SOLAR ENERGY CONVERSION AND STORAGE
Photochemical Modes

ELECTROCHEMICAL ENERGY STORAGE AND CONVERSION

Series Editor: Jiujun Zhang
National Research Council Institute for Fuel Cell Innovation
Vancouver, British Columbia, Canada

Published Titles

Electrochemical Supercapacitors for Energy Storage and Delivery: Fundamentals and Applications
Aiping Yu, Victor Chabot, and Jiujun Zhang

Proton Exchange Membrane Fuel Cells
Zhigang Qi

Graphene: Energy Storage and Conversion Applications
Zhaoping Liu and Xufeng Zhou

Electrochemical Polymer Electrolyte Membranes
Yan-Jie Wang, David P. Wilkinson, and Jiujun Zhang

Lithium-Ion Batteries: Fundamentals and Applications
Yuping Wu

Lead-Acid Battery Technologies: Fundamentals, Materials, and Applications
Joey Jung, Lei Zhang, and Jiujun Zhang

Solar Energy Conversion and Storage: Photochemical Modes
Suresh C. Ameta and Rakshit Ameta

Forthcoming Titles

Electrochemical Energy: Advanced Materials and Technologies
Pei Kang Shen, Chao-Yang Wang, San Ping Jiang, Xueliang Sun, and Jiujun Zhang

Solid Oxide Fuel Cells: From Fundamental Principles to Complete Systems
Radenka Maric

ELECTROCHEMICAL ENERGY STORAGE AND CONVERSION

SOLAR ENERGY CONVERSION AND STORAGE
Photochemical Modes

Edited by

Suresh C. Ameta
Rakshit Ameta

CRC Press
Taylor & Francis Group
Boca Raton London New York

CRC Press is an imprint of the
Taylor & Francis Group, an **informa** business

CRC Press
Taylor & Francis Group
6000 Broken Sound Parkway NW, Suite 300
Boca Raton, FL 33487-2742

© 2016 by Taylor & Francis Group, LLC
CRC Press is an imprint of Taylor & Francis Group, an Informa business

No claim to original U.S. Government works

Printed on acid-free paper
Version Date: 20150916

International Standard Book Number-13: 978-1-4822-4630-8 (Hardback)

This book contains information obtained from authentic and highly regarded sources. Reasonable efforts have been made to publish reliable data and information, but the author and publisher cannot assume responsibility for the validity of all materials or the consequences of their use. The authors and publishers have attempted to trace the copyright holders of all material reproduced in this publication and apologize to copyright holders if permission to publish in this form has not been obtained. If any copyright material has not been acknowledged please write and let us know so we may rectify in any future reprint.

Except as permitted under U.S. Copyright Law, no part of this book may be reprinted, reproduced, transmitted, or utilized in any form by any electronic, mechanical, or other means, now known or hereafter invented, including photocopying, microfilming, and recording, or in any information storage or retrieval system, without written permission from the publishers.

For permission to photocopy or use material electronically from this work, please access www.copyright.com (http://www.copyright.com/) or contact the Copyright Clearance Center, Inc. (CCC), 222 Rosewood Drive, Danvers, MA 01923, 978-750-8400. CCC is a not-for-profit organization that provides licenses and registration for a variety of users. For organizations that have been granted a photocopy license by the CCC, a separate system of payment has been arranged.

Trademark Notice: Product or corporate names may be trademarks or registered trademarks, and are used only for identification and explanation without intent to infringe.

Visit the Taylor & Francis Web site at
http://www.taylorandfrancis.com

and the CRC Press Web site at
http://www.crcpress.com

Contents

Preface vii
Editors ix
Contributors xi

Chapter 1 Introduction 1
Suresh C. Ameta

Chapter 2 Photochemical Solar Energy Conversion 7
Rakshit Ameta, Chetna Ameta, and Poonam Kumawat

Chapter 3 Basic Photoelectrochemistry 17
Purnima Dashora, Meenakshi Joshi, and Suresh C. Ameta

Chapter 4 Photoelectrochemical Cells 29
Dipti Soni, Priya Parsoya, Basant K. Menariya, Ritu Vyas, and Rakshit Ameta

Chapter 5 Organic Photovoltaic Cells 55
Meenakshi Singh Solanki, Taruna Dangi, Paras Tak, Sanyogita Sharma, and Rakshit Ameta

Chapter 6 Dye-Sensitized Solar Cells 85
Rakshit Ameta, Surbhi Benjamin, Shweta Sharma, and Monika Trivedi

Chapter 7 Photogalvanic Cells 115
Yasmin, Abhilasha Jain, Pinki B. Punjabi, and Suresh C. Ameta

Chapter 8 Hydrogen: An Alternative Fuel 139
Neelu Chouhan, Rajesh Kumar Meena, and Ru-Shi Liu

Chapter 9 Photocatalytic Reduction of Carbon Dioxide 173
Guoqing Guan, Xiaogang Hao, and Abuliti Abudula

Chapter 10 Artificial Photosynthesis 187
Neelam Kunwar, Sanyogita Sharma, Surbhi Benjamin, and Dmitry Polyansky

Chapter 11 Nanomaterials for Solar Energy 219
Mohammad Azad Malik, Sajid Nawaz Malik, and Asma Alenad

Chapter 12 Other Solar Cells 253
Rakshit Ameta

Index 265

Preface

Energy is a fundamental requirement for society. It is essential for industrialization, transportation, urbanization, food materials, and so on. The main conventional energy sources are wood, coal, petrol, diesel, kerosene, and so forth. These sources are being depleted at an ever-increasing pace, resulting in an era of energy crisis. Natural resources are limited, and it has been estimated that they will be completely exhausted in the coming few decades. Therefore, there is an urgent need to find some alternate energy sources to fulfill the energy demands of the world. In this context, researchers are stressing the use of solar energy, because it is abundant, inexhaustible, eco-friendly, and relatively low cost. Solar energy can be used to generate electricity or can be stored as chemical energy in the form of hydrogen or reduced products of carbon dioxide.

Various methods for converting light energy into electrical energy involve the use of photoelectrochemical cells, dye-sensitized solar cells, organic photovoltaic cells, photogalvanic cells, and so on. Every method has its own merits or demerits, which can be overcome by continuous and dedicated efforts. Some newer solar cells have been developed by combining or modifying these existing cells. Some examples are plasmonic solar cells, hybrid solar cells, biohybrid solar cells, perovskite solar cells, tandem solar cells, inverted tandem solar cells, and so on. Some of these cells have shown promise as future sources of energy.

Hydrogen can be obtained by photosplitting water, which is abundantly available on Earth. Hydrogen has also been advocated as the fuel of the future, because it is nontoxic and has a higher storage capacity. It generates electricity on burning in a fuel cell. The reduction of carbon dioxide to some useful synthetic fuels is another possible system for storing solar energy. Nature does this job by reducing carbon dioxide from the atmosphere in the form of biomass (carbohydrates) and regenerating oxygen. This conversion is the well-known process of photosynthesis. Efforts are being made to mimic this reaction in laboratory conditions (i.e., artificial photosynthesis). Although natural photosynthesis seems to be a simple chemical process, it is mechanistically complex. This is a challenge for chemists and biologists, in general, and photochemists in particular.

The development of nanotechnology has added many new frontiers with varied applications. Nanomaterials have smaller size, higher surface-to-volume ratio, and so on, and as a result, they may have some astonishing properties. In the last decade, nanoparticles have been extensively used in different forms (nanorods, nanowire, nanoribbon, etc.), and the time is not far off when these materials will prove their importance in solar cells, enhancing their efficiency as well as reducing water, carbon dioxide, and other materials.

In this book, the focus is on photochemical methods of converting and/or storing light energy in the form of electrical or chemical energy. Although efforts have been made to incorporate major work done in this field, due to certain limitations, some important work may have been left out. Readers are welcome to suggest any further improvement in this effort.

Suresh C. Ameta
Rakshit Ameta

Editors

Suresh C. Ameta obtained his master's degree from the University of Udaipur, India, and was awarded a Gold Medal in 1970. He secured a first position in master of philosophy in 1978 in Vikram University, Ujjain (Madhya Pradesh, India). He also obtained a PhD from this university in 1980. He served as professor and head of the Department of Chemistry, North Gujarat University Patan, India (1994) and M. L. Sukhadia University, Udaipur (2002–2005), and served as head of the Department of Polymer Science (2005–2008). He also served as dean of postgraduate studies, M. L. Sukhadia University, Udaipur (2004–2008). Now, he is serving as dean of the Faculty of Science, PAHER University, Udaipur. Professor Ameta has occupied the coveted position of president, Indian Chemical Society, Kolkata, and is now a lifelong vice president (since 2002). He has been the recipient of a number of prizes during his career, including a national prize (twice) for writing chemistry books in Hindi, the Professor M. N. Desai Award, the Professor W. U. Malik Award, the National Teacher Award, the Professor G. V. Bakore Award, and a Lifetime Achievement Award from the Indian Chemical Society. He has successfully guided 71 doctoral students. Professor Ameta has more than 300 research publications to his credit in national and international journals and has served as a reviewer. He has contributed to and authored about 40 undergraduate and postgraduate books (published by Nova Publishers, New York; Taylor & Francis Group, Oxford, United Kingdom; and Trans-Tech Publications, Pfaffikon, Switzerland), including books on green chemistry and microwave-assisted organic synthesis (published by Apple Academic Press, Waretown, New Jersey). In addition, he has delivered lectures and chaired sessions at various conferences. He completed five major research projects from different funding agencies, such as the Department of Science and Technology, the University Grants Commission, the Council of Scientific and Industrial Research, and the Ministry of Energy, Government of India. Professor Ameta has approximately 43 years of experience in teaching and research.

Rakshit Ameta obtained first position, master of science degree and was awarded a Gold Medal in 2002. He received the Fateh Singh Award from the Maharana Mewar Foundation, Udaipur, India, for his meritorious performance. He obtained a PhD in 2005 from M. L. Sukhadia University, Udaipur, India. He has worked at that university as well as at the University of Kota, Kota, India, and presently is an associate professor of chemistry at PAHER University, Udaipur. He has successfully supervised five doctoral students, and seven more are now researching various aspects of green chemistry. He has authored 70 research publications in national and international journals. Dr. Ameta has organized many national conferences, delivered a number of invited lectures, and chaired sessions at national conferences. He has been elected as scientist in charge, Industrial and Applied Chemistry Section, Indian Chemical Society, Kolkata (2014–2016), and was also elected as a council member of the Indian Chemical Society, Kolkata (2011–2013), and the Indian Council of Chemists, Agra (2012–2014). He has authored five degree-level books and contributed chapters in books (published by Nova Publishers, New York; Taylor & Francis Group, Oxford, United Kingdom; and Trans-Tech Publications, Pfaffikon, Switzerland). Two books have been published on the topics of green chemistry and microwave-assisted organic synthesis (Apple Academic Press, Waretown, New Jersey).

Contributors

Abuliti Abudula
North Japan Research Institute for
 Sustainable Energy
Hirosaki University
Aomori City
and
Graduate School of Science and Technology
Hirosaki University
Hirosaki, Aomori, Japan

Asma Alenad
School of Materials and Chemistry
University of Manchester
Manchester, United Kingdom

Chetna Ameta
Department of Chemistry
PAHER University
Udaipur, India

Rakshit Ameta
Department of Chemistry
PAHER University
Udaipur, India

Suresh C. Ameta
Department of Chemistry
PAHER University
Udaipur, India

Surbhi Benjamin
Department of Chemistry
PAHER University
Udaipur, India

Neelu Chouhan
Department of Pure and Applied Chemistry
University of Kota
Kota, India

Taruna Dangi
Department of Chemistry
PAHER University
Udaipur, India

Purnima Dashora
Department of Chemistry
PAHER University
Udaipur, India

Guoqing Guan
North Japan Research Institute for Sustainable
 Energy
Hirosaki University
Aomori City
and
Graduate School of Science and Technology
Hirosaki University
Hirosaki, Aomori, Japan

Xiaogang Hao
School of Chemistry and Chemical
 Engineering
Taiyuan University of Technology
Taiyuan, People's Republic of China

Abhilasha Jain
Department of Chemistry
St. Xavier's College
Mumbai, India

Meenakshi Joshi
Department of Chemistry
PAHER University
Udaipur, India

Poonam Kumawat
Department of Chemistry
PAHER University
Udaipur, India

Neelam Kunwar
Department of Chemistry
PAHER University
Udaipur, India

Ru-Shi Liu
Department of Chemistry
National Taiwan University
Taipei, Taiwan, Republic of China

Mohammad Azad Malik
School of Materials and Chemistry
University of Manchester
Manchester, United Kingdom

Sajid Nawaz Malik
School of Chemical and Materials
 Engineering
National University of Sciences and
 Technology
Islamabad, Pakistan

Rajesh Kumar Meena
Department of Pure and Applied Chemistry
University of Kota
Kota, India

Basant K. Menariya
Department of Chemistry
PAHER University
Udaipur, India

Priya Parsoya
Department of Chemistry
PAHER University
Udaipur, India

Dmitry Polyansky
Chemistry Department
Brookhaven National Laboratory
Upton, New York

Pinki B. Punjabi
Department of Chemistry
M. L. Sukhadia University
Udaipur, India

Sanyogita Sharma
Department of Chemistry
Pacific Institute of Technology
Udaipur, India

Shweta Sharma
Department of Chemistry
PAHER University
Udaipur, India

Meenakshi Singh Solanki
Department of Chemistry
PAHER University
Udaipur, India

Dipti Soni
Department of Chemistry
PAHER University
Udaipur, India

Paras Tak
Department of Chemistry
PAHER University
Udaipur, India

Monika Trivedi
Department of Chemistry
PAHER University
Udaipur, India

Ritu Vyas
Department of Chemistry
Pacific Institute of Technology
Udaipur, India

Yasmin
Department of Chemistry
Techno India NJR Institute of Technology
Udaipur, India

1 Introduction

Suresh C. Ameta

CONTENTS

1.1 Energy Crisis and Solar Energy 1
1.2 Solar Cells..3
1.3 Advantages..3
1.4 Disadvantages 4
1.5 Future...4

1.1 ENERGY CRISIS AND SOLAR ENERGY

Solar radiation is the most abundant natural energy source for our planet. Solar radiation can be transformed into various energy forms like heat and chemical energy. The sun provides about 120,000 terawatts to Earth's surface, which is 6000 times more than the present rate of the world's energy consumption; therefore, solar energy is going to play an important role as a future energy source. Making use of this solar energy for generating environmentally friendly chemical fuels and developing new and efficient photoconversion devices to convert this energy into the form of electricity is a challenge for chemists, in general, and photochemists, in particular. One of the advantages of a new generation of solar cells is their relatively low cost.

The solar age began in the 1950s in Bell Laboratories with the development of silicon technology. They described the first high-power silicon photovoltaic cell with 6% efficiency. This cell was much more efficient than the previous selenium solar cell. Due to the energy crisis in 1970, many technologies evolved for producing energy from renewable energy sources and photovoltaic devices. A photovoltaic device or solar cell is a solid-state electrical device that directly converts solar light energy into electricity. The light energy is transmitted in the form of small packets of photons or quanta of light, and electrical energy generated is stored.

Energy is available as heat, hydropower, electricity, chemical, wind, geothermal, tidal, biomass, liquid and gas fuels, sound, and many other forms, but there is a unity in the diversity of all of these forms of energy. The potential energy of water in a dam is used to produce electricity in a hydroelectric station, while chemical energy in fuels is converted into electrical energy or in a thermal power station battery in a generator or an inverter.

Energy is a basic necessity for life. It has provided comfort to us in its different forms. It helps us to cool down during summer and keeps us warm during winter. There has been a great increase in the global demand for energy in the last few decades as a result of industrial revolution (development) and population growth. It helps in transportation, construction, manufacturing, heat and electricity production, mining, and so on.

There are three major challenges facing mankind in the coming few decades. These are water shortages, energy challenges, and climate change. Of course, the latter depends on the first two. The simultaneous existence of all of these problems seems nearly impossible to avoid. Energy remains a great challenge to scientists, as the cost of solar energy is much higher than the cost of conventional energy technologies.

The world depends on fossil fuels at present, which are finite and not environmentally friendly. Burning oil, petrol, diesel, and so on, releases harmful greenhouse gases and sometimes even carcinogens, resulting in pollution of the environment.

There have been a number of oil shocks in the last four decades. The first was in 1973–1974, followed by those in 1979–1980, 1990–1991, 2002–2003, and the recent one in 2007–2008. It has been assumed that this trend will continue in the future.

Since the dawn of the twenty-first century (also a new millennium), the generation of energy from fuels, especially liquid fuels like diesel, petrol, and so on, with limitations on the rate of their production, has created a bottleneck resulting in an energy crisis. An energy crisis is not actually due to a shortage of energy sources, as there are enough energy sources available on Earth in some form or other; it is due to our inability to extract sufficient fuel from the globe or to produce sufficient electricity using available fuels.

The developed nations are facing such an energy crisis to meet the energy demands of the world. This is caused by uncontrolled population growth. The burning of fossil fuel produces a large quantity of the main culprit among greenhouse gases, carbon dioxide, which has an adverse effect on the environment of the earth, causing global warming. There are two conflicting opinions about carbon dioxide emissions. According to one, the world will be able to curb CO_2 emissions to 490–535 ppm by 2050, while the other is that it is able to reach 590–710 ppm. Both of these estimations are in excess of the 450 ppm targets.

As a result, there is a threat of climate change, which can bring famines, droughts, diseases, and so on. This is due to carbon emissions caused by inefficient, unclean uses of conventional energy sources, which are nonrenewable.

Prices of oil and coal are soaring, and they have increased manyfold in the last few years. The prices of other conventional energy resources are also skyrocketing. The increasing appetite for energy developed in the last few decades has been further complicated by rapidly diminishing conventional sources like coal and oil. This is added by the problem of ever-increasing demand and constrained supply.

Fossil fuels like coal and oil have definitely played a critical role as a vast energy source, which has fueled much growth in society, but now, it is agreed that these sources cannot continue to power the development of society in the future. The rising demands for oil and decreasing supply are likely to make fossil fuels economically unsustainable in years to come. Therefore, there is a pressing demand for renewable energy sources, which will decrease pollution in the environment. Solar energy enters the scene here.

Solar energy is abundantly available, and it is practically inexhaustible. Biomass, wood, and foodstuffs are derived from the sun. Even coal, petrol, diesel, and so on, are indirectly derived from the sun. The sun radiates more energy per hour than the total energy consumed globally in 1 year. The earth's surface receives as much solar energy in just 1 year, which is two times the total reserves of the earth's conventional (nonrenewable) resources of oil, coal, natural gas, and uranium combined.

Solar energy is becoming increasingly popular as we face the problems caused by the burning of conventional fuels such as wood, coal, kerosene, petrol, diesel, and so on. The popularity of solar energy supported by increased efficiency and declining cost has grown to a level at which critics of solar energy are becoming less vocal.

In an estimate made by the International Energy Agency (IEA), the energy needs of the world will be 50% more in 2030 than at present. Another forecast is that worldwide energy demand could increase almost threefold by the year 2050. Renewable energy has grown at a rapid pace in recent years. According to an estimate, renewables produced 16.5% of primary energy requirements of the world in 2005, which is a good sign as the proportion of renewable energy resources increases each day and our dependence on nonrenewable energy resources decreases.

We face two main challenges: (1) generating electricity and (2) creating solutions for its storage. Solar power accounts only for 34 TWh/y in the electricity generation mix, but this amount is projected to increase by 225 times by 2050 (i.e., somewhere between 2980 and 7740 TWh/y). It has

Introduction

been estimated that solar energy currently contributes 4% of the world's electricity. The prices of fossil fuels are increasing at a rapid pace, while the prices of solar energy production have been reduced by about 50% in the last decades. It is likely to become even cheaper in the years to come.

Common indicators of energy consumption are shares of electricity, heat, and transport fuels from renewable energy resources. The share of renewable energy sources will continue to increase. Some targets and scenarios indicate that the 20%–35% share of electricity from renewable by 2020 may increase to the range of 50%–80% by 2050.

1.2 SOLAR CELLS

Solar panels can convert sunlight into electrical energy (estimated 34% conversion efficiency, theoretically); however, only about 24% has been achieved in the majority of panels so far.

The first generation of solar cell used in most devices utilizes silicon. Silicon is highly abundant, and the best power conversion efficiency expected is 24%. But silicon solar cells are only 15%–20% efficient. This means that 80%–85% of solar energy is lost as heat.

The III–V materials are more useful for photovoltaic applications as these are better absorbers of light than silicon. These materials have the ability to stack multiple absorber layers, which provides a better match between this absorber and the solar spectra; thus, the amount of loss of sunlight as heat is reduced. As a result, the efficiency of the solar cell is increased.

The versatility and availability of different III–V materials make them quite attractive for photovoltaic applications. With these materials, one can harness different parts of the solar spectrum, resulting in an overall increase in the efficiency of solar cells.

Nanostructured photovoltaic cells make use of electrical as well as optical properties of nanomaterials that can be controlled by changing the particle size at the atomic level, but the main challenge of incorporating these nanostructures into the solar cells without disrupting the fundamental structure of the solar cell remains. There have been significant advances in the development of new materials and devices that utilize nanotechnology and nanomaterials for the conversion and storage of solar energy.

Quantum dots can be used to tune a solar cell and allow the absorption of the cell to be controlled. The Nevada Policy Research Institute (NPRI) has shown enhanced efficiency in solar cells with the use of indium (In), arsenic (As), gallium (Ga), and/or antimony (Sb) quantum dots in a standard gallium arsenide (GaAs) solar cell. They are able to show an increase in the current density.

Another approach is using concentrators in a photovoltaic cell by replacing III–V materials with less costly materials. However, higher temperatures reduce the efficiency, but this added efficiency from the concentrators outweighs the reduction in efficiency due to heat.

1.3 ADVANTAGES

- Solar energy is a clean and green source of energy, and it will never run out.
- It is estimated that all conventional energy reserves will be exhausted/depleted within a few decades, while solar energy will not be exhausted for a long time.
- Solar energy does not depend on mining raw materials as is the case with petrol, diesel, and natural gas, which causes forest and ecosystem restrictions.
- Solar energy not only benefits us financially but also protects (conserves) the environment.
- Solar energy systems are almost maintenance free, as a photovoltaic array will last for decades once it is installed. Very little maintenance is required to keep solar cells functioning smoothly, as there are no moving parts in solar cells, while other conventional sources used to generate electricity have moving parts.
- All conventional energy resources are costly, but solar energy is almost cost free except for the costs of building and other equipment. Solar energy does not require any expensive raw material, which is to be extracted, refined, and transported to the power plant.

- It is quite expensive to transmit electrical energy, while solar energy can be used efficiently even in remote areas.
- Conventional methods of generating energy cause pollution, whereas solar energy does not cause any air pollution. However, the required equipment may cause some pollution during manufacture. The major benefit of solar energy is that no greenhouse gases are generated; therefore, it will not contribute to global warming, which results in climate change, sea level rise, thinning of the ozone layer, and so on.
- Low power–consuming devices (calculators, mobile devices, watches, storage batteries, etc.) may be charged effectively by solar energy devices.
- Solar energy is available in almost every part of the world (i.e., it is decentralized in many locations), while conventional sources have deposits in some of the localized areas of the world. Even in areas that are inaccessible for power transmission by power cables, solar energy can produce electricity, no matter how remote the area, as long as sunlight is available there.
- Solar cells make absolutely no noise. There is not even a single beep when energy is harnessed, while the large machines used for extracting energy from coal, oil, petrol, and diesel are extremely noisy.
- Solar energy is renowned for its versatility; therefore, it can be used to power any device, from a small torch to a satellite.

As compared to a large number of advantages, solar energy has few disadvantages.

1.4 DISADVANTAGES

- Use of solar energy has its limits. It can be used only on sunny days. It cannot be properly utilized in adverse conditions like storms, fog, nighttime, clouds, and so on. Solar energy depends on regional, seasonal, and daily fluctuations, as solar energy is not constant in various regions of the world or certain seasons of the year, and it also is not available at night. In desert areas, large amounts of solar energy are available. In winter, the photoperiod of the day decreases as compared to summer days, and the intensity of the sun's radiation is also less.
- Large areas of land are required to capture solar energy. Areas of diffused sunlight require a large number of solar panels that utilize more space and are quite expensive.
- Batteries are to be charged if solar energy is to be used at night. These batteries are larger in size and heavy, so a space for storage is required, and these must be periodically replaced.
- At present, power generation from solar energy is quite expensive; however, the prices are coming down with time.

1.5 FUTURE

Solar energy is a completely renewable energy resource. It will surely become affordable in the future as low-cost materials are made available for producing electricity, but at present, electricity produced from solar energy is quite expensive as compared to conventional methods. Recently, in a world energy congress, it was stated that the world is going to face several significant challenges in balancing global energy needs and addressing the energy "trilemma" (a balance between energy security, social impact, and environmental sensitivity) over the next four decades.

In the future, efforts to develop new energy technology and increase the role of renewable and innovative solutions of reducing pollution and greenhouse gases should be encouraged. Solar technology is improving and its cost is decreasing, while prices of fossil fuels are constantly increasing. Improvements in the efficiency of conversion and storage of solar energy are required to reduce dependence on variability and intermittence of solar power. There is also an existing challenge to reduce the cost of renewable energy per unit, and to meet this challenge, continuous effort should be made to achieve enhanced efficiency of solar cells and bring down their costs. Energy transition is not just an imperative, it is a certainty.

A complete replacement of energy presently derived from fossil fuels with energy from alternative renewable sources is almost impossible in the short term, and it may be unrealistic to expect this transition even in the long term. If this transition is not properly managed, the consequences may be quite severe. A meaningful energy transition is still more a theory than a reality.

The sun is a known, potentially huge source of clean and renewable energy. It has been estimated that sunlight provides 10,000 times as much power as the world is using at the outset of the twenty-first century. However, still there are some major technological challenges to be met in harnessing this energy effectively. The efficiency and cost of harnessing solar power may not be ideal right now, but with improvement, the future seems quite bright and sunny.

The energy crises we experience are the result of our own choices. Our energy future will be decided by our own decisions and actions. There may be debate about the energy resources of the future, but nobody can deny the fact that solar energy is the only future energy resource.

To achieve a bright future with sufficient energy to fulfill our present and future demands, let us join the march into the solar age as soon as possible for a better world.

2 Photochemical Solar Energy Conversion

Rakshit Ameta, Chetna Ameta, and Poonam Kumawat

CONTENTS

2.1 Solar Energy Scenario 7
2.2 Some Photochemical Conversion Modes 9
References 15

2.1 SOLAR ENERGY SCENARIO

The world is currently facing energy crises, and the problem will be more grave in the future as our energy demands are ever increasing and the availability of energy sources and production of energy are limited.

A decision has been made to increase the share of renewable energy sources to about 20% of total production by 2020. No doubt, generation of energy in our present and future power plants will have its own importance, but energy sources like coal, diesel, and so on, should be carefully used and should not be wasted. These power sources are not very eco-friendly, but their significance cannot be avoided until these resources are either completely or partially replaced by some alternate renewable energy sources.

Sunlight is a clean source of energy that is inexhaustible and will remain low cost even in the future; of course, presently, it is free of cost. In spite of several benefits, there are some limitations in utilizing solar energy, such as night, fog, storms, and so on. There are also variations, such as regional and seasonal variations, that limit the use of solar energy.

The source of energy in the Sun is thermonuclear reactions involving fusion of hydrogen to form helium. About 4 billion years ago, the Earth was created, and the energy reaching the earth from the Sun was responsible for the overall development and survival of life on this planet. Humankind has used solar energy to fulfill their basic needs since ancient times.

A large amount of nuclear energy (3.89×10^{26} J s^{-1}) is released by the core of the Sun as a result of fusion reaction. This energy flux is converted in the form of thermal energy and transported toward the surfaces of different planets. The total power density emitted by the Sun is of a very high order (i.e., 64 MW m^{-2}), but only a small part of it (i.e., of the order 1.366 kW [0.001366 MW m^{-2}]) enters just above Earth's atmosphere. Therefore, with an assumption that there is no significant absorption in space, this value, 13,766 W m^{-2}, is termed the *solar constant*.

Solar radiation covers a large portion of the total spectral range. It includes radiation of nanometric and metric wavelengths. Overall solar energy flux is not equal in all ranges. It can be easily divided into three major spectral categories:

- Ultraviolet (UV) radiation ($\lambda < 400$ nm) is less than 9%.
- Visible (VIS) radiation (λ between 400 and 800 nm) is around 39%.
- Infrared (IR) radiation ($\lambda > 800$ nm) is above 52%.

The solar spectrum is quite close to the radiation of a perfect black body at 5800 K. It indicates air mass 0 (AM0) and air mass 1.5 (AM1.5), which are 1366.1 and 844 W m^{-2}, respectively. The

AM0 reference is for outside the terrestrial atmosphere; however, the radiation reaching there on the surface may be affected by many factors. The major factors are lination of the Earth's axis and the atmosphere, which are responsible for both the absorption and repulsion (albedo) of this incoming solar insolation. AM1.5 gives the influence of all these affecting parameters on solar radiation. In this spectrum, the effect of molecular alignments present in the atmosphere on light absorption is clearly visible.

The annual mean of solar radiation reaching the Earth's surface is about 180 W m^{-2}, and it is slightly lower for the oceans (170 W m^{-2}). This much solar radiation reaches the surface taking into consideration the rotation of the Earth (day and night cycle); absorption by the contents of the atmosphere; and reflection from cloud tops, oceans, and other terrestrial surfaces. Out of these, about 75% of light is received and roughly 25% is scattered by water vapor, air molecules, air Hodges, clouds, and so on.

On a clear day, the solar insolation perpendicular to the surface of the Earth is approximately 1000 W m^{-2}. This is the solar influx in clear weather at noon The solar influx is almost the same all over the globe, in spite of the increased path through the atmosphere at higher latitudes. It varies depending on the number of sun hours, clouds, and so on, in different parts of the Earth.

Sunlight is diffused as well as refreshed. This value is important for the point of solar energy conversion. Diffusion of sunlight in winter is about 35%, whereas it is only 15% in summer.

Some part of UV light from the Sun is absorbed by the ozone layer of the atmosphere. This ozone layer acts as a protective covering for life on Earth, as UV light is harmful for many biological processes. However, this layer is relatively thin over the equator. At some locations, an alarming situation has been reached due to this thinning of the ozone layer, ultimately generating an ozone hole due to various anthropogenic activities.

The Langley (L) is the conventional unit of the amount of solar radiation. One Langley is equal to 1 calorie cm^{-2}, and the incident power is normally expressed as Langleys per minute. The value of a solar constant is 1.940 L min^{-1}, which is 135.3 mW cm^{-2} or 423 Btu ft^{-2} h.

The intensity of the solar radiation reaching Earth's surface depends on the distance traveled by the light through the atmosphere. It is defined in terms of the air mass ratio (m) as

$$m = \frac{1}{\cos\theta} \tag{2.1}$$

where θ is the angle between the vertical direction and the direction of incident radiation.

- Air mass 0 (AM0) is the condition just outside the atmosphere.
- Air mass 1 (AM1) is the situation when the Sun is at the zenith ($\theta = 90° - 90° = 0°$; $\cos 0° = 1$ and $m = 1/1 = 1$).
- Air mass 1.5 (AM1.5) is used for radiation when the Sun is 41.8° to the horizon ($\theta = 90° - 41.8° = 48.2°$; $\cos 48.2° = 0.6666$ and $m = 1/0.6666 = 1.5$).
- Air mass 2 (AM2) is used for radiation when the Sun is 30° to the horizon ($\theta = 90° - 30° = 60°$; $\cos 60° = 0.5$ and $m = 1/0.5 = 2$).

The light intensity at a place is not much lower than this solar constant, if the Sun is almost overhead atmospherically and it is a bright, sunny day.

Spectral distribution of sunlight can be represented in different ways. It may be a spectral photonic flux digital, $N(\lambda)$, or a spectral irradiance, $P(\lambda)$, against wavelength.

$P(\lambda)$ and $N(\lambda)$ are related as

$$P(\lambda) = \left(\frac{h\nu}{\lambda}\right) \cdot N(\lambda) \tag{2.2}$$

Photochemical Solar Energy Conversion

This relation can also be expressed in terms of wave numbers, where $P(\bar{v})$ and $N(\bar{v})$ have the same significance but in term of wave numbers,

$$P(h\bar{v}) = (hc\bar{v}) \cdot N(\bar{v}) \tag{2.3}$$

The photonic flux density is the number of photons per unit area per unit time per unit wavelength increment, while spectral irradiance is the energy per unit area per unit time per unit wavelength increment. The total irradiation, P_{tot}, is presented as

$$P_{tot} = \int_{\lambda_m}^{\infty} P(\lambda)d\lambda = \int_0^{\bar{v}_m} P(\bar{v})\,d\bar{v} = \int_0^{\bar{v}_m} (hc\bar{v}) \cdot N(\bar{v}) \cdot d(\bar{v}) \tag{2.4}$$

The average value of the energy of a photon of white light $\langle E \rangle$ is

$$\langle E \rangle = \frac{\int_0^{\bar{v}_m} (hc\bar{v}) \cdot N(\bar{v}) \cdot d(\bar{v})}{\int_0^{\bar{v}_m} N(\bar{v}) \cdot d(\bar{v})} = \frac{P_{tot}}{N_{tot}} \tag{2.5}$$

where N_{tot} is the number of photons (at all wave numbers) incident per unit area per unit time.

2.2 SOME PHOTOCHEMICAL CONVERSION MODES

The term *solar fuel* is normally used for a system involving any chemical process that captures the light energy (photon energy), converts it, and stores it in the form of chemical energy (in the form of a chemical bond). In the last five decades or so, efforts have been made to mimic natural photosynthesis under laboratory conditions. Natural photosynthesis converts carbon dioxide of the atmosphere into useful high-energy material like glucose. One thing that can be made clear here is that in a glucose molecule, the carbon comes from carbon dioxide, and hydrogen and oxygen form water molecules. It is a misnomer to say that plants convert carbon dioxide into oxygen. Actually, this oxygen comes from water molecules, and it has been confirmed by isotopic labeling.

$$6\,CO_2 + 12\,H_2O^* \xrightarrow{h\nu}_{Chl.} C_6H_{12}O_6 + 6O_2^* + 6H_2O$$

In this process, all six carbon atoms of the carbon dioxide are incorporated in forming a glucose molecule along with its six (out of 12) oxygen atoms. The 12 hydrogen atoms forming a water molecule are utilized in the formation of a glucose molecule, while the remaining 12 oxygen atoms of water molecule and 6 oxygen atoms of carbon dioxide form 6 molecules of oxygen, which is released.

This process of photosynthesis involves chlorophyll as a sensitizer. This photosynthetic reaction looks quite simple, but it is mechanistically much more complex. Despite efforts by various chemists and photochemists, artificially mimicking photosynthesis in the laboratory remains a challenge.

It is interesting to note that in natural photosynthesis, plants utilize only about 0.023% of solar energy to feed the whole world. Looking at it this way, total solar insolation can be appreciated. The solar fuels produced by solar energy are storable, as these are generated through biomass.

A lot of effort is going into mimicking photosynthesis under laboratory conditions. A question arises in the minds of scientists or even laymen that when nature can convert carbon dioxide of the atmosphere into a useful component like biomass and produce oxygen, which is a very important component for survival of life on Earth, why can an important and beneficial process not be carried out artificially?

Chlorophyll is the natural sensitizer in photosynthesis, which can be either used as such or replaced by some other sensitizer in artificial photosynthesis. However, there are still some problems in the conversion of carbon dioxide to solar fuels, like rates of reaction, cost effectiveness, lifetime of various species involved, selectivity, overpotentials, and so on. Carbon dioxide (CO_2) can be reduced to useful synthetic fuels like $HCOOH$, $HCHO$, CH_3OH, and ultimately CH_4. However, there are some reports of conversion of CO_2 into these synthetic fuels but with very low efficiency.

It is a requirement that existing conventional energy sources be replaced by some alternate renewable, abundant, clean, and readily available energy source. This ambition can be fulfilled by artificial photosynthesis, which will utilize solar energy, CO_2, and water to generate biomass, but much more effort is required to achieve this goal.

A dye-sensitized solar cell (DSSC) is a type of thin-film solar cell. It is also known as the Gratzel cell in honor of Michal Gratzel, who invented it with Brian O'Regan. Later, it was further developed at Lausanne, Switzerland (O'Regan and Gratzel 1991). It provides a technically and economically viable concept as an alternative to p–n junction photovoltaic devices. It is alternatively abbreviated as DSSC, DSC, or DYSC. It has a number of attractive features, for instance, it is semiflexible and semitransparent, offering a variety of applications not possible with a glass-based system. In addition, most of the materials used in DSSCs are relatively low cost.

Normally, a semiconductor is responsible for both light absorption and charge carrier transport in conventional systems. These two processes are separated in a dye-sensitized solar cell. A sensitizer absorbs the light, which is anchored to the surface of a wideband semiconductor, whereas charge separation is at the interface from the dye molecule into the conduction band of the solar cell through a photoinduced electron injection. These charge carriers are then transported in the conduction band of the semiconductor, the charge collector. As the sensitizers have a broad absorption band, their use in conjunction with oxide films of a nanocrystalline nature will permit us to utilize a major portion of sunlight.

Dye-sensitized solar cells generate photocurrents from electron injection by the sensitizer dye. These cells have a photoanode and are called n-DSSCs. The reverse is the case with a photocathode (p-DSSC), which operates in the opposite manner to conventional DSSCs. Here, dye excitation is followed by a rapid electron transfer from a p-type semiconductor to the dye. It is dye-sensitized hole injection, while in n-type DSSC, it is electron injection. Some tandem solar cells (p-n-DSSCs) can be constructed using a combination of such n-DSSC and p-DSSC. The calculated/theoretical efficiency of these tandem DSSCs is far beyond that of a single-junction n-type or p-type DSSC. Any tandem cell (p-n-DSSCs) will have one n-DSSC and one p-DSSC in a simple sandwich form with an intermediate electrolyte layer. These n-DSSCs and p-DSSCs are connected in series. This implies that a resulting photocurrent will be controlled by the weakest photoelectron, and the photovoltages are additive. Therefore, it is very important to have photocurrent matching for the construction of any tandem DSSCs with high efficiency; however, fast charge recombination following dye-sensitized hole injection (unlike n-DSSC) will result in relatively low photocurrent in p-DSSCs. Thus, the efficiency of the overall device is reduced.

The first DSSC was designed in three basic parts. The first part is a transparent anode made of fluorine-doped tin oxide (FTO) deposited on the back of a glass plate. This conductive plate has a thin layer of titanium dioxide on the back, which is in the form of a highly porous structure with high surface area. It is known that titanium dioxide absorbs only a small fraction of solar insolation and that, too, in the UV range. This plate is then immersed in a mixture of a photosensitive ruthenium–polypyridine complex (also called molecular sensitizer) and a solvent. A thin layer of the dye was left covalently bonded on the surface of titanium dioxide after soaking the film in the dye solution. Then a separate plate was made with a thin layer of iodide electrolyte spread over a conductive sheet of the metal platinum. These two plates are joined and sealed together so that there is no leakage of the electrolyte. The efficiency of the cell depends on four energy levels: (1) the LUMO (conduction band), (2) the HOMO (valence band) of the photosensitizer, (3) the Fermi level of the TiO_2, and (4) the redox potential of the mediator (I^-/I_3^-) in electrolyte. Sunlight enters the DSSC through transparent FTO top contact and strikes the dye molecule on the surface of the TiO_2. Those photons that have enough energy to be absorbed excite the dye molecule to its excited state. An electron from the excited state of the dye is injected directly into the conduction band of TiO_2. This electron diffuses to the anode on top as a result of an electron concentration gradient. If another electron is not available, then the dye molecule losing an electron will decompose.

The dye molecule will abstract an electron from iodide in the electrolyte below TiO_2. This will lead to oxidation of iodide into triiodide (I_3^-). This electron transfer occurs quite rapidly compared to the time taken for the injected electron to recombine with the oxidized dye molecule. Thus, a recombination reaction is prevented, and the solar cell is short circuited. The triiodide then diffuses mechanically to the bottom of the cell, recovering its missing electron, where the counterelectrode will reintroduce the electrons after flowing to the external circuit.

Organic dyes have been used as sensitizers along with phthalocyanines, porphyrins, and so on, and high solar-to-electric power conversion efficiencies have been achieved using these sensitizers. Semiconductor quantum dots are also good sensitizers. These quantum dots are mostly II–VI and III–V types of semiconductor particles, and their absorption spectra can be adjusted by variation of their particle sizes. However, one of the problems with quantum dots is photocorrosion.

One of the main components of DSSC is electrolytes, as it can affect stability as well as the conversion efficiency of the solar cell.

There are three types of electrolytes used in DSSCs:

- Liquid electrolyte
- Solid electrolyte
- Quasi-solid-state electrolyte

Liquid electrolytes can be further classified into organic solvent electrolytes and ionic liquid electrolytes. Organic solvent can be used as liquid electrolyte (e.g., acetonitrile, ethylene carbonate, valeronitrile, etc.). The ionic liquid electrolyte can also be used. 1-Methyl-3-alkyl-imidazolium iodide or 1-methyl-3-ethyl/propyl/butyl imidazolium salts have been most commonly used as ionic liquid electrolytes. Various counterions have been used in these salts. These are I^-, BF_4^-, NCS^-, PF_6^-, and so on. Ionic liquids are commonly used in DSSCs because of their good stability, high thermal stability, high ionic conductivity, negligible vapor pressure, and so on. Various redox couples have been used as electrolytes. These are Br^-/Br_2, I_3^-/I^-, $SCN^-/(SCN)_2$, $SeCN^-/(SeCN)_2$, and so on (Boschloo and Hagfeldt 2009; Wang et al. 2010).

Some solid-state electrolytes have been used from time to time in DSSC, such as 1-methyl-3-acetyl-imidaolium iodide (Zhao et al. 2008), 4-cyano-4′-hydroxybiphenyl and imidazolium units (Cao-Cen et al., 2012), [((3-(4-vinylpyridine) propanesulfonic acid) iodide)-co-(acrylonitrile)] (Fang et al. 2011), and carbazole-imidazolium cation (Midya et al. 2010), with efficiency varying from 2.85% to 6.95%.

Similarly, some quasi-solid-state electrolytes were also used in dye-sensitized solar cells. Kumara et al. (2002) used triethylamine hydrothiocyanate as a CuI crystal growth inhibitor, while room-temperature molten salt, 1-hexyl-3-methylimidazolium iodide, iodine, and low-molecular-weight gelator were used by Kubo et al. (2002) with 5% conversion efficiency. An elastomeric copolymer of ethylene glycol and epichlorohydrin was employed in DSSC (Nogueira et al. 2001). Good efficiency of 5.4% was observed by Stathatos et al. (2003) using sol-gel nanocomposite electrolyte containing a surfactant Triton X-100, propylene carbonate, and 1-methyl-3-propylimidazolium iodide.

Although quasi-solid-state ionic liquid electrolytes have a better stability, their conversion efficiency is not as good as that of liquid electrolyte.

The concept and fundamental operation of a photogalvanic cell is quite different from that of a photovoltaic cell. These are based on chemical reactions that give rise to energy-rich products on excitation by a photon. Thereafter, these high-energy products lose energy electrochemically. Thus, photogalvanic processes are defined as different types of physical and chemical processes converting a flux of light energy into electrical power. A photogalvanic device functions as a simple transducer or it may store sufficient energy as chemical potential under an open-circuit condition and release this energy as electricity on closing the external circuit. The first photogalvanic cell was an iron–thionine cell, where a platinum electrode was illuminated and the other counterelectrode was kept in the dark.

This system has five basic steps:

1. Absorption of incident light by a colored material (dye/complex)
2. Conversion of this energy into chemical potential of a charge carrier
3. Diffusion of these charge carriers to the surface of the electrode
4. Transfer of the charge to the surface of the electrode
5. Generation of current in an external circuit

Various modifications to the photogalvanic cell have been made in the last few decades, such as modification of the electrode, synthesis, and the use of newer colored systems, exposing alternatively the two electrodes, use of micelles, and so on.

The photoelectric effect was discovered by Becquerel. Since that discovery, efforts have been made to convert sunlight to electric energy. When a photon strikes a semiconductor, it can create an electron–hole pair, and as a result, an electric potential develops.

To simplify, the mechanism of a photoelectrochemical cell is based on this conversion involving two electrodes. There are basically three possible photoelectrodes in the fabrication of a photoelectrochemical cell:

- A photoanode (n-type semiconductor) and cathode (metal)
- A photoanode (n-type semiconductor) and a photocathode (p-type semiconductor)
- A photoanode (p-type semiconductor) and a cathode (metal)

Initially, titanium dioxide and some other metal oxides were used in a photoelectrochemical cell to achieve reasonable efficiency. Later, some titanates were used, where the conduction band was mainly 3d character and the valence band was oxygen 2p character.

The bandgap of titania (approximately 3 eV) is wide enough, making it unable to absorb visible radiation more efficiently. Various modifications have been made to bring the absorption in the visible range to make it more effective. TiO_2 nanowires over porous nanocrystalline TiO_2 have also been explored for their possible use in a photoelectrochemical cell. Gallium nitride as well as other metal nitrides can also be good options because they can utilize almost the entire solar spectrum as their bandgaps are narrow enough (Wang et al. 2011). Some other nonoxides semiconductors have also been used, such as MoS_2, $MoSe_2$, WSe_2, GaAs, and so on, as n-type electrodes (Kline et al. 1981).

The field of photoelectrochemistry has been discussed by different people (Fujishima et al. 1969; Cao et al. 1996; Tyrk et al. 2000). One of the major applications of a photoelectrochemical cell is water cleavage. It provides generation of synthetic fuel, hydrogen. Hydrogen has been advocated as the fuel of the future.

The synthesis of energy-rich molecules is one of the main objectives in the chemical storage of solar energy. These energy-rich molecules may be formaldehyde, formic acid, methanol, methane, hydrazine, and so on, so that they can store energy, and their long-range distribution is also possible.

The reduction of carbon dioxide could be a good alternative as it could solve both the problems: the generation of synthetic fuels to solve the problem of energy crises and also to control the

ever-increasing amount of carbon dioxide in the atmosphere. Conventional energy fuels (i.e., diesel, petrol, kerosene, coal, etc.) add carbon dioxide to the atmosphere, resulting in global warming and ultimately natural catastrophes, while the use of synthetic fuels prepared by the reduction of carbon dioxide would not add even a single molecule of carbon dioxide to the atmosphere as these fuels are generated by photoreduction of carbon dioxide. Second, formaldehyde is the photoreduced product in some cases. It can be used as a precursor for the synthesis of glucose, which is a step toward mimicking photosynthesis.

Hydrogen is the most useful form of converted solar energy, as it can be easily substituted for petroleum-based fuels. Photodissociation of water in natural photosynthesis is an endothermic reaction, and 284 kJ mol^{-1} energy is released on decomposition of water in the form of hydrogen and oxygen. For this conversion, 1.23 eV of energy are required.

The system for generation of H_2 from the photosplitting of water consists of three components:

- Compounds include two components, that is, reductant and oxidant, which can be reduced or oxidized by quenching the excited species.
- A catalyst, which is able to collect the electrons and transfer them to the water.

However, a fourth species, electron donor, may also be required to prevent back-electron transfer or recombination of an electron–hole pair.

Hydrogen has its own importance as a synthetic fuel because of its large storage capacity, (i.e., approximately 119,000 J g^{-1}). Hydrogen can generate electricity, a source of energy, when burned in a fuel cell, and produces water as the product. Water, on photolytic decomposition, gives hydrogen and oxygen utilizing only light energy. These two chemical components, hydrogen and oxygen, are original sources and will give energy without polluting the environment. So photogeneration of hydrogen is welcome, as it can fulfill our energy demands of the future because it is renewable in nature, while other conventional fuels not only harm the environment but are limited and cannot be regenerated.

Photogeneration of another molecule is also interesting from the solar energy storage point of view. Of interest is photosynthesis of hydrogen peroxide from an oxygen molecule. It was reported that ethanol in the presence of oxygen and alkaline medium produces hydrogen peroxide in the presence of metal-free phthalocyanine or eosin. Similarly, regeneration of free halogen from halide ions has its own importance in this context. One interesting reaction is the formation of chlorine from Cl$^-$ in the presence of sulfonated anthraquinone salts. The oxidation of chloride to chlorine is also endothermic in nature. The formation of a strong oxidant may be considered as a novel way to store solar energy. Another promising system of storing solar energy is photoisomerization of norbornadiene to quadricyclane (although it has low enthalpy), dimerization of anthracene, ring-opening reaction of piperidine and pyrrolidine derivatives, and so on. The reduction of nitrogen to its reduced counterparts, like hydrazine, may prove to be a reaction of interest from an energy point of view. Photolysis of nitrosyl chloride and the decomposition of nitrogen dioxide, phosgene, and sulfur dioxide are other reactions of importance.

A photovoltaic cell is a device that converts visible light into direct current. Nanomaterial absorbs a photon, and it creates an excited state. This excited state is an electron–hole pair bound together by the electrostatic interaction. These are called *excitons*. Excitons are then broken up into free electron–hole pairs by some effective field in photovoltaic cells. An effective field is set up by creating a heterojunction between two different (dissimilar) materials. This effective field breaks up excitons as electrons fall from the conduction band of the absorber to the conduction band of the acceptor molecule. For this transfer, the conduction band of the absorber material needs to be higher than the conduction band edge of the acceptor material (McGehee and Topinka 2006; Nelson 2002; Halls and Friend 2001).

The simplest example of an organic photovoltaic cell is a monolayer or single-layer type. In these cells, a layer of organic electronic material is sandwiched between two metallic conductors.

Normally, a layer of indium tin oxide (ITO) is used, which has high work function. Another layer is of metals like magnesium, aluminum calcium, and so on, which have low work function. Because of the difference in these work functions, an electric field is set up in the organic layer. When this organic layer absorbs light photons, electrons from the highest occupied molecular orbital (HOMO) will be excited to the lowest unoccupied molecular orbital (LUMO), thus leaving behind a hole. The potential created by these two different work functions helps to split them to exciton pairs. As a consequence, electrons are pulled to the positive electrode and the holes to the negative electrodes. Kearns and Calvin (1958) reported the photovoltaic effect with a photovoltage of 200 mV. A macrocyclic compound magnesium phthalocyanine was used by them. Ghosh et al. (1974) reported an Al/phthalocyanine/Ag Schottky barrier cell. The cell showed photovoltaic cell efficiency of 0.01% on illuminating it with radiation of 690 nm.

Weinberger et al. (1982) utilized polyacetylene as the organic layer sandwiched between aluminum and graphite. They produced open-circuit voltage of 300 mV with a charge collection efficiency of 0.3%. The low quantum efficiency as well as low power conversion efficiency limit the use of single-layer organic solar cells. Another problem is that the resulting electric field between two conductive electrodes is not sufficient to split the excitons as electrons often recombine with the holes before these reach the counterelectrode.

Another type of organic photovoltaic cell is bilayer cells, with the only difference being that there are two layers sandwiched between two electrodes. One of the layers is an electron-donating layer, and the other is an electron-accepting layer. As these two layers have different electron affinity and ionization energy, electrostatic forces are generated, and the interface between these two layers is relatively large. The local electric fields are therefore stronger, which splits excitons more efficiently than single-layer photovoltaic cells. Such a structure is also known as a planar donor–acceptor hydrogen. A poly(p-phenylene vinylene) (PPV/C_{60}) photovoltaic cell was fabricated by Halls et al. (1996). They reported quantum efficiency of 9%, power conversion efficiency of 1%, and a fill factor of 0.48%. They also reported 6% quantum efficiency and a fill factor of 0.6 using a cell with a layer of bis(phenethylimido)perylene over a layer of PPV (Halls and Friend, 1997). In such a cell, the problem is diffusion of excitons to the interface of the layers and their splitting into carriers. Approximately 100 nm thickness of polymer layer is required to absorb enough light. Thus, only a limited fraction of excitons are able to reach the heterojunction interface.

Imec (Belgium) fabricated a photovoltaic cell with 8.4% efficiency using a three-layer (two electron acceptors and one donor) fullerene free stack. The open-circuit voltage of 1 V was achieved. Different kinds of heterojunctions have been used in photovoltaic cells, such as discrete, bulk, and gradient heterojunctions.

Some other types of solar cells are fabricated either by modification of existing solar cells or by a combination of some solar cells. These are hybrid solar cells, biohybrid solar cells, plastic solar cells, plasmonic solar cells, perovskite solar cells, and so on.

Biohybrid solar cells consist of a combination of inorganic matter and organic matter (photosystem I). This type of solar cell has been prepared by research workers at Vanderbilt University, Nashville, Tennessee. Ciesielski et al. (2010) prepared photosystem I biohybrid photoelectrochemical cells. In these cells, photosystem I complexes on the surface of a cathode will generate photocurrent density (approximately 2 $\mu A\ cm^{-2}$) through a photocatalytic effect. These biohybrid solar cells have remarkable stability and remain active for more than 9 months. Yehezkeli et al. (2012) used integrated photosystem II–based photobioelectrochemical cells, which generate electricity on illuminating biomaterial functionalized electrodes in aqueous solutions. The photoanode used was photosystem II functionalized and electrically wired bilirubin oxidase/carbon nanotubes modified as cathode. Here, water is oxidized to O_2 at the anode, while O_2 is reduced to water at the cathode.

Similarly, hybrid solar cells utilize organic as well as inorganic semiconductors. In this case, the organic material, which is normally conjugated polymer, absorbs light as the donor and as a consequence, before the hole is transferred. The inorganic materials are used in these cells as electron transporters. Various electron acceptors used in such devices are fullerenes, polymers, inorganic nanocrystals, and so on. Thick oxide nanoparticles dispersed in a semiconducting polymer were used by Beek et al. (2004), which act as an active layer converting 40% of incident light into electrical power at 500 nm with conversion efficiency of about 1.5%. It may be considered a step toward green electricity if eco-friendly materials are used at low temperatures. Beek et al. (2005) also reported such hybrid solar cells using crystalline zinc oxide nanoparticles as the electron acceptor and conjugated polymer blend of poly [2-methyl-5-(3′,7′-dimethyloctyloxy)-1,4-phenylene vinylene] (MDMO-PPV) as the electron donor. They also investigated the effect of degree of type of mixing of the two components as well as size and shape of nanocrystalline zinc oxide particle and observed optimized performance of about 1.6% and photon-to-current conversion efficiency ~50%.

The light energy may also be converted into electricity using plasmons. Such solar cells are called *plasmonic cells*. These cells are thin-film solar cells, where thickness of the film is about 1–2 μm. Here, lower-cost materials than silicon can be used, such as plastic, glass, steel, and so on. Catchpole and Polman (2008) have reviewed plasmonic solar cells. They reported that scattering from metal nanoparticles is a promising way of increasing light absorption in thin solar cells, which is commonly encountered in thin solar cells. Brown et al. (2010) reported the synthesis of core–shell metal insulator nanoparticles using $Au-SiO_2-TiO_2$ and plasmonic enhancement of dye-sensitized solar cells. These particles provide better efficiency than $Au-SiO_2$ plasmonic nanoparticles. It was also concluded that this is due to a near-field plasmonic effect (Sheehan et al. 2013).

The use of polymer-based photovoltaic materials makes it possible to obtain low-cost materials. The photoinduced electron transfer from donor-type semiconducting material with acceptor-type polymers, fullerenes, and so on, is used in these plastic solar cells. However, Sariciftci (2004) reported that power conversion efficiency ~5% is possible in such plastic photovoltaic devices. It was anticipated that this will increase by 8%–10% in the future.

Some inorganic and hybrid light absorbers can also be used in thin-film photovoltaic devices like quantum dots and organometal halide perovskite as these have good potential for high conversion efficiency and are also low in cost. It was demonstrated that by using meso-superstructure organometal halide perovskite solar cells with core shell $Au-SiO_2$ nanoparticles, the efficiency of the solar cell could be improved to 11.4%. This opens up an avenue for facilitated tuning of exciton binding energies in perovskite semiconductors. Chiechi et al. (2013) have reviewed the field of plastic solar cells and also the trends in polymeric nanomaterials and fullerene acceptors for hybrid polymer/quantum dot devices.

REFERENCES

Beek, W. J. E., M. M. Wienk, and R. A. J. Janssen. 2004. Efficient hybrid solar calls from zinc oxide nanoparticles and a conjugated polymer. *Adv. Mater.* 16: 1009–1013.

Beek, W. J. E., M. M. Wienk, M. Kemerink, X. Yang, and R. A. J. Janssen. 2005. Hybrid zinc oxide conjugated polymer bulk heterojunction solar cell. *J. Phys. Chem. B.* 109: 9505–9516.

Boschloo, G., and A. Hagfeldt. 2009. Characteristics of the iodide/triiodide redox mediator in dye-sensitized solar cells. *Acc. Chem. Res.* 42: 1819–1826.

Brown, M. D., T. Suteewong, R. S. S. Kumar, V. D'Innocenzo, A. Petrozza, M. Lee, et al. 2010. Plasmonic dye-sensitized solar cells using core-shell metal-insulator nanoparticles. *Nano Lett.* 11: 438–445.

Cao, F., G. Oskam, G. J. Meyer, and P. C. Searson. 1996. Electron transport in porous nanocrystalline TiO_2 photoelectrochemical cells. *J. Phys. Chem.* 100: 17021–17027.

Cao-Cen, H., J. Zhao, L. Qiu, D. Xu, Q. Li, X. Chen, et al. 2012. High performance all-solid-state dye sensitized solar cells based on cyanobiphenyl-functionalized imidazolium type ionic crystals. *J. Mater. Chem.* 22: 12842–12850.

Catchpole, K. R., and A. Polman. 2008. Plasmic solar cells. *Opt. Express.* 16: 21793–21800.

Chiechi, R. C., R. W. A. Havenith, J. C. Hummelen, L. J. A. Koster, and M. A. Loi. 2013. Modern plastic solar cells: Materials, mechanisms and modeling. *Mater. Today.* 16: 281–288.

Ciesielski, P. N., F. M. Hijazi, A. M. Scott, C. J. Beard, K. Emmett, S. J. Rosenthal, et al. 2010. Photosystem I-baded biohybrid photoelectrochemical cells. *Biosource Technol.* 101: 3047–3053.

Fang, Y., W. Xiang, X. Zhou, Y. Lin, and S. Fang. 2011. High performance novel acidic ionic liquid polymer/ionic liquid composite polymer electrolyte for dye-sensitized solar cells. *Electrochem. Commun.* 13: 60–63.

Fujishima, A., K. Honda, S. Kikuchi, and K. K. Zasshi. 1969. Recent topics in photoelectrochemistry. Achievements and future prospects. *Electrochim. Acta.* 72: 108–113.

Ghosh, A. K., D. L. Morel, T. Feng, R. F. Shaw, and C. A. Rowe Jr. 1974. Photovoltaic and rectification properties of Al/Mg phthalocyanine/Ag Schottky-barrier cells. *J. Appl. Phys.* 45: 230–236.

Halls, J. J. M., and R. H. Friend. 1997. The photovoltaic effect in a poly(p-phenylenevinylene)/perylene heterojunction. *Synth. Met.* 85: 1307–1308.

Halls, J. J. M., and R. H. Friend. 2001. In Archer M. D., and R. D. Hill (Eds.), *Clean Electricity from Photovoltaics*. London: Imperial College Press, pp. 377–445.

Halls, J. J. M., K. Pichler, R. H. Friend, S. C. Moratti, and A. B. Holmes. 1996. Exciton diffusion and dissociation in a poly(p-phenylenevinylene)/C60 heterojunction photovoltaic cell. *Appl. Phys. Lett.* 68: 3120–3122.

Kearns, D., and M. Calvin. 1958. Photovoltaic effect and photoconductivity in laminated organic systems. *J. Chem. Phys.* 29: 950–951.

Kline, G., K. Kam, D. Canfield, and B. Parkinson. 1981. Efficient and stable photoelectrochemical cells constructed with WSe_2 and $MoSe_2$ photoanodes. *Sol. Energy Mater.* 4: 301–308.

Kubo, W., T. Kiramura, K. Hanabusa, Y. Wada, and S. Yanagida. 2002. Quasi-solid-state dye-sensitized solar cells using room temperature molten salts and a low molecular weight gelator. *Chem. Commun.* 374–375.

Kumara, G. R., S. Kaneko, M. Okuya, and K. Tennkone. 2002. Fabrication of dye-sensitized solar cells using triethylamine hydrothiocyanate as a CuI crystal growth inhibitor. 18: 10493–10495.

McGehee, D. G., and M. A. Topinka. 2006. Solar cells: Pictures from the blended zone. *Nat. Mater.* 5: 675–676.

Midya, A., Z. Xie, J. X. Yang, Z. K. Chen, D. J. Blackwood, J. Wang, S. Adams, and K. P. Loh. 2010. A new class of solid state ionic conductors for application in all solid state dye sensitized solar cells. *Chem. Commun.* 46: 2091–2093.

Nelson, J. 2002. Organic photovoltaic films. *Curr. Opin. Solid State Mater. Sci.* 6: 87–95.

Nogueira, A. F., J. R. Durrant, and M. A. De Paoli. 2001. Dye-sensitized nanocrystalline solar cells employing a polymer electrolyte. *Adv. Mater.* 13: 826–830.

O'Regan, B., and M. Grätzel. 1991. A low-cost, high-efficiency solar cell based on dye-sensitized colloidal TiO_2 films. *Nature* 353: 737–740.

Sariciftci, N. S. 2004. Plastic photovoltaic devices. *Mater. Today* 7: 36–40.

Sheehan, S. W., H. Noh, G. W. Brudvig, H. Cao, and C. A. Schmuttenmaer. 2013. Plasmonic enhancement of dye-sensitized solar cells using core-shell-shell nanostructures. *J. Phys. Chem. C.* 117: 927–934.

Stathatos, E., P. Lianos, S. M. Zakeeruddin, P. Kiska, and M. Gratzel. 2003. A quasi-solid-state dye-sensitized solar cell based on a sol-gel nanocomposite electrolyte containing ionic liquid. *Chem. Mater.* 395: 583–585.

Tryk, D., A. Fujishima, and K. Honda. 2000. Recent topics in photoelectrochemistry achievements and future prospects. *Electrochim. Acta.* 45: 2363–2376.

Wang, D., A. Pierre, M. G. Kibria, K. Cui, X. Han, K. H. Guo, et al. 2011. Wafer-level photocatalytic water splitting on GaN nanowire arrays grown by molecular beam epitaxy. *Nano Lett.* 11: 2353–2357.

Wang, M., N. Chamberland, L. Breau, J.-E. Moser, R. Humphry-Baker, B. Marsan, et al. 2010. An organic redox electrolyte to rival triiodide/iodide in dye-sensitized solar cells. *Nat. Chem.* 2: 385–389.

Weinberger, B. R., M. Akhtar, and S. C. Gau. 1982. Polyacetylene photovoltaic devices. *Synth. Met.* 4: 187–197.

Yehezkeli, O., R. Tel-Vered, J. Wasserman, A. Trifonov, D. Michaeli, R. Nechushtai, et al. 2012. Integrated photosystem II-based photoelectrochemical cells. *Nat. Commun.* 3: Art. No. 742.

Zhao, Y., J. Zhai, J. He, X. Chen, L. Chen, L. Zhang, et al. 2008. High performance all-solid-state dye sensitized solar cells utilizing imidazolium-type ionic crystal as charge transfer layer. *Chem. Mater.* 20: 6022–6028.

3 Basic Photoelectrochemistry

Purnima Dashora, Meenakshi Joshi, and Suresh C. Ameta

CONTENTS

3.1 Electrochemistry of Semiconductors 17
3.2 Photoelectrochemistry 24
3.3 Photocatalysis 26
References 28

3.1 ELECTROCHEMISTRY OF SEMICONDUCTORS

Materials are normally classified into three categories, depending on energy of conduction band (CB) and valence band (VB) (Figure 3.1). The conduction band is the lowest unoccupied molecular orbital (LUMO), while the valence band is the highest occupied molecular orbital (HOMO). The conducting materials have a bandgap less than 1.0 eV or, sometimes, there may even be almost an overlap between these two bands, so that there is smooth conduction of electricity. The insulators (nonconducting material) have a wide bandgap that is more than 5.0 eV; therefore, they do not conduct electricity. There are some materials that have a bandgap of the medium order that is 1.5–3.0 eV. These are basically nonconducting in nature, but in the presence of some energy source, they become conducting. Therefore, such materials are called *semiconductors*.

A metal (M) can be easily described as the number of metal ions (M^{n+}) arranged rigidly in a crystal lattice and freely moving electrons. The number of electrons will be n times the number of metal ions. These electrons are conducting in nature and distributed in different energy levels according to the laws of quantum machines and statistical thermodynamics. At absolute zero temperature ($T = -273.15°C$ or 0 K), all the energy levels are filled up to a level called the Fermi level (E_f). This Fermi level is a characteristic of a particular material. There is a probability of finding electrons in energy levels higher than Fermi levels and an equal number of holes in energy levels lower than Fermi levels (Figure 3.2).

In the case of an insulator, all electrons are firmly attached to the atomic nuclei in the lattice or to the region between these (valence electrons). As there are no free moving electrons in an insulator, they show low or no conductance. However, if by any means it is possible to promote an electron from the VB to CB resulting in the creation of a hole in the VB, then both of these carriers, electrons and holes, are mobile under the application of the electric field. As a result, there is a possibility of a slight increase in the electrical conductivity of that insulator.

A semiconductor in a pure form (with no impurities) is called an intrinsic semiconductor. The number of electrons in the CB (per unit volume) equals the number of holes in the VB (per unit volume). If the temperature and bandgap are known, then the values of the numbers of electrons and holes can be calculated using Fermi–Dirac statistics.

The Fermi level in an intrinsic semiconductor is exactly half the way between the top of VB and the bottom of CB (Figure 3.3).

Doping an intrinsic semiconductor by some impurities with a small amount will change the property of that intrinsic semiconductor. The addition of an impurity of atoms with more valence electrons than the atom of an intrinsic semiconductor will add to the number of electrons in doped semiconductor than intrinsic semiconductor. Such doped semiconductors are called n-type or n-doped semiconductors (Figure 3.4). In such cases, electrons are more than the holes. Therefore,

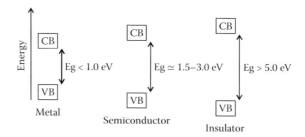

FIGURE 3.1 Classification of materials.

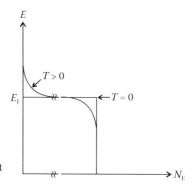

FIGURE 3.2 Distribution of electrons and holes in energy levels at different temperatures.

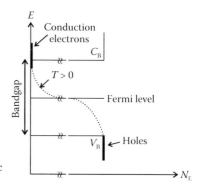

FIGURE 3.3 Fermi level, electrons and holes in intrinsic semiconductor.

electrons will act as majority carriers, and holes will act as minority carriers. Here, the Fermi level is close to the bottom of the CB, because the number of electrons is higher.

The reverse is true in the case of doping by an atom that has a number of electrons less than the number of atoms of an intrinsic semiconductor. If this is the case, then the doped semiconductor will be named p-type or p-doped semiconductor (Figure 3.5). Here, holes are majority carriers, while electrons are minority carriers because the number of holes (n_h) is greater than the number of electrons (n_e). As the number of holes is higher, the Fermi level is closer to the top of the VB.

At room temperature, the intrinsic semiconductor silicon has a number of electrons or holes equal to approximately $3.0 \times 10^{16}\,\text{m}^{-3}$, which becomes approximately $10^{23}\,\text{m}^{-3}$(n) and $10^{10}\,\text{m}^{-3}$(p) for n-doped Si; here, n ≫ p. On the other hand, if silicon is doped with gallium (which has three valence electrons only), then it gives p-doped Si, where p ≫ n. However, for the given semiconductor at a certain temperature, the product of n and p is independent of degree of doping.

If a semiconductor is illuminated with an energy, which is more than its bandgap, then it was observed that there is an increase in the conductivity of that semiconductor. This phenomenon is known as *photoconductivity*. In the case of an intrinsic semiconductor, the number of electrons

Basic Photoelectrochemistry

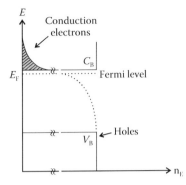

FIGURE 3.4 n-Doped semiconductor.

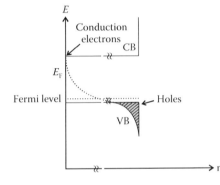

FIGURE 3.5 p-Doped semiconductor.

and holes both increase equally. This is not the case with n-type and p-type semiconductors. In such cases, there is a large increase in the number of minority carriers and a relatively insignificant increase in the number of majority carriers. The photoexcitation of the n-doped semiconductor will cause a large increase in the number of holes and a small increase in the number of electrons. The reverse is true for the p-doped semiconductor, where there is a greater increase in the number of electrons and a small increase in the number of holes.

If a metal comes in contact with an n-type semiconductor, then a contact zone is formed, which is called a Schottky barrier. As such a contact is formed, then the electron charge will move from the metal to the semiconductor. As a result, the semiconductor becomes negatively charged with respect to the metal. This transfer process continues until it reaches a state of equilibrium (i.e., the rate of electron transfer becomes the same in both directions, as the Fermi level is the same in both of these materials). There will be a space charge in the region of the semiconductor in contact with the metal. As a consequence, a band bending is observed (Figure 3.6).

Reverse band-bending is observed when a p-type semiconductor is in contact with the metal (Figure 3.7).

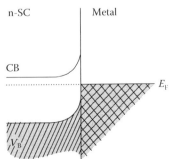

FIGURE 3.6 Band bending in n-SC/metal contact.

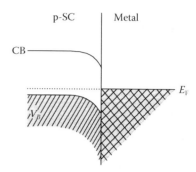

FIGURE 3.7 Band bending in p-SC/metal contact.

There will be a potential difference between the semiconductor and the metal as the charge carriers will be moved in opposite directions. If the circuit is closed by an external load, some current will pass, and the strength of this current is proportional to light intensity.

Such a contact zone is also formed if two different kinds of semiconductor (n-type and p-type) are kept in contact. This is called a case of n-p junction, and the contact zone is called an n-p barrier (Figure 3.8). Illumination of this combination also will produce the current in an external circuit, which is proportional to the intensity of light used. These cells are known as Schottky-type cells if there is contact between the semiconductor and the metal; they are known as n-p-type cells where electrodes in contact are n-doped and p-doped semiconductors.

If these is an n-type semiconductor and an electrolyte containing a redox couple ($A^{n+}\ A^{n+1}$) is in contact, then the Fermi level of the semiconductor will be at a higher energy than the redox potential of the couple (Figure 3.9).

If these two phases are brought into contact with each other, then the system will try to attain equilibrium, and the Fermi energies of two phases will become equal. There will be a transfer of electrons from the semiconductor to the electrolyte resulting in adjustment of the Fermi level. As the conductance is electronic in nature in the solid phase (semiconductor) but it is ionic in nature

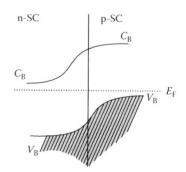

FIGURE 3.8 Band bending in n-SC/p-SC contact.

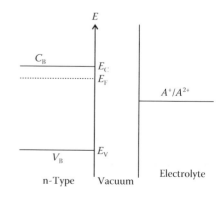

FIGURE 3.9 n-Type semiconductor/redox couple in vacuum.

Basic Photoelectrochemistry

in a solution (electrolyte), electrons cannot penetrate the interface, and transfer of the charge has to take place by a reduction process:

$$A^{(n+1)} + e^- \rightarrow A^{n+} \tag{3.1}$$

This electron transfer process produces a positive charge on the semiconductor, and a surface region is formed with a lower concentration of electrons than in the bulk of the semiconductor. This region is called the *depletion layer* or the *space charge layer*. As a result, there will be a corresponding increase in the number of electrons in the electrolyte. Thus, a negative countercharge is developed in the electrolyte. Due to this low concentration of charge carrier in the semiconductor as compared to that in electrolyte, the thickness of the depletion layer is quite higher in magnitude than the negative countercharge region in the electrolyte.

An electric field exists in this depletion layer because of this charge transfer. This electric field will change the electrostatic potential with the distance from the surface. If the potential related to a reference level is V_s at the surface, then it will have another value V_B in the bulk of the semiconductor. The difference between these two potentials, $(V_S - V_B)$, is called the *band bending*. This variation in potential corresponds to the change in energy level as potential and energy are almost identical quantities for the electron (Figures 3.10 and 3.11).

The width (W) of this depletion layer is approximately equal to

$$W = \left(\frac{2\varepsilon\varepsilon_0 V_B}{eN_D} \right)^{1/2} \tag{3.2}$$

where e is the elementary charge, N_D is the concentration of ionized donor atom, ε is the dielectric constant of the material, and ε_0 is the permittivity of free space. The semiconductor–electrolyte interface has many properties similar to the metal–semiconductor interface; therefore, it is quite commonly referred to as a Schottky-type junction.

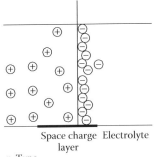

FIGURE 3.10 Space charge layer in n-type SC/electrolyte system.

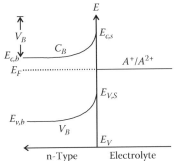

FIGURE 3.11 n-Type semiconductor/electrolyte with a redox couple.

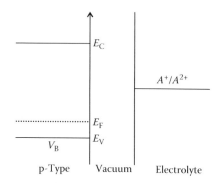

FIGURE 3.12 p-Type semiconductor/redox couple in vacuum.

In the case of a p-type semiconductor in contact with an electrolyte, the case is reversed. A Fermi level in this case is quite close to the valence band (Figure 3.12).

In the case of a p-type semiconductor in contact with the electrolyte, the redox level is above that of the Fermi level of the semiconductor. As a result, electrons will start flowing from the electrolyte into the semiconductor until the equilibrium is attained. In such a situation, a negative space charge layer is formed and the energy level is bent upward at equilibrium. This means that the energy level will be higher in bulk than at the surface. A positive countercharge is formed in the electrolyte close to the semiconductor. Here, the width of the depletion layer is

$$W = \left(\frac{2\varepsilon\varepsilon_0 V_B}{eN_A}\right)^{1/2} \quad (3.3)$$

with the only difference that N_D is replaced by N_A, the concentration of ionized acceptor ions (Figure 3.13).

The position of the Fermi level with some reference level of the solution (normally H_2/H^+ level) is called the *potential* of the semiconductor. The Fermi level of the semiconductor can be varied relative to this reference level by applying some external potential to the semiconductor electrode. Therefore, a band bending can be changed from value at equilibrium. Whenever, there is no field in the semiconductor, ($V_S = V_B$), then the potential is called the flat-band potential (V_{fb}). This flat-band potential decides the position of the conduction band of the n-type semiconductor and valence band of the p-type semiconductor, approximately.

The flat-band potential can be determined by measuring the differential capacitance (C_{dl}) of the space charge-transfer layer as a function of the potential of the semiconductor electrode. According to the Mott–Schottky equation (Myamlin and Pleskov 1967),

$$C_{dl}^{-2} = (2\varepsilon\varepsilon_0 eN_D)\left(V - V_{fb} - \frac{kT}{e}\right) \quad (3.4)$$

where k is the Boltzmann constant, T is the temperature, and C_{dl} is the differential capacity.

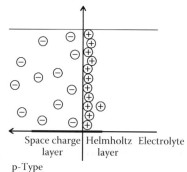

FIGURE 3.13 Space charge layer in a p-type semiconductor/electrolyte system.

Basic Photoelectrochemistry

A plot of $1/C_{dl}^2$ v/s potential (V) of the semiconductor gives a straight line. This line cuts the potential axis near the flat-band potential ($kT/e \approx 0.025$ V at room temperature). The slope of this line is determined by the concentration of the doping atom (N_D in the case of an n-type material or N_A in the case of a p-type material). The measured capacitance is a complex quantity, because it is a sum of various contributing capacitances of the system, each belonging to the space charge. Tomkiewicz (1979) has discussed the measurement of the space charge layer capacitance.

It has been observed that the flat-band potential is found to change with the pH of the solution for some semiconductors (mostly oxides). This ability of the semiconductor to change with the pH may be explained based on an assumption. If the surface of the semiconductor acts as an acid–base couple, which can either accept or donate protons, then

$$M - OH = M - O^- + H^+ \tag{3.5}$$

The potential difference between the surface of the semiconductor and the bulk of the electrolyte, which is approximately the potential drop of the Helmholtz layer (assuming that the contribution of Gouy layer is negligible) will be

$$\Delta V_H = \text{constant} - 0.059\, \text{pH} \tag{3.6}$$

When there is a vacuum between the semiconductor and the electrolyte, then the Fermi energy level of the semiconductor is termed E_F^{vac} (Figure 3.14).

As the electrolyte and semiconductor approach each other, a field is induced, which causes a change of the vacuum layer level in the region between these two phases (Figure 3.15). When this field is completely developed after the Helmholtz layer is completed, then the potential drop

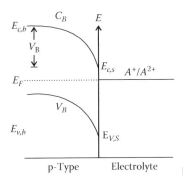

FIGURE 3.14 p-Type semiconductor/electrolyte with a redox couple.

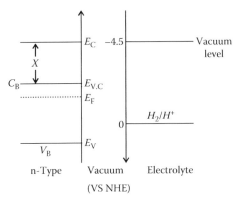

FIGURE 3.15 n-Type semiconductor in vacuum.

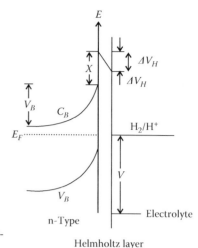

FIGURE 3.16 Effect of pH on potential drop in n-type semiconductor/electrolyte with H_2/H^+ redox couple (at lower pH).

$\Delta V_H = V_S - V_H$ must be considered in relating the energy levels of the two phases to each other. The flat-band potential is

$$V_{fb} = -(E_F^{vac}/e + 4.5 \text{ V} - \Delta V_H) \tag{3.7}$$

A change in pH also affects the electrode potential of the H_2/H^+ couple (Figure 3.16) as

$$V(H_2/H^+) = -0.059 \text{ pH} + V^0(H_2/H^+) \tag{3.8}$$

At equilibrium, the Fermi energy (i.e., the potential of the semiconductor electrode) will show similar pH dependence. It means that a band bending is independent of pH.

3.2 PHOTOELECTROCHEMISTRY

If an n-type semiconductor is in contact with an electrolyte, and it is exposed to a light of energy greater than or equal to Eg (bandgap of a semiconductor), then the photons are absorbed by the semiconductor resulting in excitation of electrons from the valence band to the conduction band (Figure 3.17). This creates electron–hole pairs. This electron–hole pair may recombine; therefore, it must be quickly separated into a free electron and a free hole. This separation is achieved in the

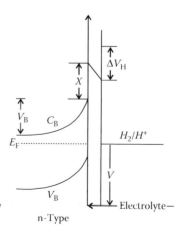

FIGURE 3.17 Effect of pH on potential drop in n-type semiconductor/electrolyte with H_2/H^+ redox couple (at higher pH).

Basic Photoelectrochemistry

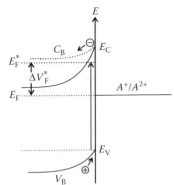

FIGURE 3.18 Effect on the band bending and Fermi level on excitation of n-type semiconductor/electrolyte system.

space charge layer, where a strong electric field is effective. The majority carriers (electrons) will pass to the bulk of the semiconductor as the minority carriers (holes) will be driven to the surface. If this excitation takes place in the region behind the depletion layer (i.e., the diffusion region), the electron–hole pairs will recombine unless the holes diffuse away into the depletion layer, as there is no field in the diffusion region. The excitation of a semiconductor in the depletion layer is an excellent means of separating the negative and positive charges quickly, and it is necessary for a system to be a good working system.

According to an assumption, no reaction at the surface can exhaust the holes; hence, a photopotential develops in the semiconductor, decreasing the band bending. As a consequence, the Fermi level will change toward its flat-band position (Figure 3.18).

This photopotential (V_{PhP}) is

$$V_{PhP} = -\frac{(E_F^* - E_F)}{e} = -\frac{\Delta E_F^*}{e} \tag{3.9}$$

where E_F is the Fermi level at equilibrium, and E_F^* is the Fermi level under illumination. The photopotential V_{PhP} can also be shown as

$$V_{PhP} \sim \left(\frac{kT}{e}\right) \ln I \tag{3.10}$$

where I represents the light intensity. As the value of

$$\frac{E_F^*(\max)}{e} = -V_{Fb} - 4.5 \text{ V}$$

the maximum photovoltage obtained will be (Gerischer 1975)

$$V_{PhP}^{\max} = V_{fb} + 4.5V + \frac{E_F}{e} \tag{3.11}$$

But in practice, this much photopotential is not obtained, maybe because of its logarithmic dependence on intensity. The flow (J) of minority carriers to the surface is the sum of the flows from the depletion layer (J_{dl}) and the diffusion region (J_{diff}). This flow is given by (Gartner 1959)

$$J = J_{dl} + J_{diff} = eI_\lambda \left[\frac{1 - \exp(-\alpha_\lambda W)}{1 + \alpha_\lambda L}\right] \tag{3.12}$$

where the wavelength of monochromatic radiation is λ, and its intensity is I_λ. If α_λ is the absorption coefficient of the semiconductor, then the diffusion length (L) of the minority carrier is

$$L = (D\tau)^{1/2} \tag{3.13}$$

where D is the diffusion coefficient, and τ is the lifetime of minority carriers.

Whether the minority carriers reaching the surface will lead to a photocurrent depends on the probability of charge transfer (S_t) and the surface in relation to the probability of recombination (S_r) with surface states. Such a relation for the generated photocurrent will be (Wilson 1977)

$$i_{PhP} = \frac{\left(\dfrac{S_t}{S_t + S_r}\right) eI_\lambda [1 - \exp(-\alpha_\lambda W)]}{1 - \alpha_\lambda W} \tag{3.14}$$

The charge transfer should preferably involve reaction with species in the electrolyte, but it may also involve a corrosion reaction leading to a photocurrent.

3.3 PHOTOCATALYSIS

Photocatalysis is an emerging and promising technology that has varied applications. The term *photocatalysis* creates confusion, as it is a combination of two terms: *photo*, meaning "light," and *catalysis*, meaning "to affect the rate of a reaction." It appears as if photocatalysis means light-catalyzed reaction, but that is not the case. This term has long been debated, but the term *photocatalysis* is still reserved for a chemical reaction in the presence of light and a semiconductor.

Photocatalysis has been classified into two types:

- Homogeneous
- Heterogeneous

Homogeneous photocatalysis involves a photocatalytic reaction, where a substance and the semiconductor are in the same phases. The best examples are dyes, coordinatation compounds, and so on. Heterogeneous photocatalysis involves the semiconductor and substrate in different phases. Common examples are insoluble semiconducting chalcogenides in binary, ternary, and sometimes, even quaternary forms.

The concept of photocatalysis (Figure 3.19) is based on photoelectrochemistry, but with a difference that here the particles of the semiconductor act as individual photoactive units. When the semiconductor is exposed to a suitable light radiation with wavelengths corresponding to equal or more than its bandgap, then an electron will be excited from its valence band to the conduction band, thus leaving behind a hole in the valence band. This electron can be utilized to reduce a substrate, or the hole can be used for oxidizing any substance, of course, depending upon the redox levels of the substrate.

It is not normally possible to have an oxidizing as well as a reducing environment simultaneously in the same system. The beauty of photocatalysis is that it provides both of these environments simultaneously in the same reaction medium. However, reduction and/or oxidation may take place depending upon a situation.

In all, there are four possibilities (Figure 3.20) depending upon the levels of conduction band, valence band, and redox levels of the substrate:

- If the reduction level of the substrate is below the CB of the semiconductor and the oxidation level is below its VB, then transfer of electrons from the semiconductor to the substrate becomes easier, resulting in reduction of the substrate.

Basic Photoelectrochemistry

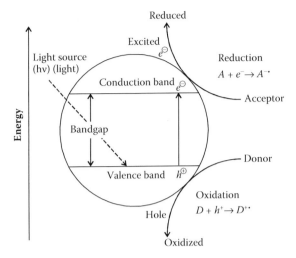

FIGURE 3.19 Photocatalysis.

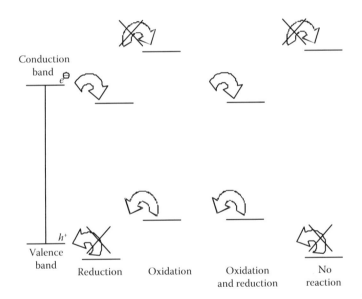

FIGURE 3.20 Various probabilities of reactions.

- If the reduction level is higher than the CB of the semiconductor, and the oxidation level of the substrate is higher than the VB of the substrate, then the electron can be easily transferred from substrate to semiconductor. In other words, the hole is transferred from semiconductor to substrate, resulting in oxidation of the substrate.
- If the reduction level of the semiconductor is lower than the CB of the semiconductor and the oxidation level is higher than the VB of semiconductor, then it appears as if the CB and VB of the semiconductor saddle over the redox levels of the substrate. In such cases, the electron and hole can be transferred from the semiconductor to the substrate. Here, both oxidation and reduction will take place.
- If the reduction level of the substrate is higher than the CB of the semiconductor, and VB is lower than the oxidation level of the semiconductor, then no reaction will take place.

Photocatalysis has provided some interesting applications like wastewater treatment, deodorization, self-cleaning glasses, disinfection, synthesis of energy-rich materials, photogeneration of hydrogen, antifogging reduction of CO_2, solar energy conversion and storage, and so on.

REFERENCES

Gartner, W. W. 1959. Depletion-layer photoeffects in semiconductors. *Phys. Rev.* 116: 84–87.

Gerischer, H. 1975. Electrochemical photo and solar cells. Principles and some experiments. *J. Electroanal. Chem.* 58: 263–274.

Myamlin, V. A., and Pleskov, V. Yu. 1967. *Electrochemistry of Semiconductors*. Plenum Press, New York.

Tomkiewicz, M. 1979. Relaxation spectrum analysis of semiconductor-electrolyte interface-TiO_2. *J. Electrochem. Soc.* 126: 2220–2225.

Wilson, R. H. 1977. A model for the current-voltage curve of photoexcited semiconductor electrodes. *J. Appl. Phys.* 48: 4292–4297.

4 Photoelectrochemical Cells

*Dipti Soni, Priya Parsoya, Basant K. Menariya,
Ritu Vyas, and Rakshit Ameta*

CONTENTS

4.1 Introduction	29
4.2 Classification	30
4.2.1 Regenerative Solar Cells	31
4.2.2 Photoelectrolytic Solar Cells	39
4.2.3 Photocatalytic Cells	42
4.3 PEC Cells as Storage Cells	47
References	48

4.1 INTRODUCTION

Photoelectrochemical (PEC) cells are the most efficient cells for converting solar energy into a more useful form of energy. These devices are quite simple to construct, and often consist of a photoactive semiconductor electrode (either n- or p-type) and a metal counterelectrode. Both of these electrodes are immersed in a suitable redox electrolyte. The PEC cells use light to carry out a chemical reaction for converting light to chemical energy. They have a solid–liquid interface, whereas photovoltaic (PV) solar cells have a solid–solid interface.

The commercial use of a PEC solar cell depends on its conversion efficiency and stability. Various efforts have been made to make PEC cells more efficient, such as electrolyte modification, surface modification of the semiconductors, photoetching of layered semiconductors, semiconductor septum-based PEC solar cells, and so on. The dream is to capture the energy that is freely available from sunlight and turn it into electric power.

The PEC cells based on III–V semiconductor electrodes have achieved high solar power conversion efficiencies in regenerative cells and in photoelectrolytic production of hydrogen. Miller (1984) discussed the corrosion chemistry associated with charge transfer at these interfaces, the influence of film formation, and the consequences for both photoanodic and photocathodic cells. Single-bandgap semiconductors in PEC cells have lower values (up to 16%) of energy conversion than the multiple bandgap cells that have significantly higher conversion efficiencies (Licht, 2001; Licht et al. 1998b,c,d).

Energy production by these PEC processes has also been reviewed by Memming (1978). Bhavani et al. (1986) studied the reactions and PECs from the standpoint of energy conversion efficiency and the possibility of energy storage. McEvoy (2005) demonstrated that in a PEC, injected electrons facilitate a current in an external circuit, returning to the redox electrolyte through a cathode that is in contact with it. The uncharged ground state of the dye is restored by electron transfer from the redox system, which completes the circuit and provides a regenerative cycle comparable with other photovoltaic devices.

French scientist Edmond Becquerel (1839) noticed a photovoltaic effect when he immersed a silver electrode in a chloride electrolyte and it was illuminated. Fujushima and Honda (1972) worked on illuminated semiconducting TiO_2, which leads to photooxidation of water to oxygen. They photoelectrolyzed water into H_2 and O_2 by using such a system. This process results in the conversion of sunlight to stored chemical energy. Metal oxide semiconductors like TiO_2, $SrTiO_3$, WO_3, and so on, are very stable and have large bandgaps, which converts only a small fraction of the solar spectrum into electrical or chemical energy. It is challenging to find a semiconductor material with a small bandgap (1.1–1.5 eV) for efficiently converting sunlight to usable energy.

The semiconductor–liquid-junction-based PEC solar cell consists of a photoactive semiconductor electrode immersed in electrolytic solution containing suitable redox couple and counterelectrode, which can be a metal or semiconductor. Irradiation of the semiconductor–electrolyte junction with light of $h\nu > E_g$ (E_g is the bandgap of SC) results in the generation and separation of charge carriers. The majority of carriers are electrons in an n-type semiconductor, which move to counterelectrode through an external circuit and take part in a counter reaction. Holes are the minority charge carriers, which in turn migrate to electrolytes and participate in electrochemical reactions.

4.2 CLASSIFICATION

The PEC cells can be classified in two major categories on the basis of change in Gibbs free energy:

1. Regenerative PEC solar cells with $\Delta G = 0$. Here, the photoenergy is converted into electric energy (Figure 4.1).
2. Photoelectrosynthetic cells with $\Delta G \neq 0$. Here, the photoenergy is used to affect chemical reactions, with nonzero free energy change in the electrolyte. These cells can be further classified into two types of cells:
    a. Photoelectrolytic cells with $\Delta G > 0$, where the photoenergy is stored as chemical energy in endergonic reactions (e.g., $H_2O \rightarrow H_2 + \frac{1}{2}O_2$) (Figure 4.2).
    b. Photocatalytic cells with $\Delta G < 0$, where photoenergy provides activation energy for exergonic reactions (e.g., $N_2 + 3H_2 \rightarrow 2NH_3$) (Figure 4.3).

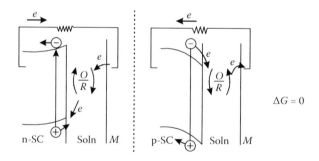

FIGURE 4.1 Regenerative solar cells.

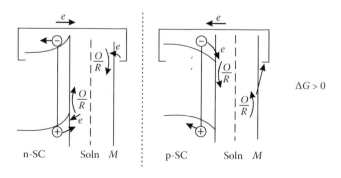

FIGURE 4.2 Photoelectrolytic solar cells.

Photoelectrochemical Cells

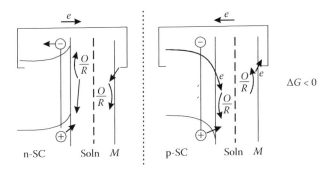

FIGURE 4.3 Photocatalytic solar cells.

Three electrical parameters determine the conversion efficiency of a liquid-junction PEC cell:

- Open-circuit voltage (V_{oc})
- Short-circuit photocurrent (i_{sc})
- The fill factor (FF)

An efficient PEC device can be constructed by optimizing these three quantities.

Solar-to-electrical conversion efficiency (η) in a semiconductor/liquid-junction solar cell is

$$\eta = \left[\frac{(i \times V)_{max}}{P}\right] \times 100 \quad (4.1)$$

where $(i \times V)_{max}$ is the maximum output power of the solar cell, and P is the optical power input. Fill factor (*FF*) of a PEC solar cell is

$$FF = \frac{i_{max} \times V_{max}}{i_{sc} \times V_{oc}} \quad (4.2)$$

The deviation of fill factor (normally less than unity) indicates the extent of departure from the ideal *i-V* behavior.

The maximum η for operation in regenerative (light to electricity) mode is 25%–30% for single electrode–based PEC cells employing semiconductors with bandgaps of 1.2–1.5 eV. If the optimum bandgap for photoelectrolysis is considered to be 1.8 eV, then the theoretical maximum efficiency is estimated to be 25%, while for double photoelectrode–based (p-n) water photoelectrolysis cells, it is almost 45%.

4.2.1 Regenerative Solar Cells

Regenerative PEC solar cells are used to convert solar energy to electricity. Regenerative PEC solar cells, which are based on a narrow-bandgap semiconductor and a redox couple, convert optical energy into electrical energy without bringing about any change in the free energy of the redox electrolyte ($\Delta G = 0$). The electrochemical reaction occurring at the counterelectrode is opposite to the photoassisted reaction occurring at the semiconductor working electrode. Thus, they are also called electrochemical photovoltaic cells. Electrochemical photovoltaic cells have the following advantages over solid photovoltaic cells:

- They are not sensitive to the defects in semiconductors.
- The solid–liquid junction is easy to form, and the production price will be much lower.

- Direct energy transfer from photons to chemical energy is possible. Unlike conventional solid-state photovoltaic cells, the potential of the working electrode can be varied with respect to the reference electrode by means of an external voltage source connected between the working electrode and the counterelectrode.

Doubly β-functionalized porphyrin sensitizers were prepared by Park et al. (2008) to study the photoelectrochemical properties of dye-sensitized nanocrystalline-TiO_2 solar cells. These porphyrin sensitizers were functionalized at meso- and β-positions with different carboxylic acid groups, which were then employed to investigate electronic and photovoltaic properties. Multiple pathways through olefinic side chains at two β-positions enhance the overall electron injection efficiency, and the moderate distance between the porphyrin sensitizer and the TiO_2 semiconductor layer will retard the charge recombination processes. Consequently, these combined effects give rise to higher photovoltaic efficiency in photovoltaic regenerative solar cells.

Kohl and Bard (1979) constructed regenerative photoelectrochemical cells by using single-crystal n-GaAs in acetonitrile solutions. Solution redox couples (anthraquinone, p-benzoquinone, dimethyl ferrocene, ferrocene, hydroxymethyl ferrocene, and tetramethyl-p-phenylenediamine) were reduced at a Pt counterelectrode and photooxidized at the semiconductor electrode converting light directly into electrical energy. A power conversion efficiency of 14% was obtained for the n-GaAs electrode in a ferrocene–ferricenium acetonitrile solution at a radiant intensity of 0.52 mW cm^{-2}.

Tributsch (1980) studied the photoelectrochemical properties of semiconducting layer-type disulfides and diselenides of transition metals belonging to groups IV, VI, and VIII, which have energy gaps ranging between 1 and 2 eV. The compounds of Mo and W can be used as stable electrodes for regenerative and fuel-producing solar cells. Neumann-Spallart and Kalyanasundaram (1981) fabricated polycrystalline CdS electrodes from CdS powders by coating a thin film of a paste formed with an aqueous solution of a surfactant and $ZnCl_2$ on a Ti substrate and sintering at 670°C in argon. Light conversion efficiencies of these electrodes have been tested in two regenerative solar cells and were found to range from 0.37% to 4.4%.

Singh et al. (1981) reported the photoelectrochemical behavior of n-GaAs electrodes used in regenerative PEC cells. Both single-crystal and polycrystalline n-GaAs electrodes were used, and it was concluded that this PEC system was capable of generating high open-circuit potentials. Kline et al. (1981) prepared and used single crystals of n-WSe_2 and n-$MoSe_2$ as the photoanodes in a regenerative PEC cell with iodide/triiodide electrolyte. The conversion efficiencies observed were 10.2% and 9.4% on selected crystals of WSe_2 and $MoSe_2$, respectively.

Kline et al. (1982) studied the PEC behavior of synthetic crystals of WS_2, MoS_2, and crystals with mixed-metal and chalcogen composition and compared that with the behavior of $MoSe_2$ and WSe_2. The composition and stoichiometry of the crystals and the composition of the electrolyte were varied, and the behavior of the materials in a regenerative liquid-junction solar cell was observed. The quantum yields and sunlight-to-electricity conversion efficiencies were also measured.

Catalytically modified n-silicon/indium tin oxide anodes were used in PEC generation of chlorine by Thompson et al. (1982). Silicon photoanodes showed stable and efficient behavior Cl_2/Cl^- with aqueous LiCl electrolytes in PEC cells. Regenerative devices using the Cl_2/Cl^- redox coupled with approximately 3% optical-to-electrical conversion efficiency have been reported. Unmodified n-Si/ITO electrodes show minimal electro- and photocatalytic activity for the generation of Cl_2.

Ang et al. (1983) made a two-photoelectrode regenerative PEC cell, which showed a combined photovoltage in excess of 1 V. The cell consists of n-CdSe as the photoanode and p-InP as the photocathode in an aqueous sulfide/polysulfide electrolyte. The photo potentials generated were sufficient for the external change of a number of redox battery couples. Rajeshwar et al. (1983) used polymer-coated n-GaAs photoanodes with aqueous electrolyte in regenerative PEC cells. The n-GaAs photoanodes were coated with films of polystyrene pendant $[Ru(bpy)_3]^{2+}$ complex, and it was called a *PSt-bpy-Ru complex*. The PSt-bpy-Ru-coated n-GaAs photoanodes were tested in aqueous redox electrolytes in regenerative PEC cells.

The behavior of regenerative PEC cells of large-grained polycrystalline n-CuInSe$_2$ films was studied by Bicelli et al. (1985). These films were synthesized by a new technique of depositing the semiconducting material on thin metal layers and then keeping it at a temperature lower than its melting point. Scanning electron microscopy was used to characterize the surface morphology of the films; systematic scanning laser spot analysis was employed to examine the output power characteristics when in contact with different redox couples. Keita and Nadjo (1984) used a single redox couple, sodium 9,10-anthraquinone-2,6-disulfonate/sodium 9,10-anthrahydroquinone-2,6-disulfonate (AQ/AQH$_2$) to illustrate various aspects of PEC cells, particularly regenerative photocells with p-type Si and p-type WSe$_2$ photocathodes and synthetic photocells. The conversion efficiencies were in the range of 2% for p-type Si electrode and 10% for layer-type materials at 632.8 nm light.

Osaka et al. (1985) studied the iron oxide/n-Si heterojunction electrode as a photoanode for a regenerative PEC cell. The effects of modifying the top layer of the electrode with Pd or RuO$_2$ were also observed. The photocurrent at the heterojunction electrode was generated by the holes, which were photoexcited in both iron oxide and n-Si. The addition of Pd or RuO$_2$ on the heterojunction electrode surface increased the optical-to-electrical conversion efficiency. The efficiencies for a stable working PEC cell were found to be 1.34% and 1.60% for a Pd- and a RuO$_2$-modified electrode, respectively. This efficiency was measured in a 0.2 M KOH solution containing 0.2 M K$_4$[Fe(CN)$_6$] and 0.01 M K$_3$[Fe(CN)$_6$] at an intensity of 55 mW cm^{-2}. The photoanode was made highly stable by the use of iron oxide, as compared to an electrode such as RuO$_2$/n-Si.

Cattarin et al. (1986) studied the photoelectrochemical behavior of copper mono- and disubstituted para-diethynylbenzene, deposited chemically and electrochemically as thin films onto copper electrodes in 0.1 M LiOH. These materials act as p-type semiconductors with bandgaps of 2.2 and 2.1 eV, respectively, in a rectifying liquid junction with a flat-band potential of 0.3 V versus SCE. High photocurrents (8.0 mA cm^{-2}) and photovoltages (0.5 V) were observed by the disubstituted compound under oxygen and 35 mW cm^{-2} white light. The monochromatic quantum yields of 12%–13% were obtained. These results displayed a remarkable improvement with respect to evaluated copper acetylide systems.

The solar-to-electricity conversion efficiency of a regenerative PEC cell was determined by Carlsson and Holmström (1986). The analysis was based on two types of laboratory measurements: (1) photocurrent as a function of photoelectrode potential at constant wavelength of illuminating light and (2) photocurrent as a function of wavelength at constant potential.

Guay et al. (1987) obtained high short-circuit photocurrents with phthalocyanine in a regenerative PEC cell. Purified phthalocyanine-AlCl was synthesized, which generated photocathodic short-circuit photocurrents in the milliampere per square centimeter (mA cm^{-2}) range under white light irradiation.

Ramaraj and Natarajan (1989) synthesized and characterized macromolecular-bound phenosafranine dyes. Photoelectrochemical investigations of these macromolecular phenosafranine dyes showed different behaviors depending on the macromolecule. Cathodic behavior with reference to an inert electrode was observed when the electrode was coated with a film of poly(acrylamidomethylphenosafranine-co-methylolacrylamide), while an electrode coated with a film of poly(acrylamidomethylphenosafranine-co-methylolacrylamide-co-vinylpyridine) exhibited anodic polarity. These polymeric phenosafranine-coated electrodes were used to operate a water-splitting regenerative cell.

Ladouceur et al. (1990) prepared two substoichiometric tungsten oxide films by plasma spray of WO$_3$ powder on Ti substrates. The films were 40 ± 20 μm thick and yellow (WO$_{2.99}$) or dark blue (WO$_{2.97}$) in color. The yellow films have been used in regenerative cells by using O$_2$/H$_2$O redox couple at pH 2.0. The WO$_{2.99}$ films have been analyzed for NH$_3$ photoproduction. Chloro- and bromoaluminum phthalocyanines were studied in cells using I$_3^-$/I$^-$ as the redox system.

Rothenberger et al. (1992) synthesized nanostructured TiO$_2$ films by sintering 15 nm diameter colloidal anatase particles on a conducting glass support. They optically determined the flat-band potential of colloidal titanium dioxide films. The ability to measure flat-band potentials in such porous electrodes helps in the optimization of interfacial charge-transfer processes and in the design of efficient regenerative PEC cells.

Redmond et al. (1994) prepared transparent nanocrystalline ZnO films on a conducting glass substrate by sol-gel techniques. They described the sensitization by adsorption of a ruthenium-based complex and subsequent incorporation as the light-harvesting unit in a regenerative PEC cell. The resulting device had a monochromatic incident photon-to-current conversion efficiency of 13% at 520 nm.

The effect of aqueous polyselenide solution modification on the photoelectrochemical behavior of an n-GaAs/aqueous polyselenide PEC cell was described by Forouzan and Licht (1995). Solution modification was achieved by changing the ratio of dissolved selenium to selenide and pH control. It was concluded that PEC photocurrent, photovoltage, and fill factor were affected by the distribution of hydroselenide, selenide, and polyselenide in solution. An optimum pH ([KOH] = 1–2 M) provides sufficient concentrations of Se^{2-} to improve PEC response. The photopotential was enhanced up to 50 mV in high-$Se°$ electrolytes.

Matthews et al. (1996) calculated the photocurrent-potential characteristics for regenerative, sensitized semiconductor electrodes. The steady-state anodic photocurrent for a sensitized semiconductor electrode has been reported, taking into account the rates of light absorption by the sensitizer, electron injection from the excited state of the sensitizer to the conduction band of the semiconductor, decay of the excited state, and reductive regeneration of the sensitizer by the redox electrolyte. In this model, the rate of recombination between the conduction band electron and the oxidized sensitizer and the reactions between the excited state and the redox couple have been assumed to be negligible.

Papageorgiou et al. (1996) investigated the physical–electrochemical properties of methylhexylimidazolium iodide (MHImI) and its mixtures with organic solvents, such as n-methyloxazolidinone and acetonitrile, and with other lower-viscosity molten salts, such as methyl-butylimidazolium triflate. Roušar et al. (1996) worked on the optimization of physical and geometrical parameters of an electrochemical photovoltaic regenerative solar cell with current leads located on opposite sides of a cell unit. The dependence of local current density on the length coordinate was expressed using dimensionless quantities, and a linear polarization curve was assumed. The optimization of the output on the ratio of the active surface to the total surface of the cell was explained, and the construction of a cell with line current collectors was also discussed.

The photoelectrochemical measurements of the modified TiO_2 electrodes in regenerative solar cells were reported by Heimer et al. (1996). The coordination compounds of the type $Ru(dmb)_2(LL)(PF_6)_2$, where dmb is 4,4'-$(CH_3)_2$-2,2'-bipyridine and LL is 4-(CH_3)-4'-$(COOH)$-2,2'-bipyridine, or 4-(CH_3)-4'-$((CH_2)_3COOH)$-2,2'-bipyridine, or 4-(CH_3)-4'-$((CH_2)_3$- $COCH_2COOC_2H_5)$-2,2'-bipyridine, were prepared for attachment to the TiO_2 surface. The optical and redox properties of these compounds in dichloromethane solution were investigated. The Langmuir adsorption isotherm model was used to analyze the binding to porous nanostructured TiO_2 films.

Garcia et al. (1998) used 4-phenylpyridine as ancillary ligand in ruthenium(II) polypyridyl complexes for sensitization of n-type TiO_2 electrodes. Two types of molecular sensitizers, cis-$[(dcbH_2)_2Ru(ppy)_2]^{2+}$ and cis-$[(dcbH_2)_2Ru(ppy)(H_2O)]^{2+}$ ($dcbH_2$ is 4,4'-$(CO_2H)_2$-2,2'-bipyridine; ppy is 4-phenylpyridine), have been prepared. One coordinated ppy, cis$[(dcbH_2)_2Ru(ppy)(H_2O)]^{2+}$, has a higher incident photon to current efficiency (IPCE) value than the corresponding derivative with two coordinated ppy, cis-$[(dcbH_2)_2Ru(ppy)_2]^{2+}$.

Falaras et al. (1998) synthesized and characterized $Ru(PPh_3)_2(dcbipy)Cl_2$ (PPh_3 is triphenylphosphine; dcbipy is 2,2'-bipyridyl-4,4'-dicarboxylate) for efficient photosensitization of titanium oxide in wet regenerative PEC cells. The broad bands in the visible spectrum as well as the reversibility of the redox couple Ru^{III}–Ru^{III} make this complex potentially beneficial for the photosensitization process. Vlachopoulos et al. (1988) used high surface area polycrystalline anatase films together with tris(2,2'-bipyridyl-4,4'-dicarboxylate)ruthenium(II), RuL_3^{4-}, as a sensitizer, and achieved efficient visible light energy conversion to electric current. In the presence of iodide as an electron donor, incident photon-to-current conversion efficiencies of 73% have been obtained at the λ_{max} of the dye. Bromide is oxidized under the same conditions with an efficiency of 56%. A regenerative cell based on the Br_2/Br^- redox system gives a monochromatic light-to-power conversion efficiency of 12% with a fill factor of 0.74.

A low power fullerene PEC solar cell, utilizing a regenerative polyiodide and ferri/ferrocyanide redox couple was studied by Licht et al. (1998a). They have demonstrated the photoelectrochemistry of illuminated and immersed single-crystal C_{60}, which has been shown to drive oxidation of several solution-phase redox couples. Utilization of a PEC solid–liquid junction, rather than a solid-state photovoltaic junction, resulted in improvement in the observed photocurrent. The spectral response and current–voltage behaviors in several electrolytes were also observed. A higher temperature and C_{60} oxygen depletion increase the photocurrent of fullerene PEC solar cells. Fullerene/iodide electrolyte PEC cells consisting of intrinsic single-crystal C_{60} in aqueous 3 M KI, 0.01 M I_2, or 0.1 M tetrabutyl ammonium iodide, 0.3 M $LiClO_4$ in acetonitrile solution, drive regenerative photoinduced iodide oxidation. The photocurrent was increased by an order of magnitude 6.4 µA cm^{-2} under 100 mW cm^{-2} illumination by an increase in the aqueous cell temperature from 24°C to 82°C. In a similar way, photocurrent was increased by O_2 depletion pretreatment (24 h at 400°C in Ar) (Licht et al. 1998e).

Argazzi et al. (1998) synthesized new Ru(dcbH)(dcbH$_2$)(L) sensitizers, where L is diethyldithiocarbamate, dibenzyldithiocarbamate, or pyrrolidinedithiocarbamate; dcbH is 4-(COOH)-4'-(COO$^-$)-2,2'-bipyridine; and dcbH$_2$ is 4,4'-(COOH)$_2$-2,2'-bipyridine. These have been used in nanocrystalline TiO_2 films for light to electrical energy conversion in regenerative PEC cells with I$^-$/I$_3^-$ acetonitrile electrolyte. Photophysical measurements show that the high photocurrent observed for cis-Ru(dcb)$_2$(NCS)$_2$/TiO_2 was due to efficient and rapid iodide oxidation.

The cis-Os(dcb)$_2$(CN)$_2$/TiO_2 PEC cells have the limiting role in iodide oxidation (Alebbi et al. 1998). In situ time-resolved diffuse reflectance measurements showed that a slow iodide oxidation rate is responsible for the low photocurrent efficiency of cis-Os(dcb)$_2$(CN)$_2$/TiO_2.

Yohannes and Inganäs (1999) studied solid-state PEC cells containing a conjugated polymer, poly[3-(4-octylphenyl)-2,2'-bithiophene], film as a photoactive electrode, a solid polymer electrolyte, poly[oxymethylene-oligo(oxyethylene)] complexed with redox couple, and a counterelectrode. The short-circuit current and open-circuit voltage generated with white light illumination at approximately one sun were 0.4 µA cm^{-2} and 240 mV, respectively. Two redox couples, I_3^-/I$^-$ and Eu$^{2+/3+}$, were used. The active junction between the conjugated polymer and the polymer electrolyte was responsible for photocurrent generation.

Cation-controlled interfacial charge injection in sensitized nanocrystalline TiO_2 was done by Kelly et al. (1999). The complex of Ru(deeb)(bpy)$_2$(PF$_6$)$_2$, where bpy is 2,2'-bipyridine and deeb is 4,4'-(COOEt)$_2$-2,2'-bipyridine, was anchored to nanocrystalline TiO_2 (anatase) or ZrO_2 films. Long-lived metal-to-ligand charge-transfer (MLCT) excited states were observed in acetonitrile (or 0.1 M tetrabutylammonium perchlorate) on both TiO_2 and ZrO_2. The addition of Li$^+$ increases both the efficiency and long wavelength sensitivity of the cell.

Garcia and Iha (2001) employed [(dcbH$_2$)$_2$ RuLL'], where dcbH$_2$ is 4,4'-(CO$_2$H)$_2$-2,2'-bipyridine and L,L' is substituted pyridines, as nanocrystalline TiO_2 sensitizers in PEC solar cells. This regenerative solar cell consists of a transparent conductive oxide (TCO) glass with a dye-sensitized TiO_2 semiconductor film as a photoanode, I_2/LiI solution in acetonitrile as a redox mediator, and a transparent Pt-coated TCO glass as a counterelectrode. In this cell, 50% IPCE was achieved at 400 and 550 nm.

Nanocrystalline titanium dioxide electrodes in regenerative PEC cells were sensitized by a series of platinum-based sensitizers of the general type Pt(NN)(SS), where NN is 4,4'-dicarboxy-2,2'-bipyridine (dcbpy) or 4,7-dicarboxy-1,10-phenanthroline (dcphen), and SS is ethyl-2-cyano-3,3-dimercaptoacrylate (ecda), quinoxaline-2,3-dithiolate (qdt), 1,2-benzenedithiolate (bdt), or 3,4-toluenedithiolate (tdt) (Islam et al. 2001a).

Islam et al. (2001b) prepared a new series of ruthenium(II) polypyridyl sensitizers with strong electron–donating dithiolate ligands Ru(dcbpy)$_2$(L) and Ru(dcphen)$_2$(L), where L is quinoxaline-2,3-dithiolate (qdt) or ethyl-2-cyano-3,3-dimercaptoacrylate (ecda) or 1,2-benzenedithiolate (bdt) or 3,4-toluenedithiolate (tdt); dcbpy is 4,4'-dicarboxy-2,2'-bipyridine; and dcphen is 4,7-dicarboxy-1,10-phenanthroline, for sensitization of nanocrystalline TiO_2 electrodes. These complexes show different sensitization to TiO_2 electrodes with increasing activity in the sequence (L = tdt, bdt, ecda, qdt) in regenerative PEC cells with I^-/I_3^- acetonitrile electrolyte. Both $Ru(dcbpy)_2(qdt)$ and $Ru(dcphen)_2(qdt)$ showed overall cell efficiency of about 3%–4% due to incident photon-to-current conversion efficiency of around 40%–45% at 500 nm. The low cell efficiency of ecda complexes may be due to slow regeneration of the dye by electron donation from iodide following charge injection into TiO_2.

New dyes of the type Ru(II)(bdmpp)(bpy) (where bdmpp is 2,6-bis(3,5-dimethyl-N-pyrazoyl) pyridine, and bpy is 2,2'-bipyridine-4,4'-dicarboxylic acid) were synthesized by Falaras et al. (2002). These compounds could be chemically anchored on TiO_2 films via ester-like linkage involving carboxylato groups. These complexes were tested to act as potential molecular antenna in dye-sensitized solar cells. The doctor blade technique was used to obtain opaque and transparent nanocrystalline TiO_2 thin-film electrodes, which were sensitized by these complexes and incorporated in a sandwich-type regenerative PEC solar cell containing 0.1 M LiI + 0.01 M I_2 in propylene carbonate. Platinized conductive glass was used as the counterelectrode. The overall energy conversion efficiency was 1.72%.

Garcia et al. (2002) prepared a transparent photoanode by immobilizing cis-[(dcbH$_2$)$_2$Ru(CNpy)(H$_2$O)]$^{2+}$ (dcbH$_2$ is 4,4'-(CO$_2$H)$_2$-2,2'-bipyridine; CNpy is 4-cyanopyridine) in a TCO substrate coated with nanocrystalline n-type TiO_2 film. Charge recombination, quenching processes, time-resolved experiments, and electron injection across the excited dye/semiconductor interface were studied.

Vorobets et al. (2002) used concentrated polysulfide solutions as electrolytes in regenerative PEC transducers. The equilibrium constants were determined, and the distribution of ion species in the solutions was measured. The effect of electrolyte solution on the output characteristics of PEC (based on cadmium selenide and cobalt sulfide) was also studied.

The pH dependence of sensitized photocurrent for porphyrin-derivatized planar TiO_2 films was studied by Watson et al. (2003). Porphyrin sensitizers like 5-(4-carboxyphenyl)-10,15,20-trimesitylporphinatozinc(II), 5-(4-carboxyphenyl)-10,15,20-trimesitylporphine, and 5-(4-carboxyphenyl)-10,15,20-trimesitylporphinatoplatinum(II) were used. All three porphyrins showed a 10-fold increase in the magnitude of sensitized photocurrent on acidification of the electrolyte from pH 12 to pH 2.

Wang et al. (2003) employed silica nanoparticles to solidify ionic liquids. These ionic liquid–based quasi-solid-state electrolytes were used in regenerative PEC cells. This yielded 7% efficiency with an amphiphilic ruthenium polypyridyl photosensitizer. Jasieniak et al. (2004) compared the photovoltaic performance of several porphyrin-derived TiO_2 films in regenerative PEC cells with the cells sensitized with a Ru(2,2'-bipyridyl-4,4'-dicarboxylate)$_2$(NCS)$_2$(N3) dye. They also studied differences in efficiencies of the porphyrin light absorbers using porphyrin sensitizers tetrakis(3',5'-di-$tert$-butylphenyl)porphyrin, tetrakis(3',5'-di-$tert$-butylphenyl)porphyrin zinc(II), and tetrakis(4'-carboxyphenyl)porphyrin.

Stergiopoulos et al. (2005) incorporated new novel compounds in sandwich-type regenerative PEC cells. A transition-metal complex with two terpy ligands [(2,2':6',2''-terpyridine-4'-iodophenyl) (2,2':6',2''-terpyridine-4'-phenylphosphonic acid)-ruthenium(II)]dichloride, was used to sensitize thin nanostructured SnO_2 film electrodes. A high molecular mass poly(ethylene) oxide electrolyte filled with titania and containing LiI and I_2 was used to transport the current of the cell at the counterelectrode. A continuous photocurrent (0.63 mA cm^{-2}) and a photovoltage (290 mV) were produced by this cell under white light illumination. Incident photon-to-current conversion efficiencies (IPCE) (16%) and energy conversion values (0.1%) were similar to those obtained with the standard N$_3$ dye under the same conditions.

Bergeron et al. (2005) implemented dye-sensitized mesoporous nanocrystalline SnO_2 electrodes and the pseudohalogen redox mediator (SeCN)$_2$/SeCN$^-$ or (SCN)$_2$/SCN$^-$ or the halogen redox mediator I_3^-/I^-, for regenerative solar cells. The sensitizers used were Ru(deeb)(bpy)$_2$(PF$_6$)$_2$, Ru(deeb)$_2$(dpp)(PF$_6$)$_2$, and Ru(deeb)$_2$(bpz)(PF$_6$)$_2$, where deeb is 4,4'-diethylester, bpy is 2-2'-bipyridine, dpp is

2,3-dipyridyl pyrazine, and bpz is bipyrazine. The donors present in the acetonitrile electrolyte were used to reduce the oxidized sensitizers. A rate constant $k > 10^8 \, s^{-1}$ was observed for sensitizer regeneration with iodide as the donor. It was found that in regenerative solar cells, the IPCE and open-circuit voltages were comparable for $(SeCN)_2/SeCN^-$ and I_3^-/I^- for all three sensitizers.

Bouroushian et al. (2006) synthesized polycrystalline (111) textured single CdSe and binary CdSe/ZnSe thin films by electrodeposition. These films were used as active electrodes in regenerative liquid-junction solar cells with aqueous sulfide–polysulfide or ferro–ferricyanide redox electrolytes. The influence of ZnSe on the PEC properties of CdSe was studied. Corrosion effects and stabilization of cells were also discussed.

Phenylenethynylene (PE) rigid linkers (para and meta) were used by Taratula et al. (2006) to attach pyrene to the surface of TiO_2 (anatase) and ZrO_2 nanoparticle thin films through the two COOH groups of an isophthalic acid unit. Photophysical properties of the compounds were influenced by the length of the PE linkers and position of substitution (para or meta). The long wavelength absorbance of the pyrene chromophore was shifted to the red with increasing conjugation, and the extinction coefficient was also increased. Pyrene excimer acts as a sensitizer.

Brugnati et al. (2007) prepared and characterized a series of bipyridine and pyridyl-quinoline Cu(I) complexes for their possible use as electron transfer mediators in regenerative PEC cells. It was observed that the best-performing mediators produced maximum IPCEs of the order of 35%–40%.

Brennan et al. (2009) integrated a triethanolamine-protected silane, 1-(3′-amino) propylsilatrane, with porphyrin- and ruthenium-based dyes and utilized it to link them to transparent semiconductor nanoparticle metal oxide films. Silatrane reacts with the metal oxide to form strong, covalent silyl ether bonds. Silatrane-functionalized dyes and analogous carboxylate-functionalized dyes were used as visible light sensitizers for porous nanoparticle SnO_2 photoanodes. The performances of the dyes were compared in nonregenerative or regenerative PEC cells. NADH (β-nicotinamide adenine dinucleotide) is used in nonregenerative cells as a sacrificial electron donor, and Hg_2SO_4/Hg is used as a sacrificial cathode. In the regenerative cell, the iodide/triiodide redox couple was used. The PEC cell efficiency was better improved with silatrane-based dyes than carboxylate-functionalized dyes. Silatranes are more capable agents for bonding organic molecules to metal oxide surfaces.

Price and Maldonado (2009) prepared macroporous GaP photoelectrodes from nondegenerately doped single-crystalline n-GaP (100). The PEC behaviors of planar and macroporous photoelectrodes were studied in nonaqueous regenerative PEC cells under potentiostatic control and using dry acetonitrile containing ferrocene/ferrocenium. The changes in short-circuit photocurrents, open-circuit photovoltages, and fill factors with increasing porosity resulted in improvement in the photoelectrode efficiency of macroporous n-GaP.

Lee et al. (2009) developed a new procedure to prepare selenide (Se^{2-}), which was used for depositing CdSe quantum dots (QDs) over mesoporous TiO_2 photoanodes by successive ionic layer adsorption and a reaction (SILAR) process in ethanol. Optimization of QD-sensitized TiO_2 films was done by using a cobalt redox couple $[Co(o-phen)_3]^{2+/3+}$ in regenerative PEC cells. Over 4% efficiency was achieved at 100 W m^{-2} with about 50% IPCE at its maximum on addition of a final layer of CdTe. It was revealed that CdTe-terminated CdSe QD cells gave better charge collection efficiencies compared to CdSe QD cells. They also prepared multilayered semiconductor (CdS/CdSe/ZnS)–sensitized TiO_2 mesoporous solar cells by the SILAR process (Lee et al. 2010). This multicomponent sensitizer (CdS/CdSe/ZnS) was evaluated in a polysulfide electrolyte solution as a redox mediator in regenerative PEC cells.

Onicha and Castellano (2010) focused on the effects of electrolyte composition, specifically the role of Li^+ and I^- ions, on the resultant photovoltaic performance of dye-sensitized solar cells based on a new Os(II) polypyridine complex, $[Os(^tBu_3tpy)(dcbpyH_2)(NCS)]PF_6$. Photophysical and electrochemical characterization of this complex confirmed the suitability of the dye to serve as a sensitizer for regenerative DSSCs on titania films. The photovoltaic performance of Os(II)-based DSSCs could be enhanced by simply modifying the composition of redox electrolytes used in the operational sandwich cells. An abundance of I⁻ played an important role in the effective regeneration of oxidized surface-bound osmium sensitizers. The power conversion efficiency for an Os(II)-based DSSC was calculated to be 4.7%.

Ruthenium(II) sensitizer (NBu$_4$)[Ru(4,7-dpp)(dcbpyH)(NCS)$_2$], (YS5) was synthesized, where NBu$_4$ is tetrabutylammonium; 4,7-dpp is 4,7-diphenyl-1,10-phenanthroline; and dcbpyH is the singly deprotonated surface anchoring derivative of 4,4′-dicarboxy-2,2′-bipyridine (dcbpyH$_2$). This was then incorporated into regenerative mesoscopic titania-based dye-sensitized solar cells (Sun et al. 2010).

Xia et al. (2010) prepared, characterized, and anchored coordination compounds [Ru(NH$_3$)$_5$(eina)](PF$_6$)$_2$, [Ru(NH$_3$)$_4$(deeb)](PF$_6$)$_2$, and [Ru(en)$_2$(deeb)](PF$_6$)$_2$, where eina is ethyl isonicotinate, deeb is 4,4′-(CO$_2$CH$_2$CH$_3$)$_2$-2,2′-bipyridine, and en is ethylenediamine, to mesoporous nanocrystalline (anatase) TiO$_2$ thin films immersed in CH$_3$CN at room temperature. The PEC performances of [Ru(NH$_3$)$_4$(deeb)](PF$_6$)$_2$ and [Ru(en)$_2$(deeb)](PF$_6$)$_2$ on TiO$_2$ in regenerative solar cells were consistent with excitation wavelength–dependent electron injection. Heuer et al. (2010) also prepared certain Ru(II) compounds [Ru(bpy)$_2$(mcbH)]$^{2+}$ and [Ru(bpy)$_2$(dafo)]$^{2+}$, where bpy is 2,2′-bipyridine; mcbH is 3-(CO$_2$H)-2,2′-bipyridine; and dafo is 4,5-diazafluoren-9-one. These compounds were anchored to nanocrystalline mesoporous TiO$_2$ thin films for excited-state and interfacial electron transfer studies. The IPCE was found to be lower for Ru(bpy)$_2$(dafo)/TiO$_2$.

Mesoporous SnO$_2$ spheres were prepared for the first time by electrochemical anodization of tin foil in basic media. Their structural elucidation indicated that these spherical particles consist of an agglomeration of SnO$_2$ nanocrystals, resulting in a high internal surface area, which would make them a potential photoanode material for use in semiconductor-sensitized solar cells. After treating SnO$_2$ nanocrystals with aqueous TiCl$_4$ solution, a thin layer of CdSe was coated by using ionic layer adsorption and a reaction method. A power conversion efficiency of ~1.91% was achieved in this regenerative PEC cell after deposition of a ZnS passivation layer. Hagedorn et al. (2010) analyzed the steady-state PEC responses of semiconductor nanowires in nonaqueous regenerative PEC cells. The responses were used to determine the effect of width of the depletion region, relative to the nanowire radius, on photogenerated carrier collection efficiency.

Johansson et al. (2011) synthesized, characterized, and contrasted three ruthenium compounds (i.e., *cis*-Ru(dcbq)$_2$(NCS)$_2$, *cis*-Ru(dcbq)(bpy)(NCS)$_2$, and *cis*-Ru(dcb)(bq)(NCS)$_2$, where bpy is 2,2′-bipyridine, dcb is 4,4′-(CO$_2$H)$_2$-2,2′-bipyridine, bq is 2,2′-biquinoline, and dcbq is 4,4′-(CO$_2$H)$_2$-2,2′-biquinoline) with the well-known N$_3$ compound (i.e., *cis*-Ru(dcb)$_2$(NCS)$_2$) in dye-sensitized solar cells. These compounds maintained the same *cis*-Ru(NCS)$_2$ core with a variation in the energy of the π* orbitals of the diimine ligand in the order bpy > dcb > bq > dcbq. The lowered π* orbitals resulted in enhanced red absorption as compared to N$_3$. Sensitization from 400 to 900 nm was realized with *cis*-Ru(dcb)(bq)(NCS)$_2$ with HCl pretreated TiO$_2$ in regenerative solar cells. Power conversion efficiencies as high as 6.5% were obtained.

The diketonato-ruthenium(II)-polypyridyl sensitizers were anchored to nanocrystalline TiO$_2$ films for light to electrical energy conversion in regenerative PEC cells by Islam et al. (2011). Lewerenz (2011) described the principal design of nanoemitter solar cells and their applicability in PEC solar cells that operate in the regenerative photovoltaic/photoelectrocatalytic mode as well as in solid-state photovoltaics. Individual steps in the preparation of photovoltaic and photoelectrocatalytic electrochemical solar cells with n- and p-type Si were discussed, and the electronic properties of nanoemitter solar cells at the solid–liquid phase boundary were also investigated. Klahr and Hamann (2011) employed the atomic layer deposition technique to grow conformal thin films of hematite on transparent conductive oxide substrates and used it as an electrode in regenerative PEC cells. An increase in the photocurrent density and photovoltage was obtained by varying the pH and redox potentials of the contacting electrolyte, which was attributed to increasing the built-in voltage.

Xiang et al. (2011) prepared p-type cuprous oxide photoelectrodes by the thermal oxidation of Cu foils. These electrodes exhibited open-circuit voltages in excess of 800 mV in nonaqueous regenerative PEC cells. Cuprous oxide gave an open-circuit voltage of 820 mV and a short-circuit current density of 3.1 mA cm^{-2} in contact with the decamethylcobaltocene$^{+/0}$ (Me$_{10}$CoCp$_2^{+/0}$) redox couple, under simulated air mass 1.5 illumination. Han et al. (2011) explored a new class of thiocyanate-free cyclometalated ruthenium sensitizers for sensitizing nanocrystalline TiO$_2$ solar cells. A power conversion efficiency of 4.76, a short-circuit photocurrent density of 11.21 mA cm^{-2}, an open-circuit voltage of 0.62 V, and a fill factor of 0.68 were obtained under standard air mass (AM) 1.5 sunlight.

In this cell, I$^-$/I$_3^-$ acetonitrile electrolyte was used. The complexes efficiently sensitized TiO$_2$ over a broad spectral range and showed an open-circuit potential of about 600 mV with a fill factor >0.70. Carli et al. (2013) compared different poly(3,4-ethylenedioxythiophene) (PEDOT)–based counterelectrodes, obtained by potentiostatic electropolymerization of 3,4-ethylenedioxythiophene (EDOT) monomer on fluorine tin oxide (FTO) surfaces, with the platinum- and gold-coated electrodes in order to evaluate the potential use of PEDOT counterelectrodes in dye-sensitized PEC cells. A series of DSSC devices utilizing Co(III)/(II) polypyridine redox mediators ([Co(bpy)$_3$]$^{3+/2+}$, [Co(phen)$_3$]$^{3+/2+}$, [Co(dtb)$_3$]$^{3+/2+}$, where bpy is 2,2′-bipyridine, dtb is 4,4′di-*tert*-butyl-2,2′-bipyridine, and phen is 1,10-phenanthroline) having distinct electrochemical characteristics were studied. Porous PEDOT/ClO$_4$ counterelectrodes have also been proven to possess sufficient electrocatalytic properties when paired with cobalt-based redox mediators, making PEDOT-based counterelectrodes attractive for their use in DSC applications.

Mi et al. (2013) studied the behavior of WO$_3$ photoanodes, in contact with a combination of four anions (Cl$^-$, CH$_3$SO$_3^-$, HSO$_4^-$, and ClO$_4^-$) and three solvents (water, acetonitrile, and propylene carbonate), to investigate the role of the semiconductor surface, the electrolyte, and redox kinetics on the current density versus potential properties of n-type WO$_3$. The internal quantum yield for 0.50 M tetra(n-butyl)ammonium perchlorate in propylene carbonate exceeded unity at excitation wavelengths of 300–390 nm, which was an indication of current doubling. A regenerative PEC cell based on the reversible redox couple B$_{10}$Br$_{10}^{-/2-}$ in acetonitrile, with a solution potential of ~1.7 V versus the normal hydrogen electrode, showed an open-circuit photovoltage of 1.32 V under 100 mW cm^{-2}.

4.2.2 Photoelectrolytic Solar Cells

The photoelectrosynthetic (photoelectrolytic or photocatalytic) cells utilize photon energy input ($E \geq E_g$) to produce a net chemical change in the electrolyte solution ($\Delta G \neq 0$) when the reaction at the counterelectrode is not exactly opposite of the hole transfer reaction at the illuminated semiconductor–liquid interface. The solar energy conversion efficiency for photoelectrolysis cells is

$$\eta = \frac{\text{Energy stored as fuel} - \text{Electrical energy supplied}}{\text{Incident solar energy}} \tag{4.3}$$

Two redox systems are present in photoelectrolytic cells. One redox system reacts with the holes at the surface of the n-semiconductor electrode, while the other reacts with the electrons entering the counterelectrode. Anodic and cathodic compartments need to be separated to prevent mixing of the two redox couples. In case of n-SC, water is oxidized to oxygen at the semiconductor photoanode and reduced to hydrogen at the cathode. The overall reaction is the cleavage of water by sunlight:

$$H_2O \rightarrow H_2 + \frac{1}{2}O_2 \tag{4.4}$$

The commercial viability of dye-sensitized photoelectrosynthesis cells (DSPECs) depends on the stability of the surface-bound molecular chromophores and catalysts. Water-stable, surface-bound chromophores, catalysts, and assemblies are essential in DSPECs for the generation of solar fuels by water splitting and CO$_2$ reduction to CO, other oxygenates, or hydrocarbons.

Photopotentials were rarely above 0.6–0.8 V in PEC cells based on single junctions of semiconductors with solutions or metals. The higher photovoltages required multijunction cells involving multilayer electrodes or a series connection of PEC cells. These principles were demonstrated with the instruments based on silicon spheres and Si p-n junctions contacting a solution via a noble metal layer. Reactions considered include generation of Cl_2 with reduction of O_2 or generation of H_2, the photobromination of phenol, and the photochlorination of cyclohexene in acetonitrile (White et al. 1985).

Photoelectrosynthesis of dihydrogen through water splitting using S_x^{2-} ($x = 1, 2, 3...$) as an anolyte, n-CdSe as the semiconductor electrode, and platinized-Pt as the cathode was reported by Bhattacharyya et al. (1996). The semiconductor electrode was prepared by depositing a thin film of CdSe by the RF sputtering technique on a stainless steel substrate. This is a first step in developing a device for a rechargeable solar electrosynthetic cell.

Wadhawan et al. (2002) studied the mechanism of photoelectrochemically induced halex reactions. They performed PEC reductions of para-bromonitrobenzene and 2,4-dibromonitrobenzene in acetonitrile solutions. It was found that p-bromonitrobenzene followed a homogeneous $EC_{rev}CE$ pathway in acetonitrile solutions containing tetrabutylammonium-based supporting electrolytes, but when changing the supporting electrolyte to a salt of the tetramethylammonium cation, the mechanism was changed qualitatively, and an ECEE pathway was observed. An ECECE mechanism was observed on PEC reduction of 2,4-dibromonitrobenzene in acetonitrile solution containing supporting electrolytes derived from the tetrabutylammonium cation. When chloride-supporting electrolytes were used, then there was a light-induced rupture of a C–Br bond to a C–Cl bond. This halogen change is a novel approach to halex reactions.

A strategy for the two-electron formation of C–C bonds with molecular catalysts anchored to semiconductor nanocrystallites was demonstrated by Ardo et al. (2011). The semiconductor used was the anatase polymorph of TiO_2 present as a nanocrystalline, mesoporous thin film, and the catalyst was cobalt meso-5,10,15,20-tetrakis(4-carboxyphenyl)porphyrin chloride, Co(TCPP)Cl. Non-Nernstian two-electron transfer photocatalysis at iron protoporphyrin chloride-TiO_2 interfaces was performed in PESC. This non-Nernstian behavior was attributed to an environmentally dependent potential drop across the molecule–semiconductor interface, and sustained photocurrents were quantified in photoelectrosynthetic solar cells under forward bias.

Song et al. (2011) reported the use of $[(Ru(bpy)_2(4,4'-(PO_3H_2)_2bpy)]^{2+}$ attached to TiO_2 nanoparticle films in a dye-sensitized photoelectrosynthesis cell for H_2 production. Photoinduced electron transfer in a chromophore-catalyst assembly anchored to TiO_2 was investigated by Ashford et al. (2012). A light-harvesting chromophore and a water oxidation catalyst were linked by a saturated bridge designed to enable long-lived charge-separated states. Following excitation of the chromophore, a rapid electron injection into TiO_2 occurs, and an intra-assembly electron transfer occurs on the subnanosecond time scale followed by microsecond–millisecond back-electron transfer from the semiconductor to the oxidized catalyst. Song et al. (2012) also worked on the solar fuel formation in DSPECs.

The photostability of $[Ru^{II}(bpy)_2(4,4'-(PO_3H_2)_2bpy)]Cl_2$ (bpy is 4,4'-bipyridine) on nanocrystalline TiO_2 and ZrO_2 films was explored by Hanson et al. (2012). They examined stability by monitoring visible light absorbance spectral changes, during 455 nm photolysis (475 mW cm^{-2}) in a variety of conditions relevant to dye-sensitized solar cells and dye-sensitized photoelectrosynthesis cells.

Enhancing the surface-binding stability of chromophores, catalysts, and chromophore-catalyst assemblies attached to metal oxide surfaces is important for the development of photoelectrosynthesis cells. Phosphonate-derivatized catalyst and molecular assembly provided a basis for water oxidation on these surfaces in acidic solution but are unstable towards hydrolysis and loss from surfaces as the pH was increased (Vannucci et al. 2013). This provided a hybrid approach to heterogeneous catalysis combining the advantages of systematic modifications possible by chemical synthesis with heterogeneous reactivity. Alibabaei et al. (2013a) also described solar water splitting in a dye-sensitized photoelectrosynthesis cell. A derivatized, core–shell nanostructured photoanode was used with the core high surface area conductive metal oxide film coated with indium tin oxide or antimony tin oxide with a thin outer shell of TiO_2 formed by atomic layer deposition. A chromophore-catalyst assembly, $[(PO_3H_2)_2bpy)_2Ru(4-Mebpy-4-bimpy) Rub(tpy)(OH_2)]^{4+}$, was attached to the TiO_2 shell, which combines both light absorber and water oxidation catalyst in a single molecule.

Visible photolysis of the resulting core–shell assembly structure with a Pt cathode resulted in water splitting into hydrogen and oxygen with an absorbed photon conversion efficiency of 4.4% at peak photocurrent.

Norris et al. (2013) synthesized phosphonic acid derivatized bipyridine ligands and their ruthenium complexes for DSPECs. This provided a stable chemical binding on metal oxide surfaces. Hanson et al. (2013) used ALD of TiO_2 on nanocrystalline TiO_2 prefunctionalized with the dye molecule $[Ru(bpy)_2(4,4'-(PO_3H_2)bpy)]^{2+}$(RuP), to stabilize surface-bound molecules. The resulting films were more photostable than untreated films, and the desorption rate constant decreased exponentially with increased thickness of ALD TiO_2 overlayers. The photodriven accumulation of two oxidative equivalents at a single site was reported by cross electron transfer on TiO_2 (Song et al. 2013a). The TiO_2 was coloaded with a ruthenium polypyridyl chromophore $[Ru(bpy)_2((4,4'-(OH)_2PO)_2bpy)]^{2+}$ ($Ru^{II}P^{2+}$, bpy is 2,2'-bipyridine and $((OH)_2PO)_2$-bpy is 2,2'-bipyridine-4,4'-diyldiphosphonic acid) and a water oxidation catalyst [Ru(Mebimpy) $((4,4'-(OH)_2PO-CH_2)_2bpy)(OH_2)]^{2+}$ ($Ru^{II}OH_2^{2+}$; Mebimpy is 2,6-bis(1-methylbenzimidazol-2-yl)pyridine; and $(4,4'-(OH)_2PO-CH_2)_2bpy)$ is 4,4'-bis-methlylenephosphonato-2,2'-bipyridine). Steady-state illumination of coloaded TiO_2 photoanodes in a DSPEC configuration resulted in the makeup of -$Ru^{III}P^{3+}$, -$Ru^{III}OH^{2+}$, and -Ru^{IV}-O^{2+}, with -Ru^{IV}-O^{2+} formation preferred at high chromophore-to-catalyst ratios.

Thompson et al. (2013) investigated an integrated hybrid approach for making solar fuels in DSPECs. Then, $[Ru(bpy)_3]^{2+*}$ and its relatives provided a basis for exploring the energy gap law for nonradiative decay, the role of molecular vibrations, and effects of solvent and medium on excited-state properties. Alibabaei et al. (2013b) used metal oxide materials in dye-sensitized photoelectrosynthesis cells to make solar fuels.

Song et al. (2013b) observed Li^+ diffusion at the TiO_2 interface in DSPEC. This cell worked in aqueous solution at pH 4.5, and the rate constants for Li^+ intercalation and release were observed as 0.22 s^{-1} and 0.014 s^{-1}, respectively. Both processes were considerably slower in the more viscous solvent propylene carbonate. Release rate constants of Li^+ were observed to be $<2.0 \times 10^{-4} s^{-1}$. Accumulation of Li^+ under these conditions shifts conduction band/trap states to less negative potentials, thus increasing electron lifetime in TiO_2.

A sensitized Nb_2O_5 photoanode for hydrogen production in a dye-sensitized photoelectrosynthesis cell was used by Luo et al. (2013). The conduction band potential is slightly positive (<0.1 eV) in a T-phase orthorhombic Nb_2O_5 nanocrystalline film, relative to that of anatase TiO_2. The H_2 quantum yield and photostability measurements showed that Nb_2O_5 was comparable but not superior to TiO_2 when ethylenediaminetetraacetate anion ($EDTA^{4-}$) was added in DSPEC as a reductive scavenger.

Dye-sensitized photoelectrosynthesis cells used in artificial photosynthesis require the assembly of a chromophore and catalyst in close proximity on the surface of a transparent, high-bandgap oxide semiconductor for integrated light absorption and catalysis. It was confirmed by controlled potential electrolysis experiments that the surface-bound assemblies function as water oxidation electrocatalysts. The electrochemical kinetics data indicated that the assemblies exhibit greater than ten-fold rate enhancements as compared to the homogeneous catalyst alone (Ryan et al. 2014), whereas Song et al. (2014) reported that light-driven dehydrogenation of benzyl alcohol to benzaldehyde and hydrogen occurred in a dye-sensitized photoelectrosynthesis cell. The photoanode consists of nano-ITO, nano-ITO/TiO_2, and mesoporous films of TiO_2 nanoparticles in the DSPEC. Excitations of chromophore and electron injection were followed by cross-surface electron transfer activation of the catalyst to -$Ru^{IV}=O^{2+}$, and in turn, benzyl alcohol was oxidized to benzaldehyde. The injected electrons are transferred to a Pt electrode for the production of H_2. Sustained absorbed photon-to-current efficiency of 3.7% was achieved for benzyl alcohol dehydrogenation at the optimized shell thickness, which amounts to an enhancement of ~10 as compared to TiO_2.

Ashford et al. (2014a) synthesized, characterized, and studied the electrochemical and photophysical properties of a series of Ru(II) polypyridyl complexes of the type $[Ru(bpy)_2(N-N)]^{2+}$ (bpy is 2,2-bipyridine and N-N is a bidentate polypyridyl ligand). The nature of the N-N ligand was changed, either through increased conjugation or by incorporation of noncoordinating heteroatoms in this series. Excited-state reduction potentials and $Ru^{3+/2+}$ potentials were assessed in the context of preparing low energy light absorbers for application in dye-sensitized photoelectrosynthesis cells.

Ashford et al. (2014b) reported water oxidation by an electropolymerized catalyst on derivatized mesoporous metal oxide electrodes. This has certain applications in DSPECs. It was observed that catalytic rate constants for water oxidation by the polymer films were similar to those for the phosphonated molecular catalyst on metal oxide electrodes, which indicates that the physical properties of the catalysts were not significantly changed in the polymer films. The results of controlled potential electrolysis indicated sustained water oxidation over multiple hours with no decrease in the catalytic current.

A helical peptide chromophore water oxidation catalyst assembly on a semiconductor surface was loaded onto nanocrystalline TiO_2 by Bettis et al. (2014). The oligoproline scaffold approach is quite appealing due to its modular nature and helical tertiary structure. It maintained the controlled relative positions of the chromophore and catalyst. Ultrafast transient absorption spectroscopy was used to analyze the kinetics of the first photoactivation step for oxidation of water in the assembly. This step in the water oxidation cycle of the chromophore-catalyst assembly anchored to TiO_2 was completed within 380 ps.

4.2.3 Photocatalytic Cells

In photoelectrocatalytic cells, the rate of reaction will increase, when $\Delta G < 0$. Aqueous suspensions composed of irradiated semiconductor particles may be considered to be an assemblage of short-circuited microelectrochemical cells operating in the photocatalytic mode, for instance, the photooxidation of organic compounds, or reactions with high activation energy.

$$N_2 + 3H_2 \rightarrow 2NH_3 \tag{4.5}$$

Photocatalytic oxidation of methyl red by TiO_2 in a PEC cell with titania thin film loaded on titanium as a photoanode has been studied by Shi et al. (2005). Degradation of methyl red was accelerated in PEC cells due to a reduced recombination of photogenerated carriers by the separation of the anodic and cathodic reactions. A thin SnO_2 film was used as an interlayer between the substrate and the TiO_2 coating $Ti/SnO_2/TiO_2$ to obtain the assembled photoanodes, leading to an increased separation efficiency of photogenerated carriers. The PEC water splitting for hydrogen production using a combination of CIGS2 solar cell and RuO_2 photocatalyst was studied by Dhere et al. (2004). A PEC setup using a multiple bandgap combination of $CuIn_{1-x}Ga_xS_2$ (CIGS2) thin-film photovoltaic cell and ruthenium oxide photocatalyst was presented.

The importance of bandgap energy and flat-band potential for application of modified TiO_2 photoanodes in water photolysis was reported by Radecka et al. (2008). The forbidden bandgap decides the absorption spectrum, and the flat-band potential affects the recombination probability on water photolysis. A three-electrode PEC cell with a TiO_2 thin-film photoanode immersed in liquid electrolyte of variable pH was used. Titanium dioxide photoanodes doped with chromium (up to 16 at %) and tin (up to 50 at %) were synthesized by RF reactive sputtering. The photoconversion efficiency of TiO_2 + 7.6 at % Cr was much smaller ($\eta_c = 0.1\%$) than that of undoped TiO_2 ($\eta_c = 1.8\%$) and TiO_2 doped with 8 at % of Sn ($\eta_c = 1.0\%$).

Jeng et al. (2010) proposed a novel PEC cell for generation of hydrogen via photocatalytic water splitting. This PEC cell is a membrane electrode assembly (MEA) integrated with Degussa P25 TiO_2 powder as a model photocatalyst for the photoanode and Pt catalyst powder for the dark cathode. It serves as an effective separator for the generated hydrogen and oxygen as well as a compact photocatalytic reactor for water splitting. This novel PEC can be operated without the addition of water in the cathode compartment, which showed improved photoconversion efficiency. Degussa P25/$BiVO_4$ mixed photocatalyst was found to significantly enhance the hydrogen generation.

Preparation of TiO_2 P-25 working electrodes on Ti substrates (TiO_2/Ti) was reported by Philippidis et al. (2010) by using pathogenic bacteria as a model in a novel batch PEC reactor. The characterization and photoelectrocatalytic activity toward the inactivation of *Escherichia coli* XL-1 blue (*E. coli*) colonies were studied. The flat-band potential ($V_{fb} = -0.54$ V) versus Ag/AgCl) was determined by differential capacitance measurements. The photoelectrocatalytic inactivation of *E. coli* colonies has been studied under artificial illumination in a novel photoelectrocatalytic reactor.

Sun et al. (2011) observed the effect of annealing temperature on the hydrogen production of TiO_2 nanotube arrays in a two-compartment PEC cell. Highly ordered TiO_2 nanotube arrays with 4 µm length were prepared by a rapid anodization process in ethylene glycol electrolyte in order to enhance the solar conversion efficiency. The highest photoconversion efficiency of 4.49% and maximum hydrogen production rate of 122 µmol/h·cm^{-2} has been observed using TiO_2 NTs annealed at 450°C.

Iodine-doped-poly(3,4-ethylenedioxythiophene)-modified Si nanowire one-dimensional (1D) core–shell arrays were investigated by Yang et al. (2012) as efficient photocatalysts for solar hydrogen. A new 1D core–shell strategy was developed for a hydrogen generation PEC cell. An encouraging solar-to-chemical energy conversion efficiency was shown by this Si/iodine-doped poly(3,4-ethylenedioxythiophene) (PEDOT) 1D nanocable array. Photocatalytic efficiency and stability of Si NW arrays could be increased by coating with iodine-doped PEDOT.

Rahman et al. (2012) studied the effects of doping (C or N) and co-doping (C + N) on the titania coating for its application in solar water splitting. The bandgap (~3.2 eV) of the TiO_2 photocatalyst is attributed to its relatively low photoactivity toward visible light. Therefore, efficient materials in PEC cells should have a smaller bandgap (approximately 2.4 eV). The effect of dopant (C or N) and co-dopant (C + N) was seen on the physical, structural, and photoactivity of TiO_2 nano-thick coating. The type and amount of doping influenced the coating growth rate, structure, surface morphology, and roughness. The photocurrent density (indirect indication of water-splitting performance) of the C-doped photoanode was approximately 26% higher than undoped photoanode. The performance of coating doped with N or N + C in the PEC cell was not found satisfactory due to their higher charge recombination properties.

Preparation and characterization of co-doped TiO_2 materials for solar light–induced current and photocatalytic applications were done by Ganesh et al. (2012). A conventional coprecipitation and sol-gel dip coating technique was used for the preparation of different amounts of co-doped TiO_2 powders and thin films. The synthesized powders and thin films were subjected to thermal treatments from 400°C to 800°C. The photocatalytic ability of the compounds was determined by the degradation of methylene blue (MB). The results indicated that the co-doped TiO_2 powder is mainly of the anatase phase and possesses reasonably high specific surface area, low bandgap energy, and flat-band potentials amenable to water oxidation in PEC cells. The 0.1 wt.% co-doped TiO_2 composition provided the higher photocurrent, n-type semiconducting behavior, and higher photocatalytic activity among various co-doped TiO_2 compositions and pure TiO_2.

Chao et al. (2012) studied visible light–driven photocatalytic and PEC properties of porous SnS_x ($x = 1, 2$) architectures. Porous SnS and SnS_2 architectures on a large scale were synthesized using a facile and template-free polyol refluxing process. It was observed that the as-synthesized SnS and SnS_2 products mainly consist of porous flower-like microstructures with reasonable BET surface areas of 66 m^{2}g^{-1} and 33 m^{2}g^{-1}, respectively. Both samples have excellent photosensitivity and response with greatly enhanced $I_{on/off}$ as high as 1.4×10^3. The potential applications of the SnS_x nanostructures in visible light–driven photocatalysis, high response photodetectors, and other optoelectronic nanodevices have been investigated.

Generation of fuel from CO_2-saturated liquids using a p-Si nanowire n-TiO_2 nanotube array PEC cell was observed by Latempa et al. (2012). Light-driven, electrically biased p-n junction PEC cells immersed in an electrolyte of CO_2-saturated 1.0 M $NaHCO_3$ were investigated for generating hydrocarbon fuels. The PEC photocathode was composed of p-type Si nanowire arrays with and without copper sensitization, while the photoanode was composed of n-type TiO_2 nanotube array films. The PEC cells convert CO_2 into hydrocarbon fuels such as methane under bandgap illumination, along with carbon monoxide and substantial rates of hydrogen generation due to water photoelectrolysis. Methane and ethylene were formed at the combined rate of 201.5 nM/cm²-h at an applied potential of −1.5 V versus Ag/AgCl with the addition of C3–C4 hydrocarbons. This technique provides a unique path for the photocatalytic reduction of CO_2 with subsequent generation of higher-order hydrocarbons and syngas constituents of carbon monoxide and hydrogen utilizing Earth-abundant materials.

The water oxidation reactions based on photocatalyst suspension and PEC cells are currently the most important field of research for sustainable energy. The effects of pH, crystallinity, and grain size of tungsten trioxide (WO_3) particles on photoconductivity were studied (Ho et al. 2012). A simple hydrothermal route was employed to synthesize Cs-loaded WO_3 (Cs-WO_3) particles to enhance the photochemical reactivity. A different photoreactivity performance in the PEC and photocatalyst suspension system was shown by the photoanodes and photocatalyst based on the Cs-WO_3. This important fundamental insight can assist in the optimization of WO_3 particles as the photocatalyst and photoanode for future hybrid photocatalysis-electrolysis water-splitting systems. Surface-modified WO_3 particles were used for efficient water-splitting catalyst due to high photocatalytic capability.

Wongwanwattana et al. (2012) fabricated PEC cells based on metal-doped (Be and Fe) and Pt-loaded nanostructured-TiO_2 films as working electrodes for solar hydrogen production. Titanium tetraisopropoxide was used as a precursor. The photocurrent density of Be-doped PEC without an external applied potential has been observed as 0.32 mA cm^{-2} under illumination of 75 mW cm^{-2}. This device produced hydrogen by water photoelectrolysis at the rate of 0.1 mL h^{-1} cm^{-2}, with a photoconversion efficiency of 0.52%, whereas the maximum photocurrent density of Fe-doped PEC was obtained as 0.80 mA cm^{-2} without an external applied potential and under the illumination of 100 mW cm^{-2} with photoconversion efficiency of 0.98%.

A facile polyol refluxing process has been used for the synthesis of tin sulfide (SnS), thick 10–20 nm nanoribbons and length up to several microns (Chao et al. 2013). The photoconductive properties of the SnS nanoribbons were determined by collecting the samples into PEC cells, exhibiting excellent photosensitivity. The photocurrent density was 87 μA cm^{-2}, which is the highest in all the SnS photoelectrodes.

Gao et al. (2013) reported reduced graphene oxide–$BiVO_4$ (GO-$BiVO_4$) composite for PEC cells and photocatalysis. The PEC cell working electrode was prepared by a doctor blade method on fluorine-doped tin oxide (FTO) coated glass. Graphene oxide–$BiVO_4$ composites were synthesized by hydrothermal reaction and, subsequently, reduced graphene oxide (RGO)-$BiVO_4$ was obtained with annealing in an N_2 atmosphere. The RGO-$BiVO_4$ films showed enhanced PEC properties under visible light compared to pure $BiVO_4$ film. It gave a high photocurrent response of 160 μA cm^{-2} and quantum efficiency of over 1.81%. The enhanced PEC activity of RGO-$BiVO_4$ could be explained by its larger recombination resistance (R_{rec}) and longer electron lifetime (τ).

Jacobsson et al. (2013) reported that CIGS ($CuIn_xGa_{1-x}Se_2$) can be utilized in the photocatalytic reduction of water into hydrogen as an efficient absorber material. The efficiency was significantly improved, and a photocurrent of 6 mA cm^{-2} was observed for the reduction reaction in the configuration of a PEC cell by utilizing a solid-state p-n junction for charge separation and a catalyst deposited on the surface. The separation between the charge carrier generation was demonstrated, which takes place in the solar cell due to the catalysis. It takes place in the electrolyte leading to improved stability, while keeping the essential functions of the processes. Photocurrents in excess of 20 mA cm^{-2} were reached for the reduction half-reaction by incorporating appropriate charge separation layers and optimizing the catalytic conditions at the surface of the electrodes.

The TiO_2 branched nanorod arrays (TiO_2 BNRs) were prepared (Su et al. 2013) with attached plasmonic Au nanoparticles on the surface. It was observed that Au/TiO_2 BNR composites exhibit high photocatalytic activity in PEC water splitting. The unique structure of Au/TiO_2 BNRs shows enhanced activity with a photocurrent of 0.125 mA cm^{-2} under visible light (≥420 nm) and

2.32 ± 0.1 mA cm^{-2} under AM 1.5 G illumination (100 mW cm^{-2}). Furthermore, the Au/TiO$_2$ BNRs achieve the highest efficiency of ~1.27% at a low bias of 0.50 V versus reversible hydrogen electrode (RHE), indicating elevated charge separation and transportation efficiencies. It was also reported that the high PEC performance was mainly due to the plasmonic effect of Au nanoparticles, which enhances the visible light absorption, together with the large surface area, efficient charge separation, and high carrier mobility of the TiO$_2$ BNRs. The carrier density of Au/TiO$_2$ BNRs was nearly six times higher than the pristine TiO$_2$ BNRs as calculated by the Mott-Schottky plot.

Yang et al. (2013) suggested that dual cocatalysts are essential for developing photocatalytic efficiency and PEC water-splitting reactions. It was observed that loading suitable dual cocatalysts on semiconductors can significantly increase the photocatalytic activities of hydrogen and oxygen evolution reactions, and it makes the overall water-splitting reaction possible.

Enhancing visible light PEC water splitting through transition-metal-doped TiO$_2$ nanorod arrays was reported by Wang et al. (2014). It was difficult to maintain high photocatalytic activity while extending the photoresponse from the ultraviolet to the visible light region. Use of a transition-metal doping treatment enhances the performance of TiO$_2$ nanorods in the visible light region for PEC water splitting. Then, Fe, Mn, and Co were used as dopants, and it was found that Fe doping is the most effective among them. The photocurrent density of the Fe-TiO$_2$ sample significantly increases with bias voltage and reaches 2.92 mA cm^{-2} at 0.25 V versus Ag/AgCl, which is five times higher than that of undoped TiO$_2$. The Fe-TiO$_2$ nanorod sample significantly improves the photoresponse, not only in the UV region but also in the visible light region. The Fe-TiO$_2$ nanorods have many applications, such as solar water splitting, dye-sensitized solar cells, and photocatalysis.

The preparation of extremely smooth and boron-fluorine co-doped TiO$_2$ nanotube arrays with enhanced PEC and photocatalytic performance was reported by Li et al. (2014). Highly ordered TiO$_2$ nanotube arrays (BF-TNTs) were prepared in the unique NH$_4$BF$_4$-based electrolyte by the anodization method. Boron and fluorine elements were simultaneously doped in the NH$_4$BF$_4$ electrolyte to obtain BF-TNTs during the anodization process. Electrolyte NH$_4$BF$_4$ provides a smoother effect on the tube walls of BF-TNTs. The BF-TNTs have better PEC properties and photocatalytic performance than do F-TNTs. Dual-template synthesis and PEC performance of hierarchical porous zinc oxide were reported by Zhao et al. (2014). The ZnO was prepared by a sol-gel method. A novel hierarchical porous ZnO was successfully synthesized through a sol-gel method. The photoelectrochemical property of this hierarchical porous ZnO was investigated.

The unveiling of two electron transport modes in oxygen-deficient TiO$_2$ nanowires and their influence on PEC operation were investigated by Chen et al. (2014). One of the most promising ways to enhance light-harvesting and photocatalytic efficiencies of PEC cells for water splitting is to introduce oxygen vacancies (V_O) into TiO$_2$ materials, among others. The nature of electron transport in V_O-TiO$_2$ nanostructures was not confirmed, especially in an operating device. A modulated photocurrent technique was used to study the electron transport property of V_O-TiO$_2$ nanowires (NWs). The electron transport in pristine TiO$_2$ displays a single trap-limited mode, whereas two electron transport modes were found in V_O-TiO$_2$ NWs. The considerably higher diffusion coefficient of the trap free transport mode grants a more rapid electron flow in V_O-TiO$_2$ NWs than that in pristine TiO$_2$ NWs.

ZnO was used as a photocatalytic electrode to TiO$_2$ for solar-powered PEC electrolysis, and H$_2$ is generated by direct water splitting in such a cell with a metal cathode and a semiconducting anode. The electrochemical potential and Fermi energy of the ZnO NR were calculated by the electrochemical current density in acid and alkaline solutions via phenomenological thermodynamic analysis. The ZnO NR has excellent potential for the storage of evolved H$_2$, and it also appears to operate as a hydrogen reservoir (Harinipriya et al. 2014).

Du et al. (2014) presented a novel biocathode-coupled PEC (Bio-PEC) integrating the advantages of photocatalytic anode and biocathode. The electrochemical anodized TiO$_2$ nanotube arrays fabricated on Ti substrate were used as Bio-PEC anodes. The TiO$_2$ nanotubes had inner diameters of 60–100 nm and wall thicknesses of about 5 nm. A pronounced photocurrent output (325 μA cm^{-2})

under xenon illumination was reported as compared to dark conditions. A comparative study was also carried out between the bio-PEC and PECs with Pt/C cathodes. The fill factors of bio-PEC and brush-PEC (50 mg) were observed to be 39.87% and 43.06%, respectively. The charge-transfer resistance of the biocathode was 13.10 ω, larger than the brush cathode with 50 mg Pt/C (10.68 ω), but smaller than the brush cathode with 35 mg Pt/C (18.35 ω), indicating comparable catalytic activity with the Pt/C catalyst. The biocathode was considered a promising alternative for the Pt/C catalyst based on the performance and cost of the PEC system.

One metal oxide (titanium dioxide) and another silicon-based compound (silicon carbide) have been given special attention as wide-bandgap semiconductor materials (Pessoa et al. 2015). Pessoa et al. (2015) also presented material characteristics, synthesis methods, and recent photocatalytic applications. The effects of the increase in efficiency of PEC devices that developed from the heterojunction of TiO_2 and SiC were also observed.

Chen and Kamat (2014) used glutathione-capped metal (gold) nanoclusters as photosensitizers for hydrogen generation in a PEC cell and a photocatalytic slurry reactor. The reversible reduction ($E^0 = -0.63$ V vs. RHE) and oxidation ($E^0 = 0.97$ and 1.51 V vs. RHE) potentials of these metal nanoclusters make them suitable for driving the water-splitting reaction. They also observed significant photocurrent activity under visible light (400–500 nm) excitation, when a mesoscopic TiO_2 film sensitized by Au_x-GSH NCs was used as the photoanode with a Pt counterelectrode in aqueous buffer solution (pH = 7). The rate of hydrogen production was 0.3 mol of hydrogen ($h^{-1} g^{-1}$) of Au_x-GSH NCs, by sensitizing Pt/TiO_2 nanoparticles with Au_x-GSH NCs in an aqueous slurry system under visible light. The rate of H_2 evolution was significantly enhanced (~5 times), when EDTA (as a sacrificial donor) was introduced into the system.

A comparative study of photoelectrocatalytic and photocatalytic inactivation was made by Nie et al. (2014). It was observed that the PEC inactivation was more effective to bacterial (*E. coli* K-12 and its mutant *E. coli* BW25113) strains than the photocatalytic (PC) process. *E. coli* BW25113 showed higher resistance than *E. coli* K-12 in both PEC and PC systems. It was found that h^+ was the major reactive species for PEC inactivation. Scanning electron microscopy images showed that the cells were severely damaged, resulting in a leakage of the intracellular components during the PEC inactivation process. The PEC inactivation efficiencies of both strains were enhanced in the presence of NaCl or NaBr.

Jin et al. (2014) prepared a flexible mesoporous TiO_2 microspheres/cellulose acetate (TCA) hybrid film as a recyclable photocatalyst with high performance, tunable size, and transparency. It was obtained by a simple method of dispersing mesoporous TiO_2 microspheres onto the surface of a freestanding cellulose acetate (CA) film at room temperature. The PEC properties of the mesoporous TiO_2 microspheres were studied by configuring them as a simple self-powered PEC cell. It was also observed that the TCA hybrid film displays excellent flexibility and favorable recyclable photocatalytic activity for the decomposition of methylene blue (MB) solution under UV light irradiation. The pH value of the solution has a more significant effect than temperature on the photoactivity of the sample.

Kim et al. (2014) reported the hierarchical In_2O_3:Sn/TiO_2/CdS heterojunction nanowire array photoanode (ITO/TiO_2/CdS-nanowire array photoanode). It provides a short travel distance for the charge carrier and a long light absorption pathway through the scattering effect. A comparison of optical properties and device performance of the ITO/TiO_2/CdS-nanowire array photoanode with the TiO_2 nanoparticle/CdS photoanode has been presented. The photocatalytic properties for water splitting were also observed in the presence of a sacrificial agent such as SO_3^{2-} and S^{2-} ions. The ITO/TiO_2/CdS-nanowire array photoanode exhibits a photocurrent density of 8.36 mA cm^{-2}, under illumination (AM 1.5 G, 100 mW cm^{-2}), at 0 V versus Ag/AgCl, which is four times higher than that of the TiO_2 nanoparticle/CdS photoanode. It was concluded that improved light-harvesting and charge collection properties, due to the increased light absorption pathway and reduced electron travel distance by ITO nanowire, lead to enhancement of the PEC performance.

Guo et al. (2014) prepared ZnO/Cu_2S core–shell nanorods from ZnO NRs for PEC water splitting through a versatile hydrothermal chemical conversion method (H-ZnO/Cu_2S core–shell NRs) and successive ionic layer adsorption and reaction method (S-ZnO/Cu_2S core–shell NRs). The photoelectrode was composed of a core–shell structure, where the core portion is ZnO NRs and the shell portion is Cu_2S nanoparticles sequentially located on the surface. It was observed that the ZnO NRs array provides a fast electron transport pathway due to its high electron mobility properties. This PEC system produced very high photocurrent density and photoconversion efficiency under 1.5 AM irradiation for hydrogen generation. It was demonstrated that H-ZnO/Cu_2S core–shell NRs exhibit a much higher photocatalytic activity than S-ZnO/Cu_2S core–shell NRs. The photocurrent density and photoconversion efficiency of H-ZnO/Cu_2S core–shell NRs were up to 20.12 mA cm^{-2} at 0.85 V versus SCE and 12.81% at 0.40 V versus SCE, respectively.

Li et al. (2015) constructed ternary CdS/reduced graphene oxide/TiO_2 nanotube array hybrids for enhanced visible light–driven PEC and photocatalytic activity. The coupling technique of electrophoretic deposition (EPD) was used for the synthesis of ternary nanocomposite photoelectrodes composed of CdS nanocrystallites, reduced graphene oxide (RGO), and TiO_2 nanotube arrays (TNTs). The ternary CdS/RGO/TNTs hybrids show more visible light–driven PEC and photocatalytic activity because the outer layer of CdS acts as a sensitizer for trapping substantial photons from the visible light. The middle layer of RGO not only serves as a transporter for suppressing the recombination of photogenerated carriers, but also acts as a green sensitizer for increasing visible light absorption. The inner TNTs with narrowed bandgap collect the hot electrons form the visible light absorption, while CdS and RGO participate in subsequent redox reactions for hydrogen production and degradation of organic pollutants.

The effect of NH_4F concentration and controlled charge consumption on the photocatalytic hydrogen generation on TiO_2 nanotube arrays was studied by Yujing et al. (2015). Electrochemical anodization in ethylene glycol–based electrolytes with various NH_4F concentrations was used for fabricating self-organized TiO_2 nanotube arrays (TiO_2 NTs) for hydrogen evolution. Hydrogen production by photocatalytic water splitting was performed in a two-compartment PEC cell, and in this process, no applied voltage was used. The impacts of NH_4F concentration on the morphological structure, PEC performance, and hydrogen evolution of TiO_2 samples were analyzed. The TiO_2 anodized with 0.50 wt% of NH_4F concentration for 60 min exhibited the highest hydrogen evolution of 2.53 mL h^{-1} cm^{-2} and maximum photoconversion efficiency of 4.39%. Another series of TiO_2 nanotube arrays samples with the equal charge consumption (designated as TiO_2 NTs-EC) was prepared. It was observed that hydrogen production and PEC properties of TiO_2 NTs-EC samples have increased. It was concluded that anodization charge density plays an important role in hydrogen generation by TiO_2 NTs.

4.3 PEC CELLS AS STORAGE CELLS

The PEC solar cells have the potential not only to convert but also store incident solar energy. Efficient photochemical storage and conversion require better functional performance of the separate cell components and system compatibility.

Some parameters should be optimized for obtaining maximum efficiency. These are

- Stability of the photoelectrode
- Stability of the counterelectrode
- Stability of the electrolyte
- Reversibility
- High photoelectrode-conversion efficiency
- Minimization of loss in light intensity reaching the photoelectrode
- High current and potential efficiency of the redox storage process
- High energy capacity of the redox storage

The PEC cells not only generate electricity, but they also split water efficiently into hydrogen and oxygen. Water splitting is a welcome step to produce hydrogen, which is being advocated as the fuel of the future. A number of PEC cells have been reported with varied materials in the form of either electrodes or electrolytes. Various efforts have been made to achieve efficient and stable PEC cells, including surface modification of the photoelectrode, electrolyte modification, photoetching of layered semiconductors, new configurations of PEC solar cells, dye sensitization, and so on. Still there are many opportunities available to develop newer materials for the future.

REFERENCES

Alebbi, M., C. A. Bignozzi, T. A. Heimer, G. M. Hasselmann, and G. J. Meyer. 1998. The limiting role of iodide oxidation in cis-Os(dcb)$_2$(CN)$_2$/TiO$_2$ photoelectrochemical cells. *J. Phys. Chem. B*. 102: 7577–7581.

Alibabaei, L., H. Luo, R. L. House, P. G. Hoertz, R. Lopez, and T. J. Meyer. 2013a. Applications of metal oxide materials in dye sensitized photoelectrosynthesis cells for making solar fuels: Let the molecules do the work. *J. Mater. Chem. A*. 1: 4133–4145.

Alibabaei, L., M. K. Brennaman, M. R. Norris, B. Kalanyan, W. Song, M. D. Losego, et al. 2013b. Solar water splitting in a molecular photoelectrochemical cell. *Proc. Nat. Acad. Sci. USA*. 110: 20008–20013.

Ang, P. G. P., A. F. Sammells, Y. Sun, A. C. Onicha, M. Myahkostupov, and F. N. Castellano. 1983. A one-volt p-InP/n-CdSe regenerative photoelectrochemical cell. *J. Electrochem. Soc*. 130: 1784–1786.

Ardo, S., D. Achey, A. J. Morris, M. Abrahamsson, and G. J. Meyer. 2011. Non-Nernstian two-electron transfer photocatalysis at metalloporphyrin–TiO$_2$ interfaces. *J. Am. Chem. Soc*. 133: 16572–16580.

Argazzi, R., C. A. Bignozzi, G. M. Hasselmann, and G. J. Meyer. 1998. Efficient light-to-electrical energy conversion with dithiocarbamate-ruthenium polypyridyl sensitizers. *Inorg. Chem*. 37: 4533–4537.

Ashford, D. L., C. R. K. Glasson, M. R. Norris, J. J. Concepcion, S. Keinan, M. K. Brennaman, et al. 2014a. Controlling ground and excited state properties through ligand changes in ruthenium polypyridyl complexes. *Inorg. Chem*. 53: 5637–5646.

Ashford, D. L., A. M. Lapides, A. K. Vannucci, K. Hanson, D. A. Torelli, D. P. Harrison, et al. 2014b. Water oxidation by an electropolymerized catalyst on derivatized mesoporous metal oxide electrodes. *J. Am. Chem. Soc*. 136: 6578–6581.

Ashford, D. L., W. Song, J. J. Concepcion, C. R. K. Glasson, M. K. Brennaman, M. R. Norris, et al. 2012. Photoinduced electron transfer in a chromophore-catalyst assembly anchored to TiO$_2$. *J. Am Chem Soc*. 134: 19189–19198.

Bergeron, B. V., A. Marton, G. Oskam, and G. J. Meyer. 2005. Dye-sensitized SnO$_2$ electrodes with iodide and pseudohalide redox mediators. *J. Phys. Chem. B*. 109: 937–943.

Bettis, S. E., D. M. Ryan, M. K. Gish, L. Alibabaei, T. J. Meyer, M. L. Waters, et al. 2014. Photophysical characterization of a helical peptide chromophore-water oxidation catalyst assembly on a semiconductor surface using ultrafast spectroscopy. *J. Phys. Chem. C*. 118: 6029–6037.

Bhattacharyya, R. G., D. P. Mandal, S. C. Bera, and K. K. Rohatgi-Mukherjee. 1996. Photoelectrosynthesis of dihydrogen via water-splitting using S_x^{2-} (x = 1, 2, 3...) as an anolyte: A first step for a viable solar rechargeable battery. *Int. J. Hydrogen Energy*. 21: 343–347.

Bhavani, N. K., D. Vijayalakshmi, and S. Seshan. 1986. Photochemical conversion and storage of solar energy. *Energy Manage*. 10: 27–32.

Bicelli, L. P., G. Razzini, N. Romeo, and V. Canevari. 1985. Photoelectrochemical characterization of n-CuInSe$_2$ films prepared by quasi-rheotaxy. *Surface Technol*. 25: 327–334.

Bouroushian, M., D. Karoussos, and T. Kosanovic. 2006. Photoelectrochemical properties of electrodeposited CdSe and CdSe/ZnSe thin films in sulphide-polysulphide and ferro-ferricyanide redox systems. *Solid State Ionics*. 177: 1855–1859.

Brennan, B. J., A. E. Keirstead, P. A. Liddell, S. A. Vail, T. A. Moore, A. L. Moore, et al. 2009. 1-(3′-amino) propylsilatrane derivatives as covalent surface linkers to nanoparticulate metal oxide films for use in photoelectrochemical cells. *Nanotechnology*. 20: 505203.

Brugnati, M., S. Caramori, S. Cazzanti, L. Marchini, R. Argazzi, and C. A. Bignozzi. 2007. Electron transfer mediators for photoelectrochemical cells based on Cu(I) metal complexes. *Int. J. Photoenergy*. 2007: 80756.

Carli, S., E. Busatto, S. Caramori, R. Boaretto, R. Argazzi, C. J. Timpson, et al. 2013. Comparative evaluation of catalytic counter electrodes for Co(III)/(II) electron shuttles in regenerative photoelectrochemical cells. *J. Phys. Chem. C*. 117: 5142–5153.

Carlsson, P., and B. Holmström. 1986. Photoelectrochemical cells: Laboratory determination of solar conversion efficiencies. *Solar Energy*. 36: 151–157.

Cattarin, S., G. Zotti, M. M. Musiani, and G. Mengoli. 1986. Photoelectrochemistry of para-diethynylbenzene copper(I) acetylides deposited as thin films onto copper electrodes. *J. Electroanal. Chem*. 207: 247–261.

Chao, J., Z. Wang, X. Xu, Q. Xiang, W. Song, G. Chen, et al. 2013. Tin sulfide nanoribbons as high performance photoelectrochemical cells, flexible photodetectors and visible-light-driven photocatalysts. *RSC Adv.* 3: 2746–2753.

Chao, J., Z. Xie, X. Duan, Y. Dong, Z. Wang, J. Xu, et al. 2012. Visible-light-driven photocatalytic and photoelectrochemical properties of porous SnS_x ($x = 1,2$) architectures. *Cryst. Eng. Comm.* 14: 3163–3168.

Chen, H., Z. Wei, K. Yan, Y. Bai, and S. Yang. 2014. Unveiling two electron-transport modes in oxygen-deficient TiO_2 nanowires and their influence on photoelectrochemical operation. *J. Phys. Chem. Lett.* 5: 2890–2896.

Chen, Y.-S., and P. V. Kamat. 2014. Glutathione-capped gold nanoclusters as photosensitizers. Visible light-induced hydrogen generation in neutral water. *J. Am. Chem. Soc.* 136: 6075–6082.

Dhere, N. G., A. H. Jahagirdar, U. S. Avachat, and A. A. Kadam. 2004. Photoelectrochemical water splitting for hydrogen production using combination of CIGS2 solar cell and RuO_2 photocatalyst. *Thin Solid Films.* 481–482: 462–465.

Du, Y., Y. Feng, Y. Qu, J. Liu, N. Ren, and H. Liu. 2014. Electricity generation and pollutant degradation using a novel biocathode coupled photoelectrochemical cell. *Environ. Sci. Technol.* 48: 7634–7641.

Falaras, P., K. Chryssou, T. Stergiopoulos, I. Arabatzis, G. Katsaros, V. J. Catalano, et al. 2002. Dye-sensitization of titanium dioxide thin films by Ru(II)-bpp-bpy complexes. *Proc. SPIE–Int. Soc. Optical Eng.* 4801: 125–135.

Falaras, P., A. P. Xagas, and A. H.-L. Goff. 1998. Synthesis and characterization of dichloro(2,2′-bipyridyl-4,4′dicarboxylate)bis(triphenylphosphine) ruthenium(II) for efficient photosensitization of titanium oxide. *New J. Chem.* 22: 557–558.

Forouzan, F., and S. Licht. 1995. Solution-modified n-GaAs/aqueous polyselenide photoelectrochemistry. *J. Electrochem. Soc.* 142: 1539–1545.

Fujishima, and K. Honda. 1972. Electrochemical photolysis of water at a semiconductor electrode. *Nature.* 238: 37–38.

Ganesh, I., A. K. Gupta, P. P. Kumar, P. S. Chandra Sekhar, K. Radha, G. Padmanabham, et al. 2012. Preparation and characterization of Co-doped TiO_2 materials for solar light induced current and photocatalytic applications. *Mater. Chem. Phys.* 135: 220–234.

Gao, L., F. Qu, and X. Wu. 2013. Reduced graphene oxide-BiVO4 composite for enhanced photoelectrochemical cell and photocatalysis. *Sci. Adv. Mater.* 5: 1485–1492.

Garcia, C. G., and N. Y. M. Iha. 2001. Photoelectrochemical solar cells using [(dcbH$_2$)$_2$ RuLL′], L,L′ = substituted pyridines, as nanocrystalline TiO_2 sensitizers. *Int. J. Photoenergy.* 3: 130–135.

Garcia, C. G., N. Y. M. Iha, R. Argazzi, and C. A. Bignozzi. 1998. 4-Phenylpyridine as ancillary ligand in ruthenium(II) polypyridyl complexes for sensitization of n-type TiO_2 electrodes. *J. Photochem. Photobiol. A: Chem.* 115: 239–242.

Garcia, C. G., A. K. Nakano, C. J. Kleverlaan, and N. Y. M. Iha. 2002. Electron injection versus charge recombination in photoelectrochemical solar cells using cis-[(dcbH$_2$)$_2$Ru(CNpy)(H$_2$O)]Cl$_2$ as a nanocrystalline TiO_2 sensitizer. *J. Photochem. Photobiol. A: Chem.* 151: 165–170.

Guay, D., R. Cote, R. Marques, J. P. Dodelet, M. F. Lawrence, D. Gravel, et al. 1987. High short-circuit photocurrents obtained with a phthalocyanine in a regenerative photoelectrochemical cell. *J. Electrochem. Soc.* 134: 2942–2943.

Guo, K., X. Chen, J. Han, and Z. Liu. 2014. Synthesis of ZnO/Cu2S core/shell nanorods and their enhanced photoelectric performance. *J. Sol-Gel Sci. Tech.* 72: 92–99.

Hagedorn, K., C. Forgacs, S. Collins, and S. Maldonado. 2010. Design considerations for nanowire heterojunctions in solar energy conversion/storage applications. *J. Phys. Chem. C.* 114: 12010–12017.

Han, L., S. P. Singh, A. Islam, and M. Yanagida. 2011. Development of a new class of thiocyanate-free cyclometalated ruthenium(II) complex for sensitizing nanocrystalline TiO_2 solar cells. *Int. J. Photoenergy.* 2011: 520848.

Hanson, K., M. K. Brennaman, H. Luo, C. R. K. Glasson, J. J. Concepcion, W. Song, et al. 2012. Photostability of phosphonate-derivatized, Ru[II] polypyridyl complexes on metal oxide surfaces. *ACS Appl. Mater. Interfaces.* 4: 1462–1469.

Hanson, K., M. D. Losego, B. Kalanyan, G. N. Parsons, and T. J. Meyer. 2013. Stabilizing small molecules on metal oxide surfaces using atomic layer deposition. *Nano Lett.* 13: 4802–4809.

Harinipriya, S., B. Usmani, D. J. Rogers, V. E. Sandana, F. H. Teherani, Lusson, et al. 2014. ZnO nanorod electrodes for hydrogen evolution and storage (Conference Paper). *J. Sol-Gel Sci. Technol.* 72: 92–99.

Heimer, T. A., S. T. D'Arcangelis, F. Farzad, J. M. Stipkala, and G. J. Meyer. 1996. An acetylacetonate-based semiconductor-sensitizer linkage. *Inorg. Chem.* 35: 5319–5324.

Heuer, W. B., H.-L. Xia, M. Abrahamsson, Z. Zhou, S. Ardo, A. A. N. Sarjeant, et al. 2010. Reaction of RuII diazafluorenone compound with nanocrystalline TiO_2 thin film. *Inorg. Chem.* 49: 7726–7734.

Ho, G. W., K. J. Chua, and D. R. Siow. 2012. Metal loaded WO_3 particles for comparative studies of photocatalysis and electrolysis solar hydrogen production. *Chem. Eng. J.* 181–182: 661–666.

Islam, A., S. P. Singh, and L. Han. 2011. Synthesis and application of new ruthenium complexes containing β-diketonato ligands as sensitizers for nanocrystalline TiO_2 solar cells. *Int. J. Photoenergy.* 2011: 204639.

Islam, A., H. Sugihara, K. Hara, L. P. Singh, R. Katoh, M. Yanagida, et al. 2001a. Dye sensitization of nanocrystalline titanium dioxide with square planar platinum(II) diimine dithiolate complexes. *Inorg. Chem.* 40: 5371–5380.

Islam, A., H. Sugihara, K. Hara, L. P. Singh, R. Katoh, M. Yanagida, et al. 2001b. Sensitization of nanocrystalline TiO_2 film by ruthenium(II) diimine dithiolate complexes. *J. Photochem. Photobiol. A: Chem.* 145: 135–141.

Jacobsson, T. J., C. Platzer-Björkman, M. Edoff, and T. Edvinsson. 2013. $CuIn_xGa_{1-x}Se_2$ as an efficient photocathode for solar hydrogen generation. *Int. J. Hydrogen Energy.* 38: 15027–15035.

Jasieniak, J., M. Johnston, and E. R. Waclawik. 2004. Characterization of a porphyrin-containing dye-sensitized solar cell. *J. Phys. Chem. B.* 108: 12962–12971.

Jeng, Ki.-T., Y.-C. Liu, Y.-F. Leu, Y.-Z. Zeng, J.-C. Chung, and T.-Y. Wei. 2010. Membrane electrode assembly-based photoelectrochemical cell for hydrogen generation. *Int. J. Hydrogen Energy.* 35: 10890–10897.

Jin, X., J. Xu, X. Wang, Z. Xie, Z. Liu, B. Liang, et al. 2014. Flexible TiO_2/cellulose acetate hybrid film as a recyclable photocatalyst. *RSC Adv.* 4: 12640–12648.

Johansson, P. G., J. G. Rowley, A. Taheri, G. J. Meyer, S. P. Singh, A. Islam, et al. 2011. Long-wavelength sensitization of TiO_2 by ruthenium diimine compounds with low-lying π* orbitals. *Langmuir.* 27: 14522–14531.

Keita, B., and L. Nadjo. 1984. Electrochemistry and photoelectrochemistry of sodium 9,10-anthraquinone-2,6. disulfonate in aqueous media. Application to rechargeable solar cells and to the synthesis of hydrogen peroxide. *J. Electroanal. Chem.* 163: 171–188.

Kelly, C. A., F. Farzad, D. W. Thompson, J. M. Stipkala, and G. J. Meyer. 1999. Cation-controlled interfacial charge injection in sensitized nanocrystalline TiO_2. *Langmuir.* 15: 7047–7054.

Kim, J. S., H. S. Han, S. Shin, G. S. Han, H. S. Jung, K. S. Hong, et al. 2014. In_2O_3:Sn/TiO_2/CdS heterojunction nanowire array photoanode in photoelectrochemical cells. *Int. J. Hydrogen Energy.* 39: 17473–17480.

Klahr, B. M., and T. W. Hamann. 2011. Current and voltage limiting processes in thin film hematite electrodes. *J. Phys. Chem. C.* 115: 8393–8399.

Kline, G., K. Kam, D. Canfield, and B. A. Parkinson. 1981. Efficient and stable photoelectrochemical cells constructed with WSe_2 and $MoSe_2$ photoanodes. *Solar Energy Mater.* 4: 301–308.

Kline, G., K. K. Kam, R. Ziegler, and B. A. Parkinson. 1982. Further studies of the photoelectrochemical properties of the group VI transition metal dichalcogenides. *Solar Energy Mater.* 6: 337–350.

Kohl, P. A., and A. J. Bard. 1979. Semiconductor electrodes-liquid junction photovoltaic cells based on n-GaAs electrodes and acetonitrile solutions. *J. Electrochem. Soc.* 126: 603–608.

Ladouceur, M., J. P. Dodelet, G. Tourillon, L. Parent, and S. Dallaire. 1990. Plasma-sprayed semiconductor electrodes: Photoelectrochemical characterization and NH3 photoproduction by substoichiometric tungsten oxides. *J. Phys. Chem.* 94: 4579–4587.

Latempa, T. J., S. Rani, N. Bao, and C. A. Grimes. 2012. Generation of fuel from CO_2 saturated liquids using a p-Si nanowire n-TiO_2 nanotube array photoelectrochemical cell. *Nanoscale.* 4: 2245–2250.

Lee, H. J., J. Bang, J. Park, S. Kim, and S.-M. Park. 2010. Multilayered semiconductor (CdS/CdSe/ZnS)-sensitized TiO2 mesoporous solar cells: All prepared by successive ionic layer adsorption and reaction processes. *Chem. Mater.* 22: 5636–5643.

Lee, H., M. Wang, P. Chen, D. R. Gamelin, S. M. Zakeeruddin, M. Grätzel, et al. 2009. Efficient CdSe quantum dot-sensitized solar cells prepared by an improved successive ionic layer adsorption and reaction process. *Nano Lett.* 9: 4221–4227.

Lewerenz, H. J. 2011. Operational principles of electrochemical nanoemitter solar cells for photovoltaic and photoelectrocatalytic applications. *J. Electroanal. Chem.* 662: 184–195.

Li, H., Z. Xia, J. Chen, L. Lei, and J. Xing. 2015. Constructing ternary CdS/reduced graphene oxide/TiO_2 nanotube arrays hybrids for enhanced visible-light-driven photoelectrochemical and photocatalytic activity. *J. Appl. Catal. B: Environ.* 168–169: 105–113.

Li, H., J. Xing, Z. Xia, and J. Chen. 2014. Preparation of extremely smooth and boron-fluorine co-doped TiO_2 nanotube arrays with enhanced photoelectrochemical and photocatalytic performance. *Electrochim. Acta.* 139: 331–336.

Licht, S. 2001. Multiple band gap semiconductor/electrolyte solar energy conversion. *J. Phys. Chem. B.* 105: 6281–6294.

Licht, S., O. Khaselev, P. A. Ramakrishnan, D. Faiman, E. A. Katz, A. Shames, et al. 1998a. Fullerene photoelectrochemical solar cells. *Sol. Energy Mater. Solar Cells.* 51: 9–19.

Licht, S., O. Khaselev, P. A. Ramakrishnan, T. Soga, and M. Umeno. 1998b. Multiple bandgap photoelectrochemistry: Bipolar semiconductor ohmic regenerative electrochemistry. *J. Phys. Chem. B.* 102: 2536–2545.

Licht, S., O. Khaselev, P. A. Ramakrishnan, T. Soga, and M. Umeno. 1998c. Multiple bandgap photoelectrochemistry: Inverted semiconductor Ohmic regenerative electrochemistry. *J. Phys. Chem. B.* 102: 2546–2554.

Licht, S., O. Khaselev, T. Soga, and M. Umeno. 1998d. Multiple bandgap photoelectrochemistry: Energetic configurations for solar energy conversion. *Electrochem. Solid State Lett.* 1: 20–23.

Licht, S., P. A. Ramakrishnan, D. Faiman, E. A. Katz, A. Shames, and S. Goren. 1998e. Photoaction, temperature and O_2 depletion effects in fullerene photoelectrochemical solar cells. *Sol. Energy Mater. Solar Cells.* 56: 45–55.

Luo, H., W. Song, P. G. Hoertz, K. Hanson, R. Ghosh, S. Rangan, et al. 2013. A sensitized Nb_2O_5 photoanode for hydrogen production in a dye-sensitized photoelectrosynthesis cell. *Chem. Mater.* 25: 122–131.

Matthews, D., P. Infelta, and M. Grätzel. 1996. Calculation of the photocurrent-potential characteristic for regenerative, sensitized semiconductor electrodes. *Sol. Energy Mater. Solar Cells.* 44: 119–155.

McEvoy, A. J. 2005. Photoelectrochemical solar cells. In T. Markvart and L. Castaner (Eds), *Solar Cells: Materials, Manufacture and Operation*. Oxford: Elsevier, pp. 395–417.

Memming, R. 1978. Energy production by photoelectrochemical processes. *Philip. Tech. Rev.* 38: 160–177.

Mi, Q., R. H. Coridan, B. S. Brunschwig, H. B. Gray, and N. S. Lewis. 2013. Photoelectrochemical oxidation of anions by WO_3 in aqueous and nonaqueous electrolytes. *Energy Environ. Sci.* 6: 2646–2653.

Miller, B. 1984. Charge transfer and corrosion processes at III-V semiconductor/electrolyte interfaces. *J. Electroanal. Chem.* 168: 91–100.

Neumann-Spallart M., and K. Kalyanasundaram. 1981. Photoelectrochemical cells with polycrystalline cadmium sulfide as photoanodes. *Ber. Bunsen. Phys. Chem.* 85: 1112–1117.

Nie, X., G. Li, M. Gao, H. Sun, X. Liu, H. Zhao, et al. 2014. Comparative study on the photoelectrocatalytic inactivation of *Escherichia coli* K-12 and its mutant *Escherichia coli* BW25113 using TiO_2 nanotubes as a photoanode. *Appl. Catal. B: Environ.* 147: 562–570.

Norris, M. R., J. J. Concepcion, C. R. K. Glasson, Z. Fang, A. M. Lapides, D. L. Ashford, et al. 2013. Synthesis of phosphonic acid derivatized bipyridine ligands and their ruthenium complexes. *Inorg. Chem.* 52: 12492–12501.

Onicha, A. C., and F. N. Castellano. 2010. Electrolyte-dependent photovoltaic responses in dye-sensitized solar cells based on an osmium(II) dye of mixed denticity. *J. Phys. Chem. C.* 114: 6831–6840.

Osaka, T., N. Hirota, T. Hayashi, and S. S. Eskildsen. 1985. Characteristics of photoelectrochemical cells with iron oxide/n-Si heterojunction photoanodes. *Electrochim. Acta.* 30: 1209–1212.

Papageorgiou, N., Y. Athanassov, M. Armand, P. Bonhôte, H. Pettersson, A. Azam, et al. 1996. The performance and stability of ambient temperature molten salts for solar cell applications. *J. Electrochem. Soc.* 143: 3099–3108.

Park, J. K., H. R. Lee, J. Chen, H. Shinokubo, A. Osuka, and D. Kim. 2008. Photoelectrochemical properties of doubly β-functionalized porphyrin sensitizers for dye-sensitized nanocrystalline-TiO_2 solar cells. *J. Phys. Chem. C.* 112: 16691–16699.

Pessoa, R. S., M. A. Fraga, L. V. Santos, M. Massi, and H. S. Maciel. 2015. Nanostructured thin films based on TiO_2 and/or SiC for use in photoelectrochemical cells: A review of the material characteristics, synthesis and recent applications. *Mater. Sci. Semicond. Proc.* 29: 56–68.

Philippidis, N., E. Nikolakaki, S. Sotiropoulos, and I. Poulios. 2010. Photoelectrocatalytic inactivation of *E. coli* XL-1 blue colonies in water. *J. Chem. Technol. Biotechnol.* 85: 1054–1060.

Price, M. J., and S. Maldonado. 2009. Macroporous n-GaP in nonaqueous regenerative photoelectrochemical cell. *J. Phys. Chem. C.* 113: 11988–11994.

Radecka, M., M. Rekas, A. Trenczek-Zajac, and K. Zakrzewska. 2008. Importance of the band gap energy and flat band potential for application of modified TiO_2 photoanodes in water photolysis. *J. Power Sources.* 181: 46–55.

Rahman, M., B. H. Q. Dang, K. McDonnell, J. M. D. MacElroy, and D. P. Dowling. 2012. Effect of doping (C or N) and co-doping (C + N) on the photoactive properties of magnetron sputtered titania coatings for the application of solar water-splitting. *J. Nanosci. Nanotechnol.* 12: 4729–4735.

Rajeshwar, K., M. Kaneko, and A.Yamada. 1983. Regenerative photoelectrochemical cells using polymer-coated n-GaAs photoanodes in contact with aqueous electrolytes. *J. Electrochem. Soc.* 130: 38–43.

Ramaraj, R., and P. Natarajan. 1989. Photoelectrochemical investigations of phenosafranine dye bound to some macromolecules. *J. Chem. Soc. Faraday Trans.* 1. 85: 813–827.

Redmond, G., D. Fitzmaurice, and M. Graetzel. 1994. Visible light sensitization by cis-bis(thiocyanato) bis(2,2′-bipyridyl-4,4′-dicarboxylato) ruthenium(II) of a transparent nanocrystalline ZnO film prepared by sol-gel techniques. *Chem. Mater.* 6: 686–691.

Rothenberger, G., D. Fitzmaurice, and M. Grätzel. 1992. Spectroscopy of conduction band electrons in transparent metal oxide semiconductor films: Optical determination of the flatband potential of colloidal titanium dioxide films. *J. Phys. Chem.* 96: 5983–5986.

Roušar, I., M. Rudolf, P. Lukášek, L. Kavan, N. Papageorgiou, and M. Grätzel. 1996. Optimization of parameters of an electrochemical photovoltaic regenerative solar cell. *Sol. Energy Mater. Solar Cells.* 43: 249–262.

Ryan, D. M., M. K. Coggins, J. J. Concepcion, D. L. Ashford, Z. Fang, L. Alibabaei, et al. 2014. Synthesis and electrocatalytic water oxidation by electrode-bound helical peptide chromophore-catalyst assemblies. *Inorg. Chem.* 53: 8120–8128.

Shi, J.-Y., W.-H. Leng, X.-F. Cheng, J.-Q. Zhang, and C.-N. Cao. 2005. Photocatalytic oxidation of methyl red by TiO_2 in a photoelectrochemical cell. *Acta Phys. Chim. Sinica.* 21: 971–976.

Singh, P., R. Singh, K. Rajeshwar, and J. DuBow. 1981. Photoelectrochemical behavior of n-GaAs electrodes in ambient temperature molten salt electrolytes: Device characterization and loss mechanisms. *J. Electrochem. Soc.* 128: 1145–1150.

Song, W., M. K. Brennaman, J. J. Concepcion, J. W. Jurss, P. G. Hoertz, H. Luo, et al. 2011. Interfacial electron transfer dynamics for $[Ru(bpy)_2((4, 4′-PO_3H_2)_2bpy)]^{2+}$ sensitized TiO_2 in a dye-sensitized photoelectrosynthesis cell: Factors influencing efficiency and dynamics. *J. Phys. Chem. C.* 115: 7081–7091.

Song, W., Z. Chen, C. R. K. Glasson, K. Hanson, H. Luo, M. R. Norris, et al. 2012. Interfacial dynamics and solar fuel formation in dye-sensitized photoelectrosynthesis cells. *Chem. Phys. Chem.* 13: 2882–2890.

Song, W., A. Ito, R. A. Binstead, K. Hanson, H. Luo, M. K. Brennaman, et al. 2013a. Accumulation of multiple oxidative equivalents at a single site by cross-surface electron transfer on TiO_2. *J. Am. Chem. Soc.* 135: 11587–11594.

Song, W., H. Luo, K. Hanson, J. J. Concepcion, M. K. Brennaman, and T. J. Meyer. 2013b. Visualization of cation diffusion at the TiO_2 interface in dye sensitized photoelectrosynthesis cells (DSPEC). *Energy Environ. Sci.* 6: 1240–1248.

Song, W., A. K. Vannucci, B. H. Farnum, A. M. Lapides, M. K. Brennaman, B. Kalanyan, et al. 2014. Visible light driven benzyl alcohol dehydrogenation in a dye-sensitized photoelectrosynthesis cell. *J. Am. Chem. Soc.* 136: 9773–9779.

Stergiopoulos, T., I. M. Arabatzis, M. Kalbac, I. Lukes, and P. Falaras. 2005. Incorporation of innovative compounds in nanostructured photoelectrochemical cells. *J. Mater. Proc. Technol.* 161: 107–112.

Su, F., T. Wang, R. Lv, J. Zhang, P. Zhang, J. Lu, et al. 2013. Dendritic Au/TiO_2 nanorod arrays for visible-light driven photoelectrochemical water splitting. *Nanoscale.* 5: 9001–9009.

Sun, Y., A. C. Onicha, M. Myahkostupov, and F. N. Castellano. 2010. Viable alternative to N719 for dye-sensitized solar cells. *ACS Appl. Mater. Interfaces.* 2: 2039–2045.

Sun, Y., K. Yan, G. Wang, W. Guo, and T. Ma. 2011. Effect of annealing temperature on the hydrogen production of TiO_2 nanotube arrays in a two-compartment photoelectrochemical cell. *J. Phys. Chem. C.* 115: 12844–12849.

Taratula, O., J. Rochford, P. Piotrowiak, E. Galoppini, R. A. Carlisle, and G. J. Meyer. 2006. Pyrene-terminated phenylenethynylene rigid linkers anchored to metal oxide nanoparticles. *J. Phys. Chem. B.* 110: 15734–15741.

Thompson, D. W., A. Ito, and T. J. Meyer. 2013. $[Ru(bpy)_3]^{2+*}$ and other remarkable metal-to-ligand charge transfer (MLCT) excited states. *Pure Appl. Chem.* 85: 1257–1305.

Thompson, L., J. DuBow, and K. Rajeshwar. 1982. Photoelectrochemical generation of chlorine on catalytically modified n-silicon/indium tin oxide anodes. *J. Electrochem. Soc.* 129: 1934–1935.

Tributsch, H. 1980. Photoelectrochemical behaviour of layer-type transition metal dichalcogenides. *Faraday Discuss. Chem. Soc.* 70: 189–205.

Vannucci, A. K., L. Alibabaei, M. D. Losego, J. J. Concepcion, B. Kalanyan, G. N. Parsons, et al. 2013. Crossing the divide between homogeneous and heterogeneous catalysis in water oxidation. *Proc. Natl. Acad. Sci. USA.* 110: 20918–20922.

Vlachopoulos, N., P. Liska, J. Augustynski, and M. Grätzel. 1988. Very efficient visible light energy harvesting and conversion by spectral sensitization of high surface area polycrystalline titanium dioxide films. *J. Am. Chem. Soc.* 110: 1216–1220.

Vorobets, V. S., S. K. Kovach, and G. Y. Kolbasov. 2002. Distribution of ion species and formation of ion pairs in concentrated polysulfide solutions in photoelectrochemical transducers. *Russ. J. Appl. Chem.* 75: 229–234.

Wadhawan, J. D., T. J. Davies, A. D. Clegg, N. S. Lawrence, J. C. Ball, O. V. Klymenko, et al. 2002. Photoelectrochemistry of bromonitrobenzenes: Mechanism and photoelectrochemically-induced halex reactions. *J. Electroanal. Chem.* 533: 33–70.

Wang, C., Z. Chen, H. Jin, C. Cao, J. Li, and Z. Mi. 2014. Enhancing visible-light photoelectrochemical water splitting through transition-metal doped TiO_2 nanorod arrays. *J. Mater. Chem. A.* 2: 17820–17827.

Wang, P., S. M. Zakeeruddin, P. Comte, I. Exnar, and M. Grätzel. 2003. Gelation of ionic liquid-based electrolytes with silica nanoparticles for quasi-solid-state dye-sensitized solar cells. *J. Am. Chem. Soc.* 125: 1166–1167.

Watson, D. F., A. Marton, A. M. Stux, and G. J. Meyer. 2003. Insights into dye-sensitization of planar TiO2: Evidence for involvement of a protonated surface state. *J. Phys. Chem B.* 107: 10971–10973.

White, J. R., F.-R. Fan, and A. J. Bard 1985. Semiconductor electrodes: LVI. Principles of multijunction electrodes and photoelectrosynthesis at Texas Instruments' p/n-Si solar arrays. *J. Electrochem. Soc.* 132: 544–550.

Wongwanwattana, P., P. Krongkitsiri, P. Limsuwan, and U. Tipparach. 2012. Fabrication and photocatalysis of nanostructured TiO_2 for solar hydrogen production. *Ceramics Int.* 38 (Suppl): S517–S519.

Xia, H.-L., F. Liu, S. Ardo, A. A. N. Sarjeant, and G. J. Meyer. 2010. Photoinduced electron transfer from Ru am(m)ine compounds with low-lying ligand field excited states to nanocrystalline TiO_2. *J. Photochem. Photobiol. A: Chem.* 216: 94–103.

Xiang, C., G. M. Kimball, R. L. Grimm, B. S. Brunschwig, H. A. Atwater, and N. S. Lewis. 2011. 820 mV open-circuit voltages from Cu_2O/CH_3CN junctions. *Energy Environ. Sci.* 4: 1311–1318.

Yang, J., D. Wang, H. Han, and C. Li. 2013. Roles of cocatalysts in photocatalysis and photoelectrocatalysis. *Acc. Chem. Res.* 46: 1900–1909.

Yang, T., H. Wang, X.-M. Ou, C.-S. Lee, and X.-H. Zhang. 2012. Iodine-doped-poly(3,4-ethylenedioxythiophene)-modified Si nanowire 1D core-shell arrays as an efficient photocatalyst for solar hydrogen generation. *Adv. Mater.* 24: 6199–6203.

Yohannes, T., and O. Inganäs. 1999. All-solid-state photoelectrochemical energy conversion with the conjugated polymer poly[3-(4-octylphenyl)-2,2′-bithiophene](FT). *Synth. Met.* 107: 97–105.

Yujing X., Y. Sun, G. Wang, K. Yan, and J. Zhao. 2015. Effect of NH_4F concentration and controlled-charge consumption on the photocatalytic hydrogen generation of TiO_2 nanotube arrays. *Electrochim. Acta.* 155: 312–320.

Zhao, Y., Y. He, D.-B. Xiong, W. Ran, Z. Liu, and F. Gao. 2014. Dual template synthesis and photoelectrochemical performance of 3-D hierarchical porous zinc oxide. *Int. J. Hydrogen Energy.* 39: 13486–13490.

5 Organic Photovoltaic Cells

Meenakshi Singh Solanki, Taruna Dangi, Paras Tak, Sanyogita Sharma, and Rakshit Ameta

CONTENTS

5.1 Introduction 55
5.2 History 57
5.3 Basic Processes 58
    5.3.1 Photon Absorption 58
    5.3.2 Exciton Diffusion 58
    5.3.3 Charge Separation 59
    5.3.4 Charge Transfer 59
    5.3.5 Charge Collection 60
5.4 Characteristics 60
5.5 Materials 61
    5.5.1 Pigments 62
    5.5.2 Dyes 62
    5.5.3 Small Molecules, Oligomers, Polymers, and Dendrimers 63
    5.5.4 Liquid Crystals 65
    5.5.5 Other Materials 66
5.6 Types of Organic Photovoltaic Cells 68
    5.6.1 Single Layer 68
    5.6.2 Bilayer 69
    5.6.3 Heterojunctions 71
5.7 Tandem Solar Cells 75
5.8 Hybrid Tandem Photovoltaic Cell 76
5.9 Inverted Tandem Solar Cell 77
5.10 Inverted Organic Photovoltaic Cells 77
References 78

5.1 INTRODUCTION

There is a continuous increase in demand for electrical energy in almost all fields (i.e., from consumer electronics, to small-scale distributed power systems, to centralized megawatt-scale power plants, in domestic use, for transportation, etc.). Conventional methods for generating electrical energy are time consuming, high in cost, low power generating, and release harmful by-products into the environment, which lead to global warming. Global warming is clear evidence of the fundamental idea of Newton that there is no action without reaction. But renewable energies produced from our natural environment, such as wind, solar, thermal, photovoltaic, geothermal, marine, and hydropower, help us to reduce our dependence on conventional methods. Therefore, solar energy, a clean alternative to traditional methods of power generation for sustainable development, is a good as well as promising choice. Direct utilization of solar radiation is similar to the ideal use of solar radiation in nature in the form of photosynthesis.

Solar electricity is a steadily growing energy technology. Among different technologies already available to convert solar light directly into electricity, photovoltaics (PVs) offer several benefits. Photovoltaic cells can produce power near the end user of the electricity, thus avoiding transmission losses and costs. Solar panels do not produce any noise or greenhouse gas emissions and require very little maintenance. Photovoltaic devices based on inorganic materials have commonly been used, but photovoltaic cells based on organic materials are garnering attention. The main advantages of organic photovoltaic (OPV) cells compared to other types of photovoltaic cells are their low weight, attractive form factor, scalability, flexibility, and low-cost fabrication.

The term *photovoltaic* is derived from *photo* meaning "light" and *voltaic* meaning "electricity"; thus, photovoltaic cells are cells that convert sunlight directly into electricity at the atomic level. Photovoltaic cells are made of special materials called *semiconductors*, such as silicon, that exhibit a property known as the photoelectric effect. Basically, when light energy strikes the cell, loosely bound electrons are knocked out from the atoms in the semiconductor material. When these free electrons are captured, the result is an electric current that can be used as electricity.

Silicon-based photovoltaic cells account for the major portion of production of solar electricity in the world. However, although the prices of these cells have been drastically reduced, they are still too expensive. Therefore, organic compounds are used to overcome such high cost problem. Organic semiconductors are a comparatively less expensive alternative to inorganic semiconductors. Moreover, they are less energy consuming. One of the challenges in realizing efficient organic photoconversion systems (organic PV cells) is that the electron–hole pair created via photon absorption must overcome the Coulomb attraction (i.e., losing the strongly bound electron and hole that result from photon exposure) to generate photocurrent. Many compelling solutions to this problem have been proposed.

The main difference between organic and inorganic materials is that in an inorganic semiconductor, the free charge carriers, electrons, and holes are created directly upon light absorption, whereas electrostatically bound charge carriers, excitons, are formed in an organic semiconductor:

- The dielectric constant and Bohr radius of carriers are less in organic semiconductors as compared to inorganic semiconductors.
- An organic semiconductor has an easy processing at 20°C–200°C, but on the contrary, an inorganic semiconductor requires high temperatures of 400°C–1400°C.

The main benefit of an organic cell over an electrochemical cell is the absence of a liquid electrolyte. The active layer thickness of an organic solar cell is only 100 nm, which is 1000 times thinner than that of a Si-solar cell and 10 times thinner than that of an inorganic thin-film solar cell. Therefore, organic solar cells have the potential for cost-efficient and large-scale applications.

Organic photovoltaics, a solar cell technology, is related to the idea of providing flexible, wearable, lightweight, and low-cost photovoltaic materials such as polymers (plastics), dyes, or certain organic electron donor and acceptor molecules. In the last few decades, this field of organic photovoltaic materials has attracted scientific and economic interest, which is triggered by a rapid increase in power conversion efficiencies. Commercially available organic solar cells have power conversion efficiencies less than 3%, while silicon-based solar cells have power conversion efficiencies of 40.7% on average. Organic photovoltaic devices have not yet reached the limits of their inorganic counterparts (i.e., approximately 10%–20%) (Harald and Sariciftci 2004). Organic solar cells are developing in a dynamic way, as they have low production costs because they are printed onto a substrate. However, silicon-based solar cell technology has the limitation of extremely high manufacturing and material costs.

Tang (1986) has reported a bilayer thin-film organic photovoltaic cell of copper phthalocyanine and a perylene tetracarboxylic derivative with efficiency up to 1%. This lower yield (almost 1%) has now increased almost 10 times in some other OPV cells. A photovoltaic conversion efficiency of about 5% was observed in a solar cell having ZnO and pentacene-based single crystals and thin-film heterojunction devices (Schon et al. 2000). However, the major objective in this field remains

Organic Photovoltaic Cells

the cost reduction of photovoltaic modules. Thin-film technology or silicon solar cells are still more efficient, but there is a pressing demand to develop technology that is eco-friendly, is readily available, and utilizes low-cost materials. Polymeric materials, particularly plastics, have great potential in these aspects.

5.2 HISTORY

Bequerel (1839) was the first to notice the photoelectric effect. He found that platinum electrode coated with silver halogen produces small amounts of electric current when irradiated with electromagnetic radiation in aqueous solution. Smith (1873) discovered the photoconductivity of selenium. The next significant photovoltaic development was the photoconductive effect in selenium placed between two metal electrodes (Adams and Day 1877). Adams and Day observed an anomaly that could be explained by the generation of internal voltages. Heated platinum contacts were pushed into opposite ends of small cylinders of vitreous selenium. They also reported that it was possible to get current in the selenium by the action of light. This was the first demonstration of the photovoltaic effect in a solid-state system. The photogenerated currents were attributed to light-induced crystallization of the outer layers of the selenium bar. The photovoltaic action of the selenium differed from its photoconductive action, where a current was produced spontaneously by the action of light.

The next significant step was taken by Fritts (1883). He coated molten selenium plate with an extremely thin layer of gold to prepare the first thin-film photovoltaic device, which was as large as 30 cm^2 in area, but power conversion efficiency was limited to less than 1%. The photoconductivity in an organic compound (anthracene) was reported by Pochettino (1906).

Grondahl (1933) reported photovoltaic cells based on a copper-cuprous oxide junction. In this model, Pb wire was used to provide firm contact to the illuminated surface of the cell. Bergmann (1931) reported a selenium device that was much more effective than a copper-based device and became the commercially dominant product. A thallous-sulfide cell of similar performance was also studied (Nix and Treptwo 1939). Excited by this outstanding achievement, the *New York Times* forecasted that solar cells would eventually harness the "limitless energy of the sun." In 1958, PV array-powered radios appeared on the U.S. Vanguard I space satellite, and this was the first time that PV technology was practically utilized. During this period, PV cells made of cadmium sulfide, gallium arsenide, cadmium telluride, and indium phosphide also appeared. However, each technology had its own disadvantages, as well as merits. Cadmium is used in CdS and CdTe PV cells.

These early PV cells of copper-cuprous oxide thin film, lead sulfide, and thallium sulfide were thin-film Schottky barrier devices, where a semitransparent layer of metal deposited on top of the semiconductor provided both asymmetric electronic junctions, which are necessary for photovoltaic action, and access to the junction for the incident light.

However, photoconductivity was what excited researchers, not the PV properties of materials like selenium. The current generated was proportional to the intensity of the incident light, and this was related to the wavelength. The photoconductive materials were ideal for photographic light meters. Barrier structures in the photovoltaic effect were an added benefit, which means the light meter could operate without a power supply. It was not until the 1950s that potentially useful quantities of power were produced by PV devices in crystalline silicon by development of good-quality silicon wafers. These were used in the new solid-state electronics devices.

In 1954, Bell Labs revealed the first high-power silicon PV cell using a p-n junction with power conversion efficiency (PCE) of 6%. Photoconductivity in the poly(*N*-vinyl-carbazole) (PVK) polymers was discovered by Hoegel (1965). The concepts for organic semiconductors were established by Chiang et al. (1977) and Shirakawa et al. (1977). Their findings were astonishing, because they highlighted the potential transition of photovoltaic substances from inorganic to organic semiconductors, which could lower production costs and ease of processing.

In the last few decades, scientists have had great interest in developing organic PV devices. During the 1990s, it was shown that the quantum efficiency of electron transfer is very high in excited polymers and C_{60}, which is promising for change carrier separation in PV cells (Sariciftci et al. 1992, 1993). Organic displays with organic light-emitting devices (OLEDs) have introduced new technology (Diaz et al. 2001).

5.3 BASIC PROCESSES

Typically, a photovoltaic cell is made up of two photoactive materials placed in between two metallic electrodes, including a transparent electrode, which is exposed to light. Photons with sufficient energy are adsorbed by the surface, and loosely bound electrons are removed. After the charge separation process, the charge carriers (electrons) are transported to the electrodes. Finally, charges are collected by the electrode, and these will produce electric current in the external circuit. These steps are important steps of any photovoltaic process (Chamberlain 1983). Basically, there are five important steps to achieve high conversion efficiency of solar radiation into electrical energy:

- Photon absorption
- Exciton diffusion
- Charge separation
- Charge transfer
- Charge collection

5.3.1 Photon Absorption

The first process in any photovoltaic device is the absorption of photons. A photosensitive semiconductor is exposed to light, leading to excitation of electrons from the highest occupied molecular orbital (HOMO) to the lowest unoccupied molecular orbital (LUMO). It may be easier for any inorganic photovoltaic devices to have better absorption, but it is slightly difficult for organic photovoltaic devices, because the bandgap is quite high for organic materials. The bandgap of an organic material should be 1.1 eV to permit sufficient absorption of solar radiation reaching Earth (about 80%). This corresponds to 1100 nm, while most of the semiconducting polymers have bandgaps more than 2.0 eV, corresponding to 600 nm. This limits the absorption of solar radiation to about 30%. Normally, the organic layer is thin, but in photovoltaic devices, the low charge carrier and exciton mobilities require slightly thick layers (~100 nm). As the absorption coefficient of organic materials is relatively higher than that of silicon, layers of 100 nm thickness can absorb 60%–90% solar radiation provided that a reflective back contact is used. Therefore, in order to increase light absorption, efforts are being made to develop small-bandgap polymers.

Ko et al. (2011) reported a threefold enhancement in photon absorption in specific regions of the solar spectrum, in part through multiple excitation resonances by the two-dimensional (2D) photonic crystal (PC) geometry in comparison to the geometry of a conventional planar cell. Photonic crystal geometry is developed using a material-agnostic process called PRINT (pattern replication in nonwetting templates) on an organic photoactive bulk heterojunction blend of P3HT (poly(3-hexylthiophene)) and PCBM (phenyl-C_{61}-butyric acid methyl ester). In this process, highly ordered arrays of nanoscale features were readily made in a single processing step over wide areas (4 cm²) that are scalable, exhibiting efficiency improvements of 70% that result not only from greater absorption, but also from electrical enhancements.

5.3.2 Exciton Diffusion

Excitons are generated on photon absorption. These excited excitons should reach a dissociation site. The diffusion length should be at least equal to the thickness of the layer for sufficient absorption,

Organic Photovoltaic Cells

as such sites are normally available at the other ends of the semiconductors. If this is not the case, then the excited excitons may recombine, and photons are wasted. The diffusion ranges in pigment and polymers are ordinarily quite low (~10 nm).

5.3.3 Charge Separation

Electrons are transferred to an electron acceptor from electron donor material in the charge separation process. The charge separation may occur at organic semiconductor/metal interfaces at oxygen as impurity or it may be between materials with quite different electron affinities and ionization potentials. If the difference between ionization potential and electron affinity is not sufficient, then excitons may just jump on the material with the lower bandgap without splitting its charges. As a result, it may recombine and not contribute changes to the photocurrent. An electron donor material has small electron affinity. On the contrary, an electron-accepting material has high electron affinity. This difference in electron affinity is the driving force that leads to exciton dissociation.

It is very important for an efficient charge generation that most of the energy of the absorbed photon is used for the charge separation process. This means that the energy of absorbed light should not be wasted in competitive processes (fluorescence or nonradiative decay). The separated charge should be stabilized in order to transfer photogenerated charge to one of the electrodes (i.e., the back-electron transfer or recombination should be minimized).

Excitons must overcome the Coulomb attraction force present between them for long-range charge separation. The formation of an excited, delocalized band state helps to obtain long-range charge separation, which has a positive effect on the efficiency of photovoltaic devices. Exposing excitons under infrared radiation (photon) for less than 1 ps results in the transfer of a charge-pair bound at a heterojunction to delocalized band states, which act as the gateway for charge separation (Bakulin et al. 2012).

The charge separation mechanism helps in increasing the performance of organic solar devices. Gelinas et al. (2014) studied the ultrafast long-range charge separation process. They found that charge separations of an electron–hole pair generated by light absorption across the donor–acceptor heterojunction are time dependent in organic solar cells. They also found that when charge separation distance is near 4 nm, then about 200 meV of electrostatic energy is created from charge separation within 40 fs of excitation. At this stage, the residual Coulomb attraction between charges is at or below thermal energies, so charges (electron and hole) are separated completely. This early behavior is consistent with charge separation through access to delocalized π-electron states in ordered regions of the fullerene acceptor material.

5.3.4 Charge Transfer

The transport of charges is affected by recombination as it moves to the electrode. This is more specific when the same material is used as the transport medium for both carriers (i.e., electrons and holes). Their interaction with other atoms or charges will lower the speed and, consequently, limit the current. Charge transport is also low due to less intrinsic mobility in organic material and by the charge trapping effects of impurities and defects. The interfacial charge pair resides at the donor–acceptor heterointerface, which is called the *charge-transfer complex*. It is responsible for the photocurrent and the open-circuit voltage (Deibel et al. 2010).

The dissociation of excitons into free electrons and holes at donor–acceptor heterointerfaces leads to photocurrent generation in OPV cells. It was also observed that the dielectric constant helps in determining the charge-transfer energy. A low dielectric constant results in strong Coulomb interactions between electron–hole pairs, which opposes the generation of free charges. The charge-transfer polarizability increases when the C_{60} crystallite size exceeds the threshold (i.e., approximately 4 nm). The charge-transfer step plays a major role in the performance of OPV cells (Piliegoa and Loi 2012).

Jailaubekov et al. (2013) used femtosecond nonlinear optical spectroscopies and nonadiabatic mixed quantum mechanics/molecular mechanics simulations in the phthalocyanine-fullerene model OPV system to explain hot charge-transfer exciton formation and relaxation. The first excitation on phthalocyanine produces hot charge-transfer excitons in 10–13 s and then relaxation to lower energies and shorter electron–hole distances of 10–12 s. Thus, this hot CT exciton cooling process and collapse of charge separation set the fundamental time limit for competitive charge separation channels, which leads to efficient photocurrent generation. Bernardo et al. (2014) investigated the charge-transfer process, which is affected by variation of fullerene fraction in small molecule–fullerene bulk heterojunction PV cells.

As organic solids have lower electron mobility, inorganic materials are also used as electron-transporting components. TiO_2 has been used to increase performance of PVs such as conjugative polymer, poly(p-phenylene vinylene) (PPV) with ultrathin TiO_2 nanocrystal (Salafsky 1999), and mesoporous titania (Coakley and McGehee 2003). Charge separation and charge transport are difficult in some organic materials like polymers. Elongated crystalline components are attractive as electron transporters if crystal size and orientation can be managed. The charge-transfer and charge separation processes in PV cells are increased by a soluble perpylene dye–polymer blend (Dittmer et al. 2000) and graphene (Chang et al. 2014).

5.3.5 Charge Collection

In most organic solar cells, the separated electrons and holes are transported to the opposite electrodes in an internal electric field created by asymmetry of the electrodes (different work functions) or in built-in potentials. Selective doping enhances the performance of the PV cell as it provides low serial resistance and creates internal electric fields for the collection of charges. The multilayer model of ITO/poly(2,5-dioctyloxy-phenylene vinylene) (OOPPV)/octaethylporphine (OEP)/C_{60}/Al has double heterojunction of OOPPV/OEP and OEP/C_{60}, which leads to charge generation by excitonic dissociation (Yoshino et al. 1997).

Good nonblocking contacts between the molecular materials and the electrodes of the cell are needed for efficient charge collection. An additional material layer between the metal and organic layer is also used to facilitate good ohmic contact in some cases, such as with LiF between conjugated polymer/methanofullerene (MDMO-PPV-PCMB) blend and aluminum (Brabec et al. 2001a; Shaheen et al. 2001a).

Charge collection is affected by the presence of inorganic film as the electron and hole collection layer. Poly(3-hexylthiophene (P3HT) and [6,6]-phenyl-C_{71}-butyric acid methyl ester ($PC_{71}BM$) as the electron donor and electron acceptor, respectively, were prepared by Vasilopoulou et al. (2014a). They used ZnO film as the electron-collecting layer and an under-stoichiometric molybdenum oxide MoO_x as the hole collection layer to improve the inverted organic PV cell efficiency. The ZnO film was prepared using two methods in this fabrication: by atomic layer deposition (ALD-ZnO) and by using the sol-gel method (sg-ZnO). Both films had the same thickness of 20 nm, but results proved that the ALD method shows significantly enhanced efficiency as it develops conformal and defect-free ZnO electron collection layers, which are responsible for high-performance organic photovoltaics.

5.4 CHARACTERISTICS

Some characteristics of the solar cell that determine its photovoltaic performance and electrical behavior are

- Open-circuit voltage (V_{oc}): This is maximum voltage across the PV cell in solar radiation, when no current is flowing in the PV cell.
- Short-circuit current (i_{sc}): This is the current that flows through an exposed PV cell when there is no external resistance. The maximum current that a device is able to produce is

Organic Photovoltaic Cells

called the *short-circuit current*. Under an external load, the current will always be less than i_{sc}.
- Fill factor (*FF*): This is the ratio of actual maximum power output to its theoretical power output (maximum current i_{sc} and voltage V_{oc}). It is the key quantity used to measure performance of the cell:

$$FF = \frac{P_{max}}{V_{oc} \times i_{sc}} = \frac{V_{max} \times i_{max}}{V_{oc} \times i_{sc}} \quad (5.1)$$

where i_{max} and V_{max} are maximum current and potential, respectively, and P_{max} is their product.
- Quantum efficiency (QE): This is the efficiency of a PV cell as a function of the energy or wavelength of the incident radiation. It specifically relates the number of charge carriers collected to the number of photons shining on the device for a particular wavelength. There are two types of quantum efficiency:
  - External quantum efficiency: This involves the losses by reflection and transmission. It is also known as incident photon-to-current efficiency (IPCE):

$$\text{Incident photon-to-current efficiency (IPCE)} = \frac{J_{sc} \times h \times c}{I \times \lambda \times e} \quad (5.2)$$

where I is the intensity of incident radiation (W cm^{-2}), λ is wavelength (μm), and J_{sc} is short-circuit current density (A cm^{-2}), while h, c, and e have their usual significance as Planck constant, velocity of light, and electronic charge, respectively.
The formula is thus

$$\text{IPCE} = \frac{1.24 J_{sc}}{I \times \lambda} \quad (5.3)$$

As h, c, and e are constant, the external photovoltaic yield (η) is the ratio of P_{max} to the product of I and surface (S) of the modules:

$$\eta = \frac{P_{max}}{I \times S} \quad (5.4)$$

  - Internal quantum efficiency: This includes losses because of reflection and transmission of light energy such that it considers processes involving absorbed photons only. By accounting for transmission and reflection processes, external quantum efficiency can be transformed into internal quantum efficiency.
- Power conversion efficiency (PCE, η): This is the ratio of power output to power input (i.e., the amount of power generated by the PV cell compared to the power available in the incident photons (P_{in}).

5.5 MATERIALS

Inorganic materials have been commonly used in photovoltaic applications, but there are certain added advantages in using organic materials for this purpose. The main advantages of using organic materials for photovoltaic application are

- Organic materials have a high optical absorption coefficient, so a large amount of light can be absorbed with a small amount of material. Therefore, the amount of organic materials required is comparatively smaller than the amount of inorganic materials required.

- Organic materials can be produced more easily than inorganic materials.
- Organic materials can be processed easily using wet processing (spin coating or doctor blade technique) or dry processing (evaporation via mask).
- Chemical/physical properties, such as solubility, valence, bandgap, and charge transfer, can be easily adjusted in the case of organic materials.
- A large variety of functionalities of the organic materials can be used, such as dyes, pigments, oligomers, polymers, dendrimers, and liquid crystals.

Organic photovoltaic materials mainly consist of three different categories, based on their mechanical or processing properties: insoluble, soluble, or liquid crystalline. They may be further categorized on the basis of conjugated π electrons such as molecules with single structural repeat units (monomer), few structural repeat units (oligomers), and more than about 10 repeat units (polymers). Oligomers and monomers that absorb in the visible light are known as chromophores and are referred to as dyes, if they are soluble, or pigments, if they are not (Petritsch et al. 2000). Basically, organic solar cells, which belong to the class of photovoltaic cells, are known as excitonic solar cells (Thompson and Fréchet 2008).

The main disadvantages related to organic materials are their low efficiency, strength, and stability.

5.5.1 Pigments

Pigments are used for coloring a material. They form polycrystalline thin film on evaporation (sublimation) from solid powder. Perylene, or perylenetetracarboxylic acid diimide, and phthalocyanine, or different metallophthalocyanines, give the structural backbone to many molecules used in efficient pigment- and dye-based organic solar cells. Fullerene (C_{60}) and pentacene are insoluble in most of the solvents; therefore, these are considered as pigments.

5.5.2 Dyes

Dyes impart color to any material on attachment. Their thin film cannot be obtained so easily from solvent, but they can be easily incorporated into a polymer host (Gautier-Thianche et al. 1998; Kido et al. 1994). Dyes can also be attached chemically to the polymer host (Cacialli et al. 1998; Jiang et al. 2000).

Photoconductivity in anthracene was observed in the beginning of the twentieth century (Pochettino 1906). Until the 1980s, the performance of a dye-sensitized solar cell was about 0.1%. A major breakthrough came when Tang (1986) showed that much higher efficiencies are attainable by producing a double-layered cell using two different dyes.

Organic dyes were also used for sensitization of solar cells. Basically, a monolayer of a dye is introduced between the donor and the acceptor to increase the absorption of light. Yoshino et al. (1997) studied a photovoltaic device made up of the three organic layers (i.e., donor–absorber–acceptor structure) to enhance photovoltaic efficiency. For a wider absorption band, there is an alternate route that replaces the electron-transporting polymer with a polymer blend with crystalline dyes. Dye crystals like anthracene or perylene were used as electron acceptors in polythiophene-based solar cells (Dittmer et al. 2000). The nature of the dye-sensitized solar cell was explained by Cahen et al. (2000).

Fanshun and Tian (2005) reported the potential use of cyanine dyes in the thin-film heterojunction of organic photovoltaic devices because it widens the absorption spectra and photoaction spectra in the visible region and, therefore, increases the performance of the cell. Walter et al. (2010) utilized organic dyes porphyrins, phthalocyanines, and related compounds as components of solar cells, including organic molecular solar cells, polymer cells, and dye-sensitized solar cells. Chen et al. (2013) reported the use of ternary organic photovoltaic systems, including polymer/

Organic Photovoltaic Cells

small molecule/functional fullerene, polymer/polymer/functional fullerene, small molecule/small molecule/functional fullerene, polymer/functional fullerene I/functional fullerene II, and polymer/quantum dot or metal/functional fullerene systems in organic photovoltaic cells. Kim et al. (2014a) were able to increase PCE to 3.08% and open-circuit voltage to 1.03 V by introducing carbazole or fluorene-containing rhodanine dyes as acceptors in OPV cells. The absorbing layer is polarized upon light absorption to drive the charges toward the appropriate transport layers.

5.5.3 Small Molecules, Oligomers, Polymers, and Dendrimers

Oligomers are typically formed during polymerization, when the number of repeat units is 2–12. The oligomer has the desired electrical and optical properties of a polymer, and their thin films are usually available in a polycrystalline state. If the repeat units are more than the limit of oligomer, then it is called a *polymer*. When the similar repeat units are present in three-dimensional (3D) form, it forms dendrimers. These dendrimers may have different shapes, such as branches of a star or an ellipsoid-like structure (Halim et al. 1999).

Conjugated polymers are attractive semiconductors for photovoltaic cells because they have large conjugative system and low bandgaps to broaden the absorption range; therefore, they are strong absorbers and can be deposited on flexible substrates at low cost. Conjugative polymers are organic molecules consisting of repeating structural units attached to each other by alternating carbon–carbon single and double bonds. Such polymers are made conductive by doping. These can be oxidized or reduced to produce p-doped or n-doped materials, respectively. The p-doped polymers are utilized in many devices, such as electrochromic devices, rechargeable batteries, capacitors, and membranes, and n-doped materials have not garnered as much attention. On the basis of molecular structure and chemical composition, polymers can be used as either electron donor or electron acceptor materials in the organic solar cells. Some common semiconducting polymers, fullerene and its derivatives are as presented in Figure 5.1.

Plastic is an insulator, but it was made a conductor by adding some impurity, and the conductivity of the polymer increased more than a billion times. This discovery and development of electrical conductive polymers snatched a Nobel Prize for Heeger, McDiarmid, and Shirakawa. Chiang et al. (1977) reported an increase in conductivity of doped polyacetylene. Shirakawa et al. (1977) treated polyacetylene with an electron acceptor such as iodine and observed higher conductivity (of the order of 10^{10} times) than the pure acetylene.

One of the most studied photoconducting polymers is poly(vinyl carbazole) (PVK). Photoconductivity of PVK was also increased by doping with fluorinated styrene chromophore (Hendrickx et al. 1999) and CdS nanocrystals (Wang et al. 2000). The conjugative polymers exhibit absorption in the blue or green region. New material combinations like polymer–fullerene were synthesized, which lower the bandgap and show absorption in the red or infrared region. The polymer–fullerene was used (Yu et al. 1995) in place of poly(*para*-phenylene vinylene) (PPV) (Jenekhe and Yi 2000), polythiophene and its derivatives (Brabec et al. 2001b; Too et al. 2001), polypyrrole/thiazadole copolymers (Dhanabalan et al. 2001), and thiophene/2-isothianaphthene copolymers (Shaheen et al. 2001b). Recent development of organic conjugated polymer–based photovoltaic elements has helped us in achieving low-cost and easy-to-produce energy from light.

The plastic solar cell (PSC) or polymer solar cell uses conductive organic polymers or small organic molecules for absorption of light and charge transport to generate electricity from sunlight (Nelson 2002). The photoactive device is based on the photoinduced charge transfer from donor-type semiconducting polymers onto acceptor-type polymers such as Buckminsterfullerene; this is the fundamental process used in photovoltaic devices. Most of the conjugated polymer can be utilized as an electron donor because on photoexcitation, electrons are promoted to the antibonding band. As the photoexcited electron is transferred to an acceptor unit, the resulting cation radical (positive polaron) species on the conjugated polymer backbone is known to be highly delocalized, mobile, and stable.

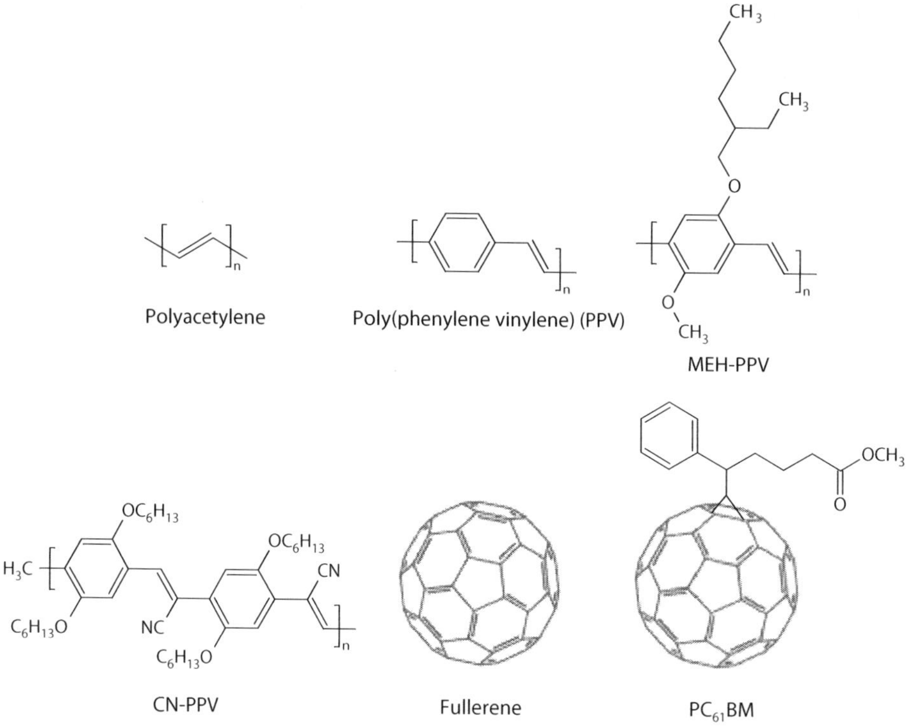

FIGURE 5.1 Some molecules used in OPV cells.

Charge carrier mobilities can be as low as 10^{-4} cm^2/V·s (approximately) in polymeric semiconductors. Because of the limit on increasing the thickness of the photoactive layer, the series resistance becomes dominant, and the short-circuit current breaks down. Using poly(p-phenylene vinylene) (PPV) with fullerenes, one can obtain an optimum thickness of ~100 nm. Therefore, instead of increasing the thickness of the photoactive layer, higher charge carrier mobility materials are required. Regioregular poly(3-alkylthiophenes) (P3AT) have better mobilities and produce higher photocurrents by increasing the thickness of the layer. Charge mobility is also sensitive to solid-state nanoscale morphology, which can be improved by different methods such as changing the solvent of the casting solution, as well as tempering the cast films. Bandgap engineering will also help in increasing the performance of the semiconducting polymer. It means different-colored semiconductor polymer can be achieved by modifying the chemical structure, which leads to a change in bandgap.

Coakley and McGehee (2003) prepared the conjugated polymer regioregular poly(3-hexylthiophene) into films of mesoporous titania by infiltration. The mesoporous titania films have pores with diameter less than 10 nm, which provide regular pathways for electrons to travel to an electrode after electron transfer. A 1.5% power conversion efficiency and an external quantum efficiency of 10% under monochromatic 514 nm light were shown. They also described that on blending polymers with electron-accepting materials such as C_{60} derivatives, cadmium selenide, and titanium dioxide, power conversion efficiencies reached 4% (Coakley and McGehee 2004).

The open-circuit voltage, which depends on the LUMO of the acceptor as well as on the HOMO of the donor, is influenced by a donor–acceptor pair and wave function of electrode. The dependence of the charge transport levels on temperature and light intensity also affect open-circuit voltage values. At normal room temperature, it is approximately 0.8–0.9 V for poly(2-methoxy-5-(3′,7′-dimethyloctyloxy)-1,4-phenylene vinylene) (MDMO-PPV) copolymer blended with [6,6]-phenyl-C_{61} butyric acid methyl ester (PC$_{61}$BM) and 0.5–0.6 V for poly-(3-hexylthiophene) (P3HT) blends with PCBM (Sariciftci 2004). Exposing an organic layer of an organic photovoltaic device to solvent vapor also affects its performance. Miller et al. (2008) reported an improvement in the performance of organic poly(3-hexylthiophene) and phenyl-C_{61}-butyric acid methyl ester devices via room temperature solvent vapor annealing.

A new conjugative polymer semiconductor was obtained by combining tetracyclic lactum monomer with thiophene (PTNT), which is a broad bandgap semicrystalline polymer with bandgap 2.2 eV. Such cells exhibited a fill factor around 0.6, high open-circuit voltage of 0.9 V, and a power conversion efficiency of 5% for more than 200 nm (or ~ 400 nm) thick active layers (Kroon et al. 2014). These values are on the higher side of those reported for a conjugated polymer with broad bandgap. Therefore, a newly developed tetracyclic unit may find some interesting applications in the future, mainly for ternary tandem PVs.

Organic photovoltaic cells must be designed in such a way that a weak electron-donating unit is used with a strong electron-withdrawing unit in order to decrease the HOMO energy level. Therefore, a conjugative polymer must be attached with a pull-and-push unit for better photon adsorption. The benzo[1,2-b:4,5-b']dithiophene is one of the most used and effective push units. It is used with pull units such as thieno[3,4-b]thiophene (TT) and benzo[2,1,3]thiodazole (BT) for form pull–push copolymer PBDTTT and PBDT-DTBT, respectively. Various organic solar devices with pull-and-push units, with their power conversion efficiency (%), are poly(3-(2′-methoxy-5′-octylphenyl) thiophene) (POMeOPT)-poly(2,5,2′,5′-tetrahexyloxy-7,8′-dicyanodi-p-phenylene vinylene) (CN-PPV), 4.5 (Roman et al. 2003); 2,5-di(thiophen-2-yl)thieno[3,2-b]thiophene and thieno[3,4-c] pyrrole-4,6-dione (PDTTPD)-[6,6]-phenyl-C_{61}-butyric acid methyl ester ($PC_{61}BM$), 5.1 (Chen et al. 2011a); benzo[1,2-b:4,5-b']dithiophene (BDT)-5,6-bis(octyloxy)benzo[c] [1,2,5]oxadiazole (BO) units, 5.7 (Jiang et al. 2011); N-alkylthieno[3,4-c]pyrrole-4,6-dione (TPD) and $PC_{61}BM$, 6.8 (Piliego et al. 2010); [6,6]-phenyl-C_{71}-butyric acid methyl ester (PBDTTT- $PC_{71}BM$), 9.2 (He et al. 2012); and so on.

The two naphthalene diimide (NDI) dimers, bis-NDI-T-EG and bis-NDI-BDTEG, bridged by thiophene and benzodithiophene, respectively, and symmetrically substituted by 2-methoxylethoxyl in the bay region were synthesized. These two NDI dimers exhibit broad absorption in the visible region of 300–650 (800) nm and display a HOMO/LUMO energy level of 5.88/3.80 and 5.46/3.78 eV, respectively. The devices with PBDTTT-C-T as donor material exhibit the best efficiency for both bis-NDI-T-EG and bis-NDI-BDT-EG of 1.31% and 1.24%, respectively (Wang et al. 2014a).

Organic materials drew attention because of their potential in providing environmentally safe, flexible, lightweight, and inexpensive electronics. The processing cost is also reduced, if the polymer is soluble. Linear polymers such as polyacetylene are much less soluble.

5.5.4 Liquid Crystals

Liquid crystals have emerged as a new category of organic solar cell materials. Their main feature is that they provide order, permitting different properties such as light absorption, charge generation, and most importantly, high charge carrier mobility and long exciton diffusion lengths (a few 100 nm), which could be beneficial for organic solar cells (Petritsch et al. 1999; Seguy et al. 2000). These materials show the phase with properties somewhere in between those of liquids and solids at certain temperatures. The liquid crystalline molecules are arranged in such a way that they look like crystalline solids but exhibit the mechanical properties of liquids (i.e., they are soft). Liquid crystals may be dyes, pigments, oligomers, or polymers. They can be processed by both wet and dry methods. They can have different shapes (smectic, nematic, columnar, etc.).

Langmuir-Blodgett films (Cimrova et al. 1996; Tokuhisa et al. 1998; Wu et al. 1996) and self-assembled monolayers (Appleyard et al. 2000) also permit molecular order control similar to that of liquid crystals. Therefore, they are used to adjust the properties of an electride (Nuesch et al. 1997). A self-organized liquid crystal organic solar cell was reported by Schmidt-Mende et al. (2001), using bilayer of liquid crystalline hexaphenyl-substituted hexabenzocoronene (HBCPhC12) as an electron donor and a perylene dicarboxylic acid diimide derivative as an electron acceptor in the active layer of the cells with high external quantum efficiency of over 34% at 490 nm.

5.5.5 Other Materials

Metals with low work function are typically used in organic solar devices because these metals are used to control carrier selectivity, transport, extraction, and blocking, as well as interface band bending. But there are some disadvantages to these materials, including that they are generally prone to reactions with water, oxygen, nitrogen, and carbon dioxide from air, leading to rapid device degradation. Therefore, lanthanides were used as new metallic cathode interlayer materials that increase device stability and still provide device efficiency similar to that achieved with a Ca interlayer (Nikiforov et al. 2013).

Chauhan et al. (2014) reported the performance of an organic solar cell with Al-doped ZnO thin films (AZO films) prepared by radio frequency (RF) sputtering at different argon pressures and low substrate temperatures of 80°C–95°C. As argon pressure was increased, their morphology, optical absorption, photoluminescence spectrum, and electrical behavior also changed. At 0.15 Pa argon pressure, Al-doped ZnO thin films show a Wurtzite-type hexagonal structure with [0001] preferred orientation, high optical transmittance of ~85% in the wavelength range 400–800 nm, and low electrical resistivity of 9.54×10^{-4} Ω cm.

Current generation in an organic photovoltaic cell mainly depends on the micro- and nanostructures in the semiconductor layers. Gilchrist et al. (2014) used high-resolution transmission electron microscopy (HRTEM) to gain unprecedented insights into the structure and composition of the molecular layers within the depth of the device structure. The technique was applied to a solar cell made of copper phthalocyanine (CuPc) and C_{60}.

Nielsen et al. (2014) developed electron-deficient truxenone derivatives as an alternative to fullerene. These derivatives are far better than PCBM because they have easily tunable absorption profiles, higher electron affinities, higher absorptivities, and highly reversible reductive characteristics. Fabrication of efficient bilayer solar cells with a subphthalocyanine (SubPc) donor supports this class of materials as promising electron acceptors in organic photovoltaic cells.

As organic materials have low charge carrier mobility, organic and inorganic material combinations have also been used to enhance performance of photovoltaic devices. Salafsky (1999) used approximately 100 nm thin titanium dioxide nanocrystals with conjugated polymer, poly(p-phenylene vinylene), where active medium was taken as Al/composite/indium tin oxide. It was observed that the ratio of inorganic materials affects the overall function of the cell.

Tamura et al. (2014) reported good fill factor and power conversion efficiency with tetrabenzoporphyrin (BP), a BP-C_{60} dyad, and PCBM for the p-, i-, and n-layers, respectively, as compared to 1:1 blend film of BP and PCBM as the i-layer in the p-i-n organic photovoltaic cell. The OPV devices made of different photoactive layers and the passivated TiO_2 electron extraction layers show significant enhancement of more than 30% in their power conversion efficiencies. Vasilopoulou et al. (2014b) prepared TiO_2 films through applying alumina (Al_2O_3) or zirconia (ZrO_2) insulating nanolayers by thermal atomic layer deposition (ALD) and used them as cathode interlayers in organic solar cells. However, this results in undesirable recombination and a high electron extraction barrier, thus reducing the open-circuit voltage and the short-circuit current of the complete OPV device.

Lee et al. (2014) fabricated a TiO_2-based solar cell using iodine-doped Cu-based metal—organic frameworks (Cu-MOFs, copper(II) benzene-1,3,5-tricarboxylate) as an active thin layer using a layer-by-layer technique. Iodine-doped MOFs gave excellent cell performance with $J_{sc} = 1.25$ mA cm^{-2} and $Eff = 0.26\%$ under illumination of 1 sun radiation, while the cell with an undoped MOF layer exhibited only $J_{sc} = 0.05$ mA cm^{-2} and $Eff = 0.008\%$.

Kim et al. (2014b) observed an increase in charge carrier mobility by heteroatom substitution of a silicon atom by a germanium atom in donor–acceptor-type low-bandgap copolymers, poly[(4,4′-bis(2-ethylhexyl)dithieno[3,2-b:2′,3′-d]silole)-2,6-diyl-alt-(2,1,3-benzothiadiazole)-4,7-diyl] (PSiBTBT) and poly[(4,4′-bis(2-ethylhexyl)dithieno[3,2-b:2′,3′-d]germole)-2,6-diyl-alt-(2,1,3-benzothiadiazole)-4,7-diyl] (PGeBTBT). Charge carrier mobility was increased because C–Ge bond length is longer in comparison to that of C–Si, which modifies the molecular conformation and leads to a more planar chain conformation in PGeBTBT than in PSiBTBT. This increase in molecular planarity leads to enhanced crystallinity and an increased preference for a face-on backbone orientation.

The ultraviolet (UV) and ozone treatment increases the power conversion efficiency and air stability of OPV cells. Le et al. (2014) prepared MoS_2 nanosheets by a simple sonication exfoliation method and used them to form a hole extraction layer (HEL). The OPV cells with MoS_2 layers show a power conversion efficiency of 1.08%, which is less than OPV cells without HEL (1.84%). After UV/ozone (UVO) treatment of the MoS_2 surface for 15 min, the efficiency value increases to 2.44%, and the work function of MoS_2 increases from 4.6 to 4.9 eV. When poly(3,4-ethylenedioxythiophene):poly(styrene sulfonate) (PEDOT:PSS) was inserted between MoS_2 and the active layer and used a hole extraction layer, the power conversion efficiency was increased to 2.81%, and air stability also increased in comparison to the device employing only PEDOT:PSS.

Cathode interfacial material (CIM) was found to be effective in improving the power conversion efficiency and long-term stability of an organic photovoltaic cell that utilizes a high work function cathode. Ultraviolet photoemission spectroscopy studies proved that CIM lowers the work function of the Ag metal as well as ITO and highly oriented pyrolytic graphite (HOPG), and facilitates electron extraction in OPV devices. Tan et al. (2014) prepared CIM by combining triarylphosphine oxide and a 1,10-phenanthrolinyl unit, which has high T_g of 116°C and attractive electron transport properties. The characterization of Ag or Al cathodes involving photovoltaic devices shows improvement in PCE as compared to the reference Ag device and compares well to that of the Ca/Al device. The CIM/Ag photovoltaic device with the active layer PTB7:$PC_{71}BM$ shows 7.51% of PCE. The PCE was further increased to 8.56% for the CIM/Al device (with J_{sc} = 16.81 mA cm^{-2}, V_{oc} = 0.75 V, and FF = 0.68).

In recent years, graphene has offered a vast range of photonic and electronic applications, such as photodetectors, optical modulators, high-speed transistors, electromagnetic wave shieldings, notch filters, linear polarizers, electrochemical energy storage, transparent and flexible electronics, displays, optoelectronics, sensors, nanoelectromechanical systems, energy technologies, and photovoltaic cells, which makes it a promising material. Its versatility provides an entirely new generation of technologies beyond the limits of conventional materials. It has a honeycomb-like structure, which is made up of a single layer of carbon atoms. Graphene is also attracting the interest of researchers because it is lightweight, flexible, atomically thin, mechanically strong, electrically tunable, visually transparent, extraordinarily high in charge mobility and saturation velocity, which helps in attaining a fast switching speed for radio frequency analog circuits. Untreated graphene is a semimetal that does not provide true off state; as a result, it typically precludes its use in digital logic electronics without bandgap engineering. It becomes highly conducting on doping.

One important characteristic of graphene is that it strongly interacts with light in the microwave range to the ultraviolet range, with spanning wavelengths of at least five orders of magnitude. Because of this exception, light interaction between light and graphene, it shows excellent electronic and mechanical properties. This is why graphene is a promising candidate for many kinds of photonic devices (Weiss et al. 2012; Xia et al. 2013). Graphene derivatives are in demand because of their excellent high electronic and thermal conductivity, high specific surface area, and optical transparency, combined with exceptionally good mechanical flexibility and environmental stability. Graphene is also used in photovoltaic, electronic, and electrochemical energy storage (Lee et al. 2013).

Graphene is a 1 atom–thick layer of graphite with a two-dimensional sp^2-hybridized carbon network. Graphene has an excellent role in improving the overall performance of OPV devices because of its peculiar properties, such as good mechanical strength, high thermal conductivity, superior transparency, large specific surface area, and tremendous charge transport properties (Chang et al. 2014).

Materials with a highly ordered phase, such as polythiophene (Too et al. 2001; Brabec et al. 2001b), offer high charge mobility, which increases the performance of the cell. Fluorination of polythiophene derivatives also affects the performance of organic photovoltaics. Recently, Jo et al. (2014) reported an increase in the power conversion efficiency by 20%–250% by fluorine atom substitution in poly(3,4-dialkylterthiophenes) (PDATs). Fluorinated PDATs show a deeper HOMO energy level than nonfluorinated ones, thus leading to higher open-circuit voltage in organic photovoltaic cells and enhanced molecular ordering.

Recently, Takao et al. (2014) used fluorinated subnaphthalocyanine derivatives as donor materials and fullerene as an acceptor for low-molecular-weight organic photovoltaic cells. Such cells have the low-lying HOMO energy levels, keeping the strong absorption band of long wavelength for improvement of open-circuit voltage without the expense of short-circuit current density. The frontier orbital energy levels can be effectively tuned by fluorination because the HOMO/LUMO energy levels of hexafluoro-, heptafluoro-, dodecafluoro-, tridecafluoro-, and the parent subnaphthalocyanine were estimated to be 5.69/3.93, 5.67/3.90, 5.96/4.19, 5.92/4.11, and 5.30/3.58 eV, respectively.

5.6 TYPES OF ORGANIC PHOTOVOLTAIC CELLS

5.6.1 Single Layer

Single-layer organic photovoltaic cells (Figure 5.2) are the simplest form, which are made up as a sandwich model, like a layer of organic electronic materials between two metallic conductors. Basically, both layers have different work functions, which help to set up an electric field in the organic layer. Typically, the first electrode is made up of an indium tin oxide layer (ITO) with high work function, and the second electrode is made up of a metal with low work function such as aluminum, magnesium, or calcium. When the organic layer absorbs light, electrons are excited from HOMO to LUMO, thereby forming excitons. Differences in work functions help to split the exciton pairs, pulling electrons to the positive electrode (an electrical conductor used to make contact with a nonmetallic part of a circuit) and holes to the negative electrode.

Ghosh et al. (1974) reported an Al/Mg phthalocyanine/Ag sandwich solar cell model with efficiency of 0.01%. Later, efficiency of 0.30% with a polyacetylene $(CH)_x$-based single-junction solar cell was reported by Weinberger et al. (1982).

Marks et al. (1994) made the single-layer device structure of OPV cells, which consists of a transparent electrode/organic photosensitive semiconductor/electrode. They used 50–320 nm thick poly(p-phenylene vinylene) (PPV) sandwiched between an indium tin oxide (ITO) anode and a low work function cathode. The quantum efficiency of this cell was reported to be approximately 0.1% under 0.1mW cm^{-2} intensity. Charge mobility in an organic semiconductor is around 10^{-3} cm^2 V^{-1}·s^{-1}, which is far less than the mobility of a single-crystalline silicon, which is about 10^3 cm^2 V^{-1}·s^{-1}, resulting in low quantum yield.

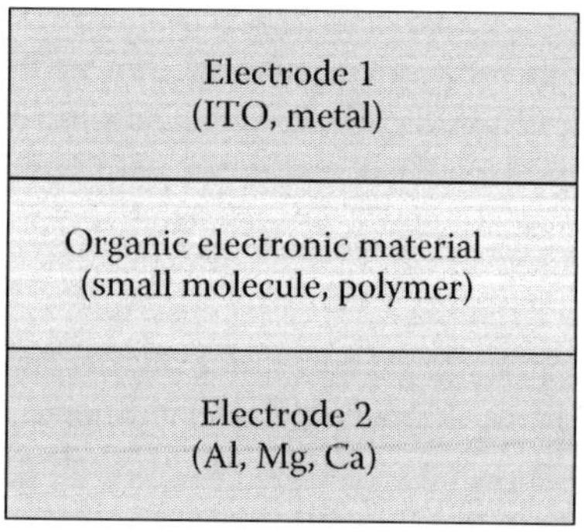

FIGURE 5.2 Structure of single-layer cells.

Single-layer organic material efficiency was reported as approximately 2%, because of its low stability of material, poor charge transport, and limitation by the low red light absorption. New materials were synthesized to solve these problems by use of material combination, optimization of molecular design, or self-assembly processing in order to control morphology, which increased performance up to 5% (Nelson 2002).

The morphology or architecture of semiconductor blends shows vast impact on the performance of the cells. It means cells should be prepared in such a way that they can capture more sunlight in order to produce more electricity or photocurrent (McGehee and Topinka 2006). Electron transfer along with effective hole transfer increase efficiency by enhancing short-circuit current. Wang et al. (2014b) showed that organic photovoltaic cells based on ambipolar donor–acceptor1–acceptor2 architectural 4-styryltriphenyl amine exhibit a long-lived charge separation state with a lifetime of 650 ns. Not only the morphology but also the conditions of preparation of organic film affect the current–voltage characteristics in a positive way. When organic film was prepared with copper phthalocyanine (CuPc) and hexadecafluoro CuPc (F_{16}CuPc) under different conditions, they then show different efficiencies of cells.

Page et al. (2014) reported that the thickness of the interlayer does not affect the performance of the single-junction solar cell. They prepared a single-junction solar cell based on the fulleropyrrolidines with amine (C_{60}-N) or zwitterionic (C_{60}-SB) substituents as cathode-independent buffer layers ranging from 5 to 55 nm thickness. It was observed that the effective work function of Ag, Cu, and Au electrodes was reduced by using a thin layer of C_{60}-N to 3.65 eV, and more than 8.5% efficiency was obtained, but this result was independent of cathode material (i.e., Al, Ag, Cu, or Au).

Beliatis et al. (2014) used PCDTBT:PC_{70}BM with solution-processed nanostructures of metal oxide–reduced graphene oxide (RGO) as electron transport layers (ETLs) exhibited the efficiency of 8% in single cells. The power conversion efficiencies were reported to be about 8% and 9%, respectively, using polymer and small molecules to synthesize a single-junction organic solar cell. Zhang et al. (2015) reported organic photovoltaic devices with oligothiophene-like small molecules, consisting of seven conjugation units as the backbone and 2-(1,1-dicyanomethylene)rhodanine as the terminal unit (i.e., DRCN7T) and [6,6]-phenyl C_{71}-butyric acid methyl ester (PC_{71}BM) as the acceptor unit. The photocurrent generation efficiency was observed as 9.30%. The DRCN7T-based cells showed exceptionally high internal quantum efficiency (100%). That may be because of a nanoscale donor–acceptor network or highly crystalline donor fibrils with approximately 10 nm diameters, which is close to diffusion length of the organic material and an efficient electron transport layer. Hybrid graphene–metal oxide materials are used as improved ETLs to improve efficiency by enhancing the charge transport process.

In a single-layer OPV cell, charge dissociates into free carriers only at one place (i.e., in the interface between semiconducting organics and a cathode). Later, it was investigated that the excitons are more efficiently dissociated at the interface between donor and acceptor, so a bilayer OPV was developed.

Single-layer organic photovoltaic cells have low quantum efficiency of less than 1% and low PCE of less than 0.1%. The major drawback with these cells is that the electric field generated from the difference between the two conductive materials is rarely enough to split an exciton. Thus, it sometimes results in electron–hole recombination.

5.6.2 Bilayer

Bilayer cells contain two layers with different electron affinities and ionization energies in between the conductive electrodes, which generate electrostatic forces at the interface between the two layers. The layer with higher electron affinity and ionization potential is the electron acceptor; the other layer is the electron donor. This structure is also called a *planar donor–acceptor heterojunction*. A bilayer OPV cell can be prepared by inserting an acceptor layer between a donor semiconducting organic and a cathode. It means the bilayer organic photovoltaic cell has an additional electron-transporting layer, which is not found in the single-layer OPV structure (Figure 5.3).

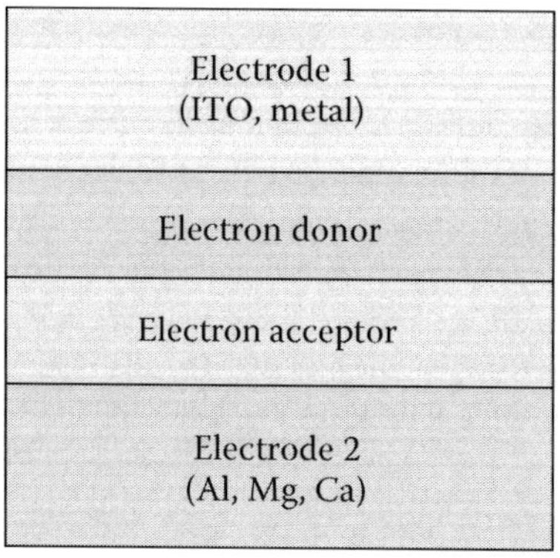

FIGURE 5.3 Structure of bilayer cells.

The interface between the two organic materials plays major role in determining the properties of the cell electrode/organic interlayer. The thin-film bilayer OPV cell structure was first realized by Tang (1986). The device consists of indium tin oxide (ITO)/copper phthalocyanine (CuPc)/perylene tetracarboxylic derivative (PV)/silver (Ag) with a power conversion efficiency of 1% under simulated AM2 conditions. This 10-fold increase in PCE resulted from improving exciton dissociation efficiency by adding an electron-transporting material that forms an offset energy band with the hole-transporting material. Antohe and Tugulea (1991) prepared a two-layer organic cell from copper phthalocyanine as a p-type organic semiconductor and 5,10,15,20-tetra(4-pyrydil)21H,23H-porphyne (TPyP) as an n-type organic semiconductor.

Sariciftci et al. (1993) prepared the bilayer organic photovoltaic cell for the first time using a conjugated polymer, where poly[2-methoxy-5-(2′-ethyl-hexyloxy)-1,4-phenylene vinylene] MEH-PPV absorbs light and transports holes to the anode, and fullerene acts as the electron-transporting material to the cathode. A power conversion efficiency of 0.04% was reported under monochromatic incident light at 514.5 nm, which was slightly higher than single polymer layer PV cells, but the performance is still low due to the intrinsically short diffusion length of excitons in organic semiconductors (Halls et al. 1996; Roman et al. 1999; Theander et al. 2000). To overcome this problem, Peumans et al. (2003) replaced perylene tetracarboxylic derivative with C_{60} as an acceptor in the device structure, which increased the device efficiency to 3.5%.

Zimmerman et al. (2013a) reported that replacing the exciton-quenching buffer layer by an exciton-blocking layer enhances the performance of the bilayer solar cell. When an exciton-blocking benzylphosphonic acid (BPA)–treated MoO_3 or NiO layer was used in place of the exciton-quenching MoO_3 anode buffer layer in bilayer organic photovoltaic cells, then the power conversion efficiency shows significant improvement. The addition of untreated MoO_3 anode buffers and BPA-treated NiO buffers in diphenylanilo-functionalized squaraine (DPSQ)/C_{60}-based bilayer devices show efficiency from 4.8% ± 0.2% to 5.4% ± 0.3% under illumination of 1 sun AM1.5G. Further addition of a highly conductive exciton-blocking bathophenanthroline (BPhen):C_{60} cathode buffer increases cell performance up to 5.9% ± 0.3%.

Use of fullerene as an electron donor increases the performance of planer heterojunction (PHJ) organic photovoltaic cells (Zhuang et al. 2013). Organic photovoltaic cells were prepared with electron donor, fullerene derivatives indene-C_{60} bisadduct (ICBA), and phenyl C_{61}-butyric acid methyl ester with fullerene C_{70} as the electron acceptor. Two processes are included: (1) fullerene bulk includes charge generation and (2) exciton dissociation at the donor–acceptor interface. Indene-C_{60} bisadduct with 5 nm thickness and C_{70} fullerene with 40 nm thickness give external quantum efficiency on an exposing long wavelength photon. This means total efficiency depends on the thickness of the donor fullerene.

A double layer of transparent conducting oxide (TCO) films increases the power conversion efficiency as compared to a single layer (Cho et al. 2014). It has been observed that indium tin oxide (ITO) and aluminum-doped zinc oxide (AZO) single-layered films are less efficient than the ITO/AZO films

Organic Photovoltaic Cells

of 500/250 nm thickness, while the double layer exhibits higher photocurrent due to higher transmittance and lower resistance.

A short rapid thermal annealing of methylammonium lead mixed halide perovskite ($CH_3NH_3PbI_{3-x}Cl_x$) at 130°C leads to the growth of large micron-sized textured perovskite domains. It also increases the short-circuit currents and power conversion efficiencies to 13.5% for the planar heterojunction perovskite solar cells (Saliba et al. 2014). Simultaneous optimization of light absorption and carrier collection increases the short-circuit current in thin planar organic photovoltaic heterojunction cells. Tsai et al. (2014) observed that an ultrathin-film solar cell with a SubPc/C_{60} photovoltaic structure shows 78% power conversion efficiency and short-circuit current of 0.790 mA cm^{-2} for a 30 nm thick cell, but only 32% power conversion efficiency for a 45 nm thick cell, for which short-circuit current is 0.980 mA cm^{-2}.

Typically, a diffusion length of organic material is 10 nm. For exciton diffusion to the interface of the layer and for generation of the charge carrier, the thickness of the layer must be equal to or in the same range as the diffusion length. However, polymer layer thickness should be 100 nm to absorb a sufficient amount of light, and at such a thickness, only a small amount of the excitons can reach the heterojunction interface.

5.6.3 Heterojunctions

Bulk heterojunction (BHJ) is made up of a nanoscale blend of donor and acceptor layers (Figure 5.4). These cells are thick enough to absorb a broad range of spectra. Yu and Heeger (1995) made a phase-separated polymer blend of donor poly[2-methoxy-5-(2′-ethyl-hexyloxy)-1,4-phenylene vinylene] (MEH-PPV) and acceptor cyano-PPV (CN-PPV). Scharber et al. (2006) prepared bulk heterojunction solar cells from conjugated polymers and a fullerene derivative. They also exhibited the relation between the open-circuit voltage and the oxidation potential for different conjugated polymers.

Li et al. (2005) showed significant enhancement in power conversion of a blend of OPV cells by using poly(3-hexylthiophene) (P3HT) as a hole-transporting polymer and PCBM as a soluble C_{60} derivative, while Kim et al. (2006) prepared a blend of organic photovoltaic cells using various P3HT and PCBM blend solutions. The P3HT solution had different regioregularities ranging from 80% to 96%. The P3HT solution with a higher order of regioregularity resulted in the crystallized fibril-like shape, which increased charge transport as well as photon absorption efficiency.

Chasteen et al. (2008) prepared photovoltaic cells consisting of layers and blends of a hole-transporting derivative of poly(p-phenylene vinylene) with different electron transporters, such as titanium dioxide, a cyano-substituted PPV, and a fullerene derivative (PCBM), to increase the performance of device. They studied time-resolved and steady-state photoluminescence phenomena and found that morphological differences such as chain conformation or domain size, often overshadowed the effect of charge transfer and also decreased losses to recombination. The charge transfer and excitons dissociation was found to increase by the addition of PCBM. But the electron-transporting polymer CN-ether-PPV does not show the same results. On increasing CN-ether-PPV

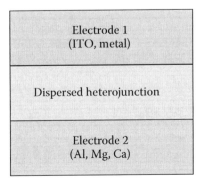

FIGURE 5.4 Structure of heterojunction cells.

in the polymer/polymer blend, large domains were created, which adversely affected the efficiency of the device. Therefore, a layered device structure is more efficient than a blended one because morphology is more easily controlled in a layered device.

The bilayer OPV cell collects a very small amount of excitons, which are created near the interface of the donor and the acceptor, leading to the development of bulk heterojunction OPV cells. In such cells, there is an intermixed composite of donors and acceptors that have a larger interface area. Ko et al. (2009) reported an enhancement in the absorption of light for OPV cells by patterning it using a photonic crystal nanostructure embossed in the photoactive bulk heterojunction layer via PRINT. This results in not only greater adsorption but also electrical enhancements. This method lends itself to a 4 cm^2–area fabrication of nanoscale features. The efficiency increased to 70%.

The efficiency of OPV devices depends on molecular ordering and lowering the HOMO level; this can be accomplished by using the proper processing method. Kim et al. (2013) synthesized conjugated polymers by the Stille polymerization reaction, in which a symmetrically branched alkyl side chain benzotriazole (DTBTz) is taken as an acceptor semiconductor, and unsubstituted or (triisopropylsilyl) ethynyl (TIPS)-substituted 2,6-bis(trimethylstannyl)benzo[1,2-b:4.5-b′] dithiophene (BDT) is taken as a donor semiconductor. The optical bandgap was reported to be 1.97 and 1.95 eV for PBDT-DTBTz and PTIPSBDT-DTBTz, respectively. The TIPS groups in the donor unit is responsible for the molecular ordering and lowering of the HOMO level, which improves the efficiency of the solar cell. Bulk heterojunction photovoltaic cells prepared from conjugative polymers (triisopropylsilyl) ethynyl (TIPS)-substituted 2,6-bis(trimethylstannyl)benzo[1,2-b:4.5-b′] dithiophene (BDT) and symmetrically branched alkyl side chain benzotriazole with power conversion efficiency of 5.5%, whereas 2,6-bis(trimethylstannyl)benzo[1,2-b:4.5-b′] dithiophene (BDT) and benzotriazole with symmetrically branched alkyl side chains show a power conversion efficiency of only 2.9%. This means that the TIPS group is responsible for higher optimal morphology and carrier mobility in OPV cells.

A bulk heterojunction is the most active layer system for and OPV cell. A BHJ is a network of domains that contains blended donor and acceptor molecules, often in pure and mixed phases (Pfannmoller et al. 2013). A bulk heterojunction with two-component p- and n-doped materials exhibits power conversion efficiencies in excess of 7% (Chen et al. 2013).

The interface structure is directly related to the performance of the cell, as it affects the recombination rate for electrons and holes at donor–acceptor heterojunctions in thin-film OPV cells. Cells must have disorder at the heterointerface and order in the bulk of the thin films (epitaxial relationships) to achieve high short-circuit current and open-circuit voltage. A squaraine donor and C_{60} acceptor heterojunction cell enhances the charge recombination through epitaxial relationships and provides interdiffusion, which reduces open-circuit voltage and improves short-circuit current (Zimmerman et al. 2013b).

Hedley et al. (2013) observed that morphology of bulk heterojunction plays an important role in the performance of OPV devices. A PTB7:PC 71 BM blend is composed of fiber-like, elongated, fullerene-rich and polymer-rich domains, which are 200–400 nm long and 10–50 nm wide. This morphology provides an efficiency of 80%, as it allows a concentration gradient for directional charge diffusion that leads to the extraction of charge pairs. But a decrease in efficiency to 45% was observed when agglomerated fullerene was used with a blend instead of elongated fullerene.

Yu and Chan (2013) prepared bulk heterojunction hybrid photovoltaic cells by incorporating a p-type NiO thin layer in between indium tin oxide (ITO)/nickel oxide (NiO)/poly(3-hexylthiophene) (P3HT): [6,6]-phenyl C_{61}-butyric(PCBM):titania (TiO_2): platinum (Pt) nanoparticles (NPs)/Ca/Al layers. It was observed that the NiO interlayer enhances stability and performance of the cell. The optimum cell performance of ITO/NiO of 5 nm thickness and P3HT:PCBM:TiO_2 of 15 wt.% : Pt of 0.03 wt.% or Ca /Al was recorded as 2.1% with an open-circuit voltage of 0.61 V, short-circuit current density of 6.22 mA cm^{-2}, and fill factor of 54.8%. Paci et al. (2013) incorporated poly(3-hexylthiophene) nanofibers in the polymer–fullerene nanostructured films to improve the structural durability.

Two electron-withdrawing fused thiadiazole units, made by fusing quinoxaline or phenazine with a benzothiadiazole unit, thiadiazolo[3,4-g]quinoxaline (DTBTQx), and thiadiazolo[3,4-i]phenazine (DTBTBPz) and a dithiophene electron-donating unit, were developed by Stille reaction. Two copolymers in the film state exhibited broad absorption spectra with a narrow bandgap of approximately 1.2 eV. A bulk heterojunction solar cell shows a power conversion efficiency of 0.52%, with a J_{sc} of 1.44 mA cm^{-2}, V_{oc} of 0.69 V, and FF of 0.40, after thermal annealing at 90°C (Hai et al. 2014).

Owczarczyk et al. (2014) carried out synthesis of push–pull conjugated copolymers based on cyclopenta[c]thiophene-4,6-dione (CTD) and benzodithiophene (BDT) as an organic photovoltaics electron donor material. Conjugated polymers have been used as electron donors and fullerene derivatives as electron acceptors. The power conversion efficiency of such a polymer solar cell has now exceeded 10%. Quinoidal units, phospholes, porphyrins, and fluorinated aromatic rings have been used as building blocks, which can now be introduced into low-bandgap conjugated polymers (Umeyama and Imahori 2014).

Shivanna et al. (2014) made a nonfullerene-based bulk heterojunction organic solar cell using a nonplanar perylene dimer (TP) as an electron acceptor and a thiophene-based donor polymer poly{[4,8-bis-(2-ethyl-hexyl-thiophene-5-yl)-benzo[1,2-6:4,5-6′] dithiophene-2,6-diyl]-alt-[2-(2′-ethyl-hexanayl)-thieno[3,4-6] thiophen-4, 6-diyl]} (PBDTTT-CT). The mixture of the donor polymer and TP at 50:50% weight ratio showed a photon-to-current conversion efficiency of 45% in the visible region and a power conversion efficiency of 3.2%.

Chan et al. (2014) synthesized a new p-type perylene diimide as a photoactive material in OPV devices. This nonplanar and three-dimensional spirobifluorene-modified perylene diimide compound can be used as a donor material in combination with fullerene to form bulk heterojunctions. This new photoactive material shows high open-circuit voltages of 0.97 V and a power conversion efficiency of up to 4%. The combination of isoindigo as the electron-deficient acceptor and 3,4-ethylenedioxythiophene as the electron-rich donor, followed by CH-arylation with different acceptors (4,7-dibromo[c][1,2,5]-(oxa, thia, and/or selena)diazole, were used to develop low-bandgap donor–acceptor–donor–acceptor (D-A-D-A) polymers. These polymers have high stability and good solubility in chlorinated solvents. Elsawy et al. (2014) utilized poly ((E)-6-(7-(benzo-[c][1,2,5]-thiadiazol-4-yl)-2,3-dihydrothieno-[3,4-b][1,4]dioxin-5-yl)-6′-(2,3-dihydrothieno-[3,4-b][1,4]-dioxin-5-yl)-1,1′-bis-(2-octyldodecyl)-[3,3′-biindolinylidene]-2,2′-dione) as the donor and PC$_{61}$BM as the acceptor in BHJ with a short-circuit current density of 8.10 mA cm^{-2}, open-circuit voltage of 0.56 V, fill factor of 35%, and PCE of 1.6%. These polymers were used as donor materials for photovoltaic applications, mainly in polymer solar cells.

Solution-processed photovoltaic cells synthesized with electron donor material 3,6-bis(N,N-dianisylamino)-fluoren-9-ylidene malononitrile (FMBDAA36) and ITO/PEDOT:PSS/ (1:3[w/w] FMBDAA36:PC$_{71}$BM)/LiF/Al show performance to 4.1%, open-circuit voltage of 0.89 V, short-circuit current of 10.35 mA cm^{-2}, and fill factor of 44.8%. Lim et al. (2014) studied the impact of ultrathin interfacial metal fluoride in the performance of bulk heterojunction organic photovoltaic cells. They used an ultrathin BaF$_2$ single-coverage layer (less than 3 nm) at the electron extraction contact, which showed a positive effect on the open-circuit voltage and power conversion efficiency of the organic photovoltaic cells, but the short-circuit current remained almost constant. An efficiency of 4.0% was obtained in the presence of the interlayer, but efficiency of only 2.1% was shown in its absence (under 100 mW cm^{-2} solar radiation). The enhancement in performance may be due to a hugely improved lifetime and lowered effective work function of the cathode caused by the large dipole moment of thin BaF$_2$ films.

Solvent-processed bulk heterojunction solar cells composed of 2,4-bis[4′-(N,N-di(4′-hydroxyphenyl)amino)-2′,6′-dihydroxyphenyl]squaraine (Sq-TAA-OH) as an active layer with configuration ITO/PEDOT:PSS/Sq-TAA-OH:PC$_{71}$BM/LiF/Al were reported (Karak et al. 2014). The bandgap and HOMO level of Sq-TAA-OH were reported to be 1.4 and −5.3 eV, respectively. The power conversion efficiency of about 4.8% was observed by a reproducible procedure using a mixture of good and poor Sq-TAA-OH-solubilizing organic solvents along with a bulk heterojunction layer made up of diiodooctane (DIO), followed by thermal annealing. The DIO is a good and poor solvent mixture that broadens the Sq-TAA-OH crystallites' size distribution in pristine films, whereas narrower size distribution is provided by thermal annealing.

Kim et al. (2014c) used difluorinated benzoselenadiazole (DFDTBSe) and ethylhexyloxy (EH)- or octyldodecyloxy (OD)-substituted benzo[1,2-b:4,5-b']dithiophene (BDT) as monomer units to form copolymers. These conjugated copolymers have low bandgaps of 1.66 and 1.69 eV and HOMO energy levels of −5.44 and −5.43 eV, respectively. The molecular packing and order of the active layer blended with [6,6]-phenyl-C_{71} butyric acid methyl ester ($PC_{71}BM$) depend on the different alkyloxy side chain in the polymer. The PODBDT-DFDTBSe:PC_{71}BM film is composed of an edge-on structure, while the PEHBDT-DFDTBSe:PC_{71}BM film has a face-on structure. It was also reported that a bulk heterojunction cell fabricated with a PEHBDT-DFDTBSe:PC_{71}BM active layer exhibited a maximum power conversion efficiency of 5.74%, which is the highest efficiency among benzoselenadiazole and BDT-derivative polymeric materials containing OPCs.

In bulk heterojunction OPV cells, the benzoselenophene derivative electron donor with $PC_{61}BM$ as the electron acceptor shows a power conversion efficiency as high as 5.8% (Park et al. 2014).

Stanculescu et al. (2014) prepared a heterojunction structure using single layers of arylene-based polymer, poly[N-(2-ethylhexyl)2.7-carbazolyl vinylene]/AMC16 and poly[N-(2-ethylhexyl)2.7-carbazolyl 1.4-phenylene ethynylene]/AMC22 with Buckminsterfullerene/C_{60} in the weight ratio of 1:2 (AMC16:C_{60}) and 1:3 (AMC22:C_{60}) by the matrix-assisted pulsed laser evaporation (MAPLE) technique. Different heterojunctions were prepared on glass/ITO/PEDOT-PSS with AMC16, AMC22, and AMC22:C_{60} layer by MAPLE. The highest performance was shown by a single layer of AMC16 polymer: glass/ITO/PEDOT-PSS/AMC16/Al.

The combination of a high fullerene-C_{70} and fullerene-C_{60} as amorphous donor and acceptor, respectively, in an optimized tandem configuration results in efficiency of 7.2%. On the contrary, a single heterojunction cell with donor–acceptor layers exhibits efficiency above 5.5%, with its material of 1.8–1.4 eV bandgap. It shows that the fill factor of the tandem stack is higher in comparison to either one of the subcells (Cheyns et al. 2014). In bulk heterojunction OPV cells, transparent conducting electrodes indium tin oxide (ITO) can be replaced by multilayer transparent electrode MTO/Ag/MTO (MAM) with a nanosized Ag thin film embedded between Mn-doped tin oxide (MTO) layers. It was prepared by employing a RF sputtering process at room temperature. The MTO/Ag/MTO multilayer electrodes are more advanced than an indium tin oxide (ITO) electrode as it exhibits transmittance of 80.1%–85.4% within a 380–780 nm wavelength range, a sheet resistance of 10.1–10.6 Ω/sq, a short-circuit current of 7.12 mA cm^{-2}, an open-circuit voltage of 0.62 V, a fill factor of 0.62, and a power conversion efficiency of 2.73% (Lee et al. 2015).

Aziz et al. (2015) used green color dye vanadyl 2,9,16,23-tetraphenoxy-29H,31H-phthalocyanine (VOPcPhO) in between a photoactive layer of VOPcPhO, which has been sandwiched between indium tin oxide (ITO) and aluminum (Al) to fabricate ITO/PEDOT: PSS/VOPcPhO/Al photovoltaic devices. The device showed a photovoltaic effect with J_{sc} of 5.26×10^{-6} A cm^{-2}, V_{oc} of 0.621 V, and FF equal to 0.33. It had a limitation in that when a single layer of vanadyl phthalocyanine derivative was incorporated in the solar cells, the efficiency decreases. On the contrary, adding a variety of donor materials to form bulk heterojunction increases the efficiency of the cell.

The photoactive layer made up of homogeneously dispersed carbon nanotubes (CNTs) in a semiconducting polymer–fullerene derivative bulk heterojunction matrix shows improvement in optical and electrical properties due to effects of the wide-band photoabsorption and high charge carrier mobility of the CNTs. The CNTs are functionalized by alkyl amide groups for high dispersion in organic media and by homogeneously mixing them with the polymer and fullerene in solution. This exhibits a 40% (3.2%–4.4%) increase in power conversion efficiency as compared to an organic solar cell using a photoactive layer without CNTs. The open-circuit voltage decreases slightly due to the small energy level variation of the active layer, when CNTs were used in the semiconducting polymer poly 3-hexylthiophene and a fullerene derivative, [6,6]-phenyl C61 butyric acid methyl ester (P3HT:PCBM) as photoactive materials. It was concluded that power conversion efficiency increases due to an increase in short-circuit current from 10.5 to 14.6 mA cm^{-2}. The photocurrent enhancement in the CNT-incorporated photoactive layer is attributed to two major causes: (1) an increase in the absorbed photons, and (2) an increase in the charge carrier mobility (Gwang et al. 2012).

The single-walled carbon nanotube (SWNT) also affects the efficiency of the cell. It may decrease or increase the performance depending on the length of the single-walled carbon nanotube. It has been reported that when full-length SWNTs were added between poly(3-hexylthiophene)-[6,6]-phenyl-C_{61}-butyric acid methyl ester (P3HT-PCBM) (1:1 w/w) active layers in the bulk heterojunction cell, then efficiency decreases (only 1.3%) at 80°C –225°C. But a cell without SWNTs shows efficiency to 2%. When shortened SWNTs were incorporated, it was noticed that efficiency reached 2% and increased by nearly 50% at 70°C. This enhancement is due to improved hole transport through the SWNTs in devices (Skupov and Adronov 2014).

The morphology or structure showed a major influence on the photocurrent power of organic photovoltaic devices. The heterojunction cell made up of zinc phthalocyanine/fullerene (ZnPc/C_{60}) with poly(3,4-ethylenedioxythiophene):poly(styrenesulfonate) and 2,5-bis(4-biphenylyl)-bithiophene (BP2T) between indium tin oxide and ZnPc layers showed much better efficiency than a conventional cell without BP2T (i.e., almost 120% higher efficiency). With the addition of BP2T in ZnPc/C_{60} bilayer films, a power current efficiency of 2.6% was obtained as a result of improved short-circuit current and fill factor. It was also noticed that when ZnPc films were tuned with long fiber-like grains, they show more photocurrent generation because the charge carriers can transport for a longer distance in comparison to the round and short fiber-like grains (Wang et al. 2014c).

Recently, new cascade heterojunction (CHJ) organic solar cells have been investigated, which are the best replacement for series-connected tandems and conventional bulk heterojunctions. Such cells have poor fill factor and minimal enhancements in power conversion efficiency, but show high internal quantum efficiency (IQE) and broad spectral coverage. The fill factor and efficiency of the cells can be increased by tuning the maximum power point voltage of the constituent parallel-connected heterojunctions and reducing the intrinsic injection barriers. It was recorded that a CHJ device with a transparent exciton dissociation layer (EDL)/interlayer/acceptor architecture with more than 99% internal quantum efficiency in the interlayer showed a 46% increase in power conversion efficiency as compared to a single heterojunction (SHJ).

5.7 TANDEM SOLAR CELLS

Tandem solar cells consist of two staked single subcells, which are connected by an interlayer. Basically, a tandem solar cell provides a new way to capture or harvest a broad range of solar radiation spectrum, as it involves combining two or more solar cells with different absorption bands.

Generally, a polymer single-layer cell has less efficiency because of the lack of a suitable low-bandgap polymer. Therefore, polymer tandem solar cells provide an innovative way to achieve high efficiency. The low-bandgap polymer decreases the performance of the cell, but a combination of such polymers shows an increase in the efficiency of the cell reported. The performance of an organic tandem cell can be increased (Hadipour et al. 2006) by electronically combining two bulk heterojunction single subcells in series to prepare a solution-processed polymer tandem cell. Two heterojunctions have maximum absorption near 850 nm and 550 nm in this cell. The cell was designed in such a way that the thickness of the bottom cell layer was tuned to maximize the optical absorption of the top cell layer. The combination concept results in a high open-circuit voltage (i.e., equal to the sum of each subcell [1.4 V]).

Tvingstedt et al. (2007) reported a doubling of the power conversion efficiency of 1.8 ± 0.3 by folding two conjugative polymer planar cells, which were spectrally different. The tandem organic solar cell performance also depends on the interlayer and illuminated light. Zhao et al. (2008) observed that a polymer–small molecule tandem cell using high transparent Al and a MoO_3 intermediate layer of 1 nm Al and 15 nm MoO_3 thickness exhibited efficiency of 2.82% on illumination under 100 mW cm^{-2}, a short-circuit current density of 6.05 mA cm^{-2}, and the open-circuit voltages of individual cells of 1.01 V. The power conversion efficiency increases to 3.88% when the tandem cell was illuminated under 300 mW cm^{-2}.

Hadipour et al. (2008) studied a multilayer tandem solar cell prepared by small molecules and polymers with their specific absorption maxima and width. The main advantage of such cells is that the voltage at which charges are collected in each subcell is closer to the energy of the photons absorbed (i.e., the photon energy is utilized more efficiently). The recombination of small- and large-bandgap polymer semiconductor cells showed efficiency of 4.9%, and the short-circuit current was observed to be far greater than the current limiting the subcell (Gilot et al. 2010).

The interlayer is a metal/semiconductor contact, which is different than the traditional tunnel junction in inorganic tandem cells. This interlayer plays an important role in deciding the performance of such devices. An interlayer of p-n type between two subcells shows a power conversion efficiency of 5.8% (Sista et al. 2010). A robust interlayer is physically quite strong, optically transparent, and electrically conductive in nature. Yang et al. (2011) reported that on exposing such cells to light, charges are collected and recombined in the interconnecting layer, while charges are generated and extracted from the tandem solar cell under dark conditions. Such an interconnecting layer cell shows high PCE (7.0%). Short-circuit current is affected by a variation in the thickness of the interlayer, which is the space between two subcells. The variations in thickness of transparent p-doped interlayer in tandem solar cells results in an enhancement in the short-circuit current but does not affect (or negligibly affects) open-circuit voltage and fill factor (Riede et al. 2011).

Graphene oxide (GO) is a two-dimensional, random copolymer with two units with distributed nanosize graphitic patches and highly oxidized domains. Tung et al. (2011) prepared a sticky interlayer by casting a mixture of GO and conducting polymer poly(3,4-ethylenedioxythiophene):poly (styrenesulfonate) (PEDOT:PSS) in water. The insulating GO makes PEDOT highly conductive.

The combination of a tandem solar cell and two vacuum-processed single heterojunctions was tried by Riede et al. (2011). The main concept behind this combination was to make a tandem solar cell, which is sensitive to the entire visible range of light. Two heterojunctions were used for this purpose. The first was C_{60} and a dicyanovinyl-capped sexithiophene derivative (DCV6T) that absorbs mainly in the green region, whereas the second incorporates C_{60} and a fluorinated zinc phthalocyanine derivative (F4-ZnPc) that absorbs mainly in the red region. Both of these together make a tandem solar cell, which is sensitive to the entire visible range of the solar spectrum.

Wang et al. (2011) investigated multilayer junction solar cells based on colloidal quantum dots (CQDs). The electronic bandgap was tuned by combining colloidal quantum dots to make it active toward a broad range of the solar spectrum. First colloidal quantum dots (CQDs) PbS, tandem solar cell was also prepared by Wang et al. in which graded recombination was used that permits electron and hole currents to meet and recombine. The open-circuit voltage equal to the sum of the two constituent single-junction devices (i.e., 1.06 V) and power conversion efficiency of about 4.2% were obtained.

Dou et al. (2012) made such a call by coupling low-bandgap conjugated polymer PBDTT and DPP with bandgap of approximately 1.44 eV. A power conversion efficiency was reported of about 8.62%, which was far better than that for a single-layer cell with the same polymer (near 6%).

Li et al. (2013) made tandem OPV cells with a solution-processed intermediate layer (IML) consisting of ZnO and neutralized poly(3,4-ethylenedioxythiophene):poly(styrenesulfonate) (PEDOT:PSS) combined in series to obtain excellent functionality, reliability, and performance. The tandem solar cells containing a processed low-temperature solution and a chemically stable interconnecting layer such as PEDOT:PSS/AZO/PEIE were far better than single-junction solar cells, because they can double the open-circuit voltage and utilize photon energy efficiently (Mitul et al. 2015).

5.8 HYBRID TANDEM PHOTOVOLTAIC CELL

A new generation of photovoltaic solar cells has evolved as a promising and renewable energy source because of their properties, such as low cost and excellence in optical and electrical properties. In the third generation of photovoltaics, semitransparent polymer material was used in hybrid tandem photovoltaic cells, and a power conversion efficiency of 5.0% was obtained by Beiley et al. (2013).

Jin et al. (2014) synthesized hybrid solar cells with $Cd_{0.75}Hg_{0.25}Te$ colloid quantum dots (CQDs) and water-soluble conjugative polymer by aqueous process. This aqueous-processed solar cell exhibited excellent sensitivity toward light because a composite of these materials provides an effective electron transfer and a power conversion efficiency of 2.7% under 100 mW cm^{-2} illumination. It showed efficiency as high as 15.04% under the Herschel infrared region (i.e., 780–1100 nm), which is reported as the highest among organic–inorganic hybrid solar cells (HSCs) made up of $Cd_xHg_{1-x}Te$ CQDs and the highest near-infrared (NIR) contribution among aqueous-processed HSCs.

5.9 INVERTED TANDEM SOLAR CELL

It is known that tandem solar cells are a combination of two or more bulk heterojunctions, active for different adsorption bands prepared by various processes. This makes the cell unable to use a wider range of solar absorption capacities. Kim et al. (2007) utilized two bulk heterojunctions made up of semiconducting polymers and fullerene derivatives by solution processes. These are separated and connected by a transparent titanium oxide (TiO_x) layer. This TiO_x layer helps in the electron transportation and collection on the first layer and provides a stable foundation that forms a second cell. Efficiency was recorded at more than 6% at illuminations of 200 mW cm^{-2}, when the front cell consists of the low-bandgap polymer–fullerene composite as the charge separating layer, and the back cell is composed of the high-bandgap polymer. This architecture is known as an inverted structure.

A metal-oxide-only MoO_3/Al/ZnO interlayer between two polymer bulk heterojunction subcells was used to make inverted-polymer tandem solar cells, which leads to a power conversion efficiency of 5.1% (Chou et al. 2011). The active layer poly(3-hexylthiophene) (P3HT) and 1-(3-methoxycarbonyl)-propyl-1-phenyl-(6,6)C_{61} (PCBM) and an intermediate layer MoO_3/Ag/Al/Ca were used to prepare an inverted tandem cell (Zhao et al. 2011). The indium tin oxide (ITO) modified by Ca was used as the cathode (electron collection) for the top subcell and Ca that form a continuous layer of 3 nm thickness, while Ag was used as the anode for hole collection. Then MoO_3 was used as the anode buffer layer, which helps to reduce exciton quenching and charge loss. A 2.89% power conversion efficiency was observed under 100 mW cm^{-2} under AM 1.5 G solar irradiation. The short-circuit current density of 4.19 mA cm^{-2}, open-circuit voltage of 1.17 V, and fill factor of 59.0% were recorded.

5.10 INVERTED ORGANIC PHOTOVOLTAIC CELLS

In spite of the high power conversion efficiency of OPV devices, there are still some limitations in terms of device stability due to the need for air-sensitive low work function metal cathodes, such as Al. The diffusion of oxygen into the organic active layer through pinholes and grain boundaries in the Al cathode may lead to the degradation of the active layer, which results in device instability in air. One approach to solving this issue is to adopt an inverted structure, where the charge separation and collection nature of the electrode are reversed.

A poly(2,7-carbazole) derivative (PCDTBT):fullerene derivative ($PC_{70}BM$) active layer was used for fabrication of highly efficient and air-stable inverted-polymer photovoltaic cells with molybdenum oxide and zinc oxide films as hole- and electron-transporting layers, respectively. It showed the highest power conversion efficiency of 6.3% (Sun et al. 2011). The solution-processed (sol-gel) vanadium oxides (VO_x) act as a hole-transporting and protecting layer in inverted OPV. The vanadium oxides (VO_x), an interfacial layer, lead to enhancement in performance (Chen et al. 2011). ZnO thin films on indium tin oxide as electron-collecting layers were fabricated along with ultrathin TiO_2 layers by atomic layer deposition. It was observed that as the thickness of TiO_2 increases, efficiency was found to decrease. The TiO_2 layer helps in preventing the electron collection on ZnO.

The electron and hole recombination on the surface of ZnO can be quenched by TiO_2, resulting in improved efficiency of the inverted organic solar cells (IOSCs) (Seo et al. 2011).

Interface treatment with 3-aminopropyltriethoxysilane (APTES) leads to an increase in power conversion efficiency of approximately 8.08% ± 0.12% with histogram skewness of −0.291 in PTB7:PC_{71}BM inverted organic photovoltaic cells. On the contrary, an untreated device shows an efficiency of 7.80% ± 0.26% with histogram skewness of −1.86. A 3-aminopropyltriethoxysilane (APTES) cathode treated interfacially provides large-area modules with high spatial performance uniformity (Luck et al. 2013).

Vasilopoulou et al. (2014a) used poly(3-hexylthiophene) (P3HT) as the electron donor and [6,6]-phenyl-C_{71}-butyric acid methyl ester (PC_{71}BM) as the electron acceptor to prepare inverted OPV cells. ZnO films were used as electron collection layers to improve its performance, while an under-stoichiometric molybdenum oxide MoO_x was used as the hole collection layer. The ZnO layer by atomic layer deposition (ALD-ZnO) with 20 nm thickness exhibited significant enhancement in the performance of OPV cells as compared to ZnO deposited by the sol-gel method (sg-ZnO) in inverted organic solar cells. The ZnO layer deposited by atomic layer deposition show reduced defect/trap concentration and a decrease in the electron extraction barrier at the ALD-ZnO/organic active layer interface.

Kim et al. (2014d) introduced ytterbium (Yb) as the electron transport layer in inverted OPVs where Yb was evaporated directly onto indium tin oxide (ITO). This electron transport layer (ETL) increases the power conversion efficiency and air stability in organic solar cells by promoting the formation of ohmic contact between the active layer and the cathode metal. The inverted OPVs made up of ITO/Yb/P3HT:PCBM/MoO_3/Ag give PCE up to 4.3% and fill factor of 71% in one sun irradiation. The Yb in inverted OPVs is vastly superior to other ETLs. Over 80% of its original PCE was retained, even after 30 days. Giansante et al. (2014) developed a solution-phase ligand exchange method, in which they replaced oleate ligands on PbS QDs with arenethiolate anions ligands in hybrid organic and inorganic quantum dot photovoltaic cells. The objective of using arenethiolate anions in place of oleate ligands was to decrease the distance between particles in PbS QD solids, which leads to enhancement of charge transport. This showed an efficiency of 1.85%, which is outstanding among single PbS QD layers.

Organic photovoltaics have wide applications because they offer many advantages, including that they are lightweight, thin, semitransparent, mechanically flexible, low cost, and not a disposal problem. There is an added advantage in that there are no restrictions on the shape, size, and color (or semitransparent) of OPV devices; therefore, they can be applied to surfaces of nearly any shape. The use of low-bandgap polymers, small molecules, or a combination of both may enhance the efficiency of OPV cells. An interesting example is that an OPV-functionalized bag is useful in charging all the portable electronic devices carried in it.

REFERENCES

Adams, W. G., and R. E. Day. 1877. The action of light on selenium. *Phil. Trans. Royal Soc. London.* 167: 313–349.

Antohe, S., and L. Tugulea. 1991. Electrical and photovoltaic properties of a two-layer organic photovoltaic cell. *Phys. Status Solidi (A).* 128: 253–260.

Appleyard, S. F. J., S. R. Day, R. D. Pickford, and M. R. Willis. 2000. Organic electroluminescent devices: Enhanced carrier injection using SAM derivatized ITO electrodes. *J. Mater. Chem.* 10: 169–173.

Aziz, F., Z. Ahmad, S. M. Abdullah, K. Sulaiman, and M. H. Sayyad. 2015. Photovoltaic effect in single-junction organic solar cell fabricated using vanadyl phthalocyanine soluble derivative. *Pigment Resin Technol.* 44: 26–32.

Bakulin, A. A., A. Rao, V. G. Pavelyev, P. H. M. van Loosdrecht, M. S. Pshenichnikov, D. Niedzialek, et al. 2012. The role of driving energy and delocalized states for charge separation in organic semiconductors. *Science.* 335: 1340–1344.

Becquerel, A. E. 1839. Recherchessur les effets de la radiation chimique de la lumieresolaire au moyen des courants electriques. *Compt. Rend. Acad. Sci.* 9: 145–149.

Beiley, Z. M., M. G. Christoforo, P. Gratia, A. R. Bowring, P. Eberspacher, G. Y. Margulis, et al. 2013. Semi-transparent polymer solar cells with excellent sub-bandgap transmission for third generation photovoltaics. *Adv. Mater.* 25: 7020–7026.

Beliatis, M. J., K. K. Gandhi, L. J. Rozanski, R. Rhodes, L. McCafferty, M. R. Alenezi, et al. 2014. Hybrid graphene-metal oxide solution processed electron transport layers for large area high-performance organic photovoltaics. *Adv. Mater.* 26: 2078–2083.

Bergmann, L. 1931. Uber eine neue Selen-sperrschicht photozelle. *Physik. Z.* 32: 286–289.

Bernardo, B., D. Cheyns, B. Verreet, R. D. Schaller, B. P. Rand, and N. C. Giebink. 2014. Delocalization and dielectric screening of charge transfer states in organic photovoltaic cells. *Nat. Commun.* 5: 3245.

Brabec, C. J., N. S. Sariciftci, and J. C. Hummelen. 2001a. Plastic solar cells. *Adv. Funct. Mater.* 11: 15–26.

Brabec, C. J., C. Winder, M. C. Scharber, N. S. Sariciftci, J. C. Hummelen, M. Svensson, et al. 2001b. Influence of disorder on the photo-induced excitations in phenyl substituted polythiophenes. *J. Chem. Phys.* 115: 7235–7244.

Cacialli, F., R. H. Friend, C.-M. Bouche, P. Le Barny, H. Facoetti, F. Soyer, et al. 1998. Napthalimide side chain polymers for organic eight-emitting diodes: Band offset engineering and role of polymer thickness. *J. Appl. Phys.* 83: 2343–2356.

Cahen, D., G. Hodes, M. Gratzel, J. F. Guillemoles, and I. Riess. 2000. Nature of photovoltaic action in dye-sensitized solar cells. *J. Phys. Chem. B.* 104: 2053–2059.

Chamberlain, G. A. 1983. Organic solar cells: A review. *Solar Cells.* 8: 47–83.

Chan, C.-Y., Y.-C. Wong, H.-L. Wong, M.-Y. Chan, and W.-W. Yam. 2014. A new class of three-dimensional, p-type, spirobifluorene-modified perylene diimide derivatives for small molecular-based bulk heterojunction organic photovoltaic devices. *J. Mater. Chem. C.* 2: 7656–7665.

Chang, D. W., H.-J. Choi, A. Filer, and J.-B. Baek. 2014. Graphene in photovoltaic applications: Organic photovoltaic cells (OPVs) and dye-sensitized solar cells (DSSCs). *J. Mater. Chem. A.* 2: 12136–12149.

Chasteen, S. V., V. Sholinb, S. A. Carterb, and Garry Rumblesc. 2008. Towards optimization of device performance in conjugated polymer photovoltaics: Charge generation, transfer and transport in poly(p-phenylene-vinylene) polymer heterojunctions. *Sol. Energy Mater. Solar Cells.* 92: 651–659.

Chauhan, R. N., R. S. Anand, and J. Kumar. 2014. RF-sputtered Al-doped ZnO thin films: Optoelectrical properties and application in photovoltaic devices. *Phys. Status Solidi (a).* 211: 2514–2522.

Chen, C.-P., Y.-D. Chen, and S.-C. Chuang. 2011. High-performance and highly durable inverted organic photovoltaics embedding solution-processable vanadium oxides as an interfacial hole-transporting layer. *Adv. Mater.* 23: 3859–3863.

Chen, G.-Y., Y.-H. Cheng, Y.-J. Chou, M.-S. Su, C.-M. Chen, and K.-H. Wei. 2011a. Crystalline conjugated polymer containing fused 2,5-di(thiophen-2-yl)thieno[2,3-b] thiophene and thieno[3,4-c]pyrrole-4,6-dione units for bulk heterojunction solar cells. *Chem. Commun.* 47: 5064–5066.

Chen, Y. C., C. Y. Hsu, R. Y. Lin, K. C. Ho, and J. T. Lin. 2013. Materials for the active layer of organic photovoltaics: Ternary solar cell approach. *ChemSusChem.* 6: 20–35.

Cheyns, D., M. Kim, B. Verreet, and B. P. Rand. 2014. Accurate spectral response measurements of a complementary absorbing organic tandem cell with fill factor exceeding the subcells. *Appl. Phys. Lett.* 104: 093302.

Chiang, C. K., C. R. Fincher, Jr., Y. W. Park, A. J. Heeger, H. Shirakawa, E. J. Louis, et al. 1977. Electrical conductivity in doped polyacetylene. *Phys. Rev. Lett.* 39: 1098.

Cho. J. M., J. Kim, H. Kim, M. Kim, S.-J. Moon, and W. S. Shin. 2014. ITO/AZO double-layered transparent conducting oxide films for organic photovoltaic cells. *Mol. Cryst. Liq. Cryst.* 597: 1–7.

Chou, C.-H., W. L. Kwan, Z. Hong, L.-M. Chen, and Y. Yang. 2011. A metal-oxide interconnection layer for polymer tandem solar cells with an inverted architecture. *Adv. Mater.* 23: 1282–1286.

Cimrova, V., M. Remmers, D. Neher, and G. Wegner. 1996. Polarized light emission from LEDs prepared by the Langmuir-Blodgett technique. *Adv. Mater.* 8: 146–149.

Coakley, K. M., and M. D. McGehee. 2003. Photovoltaic cells made from conjugated polymers infiltrated into mesoporous titania. *Appl. Phys. Lett.* 83: 3380.

Coakley, K. M., and M. D. McGehee. 2004. Conjugated polymer photovoltaic cells. *Chem. Mater.* 16: 4533–4542.

Deibel, C., T. Strobell, and V. Dyakonov. 2010. Role of the charge transfer state in organic donor–acceptor solar cells. *Adv. Mater.* 22: 4097–4111.

Dhanabalan, A. J., K. J. Van Duren, P. A. Van Hal, J. L. J. van Dongen, and R. A. J. Janssen. 2001. Synthesis and characterization of a low bandgap conjugated polymer for bulk heterojunction photovoltaic cells. *Adv. Funct. Mater.* 11: 255–262.

Diaz, F. R., J. C. Bernède, M. A. Del Valle, and V. Jousseaume. 2001. From the organic electroluminescent diodes to the new organic photovoltaic cells. *Curr. Trends Polym. Sci.* 6: 135–155.

Dittmer, J. J., E. A. Marseglia, and R. H. Friend. 2000. Electron trapping in dye/polymer blend photovoltaic cells. *Adv. Mater.* 12: 1270–1274.

Dou, L., J. You, J. Yang, C.-C. Chen, Y. He, S. Murase, et al. 2012. Tandem polymer solar cells featuring a spectrally matched low-bandgap polymer. *Nat. Photonics.* 6: 180–185.

Elsawy, W., H. Kang, K. Yu, A. Elbarbary, K. Lee, and J.-S. Lee. 2014. Synthesis and characterization of isoindigo-based polymers using CH-arylation polycondensation reactions for organic photovoltaics. *J. Polym. Sci. Part A.* 52: 2926–2933.

Fanshun, M., and H. Tian. 2005. Solar cells based on cyanine and polymethine dyes. In N. S. Sariciftci and S. S. Sun (Eds), *Organic Photovoltaics: Mechanisms, Materials and Devices*. Boca Raton: CRC Press.

Fritts, C. E. 1883. On a new form of selenium photocell. *Am. J. Sci.* 26: 465–472.

Gautier-Thianche, E., C. Sentein, A. Lorin, C. Denis, P. Raimond, and J. M. Nunzi. 1998. Effect of coumarin on blue light-emitting diodes based on carbazol polymers. *J. Appl. Phys.* 83: 4236–4241.

Gelinas, S., A. Rao, A. Kumar, S. L. Smith, A. W. Chin, J. Clark, et al. 2014. Ultrafast long-range charge separation in organic semiconductor photovoltaic diodes. *Science.* 343: 512–516.

Ghosh, A. K., D. L. Morel, T. Feng, R. F. Shaw, and C. A. Rowe, Jr. 1974. Photovoltaic and rectification properties of Al/Mg phthalocyanine/Ag Schottky-barrier cells. *J. Appl. Phys.* 45: 230–236.

Giansante, C., L. Carbone, C. Giannini, D. Altamura, Z. Ameer, G. Maruccio, et al. 2014. Surface chemistry of arenethiolate-capped PbS quantum dots and application as colloidally stable photovoltaic ink. *Thin Solid Films.* 560: 2–9.

Gilchrist, J. B., T. H. Basey-Fisher, S. C. Chang, F. Scheltens, D. W. Mccomb, and S. Heutz. 2014. Uncovering buried structure and interfaces in molecular photovoltaics. *Adv. Funct. Mater.* 24: 6473–6483.

Gilot, J., M. M. Wienk, and R. A. J. Janssen. 2010. Optimizing polymer tandem solar cells. *Adv. Mater.* 22: E67–E71.

Grondahl, L. O. 1933. The copper-cuprous-oxide rectifier and photoelectric cell. *Rev. Modern Phys.* 5: 141–168.

Gwang, H. J., S. J. Jin, S. H. Park, S. Jeon, and S. H. Hong. 2012. Highly dispersed carbon nanotubes in organic media for polymer: Fullerene photovoltaic devices. *Carbon.* 50: 40–46.

Hadipour, A., B. De Boer, and P. W. M. Blom. 2008. Organic tandem and multi-junction solar cells. *Adv. Funct. Mater.* 18: 169–181.

Hadipour, A., B. De Boer, J. Wildeman, F. B. Kooistra, J. C. Hummelen, M. G. R. Turbiez, et al. 2006. Solution-processed organic tandem solar cells. *Adv. Funct. Mater.* 16: 1897–1903.

Hai, J., W. Yu, E. Zhu, L. Bian, J. Zhang, and W. Tang. 2014. Synthesis and photovoltaic characterization of thiadiazole based low bandgap polymers. *Thin Solid Films.* 562: 75–83.

Halim, M., J. N. G. Pillow, I. D. W. Samuel, and P. L. Burn. 1999. Conjugated dendrimers for light-emitting diodes: Effect of generation. *Adv. Mater.* 11: 371–374.

Halls, M., J. J. N. K. Pichler, R. H. Friend, S. C. Morttigard, and A. B. Holmes. 1996. Excitons diffusion and dissociation in a poly(p-phenylene vinylene)/C60 heterojunction photovoltaic cell. *Appl. Phys. Lett.* 68: 3120–3122.

Harald, H., and N. S. Sariciftci. 2004. Organic solar cells: An overview. *J. Mater. Res.* 19: 1924–1945.

He, Z., C. Zhong, S. Su, M. Xu, H. Wu, and Y. Cao. 2012. Enhanced power-conversion efficiency in polymer solar cells using an inverted device structure. *Nat. Photonics* 6: 591–595.

Hedley, G. J., A. J. Ward, A. Alekseev, C. T. Howells, E. R. Martins, L. A. Serrano, et al. 2013. Determining the optimum morphology in high-performance polymer-fullerene organic photovoltaic cells. *Nat. Commun.* 4: 2867. doi:10.1038/ncomm53867.

Hendrickx, E., Y. Zhang, K. B. Ferrio, J. A. Herlocker, J. Anderson, N. R. Armstrong, et al. 1999. Photoconductive properties of PVK-based photorefractive polymer composites doped with fluorinated styrene chromophores. *J. Mater. Chem.* 9: 2251–2258.

Hoegel, H. 1965. On photoelectric effects in polymers and their sensitization by dopants. *J. Phys. Chem.* 69: 755–766.

Jailaubekov, A. E., A. P. Willard, J. R. Tritsch, W.-L Chan, N. Sai, R. Gearba, et al. 2013. Hot charge-transfer excitons set the time limit for charge separation at donor/acceptor interfaces in organic photovoltaics. *Nat. Mater.* 12: 66–73.

Jenekhe, S. A., and S. Yi. 2000. Efficient photovoltaic cells from semiconducting polymer. *Appl. Phys. Lett.* 77: 2635–2637.

Jiang, J.-M., P.-A. Yang, H.-C. Chen, and K.-H. Wei. 2011. Synthesis, characterization, and photovoltaic properties of a low-band gap copolymer based on 2,1,3-benzooxadiazole. *Chem. Commun.* 47: 8877–8879.

Jiang, X., R. A. Register, K. A. Killeen, M. E. Thompson, F. Pschenitzka, and J. C. Sturm. 2000. Statistical copolymers with side-chain hole and electron transport groups for single layer electroluminescent device application. *Chem. Mater.* 12: 2542–2549.

Jin, G., H.-T. Wei, T.-Y. Na, H.-Z. Sun, H. Zhang, and B. Yang. 2014. High-efficiency aqueous-processed hybrid solar cells with an enormous Herschel infrared contribution. *ACS Appl. Mater. Interfaces.* 6: 8606–8612.

Jo, J. W., J. W. Jung, H.-W. Wang, P. Kim, T. P. Russell, and W. H. Jo. 2014. Fluorination of polythiophene derivatives for high performance organic photovoltaics. *Chem. Mater.* 26: 4214–4220.

Karak, S., P. J. Homnick, A. M. D. Pelle, Y. Bae, V. V. Duzhko, F. Liu, et al. 2014. Crystallinity and morphology effect on a solvent-processed solar cell using a triarylamine-substituted squaraine. *ACS Appl. Mater. Interfaces.* 6: 11376–11384.

Kido, J., K. Hongawa, K. Okuyama, and K. Nagai. 1994. White light-emitting organic electroluminescent devices using the poly(N-vinylcarbazole) emitter layer doped with three fluorescent dyes. *Appl. Phys. Lett.* 64: 815–817.

Kim, G. M., I. S. Oh, A. N. Lee, and S. Y. Oh. 2014d. Applications of ytterbium in inverted organic photovoltaic cells as high-performance and stable electron transport layers. *J. Mater. Chem. A.* 2: 10131–10136.

Kim, J., K. G. Chitibabu, M. J. Cazeca, W. Kim, J. Kumar, and S. K. Tripathy. 1997. Fabrication of polymer light-emitting diodes by layer-by-layer complexation technique In optical, and magnetic properties of organic solid-state materials V. *Mater. Res. Soc. Symp. Proc.* 488: 527–532.

Kim, J.-H., S. A. Shin, J. B. Park, C. E. Song, W. S. Shin, H. Yang, Y. Li, et al. 2014c. Fluorinated benzoselenadiazole-based low-band-gap polymers for high efficiency inverted single and tandem organic photovoltaic cells. *Macromolecules.* 47: 1613–1622.

Kim, J.-H., C. E. Song, N. Shin, H. Kang, S. Wood, I.-N. Kang, et al. 2013. High-crystalline medium-band-gap polymers consisting of benzodithiophene and benzotriazole derivatives for organic photovoltaic cells. *ACS Appl. Mater. Interfaces.* 5: 12820–12831.

Kim, J. S., Z. Fei, S. Wood, D. T. James, M. Sim, K. Cho, et al. 2014b. Germanium- and silicon-substituted donor-acceptor type copolymers: Effect of the bridging heteroatom on molecular packing and photovoltaic device performance. *Adv. Energy Mater.* 4. doi :10.1002/aenm.201400527.

Kim, J. Y., K. Lee, N. E. Coates, D. Moses, T.-Q. Nguyen, M. Dante, et al. 2007. Efficient tandem polymer solar cells fabricated by all-solution processing. *Science.* 317: 222–225.

Kim, Y., S. Cook, S. M. Tuladhar, S. A. Choulis, J. Nelson, J. R. Durrant, et al. 2006. A strong regioregularity effect in self-organizing conjugated polymer films and high-efficiency polythiophene:fullerene solar cells. *Nat. Mater.* 5: 197–203.

Kim, Y., C. E. Song, S.-J. Moon, and E. Lim. 2014a. Rhodanine dye-based small molecule acceptors for organic photovoltaic cells. *Chem. Commun.* 50: 8235–8238.

Ko, D.-H., J. R. Tumbleston, W. Schenck, R. Lopez, and E. T. Samulski. 2011. Photonic crystal geometry for organic polymer: Fullerene standard and inverted solar cells. *J. Phys. Chem. C.* 115: 4247–4254.

Ko, D.-H., J. R. Tumbleston, L. Zhang, S. Williams, J. M. DeSimone, R. Lopez, et al. 2009. Photonic crystal geometry for organic solar cells. *Nano Lett.* 9: 2742–2746.

Kroon, R., A. Diaz De Zerio Mendaza, S. Himmelberger, J. Bergqvist, O. Bäcke, G. C. Faria, et al. 2014. A new tetracyclic lactam building block for thick, broad-bandgap photovoltaics. *J. Am. Chem. Soc.* 136: 11578–11581.

Le, Q. V., T. P. Nguyen, H. W. Jang, and S. Y. Kim. 2014. The use of UV/ozone-treated MoS_2 nanosheets for extended air stability in organic photovoltaic cells. *Phys. Chem. Chem. Phys.* 16: 13123–13128.

Lee, C.-H., R. Pandey, B.-Y. Wang, W.-K. Choi, D.-K. Choi, and Y.-J. Oh. 2015. Nano-sized indium-free MTO/Ag/MTO transparent conducting electrode prepared by RF sputtering at room temperature for organic photovoltaic cells. *Sol. Energy Mater. Sol. Cells.* 132: 80–85.

Lee, D. Y., D. V. Shinde, S. J. Yoon, K. N. Cho, W. Lee, N. K. Shrestha, et al. 2014. Cu-based metal–organic frameworks for photovoltaic application. *J. Phys. Chem. C.* 118: 16328–16334.

Lee, S.-K., K. Rana, and J.-H. Ahn. 2013. Graphene films for flexible organic and energy storage devices. *J. Phys Chem. Lett.* 4: 831–841.

Li, G., V. Shrotriya, J. Huang, Y. Tao, T. Moriarty, K. Emery, et al. 2005. High-efficiency solution processable polymer photovoltaic cells by self-organization of polymer blends. *Nat. Mater.* 4: 864–868.

Li, N., D. Baran, K. Forberich, M. Turbiez, T. Ameri, F. C. Krebs, and C. J. Brabec. 2013. An efficient solution-processed intermediate layer for facilitating fabrication of organic multi-junction solar cells. *Adv. Energy Mater.* 3: 1597–1605.

Lim, K.-G., M.-R. Choi, J.-H. Kim, D. H. Kim, G. H. Jung, Y. Park, et al. 2014. Role of ultrathin metal fluoride layer in organic photovoltaic cells: Mechanism of efficiency and lifetime enhancement. *ChemSusChem.* 7: 1125–1132.

Luck, K. A., T. A. Shastry, S. Loser, G. Ogien, T. J. Marks, and M. C. Hersam. 2013. Improved uniformity in high-performance organic photovoltaics enabled by (3-aminopropyl)triethoxysilane cathode functionalization. *Phys. Chem. Chem. Phys.* 15: 20966–20972.

Marks, R. N., J. J. M. Halls, D. D. C. Bradley, R. H. Friend, and A. B. Holmes. 1994. The photovoltaic response in poly(p-phenylene vinylene) thin-film devices. *J. Phys.: Condens. Matter.* 6: 1379–1394.

McGehee. M. D., and M. A. Topinka. 2006. Solar cells: Pictures from the blended zone. *Nat. Mater.* 5: 675–676.

Miller, S., G. Fanchini, Y.-Y. Lin, C. Li, C.-W. Chen, W.-F. Su, et al. 2008. Investigation of nanoscale morphological changes in organic photovoltaics during vapour annealing. *J. Mater. Chem.* 18: 306–312.

Mitul, A. F., L. Mohammad, S. Venkatesan, N. Adhikari, S. Sigdel, Q. Wang, et al. 2015. Low temperature efficient interconnecting layer for tandem polymer solar cells. *Nano Energy.* 11: 56–63.

Nelson, J. 2002. Organic photovoltaic films. *Curr. Opin. Solid State Mater. Sci.* 6: 87–95.

Nielsen, C. B., E. Voroshazi, S. Holliday, K. Cnops, D. Cheyns, and I. McCulloch. 2014. Electron-deficient truxenone derivatives and their use in organic photovoltaics. *J. Mater. Chem. A.* 2: 12348–12354.

Nikiforov, M. P., J. Strzalka, Z. Jiang, and S. B. Darling. 2013. Lanthanides: New metallic cathode materials for organic photovoltaic cells. *Phys. Chem. Chem. Phys.* 15: 13052–13060.

Nix, F. C., and A. W. Treptwo. 1939. A thallous sulphide photo EMF cell. *J. Opt. Soc. Am.* 29: 457–462.

Nuesch, F., L. Si-Ahmed, B. Francois, and L. Zuppiroli. 1997. Derivatized electrodes in the construction of organic light emitting diodes. *Adv. Mater.* 9: 222–225.

Owczarczyk, Z. R., W. A. Braunecker, S. D. Oosterhout, N. Kopidakis, R. E. Larsen, D. S. Ginley, et al. 2014. Cyclopenta[c]thiophene-4,6-dione-based copolymers as organic photovoltaic donor materials. *Adv. Energy Mater.* 4. doi: 10.1002/ aenm.201301821.

Paci, B., A. Generosi, V. R. Albertini, and R. De Bettignes. 2013. Improved structural/morphological durability for organic solar cells based on poly(3-hexylthiophene) fibers photoactive layers. *Chem. Phys. Lett.* 587: 50–55.

Page, Z. A., Y. Liu, V. V. Duzhko, T. P. Russell, and T. Emrick. 2014. Fulleropyrrolidine interlayers: Tailoring electrodes to raise organic solar cell efficiency. *Science.* 346: 441–444.

Park, Y. S., T. S. Kale, C.-Y. Nam, D. Choi, and R. B. Grubbs. 2014. Effects of heteroatom substitution in conjugated heterocyclic compounds on photovoltaic performance: From sulfur to tellurium. *Chem. Commun.* 50: 7964–7967.

Petritsch, K., J. J. Dittmer, E. A. Marseglia, R. H. Friend, A. Lux, G. G. Rozenberg, et al. 2000. Dye based donor/acceptor solar cells, *Sol. Energy Mater. Solar Cells.* 61: 63–72.

Petritsch, K., R. H. Friend, A. Lux, G. Rozenberg, S. C. Moratti, and A. B. Holmes. 1999. Liquid crystalline phthalocyanines in organic solar cells. *Synth. Met.* 102: 1776–1777.

Peumans, P., A. Yakimov, and S. R. Forrest. 2003. Small molecular weight thin-film photodetectors and solar cells. *J. Appl. Phys.* 93: 3693–3723.

Pfannmoller, M., W. Kowalsky, and R. R. Schroder. 2013. Visualizing physical, electronic, and optical properties of organic photovoltaic cells. *Energy Environ. Sci.* 6: 2871–2891.

Piliegoa, C., and M. A. Loi. 2012. Application charge transfer state in highly efficient polymer-fullerene bulk heterojunction solar cells. *J. Mater. Chem.* 22: 4141–4150.

Piliego, C., T. W. Holcombe, J. D. Douglas, C. H. Woo, P. M. Beaujuge, and J. M. J. Fréchet. 2010. Synthetic control of structural order in N-alkylthieno[3,4-c]pyrrole-4,6-dione-based polymers for efficient solar cells. *J. Am. Chem. Soc.* 132: 7595–7597.

Pochettino, A. 1906. Sul comportamento foto-elettrico dell'antracene. *Acad. Lincei Rend.* 15: 355–368.

Riede, M., C. Uhrich, J. Widmer, R. Timmreck, D. Wynands, G. Schwartz, et al. 2011. Efficient organic tandem solar cells based on small molecules. *Adv. Funct. Mater.* 21: 3019–3028.

Roman, L. S., A. C. Arias, M. Theander, M. R. Andersson, and O. Inganas. 2003. Photovoltaic devices based on photo induced charge transfer in polythiophene:CN-PPV blends. *Braz. J. Phys.* 33: 376–381.

Roman, L. S., L. A. A. Patterson, and O. J. Inganas. 1999. Modeling photocurrent action spectra of photovoltaic devices based on organic thin films. *J. Appl. Phys.* 86: 487–496.

Salafsky, J. S. 1999. Exciton dissociation, charge transport, and recombination in ultrathin, conjugated polymer–TiO_2 nanocrystal intermixed composites. *Phys. Rev. B.* 59: 10885–10894.

Saliba, M., K. W. Tan, H. Sai, D. T. Moore, T. Scott, W. Zhang, et al. 2014. Influence of thermal processing protocol upon the crystallization and photovoltaic performance of organic-inorganic lead trihalide perovskites. *J. Phys. Chem. C.* 118: 17171–17177.

Sariciftci, N. S. 2004. Plastic photovoltaic devices. *Mater. Today.* 7: 36–40.

Sariciftci, N. S., D. Braun, C. Zhang, V. I. Srdanov, A. S. Heeger, G. Stucky, et al. 1993. Semiconducting polymer-buckminsterfullerene heterojunctions: Diodes, photodiodes, and photovoltaic cells. *Appl. Phys. Lett.* 62: 585–587.

Sariciftci. N. S., L. Smilowitz, A. J. Heeger, and F. Wudl. 1992. Photoinduced electron transfer from a conducting polymer to buckminsterfullerene. *Science.* 258: 1474–1476.

Scharber, M. C., D. Mühlbacher, M. Koppe, P. Denk, C. Waldauf, A. J. Heeger, and C. J. Brabec. 2006. Design rules for donors in bulk-heterojunction solar cells—Towards 10% energy-conversion efficiency. *Adv. Mater.* 18: 789–794.

Schmidt-Mende, L., A. Fechtenkötter, K. Müllen, E. Moons, R. H. Friend, and J. D. MacKenzie. 2001. Self-organized discotic liquid crystals for high-efficiency organic photovoltaics. *Science.* 293: 1119–1122.

Schon, J. H., K. Ch, and B. Batlogg. 2000. Efficient photovoltaic energy conversion in pentacene-based heterojunction. *Appl. Phys. Lett.* 77: 2473–2475.

Seguy, I., P. Destruel, and H. Bock. 2000. An all-columnar bilayer light-emitting diode. *Synth. Met.* 15: 111–112.

Seo, H. O., S.-Y. Park, W. H. Shim, K.-D. Kim, K. H. Lee, M. Y. Jo, et al. 2011. Ultrathin TiO_2 films on ZnO electron-collecting layers of inverted organic solar cell. *J. Phys. Chem. C.* 115: 21517–21520.

Shaheen, S. E., C. J. Brabec, N. S. Sacriciftci, F. Padinger, T. Fromherz, and J. C. Hummelen. 2001a. 2.5% Efficient organic plastic solar cell. *Appl. Phys. Lett.* 78: 841–843.

Shaheen, S. E., D. Vangeneugden, R. Kiebooms, D. Vanderzande, T. Fromherz, F. Padinger, et al. 2001b. Low band-gap polymeric photovoltaic devices. *Synth. Met.* 121: 1583–1584.

Shirakawa, H., E. J. Louis, A. G. MacDiarmid, C. K. Chiang, and A. J. Heeger. 1977. Synthesis of electrically conducting organic polymers: Halogen derivatives of polyacetylene, $(CH)_x$. *J. Chem. Soc. Chem. Commun.* 578–580.

Shivanna, R., S. Shoaee, S. Dimitrov, S. K. Kandappa, S. Rajaram, J. R. Durrant, et al. 2014. Charge generation and transport in efficient organic bulk heterojunction solar cells with a perylene acceptor. *Energy Environ. Sci.* 7: 435–441.

Sista, S., M.-H. Park, Z. Hong, Y. Wu, J. Hou, W. L. Kwan, G. Li, and Y. Yang. 2010. Highly efficient tandem polymer photovoltaic cells. *Adv. Mater.* 22: 380–383.

Skupov, K., and A. Adronov. 2014. Effect of carbon nanotube incorporation into polythiophene-fullerene-based organic solar cells. *Can. J. Chem.* 92: 68–75.

Smith, W. 1873. Effect of light on selenium during the passage of an electric current. *Nature.* 7: 303.

Stanculescu, F., O. Rasoga, A. M. Catargiu, L. Vacareanu, M. Socol, C. Breazu, et al. 2014. MAPLE prepared heterostructures with arylene based polymer active layer for photovoltaic applications. *Appl. Surface Sci.* 336: 240–248.

Sun, Y., J. H. Seo, C. J. Takacs, J. Seifter, and A. J. Heeger. 2011. Inverted polymer solar cells integrated with a low-temperature-annealed sol-gel-derived ZnO film as an electron transport layer. *Adv. Mater.* 23: 1679–1683.

Takao, Y., T. Masuoka, K. Yamamoto, T. Mizutani, F. Matsumoto, K. Moriwaki, et al. 2014. Synthesis and properties of novel fluorinated subnaphthalocyanine for organic photovoltaic cells. *Tetrahedron Lett.* 55: 4564–4567.

Tamura, Y., H. Saeki, J. Hashizume, Y. Okazaki, D. Kuzuhara, M. Suzuki, et al. 2014. Direct comparison of a covalently-linked dyad and a 1:1 mixture of tetrabenzoporphyrin and fullerene as organic photovoltaic materials. *Chem. Commun.* 50: 10379–10381.

Tan, W.-Y., R. Wang, M. Li, G. Liu, P. Chen, X.-C. Li, et al. 2014. Lending triarylphosphine oxide to phenanthroline: A facile approach to high-performance organic small-molecule cathode interfacial material for organic photovoltaics utilizing air-stable cathodes. *Adv. Funct. Mater.* 24: 6540–6547.

Tang, C. W. 1986. Two-layer organic photovoltaic cell. *Appl. Phys. Lett.* 48: 183–185.

Theander, M., A. Yartsev, D. Zigmantas, V. Sundstrom, W. Mammo, M. R. Andersson, et al. 2000. Photoluminescence quenching at a polythiophene/C60 heterojunction. *Phys. Rev. B.* 61: 12957–12963.

Thompson, B. C., and J. M. J. Fréchet. 2008. Polymer-fullerene composite solar cells. *Angew. Chem. Int. Ed.* 47: 58–77.

Tokuhisa, H., M. Era, and T. Tsutsui. 1998. Polarized electroluminescence from smectic mesophase. *Appl. Phys. Lett.* 72: 2639–2641.

Too, C. O., G. G. Wallace, A. K. Burrell, G. E. Collis, D. L. Officer, E. W. Boge, et al. 2001. Photovoltaic devices based on polythiophenes and substituted polythiophenes. *Synth. Met.* 123: 53–60.

Tsai, C.-C., R. R. Grote, J. H. Beck, I. Kymissis, R. M. Osgood, Jr., and D. Englund. 2014. General method for simultaneous optimization of light trapping and carrier collection in an ultra-thin film organic photovoltaic cell. *J. Appl. Phys.* 116: 023110. doi:10.1063/1.4890275.

Tung, V. C., J. Kim, L. J. Cote, and J. Huang. 2011. Sticky interconnect for solution-processed tandem solar cells. *J. Am. Chem. Soc.* 133: 9262–9265.

Tvingstedt, K., V. Andersson, F. Zhang, and O. Inganas. 2007. Folded reflective tandem polymer solar cell doubles efficiency. *Appl. Phys. Lett.* 91: 123514. doi:10.1063/1.2789393.

Umeyama, T., and H. Imahori. 2014. Design and control of organic semiconductors and their nanostructures for polymer-fullerene-based photovoltaic devices. *J. Mater. Chem. A.* 2: 11545–11560.

Vasilopoulou, M., D. G. Georgiadou, A. Soultati, N. Boukos, S. Gardelis, L. C. Palilis, et al. 2014b. Atomic-layer-deposited aluminum and zirconium oxides for surface passivation of TiO_2 in high-efficiency organic photovoltaics. *Adv. Energy Mater.* 4. doi: 10.1002/ aenm.201400214.

Vasilopoulou, M., N. Konofaos, D. Davazoglou, P. Argitis, N. A. Stathopoulos, S. P. Savaidis, et al. 2014a. Organic photovoltaic performance improvement using atomic layer deposited ZnO electron-collecting layers. *Solid-State Electron.* 101: 50–56.

Walter, M. G., A. B. Rudine, and C. C. Wamser. 2010. Porphyrins and phthalocyanines in solar photovoltaic cells. *J. Porphyrins Phthalocyanines.* 14: 759–792.

Wang, S., S. Yang, C. Yang, Z. Li, J. Wang, and W. Ge. 2000. Poly(N-vinylcarbazole) (PVK) photoconductivity enhancement induced by doping with CdS nanocrystals through chemical hybridization. *J. Phys. Chem. B.* 104: 11853–11858.

Wang, T., K. C. Weerasinghe, D. Lui, W. Li, X. Yan, X. Zhou, and L. Wang. 2014b. Ambipolar organic semiconductors with cascades of energy levels for generating long-lived charge separated states: A donor-acceptor1-acceptor2 architectural triarylamine dye. *J. Mater. Chem. C.* 2: 5466–5470.

Wang, X., J. Huang, Z. Niu, X. Zhang, Y. Sun, and C. Zhan. 2014a. Dimeric naphthalene diimide based small molecule acceptors: Synthesis, characterization, and photovoltaic properties. *Tetrahedron.* 70: 4726–4731.

Wang, X., G. I. Koleilat, J. Tang, H. Liu, I. J. Kramer, R. Debnath, et al. 2011. Tandem colloidal quantum dot solar cells employing a graded recombination layer. *Nat. Photon.* 5: 480–484.

Wang, Z., T. Miyadera, A. Saeki, Y. Zhou, S. Seki, Y. Shibata, et al. 2014c. Structural influences on charge carrier dynamics for small-molecule organic photovoltaics. *J. Appl. Phys.* 116: 013105.

Weinberger, B. R., M. Akhtar, and S. C. Gau. 1982. Polyacetylene photovoltaic devices. *Synth. Met.* 4: 187–197.

Weiss, N. O., H. Zhou, L. Liao, Y. Liu, S. Jiang, Y. Huang, and X. Duan. 2012. Graphene: An emerging electronic material. *Adv. Mater.* 24: 5782–5825.

Wu, A., T. Fujuwara, M. Jikei, M.-A. Kakimoto, Y. Imai, T. Kubota, et al. 1996. Electrical properties and electroluminescence of poly(p-phenylene vinylene) Langmuir-Blodgett film. *Thin Solid Films.* 284–285: 901–903.

Xia, F., H. Yan, and P. Avouris. 2013. The interaction of light and graphene: Basics, devices, and applications. *Proc. IEEE.* 101: 1717–1731.

Yang, J., R. Zhu, Z. Hong, A. Kumar, Y. Li, and Y. Yang. 2011. A robust inter-connecting layer for achieving high performance tandem polymer solar cells. *Adv. Mater.* 23: 3465–3470.

Yoshino, K., K. Tada, A. Fujii, E. M. Conwell, and A. A. Zakhidov. 1997. Novel photovoltaic devices based on donor-acceptor molecular and conducting polymer systems. *IEEE Trans. Electron Dev.* 44: 1315–1324.

Yu, G., and A. J. Heeger. 1995. Charge separation and photovoltaic conversion in polymer composites with internal donor/acceptor heterojunctions. *J. Appl. Phys.* 78: 4510.

Yu, G., J. Gao, J. C. Hummelen, F. Wudl, and A. J. Heeger. 1995. Polymer photovoltaic cells: Enhanced efficiencies via a network of internal donor-acceptor heterojunctions. *Science.* 270: 1789–1791.

Yu, Y.-Y., and S.-H. Chan. 2013. Effects of metal oxide as an anode interlayer for organic photovoltaics. *Thin Solid Films.* 546: 231–235.

Zhang, Q., B. Kan, F. Liu, G. Long, X. Wan, X. Chen, et al. 2015. Small-molecule solar cells with efficiency over 9%. *Nat. Photon.* 9: 35–41.

Zhao, D. W., L. Ke, Y. Li, S. T. Tan, A. K. K. Kyaw, H. V. Demir, et al. 2011. Optimization of inverted tandem organic solar cells. *Sol. Energy Mater. Solar Cells.* 95: 921–926.

Zhao, D. W., X. W. Sun, C. Y. Jiang, A. K. K. Kyaw, G. Q. Lo, and D. L. Kwong. 2008. Efficient tandem organic solar cells with an Al/ MoO_3 intermediate layer. *Appl. Phys. Lett.* 93: 083305.

Zhuang, T. X.-F. Wang, T. Sano, Z. Hong, Y. Yang, and J. Kido. 2013. Fullerene derivatives as electron donor for organic photovoltaic cells. *Appl. Phys. Lett.* 103: 203301–203304.

Zimmerman, J. D., B. E. Lassiter, X. Xiao, K. Sun, A. Dolocan, R. Gearba, et al. 2013a. Control of interface order by inverse quasi-epitaxial growth of squaraine/fullerene thin film photovoltaics. *ACS Nano.* 7: 9268–9275.

Zimmerman, J. D., B. Song, O. Griffith, and S. R. Forrest. 2013b. Exciton-blocking phosphonic acid-treated anode buffer layers for organic photovoltaics. *Appl. Phys. Lett.* 103: 243905.

6 Dye-Sensitized Solar Cells

Rakshit Ameta, Surbhi Benjamin, Shweta Sharma, and Monika Trivedi

CONTENTS

6.1 Introduction	85
6.2 Principle	87
6.3 Characterization	88
6.4 Incident Photon-to-Current Conversion Efficiency	89
6.5 Components of DSSC	89
6.6 Sensitizer	89
6.6.1 Metal Complexes	89
6.6.1.1 Ru Metal Complex Dyes	90
6.6.1.2 Other Metal Complexes	90
6.6.2 Dyes	91
6.6.2.1 Metal Complex Free Dyes	91
6.6.3 Natural Pigments	92
6.7 Electrolytes	93
6.7.1 Liquid Electrolytes	93
6.7.2 Polymer Gel Electrolytes	94
6.7.3 Other Electrolytes	94
6.8 Electrodes	95
6.8.1 Nanocrystalline Semiconductor Working Electrode	95
6.8.1.1 Naïve Semiconductors	95
6.8.1.2 Modified Semiconductors	97
6.8.1.3 Doped Semiconductors	98
6.8.1.4 Other Nanomaterials	99
6.9 Counterelectrode	100
6.10 Other Types of Dye-Sensitized Solar Cell	101
6.10.1 Solid-State DSSC	101
6.10.2 Quasi-Solid-State DSSC	103
6.10.3 Quantum Dot DSSC	104
6.11 Applications of DSSC	105
References	105

6.1 INTRODUCTION

Solar insolation is the most abundant natural energy source for Earth. Solar radiation can be transformed into various energy forms such as heat and chemical energy (e.g., via photosynthesis). Natural energy sources will be depleted in the near future due to their rapid and uncontrolled consumption. In this regard, solar energy may play a very important role as an alternative energy source. The Sun provides about 120,000 TW of energy toward Earth's surface. This amount is about 6000 times the present rate of the world's energy consumption. It is a challenge to use solar energy as environmentally friendly chemical fuels and photoconversion devices are necessary to convert it to electricity. The main advantage of the new generation of solar cells is their low cost (less than $1/peak watt) and potential (Gratzel 2003).

A solar cell is a solid-state electrical device, which directly converts solar energy into electricity. Solar energy is transmitted in the form of small packets of photons or quantums of light, and the generated electrical energy is stored in electromagnetic fields. The solar age began in the 1950s at Bell Laboratories with the development of silicon technology. They described the first high-power silicon photovoltaic cell with 6% efficiency. The efficiency of this cell was much better than the previous selenium solar cell. Due to the energy crisis in 1970, many researchers developed technologies for producing energy from renewable energy sources and photovoltaic devices. In 1997, 33% of electrical energy was generated from photovoltaic solar devices, which is an outstanding achievement.

The first-generation solar cells were single-junction devices. The main problems of this type of solar cells include that they are labor intensive, require a cumbersome method to produce energy, and have high manufacturing costs, but their electrical output is relatively high. The second-generation solar cell was developed to solve this problem. In a thin-film solar cell, cadmium telluride (CdTe), copper indium gallium selenide (ClGS), copper indium diselenide (CIS), and polycrystalline and amorphous silicon were applied. Their costs were much lower than those of the previous solar cells. Then the third-generation solar cells were designed to enhance the electrical performance of thin-film solar cells, such as polymer solar cells, dye-sensitized solar cells (DSSCs), and nanocrystalline solar cells.

This new generation of solar cell was based on nanotechnology to lower costs. The mechanism of conversion of light energy into electrical energy is based on photogeneration of charge carriers (electrons and holes) on semiconductor material (light-absorbing material) and charge separation to produce electrical energy. These solar cells are a promising renewable energy source, which are environmentally friendly based on specific inorganic and organic compounds.

The history of photovoltaics started with the work of French physicist Alexandre-Edmond Becquerel (1839). He observed that when two platinum electrodes were immersed in an illuminated solution, electric current was generated. The electrodes were coated with light-sensitive metal halide salt (e.g., AgCl or AgBr). It was observed that as the light intensity was increased, electricity also increased. The photovoltaic effect in selenium was observed by Smith (1873). Adams and Day (1876) investigated the electrical behavior of selenium and, especially, its sensitivity to light. They discovered the photovoltaic effect of the platinum and selenium junction. Moser (1887) discovered the photosensitization effect on dyes. He used silver halide electrode painted with red dye. Gerischer et al. (1968) explained the mechanism of dye adsorption on a semiconductor to produce electricity under certain conditions.

Dye-sensitized electrochemical photocells have been used to convert light energy into electrical energy. Due to poor light harvesting and the instability of dyes, the efficiency of this device was very low (i.e., near 1% or less). Matsumura et al. (1980) reported the dye sensitization on ZnO, CdS, and TiO_2 in an electrochemical photocell. They used cationic, anionic, and zwitterionic dyes. A dye-sensitized photocell with 2.5% cell efficiency was obtained using aluminum-doped porous ZnO electrode with rose Bengal (anionic dye).

O'Regan and Grätzel (1991) described the heterojunction three-dimensional fabrication of a dye-sensitized solar cell. The basic components of this device were a semiconductor film containing nanometer-sized TiO_2 particles, with newly developed charge-transfer dyes. They reported more than 7% efficiency in the presence of direct sunlight. DSSCs are different from conventional semiconductor devices where an n-type semiconductor material such as TiO_2 was used. It generates current while dye molecules absorbed photon from light and injected electrons into the conduction band. DSSCs are photoelectrochemical solar cells. Their mechanism is based on photoinduced charge separation at a dye-sensitized interface between a nanocrystalline, mesoporous metal oxide electrode and a redox electrolyte (O'Regan and Durrant 2009).

Dye-Sensitized Solar Cells

The DSSCs are also known as "Gratzel cells." In the last decade, the DSSC has become a more active and interesting research topic because of its low cost, easy fabrication, low energy payback time, flexibility, and multicolor option. These cells are technically and economically as well as environmentally reliable alternatives to existing p-n junction photovoltaic devices. However, some improvements are still needed in their properties and durability for their future development.

6.2 PRINCIPLE

A DSSC has some basic steps for generating electrical energy at the cost of light energy. These steps are

- Nanocrystalline TiO_2 is deposited on the conducting electrode (photoelectrode) to provide a large surface area on which to absorb dye molecules (D) (sensitizer):

$$D + Photon \rightarrow D^* \rightarrow Excitation\ process \tag{6.1}$$

$$D^* + TiO_2 \rightarrow e^-\ (TiO_2) + D^+ \rightarrow Injection\ process \tag{6.2}$$

$$e^-(TiO_2) + C.E. \rightarrow TiO_2 + e^-\ (C.E.) + Electrical\ energy \rightarrow Energy\ generation \tag{6.3}$$

$$D^+ + \frac{3}{2}I^- \rightarrow D + \frac{1}{2}I_3^- \rightarrow Regeneration\ of\ dye \tag{6.4}$$

$$\frac{1}{2}I_3^- + e^-\ (C.E.) \rightarrow \frac{3}{2}I^- + C.E. \rightarrow e^-\ Recapture\ reaction \tag{6.5}$$

- When the photons are absorbed, dye molecules are excited from the highest occupied molecular orbital (HOMO)/valence band (VB) to the lowest unoccupied molecular orbital (LUMO)/conduction band (CB) states (Equation 6.1).
- Once an electron is injected into the conduction band of the wide-bandgap semiconductor, like a nanostructured TiO_2 film, the excited dye molecules (D^*) were oxidized to D^+ (Equation 6.2).
- The injected electron is transported between the TiO_2 nanoparticles and then extracted to a load, where the work done is delivered as electrical energy (Equation 6.3).
- Electrolytes containing an I^-/I_3^- redox couple system are used as an electron mediator between the TiO_2 photoelectrode and the carbon-coated counterelectrode. Therefore, the oxidized dye molecules are regenerated by receiving electrons from the I^- ions redox mediator, which is oxidized to I_3^- (triiodide ions) (Equation 6.4).
- The I_3^- ion accepts the internally donated electron from the external load and is reduced back to I^- ion (Equation 6.5).

The movement of electrons in the conduction band of the semiconductor is accompanied by the diffusion of charge-compensating cations in the electrolyte layer close to the nanoparticle surface. Therefore, generation of electric power in the DSSC causes no permanent chemical change or transformation (Gratzel 2005).

6.3 CHARACTERIZATION

The photovoltaic cell is a device that generates electrical power from conversion of incident light into electrical energy. This process includes producing voltage in the presence of an external load and supplying it through any load at the same time. This is presented by the current–voltage (i–V) characteristic curve of the cell at different illuminations and temperatures (Figure 6.1).

When the cell is short-circuited under illumination, the maximum current i_{sc} (the short-circuit current) is generated. Under open-circuit conditions, no current can flow, and the voltage obtained is at its maximum, which is called the open-circuit voltage (V_{OC}). The point in the i–V curve yielding the maximum product of current and voltage (i.e., power) is called the *power point*. Here, the electrical power is maximum ($P_{max} = i_{max} \times V_{max}$). Another important characteristic of the solar cell performance is its fill factor (*FF*), which is defined as

$$FF = \frac{i_{max} \times V_{max}}{i_{sc} \times V_{oc}} \tag{6.6}$$

where:

- i_{max} is the current per unit area at the maximum output power point
- V_{max} is the voltage per unit area at the maximum output power point
- i_{sc} is the short-circuit photocurrent density
- V_{oc} is the open-circuit voltage
- P_{in} is the light power (intensity of light) per unit area
- *FF* is the fill factor of the cell

The solar energy to electricity conversion efficiency (η) of DSSC is calculated by the following equation:

$$\eta(\%) = \frac{i_{sc} \times V_{oc} \times FF}{P_{in}} \times 100 \tag{6.7}$$

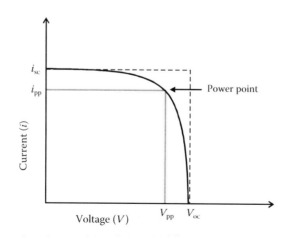

FIGURE 6.1 Current–voltage (i–V) characteristics of the cell.

Dye-Sensitized Solar Cells

6.4 INCIDENT PHOTON-TO-CURRENT CONVERSION EFFICIENCY

Incident photon-to-current conversion efficiency (IPCE) is defined as the ratio of the number of electrons in the external circuit produced by an incident photon at a given wavelength:

$$\text{IPCE } (\%) = \frac{100 \times 1240 \times i_{sc}}{(w_{in} \times \lambda)} \qquad (6.8)$$

where:
- i_{sc} is the short-circuit photocurrent (mA cm^{-2})
- w_{in} is the incident light intensity (W cm^{-2})
- λ is wavelength (nm)

6.5 COMPONENTS OF DSSC

The DSSC consists of two electrodes: photoactive working electrode and counterelectrode (Figure 6.2). These electrodes are contacted by a liquid redox electrolyte. They are explained as follows:

- Working electrode: Porous nanostructures semiconductor (TiO$_2$) attached to a conducting fluorine-doped tin oxide (FTO) glass
- Counterelectrode: Platinized conducting substrate
- Light-absorbing layer: Adsorbed dye as a sensitizer
- Redox system: A liquid electrolyte containing the redox couple system (iodide/triiodide)

6.6 SENSITIZER

6.6.1 METAL COMPLEXES

Transition-metal complexes such as derivatives of polypyridine complexes and metalloporphyrins were the first choice as photosensitizers due to their low-lying metal-to-ligand charge-transfer (MLCT) and ligand-centered excited states for maximum electron transfer. Therefore, these can also be used as photochemical systems for direct conversion and storage of solar energy. The metal

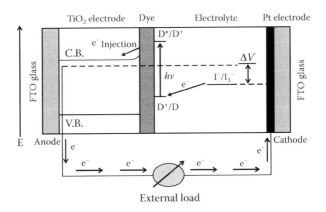

FIGURE 6.2 Dye-sensitized solar cell.

complexes attached to the surfaces of some mesoporous membrane thin films with large surface area were prepared by nanosized colloidal semiconductor dispersions that improve the energy conversion efficiency of photovoltaic devices such as dye-sensitized solar cells (Kalyanasundaram and Gratzel 1998).

6.6.1.1 Ru Metal Complex Dyes

Anderson et al. (1979) reported that the attachment of the $[Ru(Bipy)_3]^{2+}$ derivative on the single-crystal n-type TiO_2 electrode produced anodic photocurrents in the presence of visible light. The efficient sensitizer based on the *N*-bound isothiocyanato complex K[Ru(bmipy)(dcbpy)(NCS)] on a nanocrystalline TiO_2 surface was suggested by Kohle et al. (1996). A black trithiocyanatoruthenium(II) terpyridyl complex was found to be quite efficient for the sensitization of the nanocrystalline TiO_2 solar cell (Nazeeruddin et al. 1997), whereas Imahori et al. (2009) reported that the ruthenium(II) bipyridyl complexes were the most efficient complexes in the TiO_2-sensitized solar cell. The large π-aromatic molecules such as porphyrins, phthalocyanines, and perylenes also play an important role as good sensitizers to give high conversion efficiency in DSSC. Porphyrin and phthalocyanine sensitized DSSCs have shown poor light-harvesting tendencies compared to ruthenium complexes; therefore, some types of push–pull porphyrin sensitizers were discovered to further improve power conversion efficiency to the order of 6%–7%.

The other derivatives of the Ru(II) complex were also used as the sensitizers in DSSCs. Dare-Edwards et al. (1980) used Ru(bipy)2(bpca)–sensitized n-TiO_2, n-$SrTiO_3$, and n-SnO_2 electrodes in solar cells. The N719-sensitized multilayered mono- and double-layered film DSSCs were described by Wang et al. (2004a). The solar-to-electric energy conversion efficiency has been increased from 7.6% to 9.8% while changing monolayer to multilayer films, and 10.2% efficiency was achieved by using an antireflection multilayered film in the presence of AM 1.5 solar light (100 mW cm^{-2}).

A new ion-coordinating ruthenium polypridyl sensitizer dye [K68, NaRu(4-carboxylic acid-4′-carboxylate)(4,4′-bis[(triethyleneglycolmethylether) heptylether]-2,2′-bipyridine)(NCS)$_2$] has been synthesized and used by Kuang et al. (2008). Binary ionic liquid electrolytes 1-propyl-3-methyl-imidazolium iodide (PMII) and 1-ethyl-3 methyl-imidazoliumtetracyanoborate (EMIB)(CN)$_4$) have been used. They achieved 7.7% conversion efficiency for DSSC. The other amphiphilic heteroleptic ruthenium(II) sensitizer series, such as [Ru(H$_2$dcbpy)(dhbpy)(NCS)$_2$], [Ru(H$_2$dcbpy)(bccbpy) (NCS)$_2$], [Ru(H$_2$dcbpy) (mpubpy)(NCS)$_2$], and [Ru(H$_2$dcbpy)(bhcbpy)(NCS)$_2$], have been synthesized for DSSC where the latter sensitizer gave 7.4% efficient cells from all other dyes (Lagref et al. 2008). The DSSC has been sensitized with black dye [(C$_4$H$_9$)$_4$N]$_3$[Ru(Htcterpy)(NCS)$_3$] (tcterpy = 4,4′,4″-tricarboxy-2,2′,2″-terpyridine), and maximum power conversion efficiency (10.5%) was achieved from the pretreatment of HCl on the TiO_2 surface (Wang et al. 2005a). Ruthenium(II) complexes containing [Ru(dcbpyH$_2$)$_2$(NCS)$_2$] (N3) and [Ru(dcbpyH$_2$) (tdbpy)(NCS)$_2$] (N621) dye-absorbed TiO_2 films in their monoprotonated and diprotonated states have approached high power conversion efficiencies of 11.18% and 9.57%, respectively, where (dcbpyH$_2$ is 4,4′-dicarboxy-2,2′-bipyridine; and tdbpy is 4,4′-tridecyl-2,2′-bipyridine) (Nazeeruddin et al. 2005). Heteroleptic polypyridine ruthenium complex *cis*-Ru(L1)(L2)(NCS)$_2$ (where L1 is 4,4′-dicarboxylic acid-2,2′-bipyridine (dcbpy); and L2 is 4,4′-bis[*p*-diethylamino]-α-styryl]-2,2′-bipyridine) has a high molar extinction coefficient as compared to organic dye and 8.65% conversion efficiency (Jiang et al. 2008). Novel Ru(II) phenanthroline complex [(Ru(II)(4,4′,4″-tri-*tert*-butyl-2,2′:6′,2″-terpyridine)-(4,7-diphenyl-1,10-phenanthroline-disulfonicaciddisodiumsalt) (thiocyanate)] (K328) has been synthesized by Erten-Ela et al. (2014).

6.6.1.2 Other Metal Complexes

A DSSC based on a TiO_2 film using an I$^-$/I$_3^-$ liquid electrolyte was described by Wang et al. (2005b). This cell produced 85% incident photon-to-current efficiency with a highly efficient (5.6%) solar cell in the case of porphyrins-based dyes. Iron(II) complex di(aqua)bis(oxalato)iron(II), [Fe(II)(H$_2$O)$_2$(C$_2$O$_4$)$_1$]$^{2-}$, coated with bromopyrogallol ligand has shown improved photovoltaic property when compared with a bromopyrogallol ligand–coated photovoltaic device. Bromopyrogallol ligand complexed with iron(II) shows higher stability for photodegradation and photocurrent conversion efficiency (Jayaweera et al. 2001). Carboxylate-derivatized {Cu^IL_2} complexes have been used in DSSCs, and they show long-term stability and are amazingly efficient compared to ruthenium complex dye (Bessho et al. 2008). Ir(III) complexes [Ir(C^N^C)(ptpy-COOH)]$^+$ (C^N^C = 2,6-diphenylpyridinato; 2,4,6-triphenylpyridinato, ptpy-COOH = 4′-(4-carboxyphenyl)-2,2′:6′,2″-terpyridine), and [Ir(C^N^C) (tpy-COOH)]$^+$ (C^N^C = 2,6-diphenylpyridinato; 2,6-ditolylpyridinato, tpy-COOH = 4′-carboxy-2,2′:6′,2″-terpyridine) have been synthesized as photosensitizers and compared with cis-diisothiocyanatobis(4,4′-dicarboxy-2,2′-bipyridine)ruthenium(II) (N3) photosensitized DSSCs. The Ir(III) complex shows a 2.16% power conversion efficiency and a longer lifetime (Shinpuku et al. 2011). The Pt(II)-based dye 4,4′-dicarboxy-2,2′-bipyridine and quinoxaline-2,3-dithiolate ligands were used in DSSCs, and it gave a 2.6% power conversion efficiency (Islam et al. 2000).

The Os complex [Os(II)H$_3$tcterpy)(CN$^-$)]$^-$ (where H$_3$tcterpy is 4,4′,4″-tricarboxy-2,2′:6′,2″-terpyridine) showed much better photochemical stability as compared to the Ru sensitizer for DSSCs (Argazzi et al. 2004). Mordant dye has a salicylate chelating group, which is effective for the surface of a nanocrystalline TiO_2. Forty-nine types of mordant dyes have been used in DSSCs. When mordant dye was compared with N3 dye, it showed more than 0.2 mA photocurrent, and it bound more strongly with the TiO_2 surface (Millington et al. 2007)

6.6.2 Dyes

6.6.2.1 Metal Complex Free Dyes

The Ru(II) complex dyes have shown the best results in DSSCs, but these are very costly because of ruthenium, which may increase the overall cost of the cell. Some researchers developed cost-efficient dyes for dye-sensitized solar cells to overcome this problem.

The DSSC sensitized with benzothiazolemerocyanine(3-carboxyalkyl-5-[2-(3-alkyl-2-benzothiazolinyldene)ethylidene]-2-thioxo-4-thiazolidinone dye has given 4.5% cell efficiency, which was higher than that of other organic dyes. Dye molecules were aggregated on a TiO_2 surface. The J aggregation of the dye with long alkyl chains showed excellent capability for sensitization on the TiO_2 electrode. Power conversion efficiency and incident photon-to-current conversion efficiency increase when alkyl chain attached to the benzothiazole ring increases. The dye molecules were fixed by a chelate-like linkage of the carboxylate group on the TiO_2 surface (Sayama et al. 2002). The indoline dye has given the highest power conversion efficiency as compared with the metal-free organic dyes. Using 4-tert-butyl pyridine as the electrolyte and cholic acid as a co-adsorbent, 8.00% efficiency was obtained, whereas it gives 6.51%, 7.89%, and 8.26% power conversion efficiency with N3 dye and N719 dye in the same conditions (Horiuchi et al. 2004). The coumarin derivatives were found to work as excellent photosensitizers for the DSSCs. These cells performed with 5.6% power conversion efficiency (Hara et al. 2001). Other coumarin derivatives such as thiophene have been used in DSSCs, and 7.7% solar-to-electrical power conversion efficiency was achieved by Hara et al. (2003). Oligothiophene-containing coumarin dyes (NKX-2677) have also been developed by Hara et al. (2005), and 7.4% conversion efficiency of the cell was achieved. Polypyridine rings containing dyes were used as sensitizer in solar cells consisting of nanocrystalline titania, and they were examined by in situ micro and macro Raman spectroscopy (Kontos et al. 2008).

Organic chromophore dyes (3,4-ethyldioxythiophene and thienothiophene) were designed for DSSCs by Zhang et al. (2009). This dye has shown 9.8% power conversion efficiency and excellent stability of the solar cells. Some imidazole derivatives CD-4 (4.11%) and CD-6 (1.15%) triphenylamine-based organic dyes were used by Chen et al. (2014b). The JK-16 and JK-17 organic dyes containing bis-dimethylfluorenyl amino benzo[b]thiophene were synthesized and gave 7.43% and 5.49% power conversion efficiency in DSSCs, respectively (Choi et al. 2007). Chang et al. (2011)

introduced a highly efficient porphyrin sensitizer containing two ortho-substituted long alkoxyl chains for dye-sensitized solar cells. They achieved 10.17% conversion efficiency.

Metal-free dye was a better option than organic metal-containing dyes for further improvement of the sensitizer. These dyes gave almost the same output for DSSCs, and with a low fabrication cost. The metal-free organic dyes containing thienothiophene and thiophene segments were synthesized for DSSCs with a 6.23% solar-to-electricity conversion efficiency (Li et al. 2006). Three metal-free organic dyes (H11–H13) containing 3,6-disubstituted carbazole, benzothiadiazole, and cyanoacrylic acid were used by Lee et al. (2013). The H13 dye has given the best results when used with co-adsorbed chenodeoxycholic acid (CDCA) for sensitization of the TiO_2 nanocrystalline surface. The cell has displayed open-circuit voltage of 0.71 V, short-circuit current density of 12.69 mA cm^{-2}, fill factor of 0.71, and reasonably good efficiency of 6.32%.

Two thiophene-(N-aryl)pyrrole-thiophene (TPT)-based metal-free dyes have been introduced by Tamilavan et al. (2014). The performances of solar cells with these dyes in the presence and absence of co-adsorbent were reported as 7.06% and 6.85%, respectively. Cosensitized DSSC with metal-free squaraine dye (SQ2) and diarylaminofluorene-based organic dye (JD1) with power conversion efficiency 6.36% has been reported by Lin et al. (2014).

6.6.3 Natural Pigments

Ru(II) complex dyes, other organic dyes, and metal-free dyes performed very well as sensitizers in DSSCs and gave good efficiency, but these dyes are limited in number and are costly. Natural pigments (plant pigments) can be used as sensitizers for low-cost fabrication of dye-sensitized solar cells. Chlorophyll, carotene, cyanin, and cyanidine are freely available in plant leaves, flowers, and fruits (Tennakone et al. 1997). Natural pigments have the advantage of being environmentally friendly freely available in nature.

Natural pigments extracted from rosella, blue pea, and a mixture of both extracts have been reported for the fabrication of DSSCs by Wongcharee et al. (2007). The mixture of both extracts adsorbed by the TiO_2 surface did not show any synergistic light absorption and photosensitization when compared with each extract alone. The power conversion efficiencies of the DSSCs were 0.37%, 0.05%, and 0.15% when sensitized with rosella extract, blue pea extract, and the mixed extract, respectively. The efficiency of DSSCs photosensitized with rosella extract was increased from 0.37% to 0.70%, when the aqueous solution of dye was extracted at 100°C, and the pH of the extract was varied from 3.2 to 1.0. The extracts of natural fruits such as black rice, capsicum, *Erythrina variegata* flower, *Rosa xanthina*, and kelp have also been used as sensitizers in the fabrication of DSSCs (Hao et al. 2006). The black rice extract performed as the best sensitizer due to better interaction between the carbonyl and hydroxyl groups of anthocyanin molecule. Water-soluble natural pigment betalain (from beet roots) has good oxidizing capacity with strong visible light absorption. The cell was assembled with nanocrystalline TiO_2 as a photoanode treated with ethanolic HCl solution and sensitized with betanin solution. Then 2.42 mA cm^{-2} photocurrent and 0.44 V photovoltage were achieved using methoxypropionitrile containing I^-/I_3^- redox electrolyte (Zhang et al. 2008). Zhou et al. (2011) used 20 different natural dye extracts from natural materials such as flowers, leaves, fruits, traditional Chinese medicines, and beverages. The open-circuit voltages varied from 0.337 to 0.689 V, and short-circuit photocurrent varied from 0.14 to 2.69 mA cm^{-2}. They achieved maximum photo-to-electric conversion efficiency from mangosteen pericarp (1.17%).

Yusoff et al. (2014) reported the use of anthocyanin pigments from 35 native plants as potential sensitizers for DSSC. *Melastoma malabathricum* (fruit pulp), *Hibiscus rosa-sinensis* (flower), and *Codiaeum variegatum* (leaves) have given the highest absorption peaks. *Melastoma malabathricum* has the highest amount of anthocyanin pigments (8.43 mg/L), and it gave the power conversion efficiency of 1.16%, while with *Hibiscus rosa-sinensis* and *Codiaeum variegatum*, 0.16% and 1.08% power conversion efficiencies have been achieved, respectively.

6.7 ELECTROLYTES

Electrolytes are a most effective and important component in DSSCs. The stability and conversion efficiency of cells depend on the performance of the electrolytes. Three types of electrolytes have been used in DSSCs: liquid electrolyte (organic solvent and ionic liquid electrolyte), quasi-solid-state electrolyte, and solid-state electrolyte.

The basic function of a redox couple in the electrolyte is to regenerate the dye molecule after electron injection into a conduction band of a semiconductor and to transport the positive charge (holes) toward the counterelectrode. Mostly, a low-volatility solvent-based liquid electrolyte, I_3^-/I^- redox couple in acetonitrile, has been used in DSSCs.

6.7.1 Liquid Electrolytes

The iodide/triiodide redox shuttle was widely used in DSSCs with ruthenium dye (H_2-dcbpy) Ru(NCS)$_2$ (H_2-dcbpy = 2,2′-bipyridine-4,4′-dicarboxylic acid) as sensitizer, TiO$_2$ as a working electrode, and a platinum quasi-reference electrode. It was observed that in the presence of Li$^+$, the photocurrent was five times greater than when $(C_4H_9)_4N^+$ was used. Some other redox shuttles used include ferrocene, thiocyanate, triiodide, and bromide, but an iodide/triiodide redox couple was found to be the most dominant (Wolfbauer et al. 2001). Paulsson et al. (2003) investigated trialkylsulfonium iodides–based electrolytes using liquid salts of (Et$_2$MeS)I, (Bu$_2$MeS)I, and (Bu$_2$EtS)I, where iodine was present in low concentrations. Among them, iodine-doped (Bu$_2$MeS)I electrolyte (3.7%) has shown the best performance for DSSCs. Other types of ionic liquids such as 1-ethyl-3-methylimidazolium bis(trifluoromethane sulfonyl)imide (EMImTFSI), 1-ethyl-3-methylimidazolium tetrafluoroborate (EMImBF$_4$), 1-butyl-3-methylimidazolium hexafluorophosphate (BMImPF$_6$), 1-ethyl-3-methylimidazolium dicyanamide (EMImDCA), and 1-butylpyridinium bis(trifluoromethane sulfonyl)imide (BPTFSI), where 1-ethyl-3-methylimidazolium iodide (EMImI) and I$_2$ were dissolved as a redox couple, were used by Kawano et al. (2004). The EMImDCA (1-ethyl-3-methylimidazolium dicyanamide) and I$_2$ electrolyte have given 5.5% efficiency in a cell, when $[I^-] + [I_3^-] = 2$ M and $[I^-]:[I_2] = 10:1$ were used with 4-t-butylpyridine and LiI. The fabrication of large (100 mm × 100 mm) dye-sensitized solar cells was accomplished by Matsui et al. (2004). Liquid electrolyte containing EMIm–TFSA (1-ethyl-3-methylimidazolium bis(trifluoromethanesulfonyl) amide) was used as the redox system. The power conversion efficiency of 2.7% in a large-sized DSSC was achieved, while a small-sized (9 mm × 5 mm) DSSC showed 4.5% energy conversion efficiency.

Wang et al. (2003a) introduced the solvent-free ionic liquid electrolyte based on 1-methyl-3-propylimidazolium iodide, 1-methyl-3-ethylimidazolium dicyanamide, and LiI. The cell was sensitized with an amphiphilic polypyridyl ruthenium dye. The addition of LiI increases the dye regeneration rate and electron injection yield, and the cell performed with 6.6% power conversion efficiency at high light irradiation, while it showed 7.1% conversion efficiency at lower light intensities. The multicolored solar cell was fabricated by using nanocrystalline TiO$_2$ electrodes sensitized with organic dyes and platinum as a counterelectrode. A mixture of iodine and 1-butyl-3-methylimidazolium iodide in a 0.2:10 molar ratio was used. This multicolor DSSC achieved 2.1% efficiency (Otaka et al. 2004).

The low-viscosity binary ionic liquid PMII (1-propyl-3-methyl-imidazolium iodide) and EMIB(CN)$_4$ (1-ethyl-3-methyl-imidazolium tetracyanoborate) electrolyte have also been used in the DSSCs. Ru complex, Ru(2,2′-bipyridine-4,4′-dicarboxylic acid)(4,4′-bis(2-(4-*tert*-butyloxy-phenyl)ethenyl)-2,2′-bipyridine) (NCS)$_2$ has been used as the photosensitizer. Different ratios of PMII/EMIB(CN)$_4$ liquid electrolytes have been used, and 7.6% power conversion efficiency was obtained (Kuang et al. 2007).

Ionic liquid electrolyte contains nitrile and vinyl function groups attached on imidazolium cations with certain anions, for example, iodide bis[(trifluoromethyl)sulfonyl]imide ([TFSI]$^-$) or dicyanamide ([N(CN)$_2$]$^-$) has been prepared by Mazille et al. (2006). Jovanovski et al. (2006) presented the use of trimethoxysilane derivative of propylmethylimidazolium iodide as gel electrolyte for DSSCs. The overall efficiency of 3.2% was achieved when these derivatives mixed with iodine were used. High-performance dye-sensitized solar cells with electrochemically stable nonvolatile electrolytes were studied by Dai et al. (2006). Ethylmethylimidazolium-based ionic liquid electrolyte with dicyanamide anion was found to be most promising. Therefore, two different anions, tricyanomethanide and thiocyanate, were also used. An increased conversion efficiency (2.1%) was observed in the case of tricyanomethanide, while it was 1.7% for thiocyanate anion. The mixed solvent of tetrahydrofuran (THF) and acetonitrile gave a remarkable (9.74%) conversion efficiency (Fukui et al. 2006). To overcome the problem of stability and fabrication with ionic liquid electrolytes, some other types of electrolytes were developed, which have good stability and are efficient in photoconversion.

6.7.2 Polymer Gel Electrolytes

Ionic liquid EMIm-TFSI (1-ethyl-3-methylimidazolium bis(trifluoromethylsulfonyl) imide) was used as the dispersion medium for carbon nanotubes, titanium dioxide, and other carbon nanoparticles (Usui et al. 2004). It was ground and centrifuged to give an ionic nanocomposite gel liquid. They observed enhanced conversion efficiency as compared to DSSCs with bare ionic liquid electrolytes. A poly(vinylidenefluoride-*co*-hexafluoropropylene) (PVDF-HFP)-based polymeric solid electrolyte film was used by Asano et al. (2004) in a DSSC. The use of such a film was accompanied by a decrease in short-circuit current density. The cell gap of the DSSC and the diffusion coefficient of triiodide were found to play a major role in increasing this current density. It was observed that short-circuit current density of a polymeric solid electrolyte–based DSSC with 20 μm cell gap was almost 97% to that of a liquid electrolyte–based DSSC. This type of PVDF-HFP–based gel polymeric solid electrolyte was also used by Nishikitani et al. (2006). They observed that short-circuit current was found to increase by either increasing the diffusion coefficient of the triiodide and/or iodide ion or by decreasing the cell gap, or both. In general, the narrowing of the cell gap is a simple and effective way to increase short-circuit current density on account of back-electron transfer from titania to an oxidized dye, dependent on short-circuit current density. The photon-to-electron conversion efficiency of 2.93% was obtained when the semisolid ionic gel electrolyte was prepared from agarose, natural polysaccharide, and 1-alkyl-3-methyl-imidazolium salts (Suzuki et al. 2006). A polymer gel electrolyte, poly(vinylidenefluoride-*co*-hexafluoropropylene) was used by Lee et al. (2008a) in DSSC with graphite and titania nanoparticles as fillers. Graphite nanoparticles were found to be better fillers than titania. A DSSC with polymer gel electrolyte with graphite nanoparticles has cell efficiency of 6.04% as compared to 4.69% (without filler), which was close to that of a liquid system.

Park et al. (2012) introduced a new type of UV-cured resin polymer gel with the addition of GBL (γ-butyrolactone). This electrolyte has long-term stability and high photo-to-electric conversion efficiency. The iodide-conducting plastic crystal electrolytes based on N,N-dimethyl-2-(methylsilyloxy) ethanaminium cations and I^-/I_3^- anions was reported by Bertasi et al. (2014). Arof et al. (2014) developed two mixed iodide salt systems: the mixture made of potassium iodide (KI) with a small K^+ cation and tetrapropylammonium iodide (Pr_4NI) with a bulky Pr_4N^+ cation. When KI and Pr_4NI electrolytes were used alone for the cell, they gave 2.37% and 2.90% efficient cells, respectively, whereas the mixed iodide system of KI:Pr_4NI gave a 3.92% efficient cell. The PVDF-*co*-HFP (poly(vinylidenefluoride-*co*-hexafluoropropylene):side-chain liquid crystal polymer (SCLCP) electrolytes have been developed for use in DSSCs and exhibited 5.63% efficiency. This electrolyte may increase the redox couple reduction and reduction charge recombination as compared to PVDF-*co*-HFP-based polymer electrolyte (Cho et al. 2014).

6.7.3 Other Electrolytes

Wang et al. (2004b) presented solvent-free ionic electrolyte as an alternative to iodide/triiodide for nanocrystalline solar cells. The electrolyte consists of a $SeCN^-/(SeCN)_3^-$ redox couple, and the cell reached 7.5%–8.3% power conversion efficiency. After that, Wang et al. (2005c) introduced a more stable Br^-/Br_3^- redox couple in a dye-sensitized solar cell. These cells were sensitized with eosin Y with liquid electrolyte (prepared with the combination of $LiBr + Br_2$ in acetonitrile) and gave 2.61% power conversion efficiency, while $LiI + I_2$ gave only 1.67%. An increase in power conversion efficiency to 56% was observed when I_3^-/I^- was replaced with Br^-/Br_3^-. Bergeron et al. (2005) used $(SeCN)_2/SeCN^-$ or $(SCN)_2/SCN^-$ or I_3^-/I^- and mesoporous nanocrystalline SnO_2 as a redox couple pseudohalogen/halogen electrolyte and semiconductor in regenerated solar cells. They used three sensitizers, $Ru(deep)(bpy)_2(PF_6)_2$, $Ru(deep)_2(dpp)(PF_6)_2$, and $Ru(deep)_2(bpz)_2(PF6)_2$ (deep is 4,4′-diethylester-2,2′-bipyridine; dpp is 2,3-dipyridyl pyrazine; and bpz is bipyrazine). The current conversion efficiencies and open-circuit voltages were found comparable for $(SeCN)_2/SeCN^-$ and I_3^-/I^-. The binding of sensitizers (adsorption) to the SnO_2 surface was described by the Langmuir model. Sapp et al. (2002) made use of some substituted polypyridine complexes of cobalt(II/III) as efficient electron transfer mediators in DSSC. Tris(4,4′-di-*tert*-butyl-2,2′-dpyridyl)cobalt(II/III) perchlorate was found to be the best mediator with 80% efficiency, better than that of iodide/triiodide-based DSSCs. This ligand is simple and readily available; therefore, its complex with cobalt as the mediator can be considered as a practical alternative for an efficient electron transfer mediator in DSSC. It is a nonvolatile as well as noncorrosive complex. It was observed that gold and carbon are superior cathode materials to platinum with no evidence of corrosion.

Other metal-based electrolytes like tris(bipyridine) complex of cobalt(II/III) have been used as redox couple electrolytes (Pazoki et al. 2014). The effect of dye coverage of mesoporous titania electrode was also observed using D35 dye. The adsorption of the dye on TiO_2 was reasonably fit as per the Langmuir adsorption isotherm. It was observed that the electron lifetime in the DSSC increased remarkably with an increase in dye coverage on the TiO_2 electrode. As a result, there was a dramatic increase in solar cell efficiency. Maruthamuthu et al. (2006) designed a lipid (detergent)-based composite dye-sensitized solar cell without using liquid or solid electrolyte. They used a (*cis*-Ru(II) $(LH_2)_2(NCS)_2)$-TiO_2 layer exposed to Triton X $(C_{16}H_{26}O_2)_n$, ion-conducting materials, and dissolved iodine with surface-treated powder of carbon. Photovoltage (0.45 V), photocurrent (1.5 mA/cm²), and energy conversion efficiency (0.22%) have been observed under solar light.

Recently, Luz et al. (2013) have prepared BiOCl and BiOBr nanodiscs via water-based nucleation and purified these samples by a phase transfer reaction. The nanodiscs were finally deposited as a porous p-type semiconductor layer. Coumarin 343 was used as a sensitizer with $BiOCl/KI$-I_2 system, while eosin Y was used as sensitized with a $BiOBr/[C_4MPyr]_2[Br_{20}]$ ($C_4MPyr = N$-butyl-N-methylpyrrolidinium) system. They have applied such a polybromide electrolyte for the first time in a DSSC. These systems show 40.6% and 28.6% fill factor and 0.003% and 0.0005% conversion efficiency, respectively. The ionic liquid crystals C12EImI (1-dodecyl-3-ethylimidazolium iodide) and C10EImI (1-decyl-3-ethylimidazolium iodide) have been synthesized by Xu et al. (2013). Both electrolytes have the same electrochemical and thermal stability, but the I^- diffusion coefficient and photon conversion efficiency of C12EImI were higher than that of C10EImI-based DSSCs.

6.8 ELECTRODES

6.8.1 Nanocrystalline Semiconductor Working Electrode

The semiconductor is the heart of the DSSC. It is composed with nanoparticles of mesoporous oxide thin layer. Generally in a DSSC, oxides like TiO_2, ZnO, SnO_2, Nb_2O_5, In_2O_3, WO_3, Ta_2O_5, and ZrO_2 are used as semiconductor materials. The TiO_2 is used widely as a semiconductor due to its availability in the market, nontoxicity, low cost, environmentally friendly behavior, and biocompatibility.

6.8.1.1 Naïve Semiconductors

Nanocrystalline TiO_2 particles were designed with high surface area (10–20 nm diameter) to improve light absorption by the semiconductor material. The mesoporous morphology of thin film plays an important role in collecting sunlight on an oxide film. When light penetrates the dye-covered oxide "sponge," it crosses hundreds of adsorbed dye monolayers. Photons with energy close to the absorption maximum of the dye were absorbed completely. Hence, such a mesoporous material mimics the process of light absorption by green leaves as in the case of photosynthesis. Different methods of preparing TiO_2 nanocrystalline semiconductors were proposed by Li et al. (1999). They introduced the sol-gel method of preparing a TiO_2 (100 nm diameter) nanoparticle thin layer with the help of spin-coating or doctor-blading techniques. Modification in the sizes of the nanoparticles of the TiO_2 thin layer (25 nm diameter) and fabrication with screen printing have resulted in maximum power conversion efficiencies, due to the large internal surface areas of nanocrystalline particles.

The colloidal TiO_2 film has been prepared in the presence of flexible substrates without using any organic surfactant by sintering at 100°C, and its conversion efficiency was obtained as 1.2% (Pichot et al. 2000). The comparative study of rutile and anatase TiO_2-based dye-sensitized solar cells has been presented by Park et al. (2000). The scanning electron microscopy (SEM) study shows that the rutile films consist of homogeneously distributed rod-shaped particles with an average dimension of 20 × 80 nm. The photovoltage shows a small change for anatase and rutile TiO_2 film–based solar cells, whereas the photocurrent of a rutile-based cell was 30% lower than that of an anatase-based solar cell. The overall power conversion efficiency was close for both the dye-sensitized rutile and anatase TiO_2 films at the same intensity. Lindstrom et al. (2001) demonstrated a new method for preparing a nanostructured porous layer of TiO_2 film on conductive glass or plastic substrate. The cell generated 5.2% solar-to-electrical power conversion efficiency when plastic film was used.

Miyasaka et al. (2002) presented a new nonsintering method for preparing a nanoporous semiconductor thin film, where TiO_2 particles were electophoretically deposited on the electrode. The nanoporous TiO_2 was post-treated with the help of chemical vapor deposition of Ti alkoxide and microwave irradiation. It shows the conversion efficiency of 4.1% under irradiation of AM 1.5 light (100 mW cm^{-2}). A wide-bandgap nanocrystalline TiO_2 film was electrophoretically deposited without using any surfactant or post-thermal treatment (Yum et al. 2005). This film was used in flexible dye-sensitized solar cell on the electrode film. The fill factor and energy conversion efficiency of such a flexible dye-sensitized solar cell was 50.0% and 1.03%, which was raised to 56.3% and 1.66%, respectively, on compression.

Ito et al. (2006) described the fabrication of a highly efficient, flexible, dye-sensitized solar cell with a maximum 7.2% conversion efficiency in a plastic base DSSC. This cell was fabricated with Ti metal foil substrate as the photoanode and Pt-electrodeposited as counterelectrode on ITO/PEN (polyethylene naphthalate). Koo et al. (2008) investigated the effect of light scattering on the conversion efficiency of a DSSC. They used two different types of rutile TiO_2 nanoparticles with diameters of 0.3 μm (G1) and 0.5 μm (G2). The conversion efficiency of G1 and G2 TiO_2 with a 7 μm thick main layer film was improved from 7.55% to 8.94%, but when a 14 μm thick main layer film was used, the power conversion efficiency improved from 8.60% to 9.09% and 9.15%, respectively, upon depositing G1 and G2 particulate overlayers.

The fabrication of dye-sensitized solar cells based on homemade ZnO nanoparticles was reported by Lu et al. (2010), and its conversion efficiency was obtained as 3.92%. Anta et al. (2012) analyzed the performance of a ZnO-based, dye-sensitized solar cell. ZnO has the unique combination of high bulk electron mobility and the richest variety of nanostructures with a wide range of synthesis routes. The comparative study of Nb_2O_5- and TiO_2-based solar cells was reported by Sayama et al. (1998). The cell was sensitized with ruthenium(II) *cis*-bis(thiocyanato)bis(2,2'-bipyridyl-4,4'-dicarboxylic acid) complex. The Ru dye was attached to the surface of Nb_2O_5 by ester linkage. The IPCE performance of the Nb_2O_5 photoelectrode was improved to 32% at 548 nm by the treatment of Nb alkoxide, and the overall solar-to-electric energy conversion efficiency obtained was 2%.

A DSSC was fabricated by Kumara et al. (2006) using a highly conductive and optically transparent glass consisting of an inner layer of indium tin oxide (ITO) and an outer layer of fluorine-doped tin oxide (FTO). The nanocrystalline films of titania with large (30 nm) and small (5 nm) sized particles can promote porosity and light scattering. It was observed that incorporation of trace quantities of MgO into TiO$_2$ increases the efficiency of the solar cell. There was an increase in the efficiency of the solar cell from 5.6% in the absence of magnesium oxide to 7.2% in the presence of magnesium oxide. They have also discussed the mechanism of improvement of cell performance on incorporating magnesium oxide.

6.8.1.2 Modified Semiconductors

Dye-sensitized solar cells are made of nanocrystalline films of some oxide materials, but the deposition of some ultrathin shells of insulators or high bandgap containing semiconductors on crystallites can make dramatic changes (Tennakone et al. 2002). The modified TiO$_2$ nanoparticles coated with Mg(OH)$_2$ were prepared by Jung et al. (2005). The specific surface area of an electrode was prepared with the help of a nanocrystalline core–shell-structured TiO$_2$. The cell performance was increased to 45% as compared to uncoated TiO$_2$ electrode. Thus, it can be concluded that the specific surface area of nanoparticles increases the conversion efficiency.

The performance of DSSC was compared before and after treatment of TiO$_2$ electrode with an electrochemical reaction in the presence of metals such as Mg, Zn, Al, and La in 2-propanol (Yum et al. 2006). On coating metal hydroxides without a sintering process, the magnesium hydroxide gave the maximum enhancement in photovoltage, improved fill factor, and overall conversion efficiency of the solar cell. The overcoating of TiO$_2$ with magnesium hydroxide functions as the blocking layers at the fluorine-doped tin oxide and TiO$_2$ interfaces.

Lee et al. (2006) fabricated the working electrode with core–shell-type nanoparticles. Here, TiO$_2$ and CaCO$_3$ were used as cores and shells, respectively. This type of overlayering makes this cell more efficient than an uncoated TiO$_2$ layered electrode. It created maximum dye adsorption due to the presence of a high isolated point of the coated layer and anticipation of the back-electron transfer by the insulating nature of the overcoating of the layer. Three different layered TiO$_2$ photoelectrodes were designed and investigated by Hu et al. (2007). These layers were constructed in three different parts like small pore–size films, large pore–size films, and light-scattering particles on conducting glass according to required thickness. Thus, the photovoltaic performance of a cell can be affected by different porousness, pore-size distribution, and surface area. Good power conversion efficiency (6%) was obtained with 15 × 20 cm^2 surface area in AM 1.5 sunlight (100 mW cm^{-2}).

The Al$_2$O$_3$-coated nanoporous TiO$_2$ thin-film dye-sensitized solar cell showed retardation of the interfacial charge recombination and improved efficiency of the solar device 30% (Palomares et al. 2002), whereas an MgO-coated TiO$_2$ electrode was prepared by the surface modification method in the presence of DC magnetron sputtering (Wu et al. 2008). By this modification, dye adsorption was increased, which decreases in trap states and clamp interfacial recombination. The efficiency was increased from 6.45% to 7.57%.

Senevirathna et al. (2007) demonstrated SnO$_2$/MgO film dye-sensitized solar cells. In this type of cell, an SnO$_2$ crystalline surface was covered with an ultrathin shell of MgO; this electrode protests dye and electrolytic degradation better than TiO$_2$-based solar cells. Li et al. (2013) introduced a one-step electrophilic deposition method for the preparation of MgO-coated SnO$_2$ photoanodes for dye-sensitized solar cells. The MgO-coated SnO$_2$ semiconductor was prepared from a facile method. The photoanode was hydrolyzed at 413 K and irradiated under UV light. The post-treatment makes the photoanode more flexible. The overall power conversion efficiency in this case reached 5.48%.

The fabrication of a DSSC with core–shell MgO-TiO$_2$ electrode was done by a simple chemical bath deposition method to coat a thin MgO film around TiO$_2$ nanoparticles (Li et al. 2010). The TiO$_2$-MgO electrode showed excellent performance as compared to a noncoated TiO$_2$ electrode. The i_{sc}, V_{oc}, and fill factor were found to be 8.80 mA cm^{-2}, 646 mV, and 0.69, respectively, and power conversion efficiency increased from 4.32% to 5.26%. The surface modification of TiO$_2$/MnTiO$_3$ was made by Shaterian et al. (2014). MnTiO$_3$ nanoparticles were prepared by sol-gel method by adding manganese acetate and tetrabutyltitanate as Mn and Ti sources and stearic acid as complexing agent. The modified photoanode shows increased power conversion efficiency and improvement in photovoltage due to increasing electron lifetime on the TiO$_2$/MnTiO$_3$ interface.

6.8.1.3 Doped Semiconductors

Various methods have been developed for improving the photocatalytic activity of TiO_2 and other semiconductors in the presence of visible light, like the doping of oxide with metals, and co-doping with metals and nonmetals.

Nitrogen-doped TiO_2 electrode was used in a dye-sensitized solar cell by Ma et al. (2005). In nitrogen-doped titania, the substitution of oxygen sites with nitrogen atom was confirmed by x-ray photoemission spectroscopy. This powder and film showed an absorption in the visible range (400–535 nm). Such fabricated cells with nitrogen-doped titania exhibited good stability and 8% overall conversion efficiency. Tian et al. (2010) studied the photovoltaic performance and charge recombination in nitrogen-doped TiO_2 solar cells. The flat-band potential of such a film had a negative shift, which was attributed to the formation in the O-Ti-N bond. As a consequence, the open-circuit voltage improved. This cell exhibited good performance as well as stability as compared to a cell with undoped TiO_2. It was observed that the replacement of oxygen-deficient titania by nitrogen-doped TiO_2 makes the DSSC system more stable.

Mn^{2+}- and Co^{2+}-doped TiO_2 nanoparticles were prepared using a hydrothermal method by Shalan and Rashad (2013). The trap densities were increased by the doping of Mn and Co. The power conversion efficiency decreased due to the high change of flat-band edge. The doping of P25, N-TiO_2, and CN-TiO_2 gave 1.61%, 2.44%, and 3.31% power conversion efficiency, respectively (Sun et al. 2013), while iodine-doped ZnO nanoparticles have shown 3.6%–4.6% cell efficiency (Zheng et al. 2014). Electron injection and electron transportation could be controlled by introducing Cr^{3+} and CNTs into a TiO_2 photoanode, respectively. The presence of Cr^{3+} improved the morphology of the CNT-TiO_2 electrode. The solar cell was composed of 3 at.% Cr^{3+} and 0.025 wt% CNTs, which showed 7.47% efficiency (Massihi et al. 2013). A successive embedding of N-doping and surface modification of TiO_2 photoelectrodes with $CaCO_3$ for dye-sensitized solar cells was reported by Park et al. (2013). It was observed that their combined effect resulted in an increase in short-circuit current, open-circuit voltage, as well as photoelectric conversion efficiency of the cell. The efficiency of this cell was improved to 9.03% from 5.42% and 7.47% for unmodified N-doped electrode and N-doped $CaCO_3$ surface-modified electrode, respectively. Surface modification of an N-doped electrode by $CaCO_3$ formed a barrier, which suppresses the charge recombination of photogenerated electrons to dye or the electrolyte. This extends their lifetime in the electrode and, consequently, the cell efficiency.

Liu (2014a) reported DSSC based on Fe-doped titania electrode. Pure and Fe-doped TiO_2 electrodes were prepared by the hydrothermal method. The Fe doping was done using Fe salt [$Fe(C_2O_4)_3 \cdot 5H_2O$]. The flat-band potential was shifted positively after Fe doping as evident from the Mott-Schottky plot. This shift was considered the driving force for injected electrons from the valence band of the dye to the conduction band of the semiconductor. The photovoltaic efficiency of cell and fill factor increased from 6.07% and 0.53 to 7.46% and 0.63, respectively, as compared to pure TiO_2. Liu (2014b) also tried to improve the efficiency of the DSSC by using Mg-doped TiO_2 thin films. The hydrothermal method was used for the preparation of Mg-doped titania electrode by using Mg salt [$Mg(NO_3)_2 \cdot 6H_2O$]. Fluorine-doped tin oxide glass was coated with this Mg-doped slurry and sintered at 450°C. The doped Mg ions were present in the form of Mg^{2+}, which could act as a trap for e^- or h^+, and as a result, the e^-/h^+ pair recombination rate is reduced. Incorporation of Mg^{2+} increases the photovoltaic efficiency from 5.62% to 7.12% and short-circuit current from 14.9 to 19.1 mA as compared to undoped TiO_2 thin-film electrode.

The high efficiency of a dye-sensitized solar cell depends on effective light scattering and harvesting, rapid electron transfer and transport, and reduced charge recombination. Wang et al. (2014a) attempted to study the effect of the addition of ZnO into a TiO_2 electrode. They observed that the addition of only 2% zinc oxide increased dye adsorption, light scattering, and electron transfer. This modification has enhanced conversion efficiency 9.53%, which was 12% higher than with TiO_2 without the addition of ZnO.

A NiO photocathode–based dye-sensitized solar cell was investigated by Wang et al. (2014b). NiO was synthesized by the microwave-assisted hydrothermal method. Lithium-doped NiO has also been used as a photocathode, and it has shown better IPCE as well as conversion efficiency as compared to undoped NiO photocathode. A silica layer coated with highly uniform, monodispersed, hexagonal prisms was synthesized by a hydrothermal route. These NYFYE@SiO_2 (NYFYE = β-NaYF4:Yb^{+3}/Er^{3+}) crystallites were incorporated into TiO_2 porous thin film by Guo et al. (2014). Here, with the addition of NYFYE@SiO_2, improvement was seen in light scattering and near-infrared light harvesting of the composite photoanode resulting in increased short-circuit current and photoelectric conversion efficiency by 14% and 29%, respectively, as compared to pure TiO_2 photoanode alone. The cell performance of DSSC in this case was 7.28%.

A core–shell semiconductor was also used in DSSC by Törngren et al. (2014), like plasmonic core–shell Au@SiO_2 nanoparticles. A thin silica coating provides better stability during thermal treatment and also against corrosive electrolyte. Stability was further assessed using temperatures up to 500°C for thermal stability and iodide/triiodide electrolyte for chemical stability. Mercaptosilane was used as a linker to maintain a complete silica coating on gold cores. There was almost a 10% increase in the efficiency of DSSC on plasmon incorporation as compared to DSSC without core–shell nanoparticles.

Submicron-sized YVO4:Eu^{3+}, Bi^{3+}@SiO_2 core–shell particles were prepared by Lai et al. (2014) using the hydrothermal method. It was interesting to note that these particles not only enhanced light scattering with photoanode but were also able to convert ultraviolet light to visible light. By embedding such particles, the cell efficiency of DSSC was improved from 3.6% to 5.9%, which is a 64% enhancement.

6.8.1.4 Other Nanomaterials

The photo-to-electric conversion efficiency of DSSC was limited due to electron–hole charge recombination at the electrode–electrolyte interface. Therefore, current development in this research area is based on alternative nanomaterials such as nanofibers, nanowires, nanorods, nanotubes, and so on. These can be used as a photoanodes having large surface areas, low recombination rates, and high energy conversion efficiencies.

A highly efficient dye-sensitized solar cell was fabricated by using a single-crystalline TiO_2 nanotube as a thin film with 5% cell efficiency (Adachi et al. 2003). The ZnO nanorod, which was annealed in N_2/H_2 or O_2, enhanced dye adsorption due to high OH concentration on the hydrophilic surface. After the annealing of ZnO nanorods, fill factor and cell efficiency were also increased (Chung et al. 2010). ZnO nanotubes were synthesized onto an ITO glass, which gave 1.01% cell conversion efficiency (Ranjusha et al. 2011). The hybrid TiO_2-coated SnO_2 nanotube has been used in the fabrication of DSSC. The cell gave 3.53% cell efficiency, and its electron recombination rate is significantly better than that of TiO_2 nanotubes, ZnO nanowires, and TiO_2 nanoparticles thin films (Desai et al., 2013). Bamboo-like shaped TiO_2 nanotubes achieved 7.36% cell efficiency (Wang et al. 2014c).

A TiO_2-coated SnO_2 hybrid nanorod with 40 nm diameter has been synthesized by using a modified flame spray pyrolysis (FSP). The conversion efficiency was reached at 6.93%, which was much higher than SnO_2 nanorod (3.95%) and P25 (5.27%) electrodes (Huo et al. 2014). The homogeneous composition electrode of a TiO_2 macroporous material was used where TiO_2 nanowire for DSSC showed an impressive conversion efficiency of 9.51% (Wu et al. 2014). These dye-sensitized solar cells are low in cost, environmentally friendly, and a green source of energy to meet future energy requirements.

Lotey and Verma (2014) synthesized $BiFeO_3$ nanowires (average diameter of 20 nm) by the colloidal dispersion capillary force induced template-assisted technique. The synthesized nanowires possess rhombohedral structures as supported by x-ray diffraction study and Fourier transform infrared data. The bandgap was in the desired range (2.5 eV). A positive shift in flat-band potential is responsible for high electron injection efficiency. The high energy conversion efficiency of 3.02% has been achieved in this fabricated DSSC. The hybrid photoanodes containing ZnO nanoparticle-nanowire were fabricated by Yodyingyong et al. (2010) for use in DSSC. This hybrid photoanode provides a direct pathway for rapid electron transport, while nanoparticles dispersed between nanowires offer a high specific surface area for adsorption for the dye. This type of modification enhances photoconversion efficiency to a much higher percentage (4.2%) as compared to ZnO nanoparticles (1.58%) and ZnO nanowires (1.31%).

Krysova et al. (2014) have developed a simplified method for the preparation of thick mesoporous nanofiber electrodes. Four types of TiO_2 nanofibers were used. These were electrospun from polyvinylpyrrolidone (PVP) and hydroxypropylcellulose (HPC) as templates. The diameter and surface area ranged between 100–280 nm and 9–100 m^2/g, respectively. Mesoporous TiO_2 films were prepared by supramolecular templating with a block copolymer Pluronic P123. Then 2% or 15% nanofibers were added to the mesoporous film. This resulted in a decrease in dye N719 adsorption but simultaneously an increase in roughness factor. This enhances the photoelectric conversion efficiency of the cell to 5.51%, more than that of mesoporous film (4.96%).

A hybrid photoelectrode was reported by Kong et al. (2014), which was fabricated using single-crystalline TiO_2 nanowires (NWs) inlaid with TiO_2 (anatase) nanoparticles (NPs). Then 4 mm thick vertically aligned NWs were synthesized on fluorine-doped tin oxide glass by a solvothermal route. Nanowires provide a faster pathway for electron transport and the desired light-scattering effect. Sidewise, inlaid NPs give an extra space for the uptake of dye. They observed 6.2% conversion efficiency, which was about 48% more enhancement than DSSC with only NPs. This improvement was attributed to the synergistic effect of enhanced light confinement, charge collection, and loading of the dye.

6.9 COUNTERELECTRODE

The Pt electrode plays an important role in the regeneration of dye molecules by catalysis of the I^-/I_3^- redox couple. The Pt becomes a good choice for developing a counterelectrode due to its rough surface area and nature toward an exposed face, but Pt is a costly noble metal, so efforts are being made to find a low-cost alternative for production of counterelectrodes (Thomas et al. 2014).

Thus, other materials should be applied for obtaining low-cost solar cells for further development of counterelectrodes. Graphite powder and carbon black (Kay and Gratzel 1996) and single-wall carbon nanotubes (SWCNTs) (Suzuki et al. 2003) have been used to replace the Pt electrode. Other carbon materials, inorganic materials, multiple compounds, polymers and composites (Wu and Ma 2014), or metal, such as platinum, gold, and nickel, and various nanostructured carbon materials, conductive polymers (Jing et al. 2011) were also used in DSSC.

Murakami and Gratzel (2008) explained that different types of materials can be used in the preparation of counterelectrodes like platinum, graphite, activated carbon, carbon black, single-wall carbon nanotubes, poly(3,4-ethylenedioxythiophene) (PEDOT), polypyrrole, and polyaniline. These may be used as catalysts for the reduction of triiodide. They prepared carbon black–loaded stainless steel electrodes as low-cost counterelectrodes and got excellent 9.15% power conversion efficiency. The SUS-316 stainless steel gave equivalent performance as FTO glasses.

Olsen et al. (2000) investigated the stability and performance of FTO glasses coated with vapor deposition of Pt and Pd electrocatalyst, which were used as counterelectrodes. Saito et al. (2004) suggested the porous PEDOT-TsO as a counterelectrode, which was low cost and simple to fabricate.

The potential of DSSC is further increased due to catalytic activity of the counterelectrode. Nanosized carbon material was reported as counterelectrode with 7.56% cell efficiency (Lee et al. 2008b). Graphene nanosheets (GNs) were synthesized for DSSC with 6.81% power conversion efficiency (Zhang et al. 2011). On the other hand, some nanolayers were prepared to increase catalytic performance of the cell. PEDOT (Poly (3,4-ethylenedioxythiophene) nanoporous layers were prepared by Ahmad et al. (2010). The PEDOT counterelectrode has shown 8.0% power conversion efficiency, which is very close to that of the Pt counterelectrode (8.7%).

Dye-Sensitized Solar Cells

The fabrication of DSSC with graphene-PVP (polyvinyl pyrrolidone) counterelectrode coated on FTO glasses was reported by Li et al. (2012). The PVP is linked with graphene by formation of an ester bond (-C-O-OC-). Due to the high electrocatalytic activity of the electrode, it might show low charge-transfer resistance and high redox current density, and it displayed 3.01% power conversion efficiency.

Wang et al. (2013a) proposed that ZnO could be used as a counterelectrode with PEDOT:PSS (poly(3,4-ethylenedioxythiophene):polystyrene sulfonate), and it could be an efficient counterelectrode with a maximum power conversion efficiency of 8.17%. Chen et al. (2013) reported an SnS low-cost electrode for DSSC. They prepared SnS nanosheets, SnS nanowires, and SnS_2 nanosheets as alternatives to the Pt electrode. They determined power conversion efficiencies of 7.56% for Pt, 6.56% for SnS NS, 5.00% for SnS NW, and 5.14% for SnS_2 NS. The remarkable performance of SnS-based DSSC has shown that it had excellent catalytic activity for the reduction of triiodide to iodide.

Pt-free electrodes have been synthesized by Ahmad et al. (2014). They introduced counterelectrodes by using graphene nanoplatelets (GNPs) or multiwall carbon nanotubes (MWCNTs). Hou et al. (2013) introduced rust (α-Fe_2O_3) as a new counterelectrode, and Yue et al. (2013) prepared PEDOT:PSS [poly(3,4-ethylenedioxythiophene):poly(styrenesulfonate)] coated counterelectrode with high transparency and low temperature required for coating on FTO glasses. The combination of carbon nanotubes with carbon black (carbon nanocomposite) has been used as counterelectrode (Chen et al. 2012). Noh and Song (2014) innovated a new type of Ru/Ti bilayer on flat glass counterelectrode, which gave 2.4% power conversion efficiency. A new type of MoS_2 transparent counterelectrode was presented by Zhang et al. (2014). To improve the catalytic activity of the electrode, they created an artificial active edge site by patterning holes on MoS_2 atomic layers. After the hole patterning on MoS_2 layers, DSSC has shown remarkable improvement in power conversion efficiency from 2.0% to 5.8%. Nickel sulfide (6.81%) and cobalt sulfide (6.59%) coated FTO glasses were also used as counterelectrodes (Yang et al. 2014). The $MoSe_2$/Mo counterelectrode improved the efficiency of DSSC (8.13%) as compared to Pt electrode–based DSSC (8.06%) (Chen et al. 2014a).

The rigid monolithic coal-based carbon counterelectrode has been synthesized by Wang et al. (2014d). Coal-based carbon electrode has shown low internal resistance with 6.38% power conversion efficiency. The surface modification with a mesoporous carbon catalytic layer could give 8.73% power conversion efficiency.

A polypyrrole nickel-coated cotton fabric counterelectrode was proposed by Xu et al. (2014) with 3.3% power conversion efficiency. $CuSbS_2$ was also used as a counterelectrode (Ramasamy et al. 2014).

6.10 OTHER TYPES OF DYE-SENSITIZED SOLAR CELL

6.10.1 Solid-State DSSC

In solid-state DSSC (SS-DSSC), redox electrolyte is replaced by a nanocrystalline solid. A p-type semiconductor TiO_2 was used because of the independent hole transport ability of the p-type semiconductor, which could permit charge neutralization of dye molecules after electron injection.

Dye-sensitized solar cells have achieved 11% power conversion efficiency on using liquid electrolytes, whereas solid-state dye-sensitized solar cells gave 5% photon-to-electric conversion efficiency by replacing the liquid electrolyte. In order to further improve the efficiency, a variation in metal oxides, organic and inorganic sensitizers, molecular, polymeric and electrolytic hole transporter materials, and blocking layer was made (Yum et al. 2008). SS-DSSC is a low-cost and easily fabricated alternative for amorphous silicon solar cells. It is sensitized with indoline dye, and it gave 4% efficiency (Schmidt-Mende et al. 2005).

Liquid electrolytes were replaced with solid-state materials like organic and inorganic p-type semiconductors, but the power conversion efficiency of solid-state-based solar cells remained low compared to liquid electrolyte–based DSSC. Bach et al. (1998) introduced a dye-sensitized heterojunction TiO_2 with amorphous organic hole transport material, 2,2′,7,7′-tetrakis(N,N-di-p-methoxyphenyl-amine) 9,9′-spirobifluorene (OMeTAD). They got 33% yield of photon-to-current conversion efficiency. The fabrication of highly efficient SS-DSSC with porous dye-filled TiO_2 layer with ZnO coating and molten salt capped CuI crystals was done by Meng et al. (2003). The cell achieved long-term stability and 3.8% efficiency.

Dye-sensitized solid-state solar cells have shown low energy due to a high recombination rate at the n-type semiconductor/dye/p-type semiconductor interfaces, when n-type semiconductor TiO_2 and p-type semiconductor CuI were used. The modification of n-TiO_2 crystallites with the deposition of an ultrathin layer of MgO could improve power conversion efficiency (Kumara et al. 2004).

Karthikeyan and Thelakkat (2008) reported that the pore size is more important than a large surface area for filling the mesoporous layer with a solid-state hole conductor. The heteroleptic Ru(II) complexes [N,N′-bis(phenyl)-N,N′-bis(3-methylphenyl)-1,1′-biphenyl-4,4′-diamine (TPD)] have shown better efficiency due to their easy filling into pores of nc-TiO_2 layer. Recently, novel hole conductors have been synthesized with spiro-bifluorene-triphenylamine core for transporting holes and tetraethylene glycol side chain for binding lithium ion for improvement in SS-DSSC. The addition of Li salt could improve the performance of SS-DSC, because Li^+ salt is required for the interfacing of TiO_2 and dye.

Abrusci et al. (2011) replaced spiro-OMeTAD with poly(3-hexylthiophene) (P3HT) light-adsorbing polymer with the addition of lithium bis(trifluoromethylsulfonyl) imide salts (Li-TFSI) in SS-DSSC sensitized with indolene-based D102. Burschka et al. (2011) doped spiro-OMeTAD with p-type Co(III). It worked as a hole conductor for SS-DSSC and has good power conversion efficiency (7.2%). Solid-state dye-sensitized solar cells assembled with $CsSnI_{2.95}F_{0.05}$ doped with SnF_2, nanoporous TiO_2, and sensitized with N719 dye gave excellent power conversion efficiency (10.2%). CsSnI was used for hole conduction in place of a liquid electrolyte by Chung et al. (2012).

Michaleviciute et al. (2014) synthesized a star-shaped carbazole molecule, TMPCA [tri(9-(methoxyphenyl) carbazol-3-yl) amine] and fabricated the solid-state dye-sensitized solar cells based on ITO/TiO_2/D102/T4MPCA/Au. This cell has shown 2.23% power conversion efficiency. Chiang et al. (2013) prepared a new tandem solid-state dye-sensitized solar cell with 3.3% power conversion efficiency. Roh et al. (2014) used a pine tree–like TiO_2 nanotube (PTT) and sensitized it with N719 for fabricating SS-DSSC. They got an outstanding energy conversion efficiency of 8.0% at 100 mW cm^{-2}.

Solid electrolyte prepared with 2-hydroxyethyltrimethylammonium iodide (choline iodide) was used in a solid-state, dye-sensitized solar cell. This solid electrolyte has a strong quaternary ammonium cation, 2-hydroxyethyltrimethylammonium $(HETA)^+$ as a cation donor. In this cell, 2.5% overall power conversion efficiency was achieved (Wang et al. 2013b).

Solid-state dye-sensitized solar cells fabricated with hyperbranched nanostructured TiO_2 on FTO glass and sensitized with D102 dye were described by Passoni et al. (2013). This device has shown the highest 3.96% efficiency and 66% increased efficiency with respect to a reference mesoporous photoanode. Fabrication of SS-DSSC with polymer substrates was done by Xue et al. (2014) with 1.93% efficiency.

Akhtar et al. (2013) prepared a solid electrolyte with the combination of graphene (Gra) and polyethylene oxide (PEO). Gra–PEO composite electrolyte has shown the large-scale generation of iodide ions in a redox couple and reduced its velocity by adding LiI and I_2. They achieved a 5.23% overall power conversion efficiency. Fabrication of a SS-DSSC with P3HT [poly(3-hexilthiophene)] as the hole transport material for the dye regeneration process with 2% power conversion efficiency was reported by Matteocci et al. (2014).

Li et al. (2015) synthesized I^-/I_3^--doped 3-hydroxypropionitrile/polyaniline (HPN/PANi) solid electrolytes for the fabrication of SS-DSSC. The I_3^- reduction reaction from the electrolyte/

Dye-Sensitized Solar Cells

counterelectrode interface to solid electrolyte system was enhanced, and the charge diffusion path length was decreased. The HPN/PANi SS-DSSC shows 3.70% cell efficiency higher than the primitive HPN-based DSSC.

6.10.2 Quasi-Solid-State DSSC

A DSSC has achieved 7% efficiency with liquid electrolyte. But liquid electrolyte creates some technical problems related to long-term stability, and the voltaic nature of organic solvents does not allow tight sealing of the cell. From this point of view of long-term stability of the cell, liquid electrolyte could be replaced with quasi/full solid-state electrolyte for better performance. This type of solid electrolyte was developed and gave better long-term stability and higher efficiency than a liquid electrolyte.

A quasi-solid-state electrolyte has been used as a substitute for liquid electrolyte in quasi-solid-state dye-sensitized solar cell (QSS-DSSC). Kubo et al. (2002) fabricated a dye-sensitized solar cell with 1-hexyl-3-methylimidazolium iodide, iodine, and low-molecular-weight gelator as a quasi-solid-state electrolyte. They achieved 5.0% power conversion efficiency and high-temperature stability. Wang et al. (2002) used ionic liquid polymer gel electrolyte for QSS-DSSC with 5.3% power conversion efficiency. The polymer gel electrolyte consists of 1-methyl-3-propylimidazolium iodide (MPII) and poly(vinylidenefluoride-co-hexafluoropropylene) (PVDF–HFP). Ionic liquid–based quasi-solid-state electrolyte has been made from solidification of silica nanoparticles and was sensitized with ruthenium polypyridyl dye. The cell gave 7% power conversion efficiency (Wang et al. 2003b).

Mohmeyer et al. (2006) prepared a stable quasi-solid-state electrolyte for QSS-DSSC with 6.3% efficient cell by using organogelator. Cross-linked gelators containing gel electrolyte have been used in QSS-DSSC. This electrolyte consists of ionic liquids: (1-methyl-3-propylimidazolium iodide) solidified with polyvinyl pyridine and 1,2,4,5-tetra(bromomethyl) benzene). The solidification process occurred after injecting the gel electrolyte into the cell (Sakaguchi et al. 2004). A new type of ionic polymer electrolyte has been synthesized by copolymerization of alkyl-bis(imidazole)s polymer, and diiodo alkyls were used in quasi-solid-state electrolyte in QSS-DSSC (Suzuki et al. 2004). Komiya et al. (2004) introduced polymer electrolyte, which has only 7% concentration of polymer. This polymer electrolyte was used in quasi-solid-state DSSC, and it gave 8.1% power conversion efficiency. Poly(methyl acrylate)/poly(ethylene glycol)-based polymer gel electrolyte with 5.76% efficiency has been reported by Shi et al. (2009), while Joseph et al. (2006) reported hybrid gel electrolyte based on tetraethyl orthosilicate (TEOS) and poly(ethylene glycol) with 4.1% efficiency.

1-Butyl-3-methylimidazolium iodide, polyvinyl pyrrolidone, potassium iodide, and iodine-containing ionic liquid electrolyte have been used in quasi-solid-state DSSC (5.41%) (Fan et al. 2010). Poly(HEMA/GR) [poly(hydroxyethyl methacrylate/glycerin)]/PANi (polyaniline) gel electrolyte has also been used for dye-sensitized solar cell. The combination of poly(HEMA/GR) and PANi creates low charge-transfer resistance and higher electrolytic activity for the I_3^-/I^- redox reaction, and it gave 6.63% power conversion efficiency (Li et al. 2014). Block copolymer electrolyte based on PS_n-b-PEO_m-b-PS_n (PS = polystyrene, PEO = poly(ethylene oxide) as solid hosts for iodine/iodide was used with 6.7% efficiency in DSSC by Manfredi et al. (2014). Dkhissi et al. (2014) reported poly(vinylidenefluoride-co-hexafluoropropylene)–based gel electrolytes for QS-DSSC with 6.4% cell efficiency.

Fan et al. (2014) introduced metal–organic skeleton-based gel electrolytes for high-efficiency quasi-solid-state dye-sensitized solar cell (QSS-DSSC). Metal–organic gel was prepared by coordination of Al^{3+} and 1,3,5-benzenetricarboxylate (H_3BTC). The cell fabricated with gel electrolyte showed 8.60%, which is lower than the liquid electrolyte–based solar cell.

The novel polymeric ionic liquid gel electrolyte was synthesized from poly(oxyethylene)-imide-imidazole complex coupled with iodide anions (POEI-II) for quasi-solid-state dye-sensitized solar (Chang et al. 2014). POEI-II simultaneously acts as a redox mediator in the electrolyte and as a polymer for the gelation of an organic solvent-based electrolyte, while cell durability was improved. The QSS-DSSC with the POEI-II gel electrolyte reaches a high cell efficiency of 7.19%, while with the POEI-II/MWCNT gel electrolyte, 7.65% power conversion efficiency was obtained.

Shi et al. (2015) designed quasi-solid-state dye-sensitized solar cells with three-dimensional network consisting poly(adipic acid pentaerythritol ester) (PAAPE). The PAAPE was prepared by the esterification of pentaerythritol and adipic acid. This three-dimensional network gives highly efficient ion-transporting channels for iodide/triiodide (I^-/I_3^-). This gel electrolyte shows 4.03 mS cm^{-1} conductivity at 25°C, and the solar cell shows 6.81% power conversion efficiency. Many types of gel polymer electrolytes were used in DSSC, including I^-/I_3^- redox couple-contain, ethylene carbonate (EC) and propylene carbonate (PC)-plasticized, polyacrylonitrile (PAN)-based gel polymer electrolyte. Jayaweera et al. (2015) reported four different types of DSSC: (1) only plasticized gel polymer electrolyte was used in between two electrodes of the cell; (2) the same electrolyte was used, but the gel electrolyte was deeply penetrated in the matrix of the nanocrystalline TiO_2 electrode; (3) an I^-/I_3^- redox couple electrolyte was used with sealing the pores by PAN gel electrolyte; and (4) an I^-/I_3^- redox couple of liquid electrolyte was used. These cells show power conversion efficiencies for (1) 4.1%, (2) 5.2%, (3) 8.4%, and (4) 9.8%, respectively.

6.10.3 Quantum Dot DSSC

Quantum dot dye-sensitized solar cells (QD-DSSC) have received tremendous attention as one of the potential low-cost alternatives for p-n junction silicon solar cells. The basic difference between a QD-DSSC and a DSSC is the material used to harvest the visible portion of the solar spectrum.

A quantum dot cell is designed with electron conductor/QD monolayer/hole conductor junctions with high optical absorbance. Nanocrystalline semiconductor and redox electrolyte, solid-state hole conductors were borrowed from standard DSSCs (Rühle et al. 2010).

Cadmium chalcogenide (CdX, X = S, Se, or Te)-based QD-DSSC has given the best performance, but its performance could not compete with the DSSC. The overall performance of a cell depends on the fabrication method of the quantum dots, morphology of the photoanode, type of electrolyte, and choice of counterelectrode (Jun et al. 2013).

The ZnO nanorods coated with TiO_2 nanosheets were designed for cosensitized CdS/CdSe quantum dot solar cells. TiO_2 nanosheets improved the surface area of the ZnO nanorod and allowed more adsorption for quantum dots, which gave a high short current density, an energy barrier for electron hindrance in ZnO from being returned to the electrolyte, and reduced rate of charge recombination. The power conversion efficiency obtained was 2.7% (Tian et al. 2013). The hybrid GaAs-based solar cell with colloidal CdS quantum dots was demonstrated by Lin et al. (2012). CdS quantum dots have antireflective features at long wavelengths and down-conversion in a UV regime, so they can affect overall power conversion efficiency as compared with a traditional GaAs-based device.

Sarkar et al. (2012) reported different-sized 3-mercaptopropionic acid (MPA) stabilized CdTe quantum dots (QDs) for the fabrication of QDs in ZnO nanorod (NR)-based DSSCs. They highlight two major pathways: (1) a direct injection of charge carriers from QDs to ZnO semiconductor via photoinduced electron transfer (PET) and (2) an indirect excitation of sensitized dye (N719) molecules by funneling harvested light via Förster resonance energy transfer (FRET). The QD-assembled DSSCs shows higher photoconductivity and high short-circuit density as compared to DSSC fabricated with N719 sensitizer.

The heterostructured cerium oxide quantum dots (CeO_2 QDs) equipped with zinc oxide nanorods (ZnO NRs) have been used in DSSC (Rai et al. 2014). The CeO_2 QDs fitted with zinc oxide nanorods were prepared by the combination of solvothermal and chemical bath deposition methods. The solar cell assembled with CeO_2 QDs/ZnO NRs shows improved open-circuit voltage, fill factor, and maximum power conversion efficiency (2.65%). The CeO_2 QDs/ZnO NRs shows high light-harvesting efficiency due to the formation of a barrier layer and hindrance in back-electron transfer on the surface of the photoanode.

Rho et al. (2014) reported silica-coated quantum dot–embedded silica nanoparticles for the fabrication of DSSC. This solar cell was compared with unmodified DSSC. The power conversion efficiency was improved from 3.92% to 4.82% for SiO_2/QD/SiO_2 NPs DSSCs.

Fang et al. (2014) introduced graphene quantum dots (GQDs) in the fabrication of dye-sensitized solar cells. It was observed that the amount of dye adsorption first decreased and then increased with an increasing amount of GQDs in the photoanodes, while the photovoltage, photocurrent, and conversion efficiency of these solar cells were first increased and then decreased. The role of GQDs not only improves the properties of DSSC, but it also reduces the use of dyes. It has significant importance in the fabrication of environmentally friendly and low-cost DSSCs.

Graphene quantum dots have also been used as cosensitizers in hybrid dye-sensitized solar cells (Mihalache et al. 2015). GQDs were synthesized by a hydrothermal method, which permits the tuning of electronic levels and optical properties by using suitable conditions for synthesis and proper precursors. The solar cell was assembled by using TiO_2/GQD/N3 Ru dye, where GQDs were used as cosensitizer together with N3 dye. The power conversion efficiency was improved due to the energy transfer from GQDs to N3 Ru dye by the overlapping between GQD photoluminescence and N3 Ru-dye absorption spectra. The inhibition of back-electron transfer to the electrolyte by the GQDs improves the reduction of electron recombination to the redox couple.

6.11 APPLICATIONS OF DSSC

A DSSC is more effectively used in the field of energy harvesting because of its capability to operate under low light irradiation. Almost half of the sunlight spectrum belongs to the infrared region. The silicon cell and CLGC solar cells adsorb sunlight at over 1000 nm, but DSSC adsorbs light at 775 nm for N719, red dye, and 900 nm for N749, black dye.

The common silicon-based solar cell can be replaced by a DSSC because of its good performance and high efficiency (10%). In 2010, DSSCs have been fabricated with efficiency of almost 10%; hence, the commercialized DSSC modules are now attainable.

In the field of BIPV (building integrated photovoltaics), DSSCs are widely applicable in power window and shingles or pebbles because of their multicolor options (depending on the dye). The first building has been equipped with a wall of electric power–producing glass tiles for large-scale testing of DSSCs. The electric power–producing glass tiles were also manufactured. DSSC can also be used for room and outdoor calculators, gadgets, and mobile devices due to their lightweight flexibility. DSSCs can be designed for indoor decorative elements due to different-colored dyes. They can be used in many portable devices, including baggage, gears, and outfits.

REFERENCES

Abrusci, A., R. S. S. Kumar, M. Al-Hashimi, M. Heeney, A. Petrozza, and H. J. Snaith. 2011. Influence of ion induced local coulomb field and polarity on charge generation and efficiency in poly(3-hexylthiophene)-based solid-state dye-sensitized solar cells. *Adv. Funct. Mater.* 21: 2571–2579.

Adachi, M., Y. Murata, I. Okada, and S. Yoshikawa. 2003. Formation of titania nanotubes and applications for dye-sensitized solar cells. *Electrochem. Soc.* 150: G488–G493.

Adams, W. G., and R. E. Day. 1876. The action of light on selenium. *Proc. Roy. Soc. Lond. A* 25: 113–117.

Ahmad, I., J. E. McCarthya, M. Barib, and Y. K. Gun'koa. 2014. Carbon nanomaterial based counter electrodes for dye-sensitized solar cells. *Solar Energy.* 102: 152–161.

Ahmad, S., J.-H. Yum, Z. Xianxi, M. Grätzel, H.-J. Butt, and M. K. Nazeeruddin. 2010. Dye-sensitized solar cells based on poly (3,4-ethylenedioxythiophene)counter electrode derived from ionic liquids. *J. Mater. Chem.* 20: 1654–1658.

Akhtar, M. S., S. J. Kwon, F. J. Stadler, and O. B. Yang. 2013. High efficiency solid state dye sensitized solar cells with graphene–polyethylene oxide composite electrolytes. *Nanoscale.* 5: 5403–5411.

Anderson, S., E. C. Constable, M. P. Dare-Edwards, J. B. Goodenough, A. Hamnett, K. R. Seddon, et al. 1979. Chemical modification of a titanium (IV) oxide electrode to give stable dye sensitization without a super sensitiser. *Nature.* 280: 571–573.

Anta, J. A., E. Guillén, and R. Tena-Zaera. 2012. ZnO-based dye-sensitized solar cells. *J. Phys. Chem. C.* 116: 11413–11425.

Argazzi, R., G. Larramona, C. Contado, and C.-A. Bignozzi. 2004. Preparation and photoelectrochemical characterization of a red sensitive osmium complex containing 4,4′,4″-tricarboxy-2,2′:6′,2″-terpyridine and cyanide ligands. *J. Photochem. Photobiol. A: Chem.* 164: 15–21.

Arof, A. K., M. F. Aziz, M. M. Noor, M. A. Careem, L. R. A. K. Bandara, C. A. Thotawatthage, et al. 2014. Efficiency enhancement by mixed cation effect in dye-sensitized solar cells with a PVdF based gel polymer electrolyte. *Int. J. Hydrogen Energy.* 39: 2929–2935.

Asano, T., T. Kubo, and Y. Nishikitani. 2004. Electrochemical properties of dye-sensitized solar cells fabricated with PVDF-type polymeric solid electrolytes. *J. Photochem. Photobiol. A: Chem.* 164: 111–115.

Bach, U., D. Lupo, P. Comte, J. E. Moser, F. Weissörtel, J. Salbeck, et al. 1998. Solid-state dye-sensitized mesoporous TiO2 solar cells with high photon-to-electron conversion efficiencies. *Nature.* 395: 583–585.

Becquerel, A. E. 1839. Memory on the electrical effects produced under the influence of radiation solarire. *C. R. Acad. Sci. Paris.* 9: 561–567.

Bergeron, B. V., A. Marton, G. Oskam, and G. J. Meyer. 2005. Dye-sensitized SnO_2 electrodes with iodide and pseudohalide redox mediators. *J. Phys. Chem. B.* 100: 937–943.

Bertasi, F., E. Negro, K. Vezzu, and V. D. Noto. 2014. Iodide-conducting plastic crystals based on N,N-dimethyl-2-(methylsilyloxy) ethanaminium cations (MESEAn^+) for application in dye-sensitized solar cells. *Int. J. Hydrogen Energy.* 39: 2896–2903.

Bessho, T., E. C. Constable, M. Graetzel, A. H. Redondo, C. E. Housecroft, W. Kylberg, et al. 2008. An element of surprise-efficient copper-functionalized dye-sensitized solar cells. *Chem. Commun.* 3717–3719.

Burschka, J., A. Dualeh, F. Kessler, E. Baranoff, N. L. Cevey-Ha, C. Yi, et al. 2011. Tri(2-(1H-pyrazol-1-yl) cobalt (III) as p-type dopant for organic semiconductors and its application in highly efficient solid-state dye-sensitized solar cells. *J. Am. Chem. Soc.* 133: 18042–18045.

Chang, L.-Y., C.-P. Lee, C.-T. Li, M.-H. Yeh, K.-C. Ho, and J.-J. Lin. 2014. Synthesis of a novel amphiphilic polymeric ionic liquid and its application in quasi-solid-state dye-sensitized solar cells. *J. Mater. Chem. A.* 2: 20814–20822.

Chang, Y.-C., C.-L. Wang, T.-Y. Pan, S.-H. Hong, C.-M. Lan, H.-H. Kuo, et al. 2011. A strategy to design highly efficient porphyrin sensitizers for dye-sensitized solar cells. *Chem. Commun.* 47: 8910–8912.

Chen, H., Y. Xie, H. Cui, W. Zhao, X. Zhu, Y. Wang, et al. 2014a. In situ growth of a $MoSe_2$/Mo counter electrode for high efficiency dye-sensitized solar cells. *Chem. Commun.* 50: 4475–4477.

Chen, X. Y., Y. Hou, B. Zhang, X. H. Yang, and H. G. Yang. 2013. Low-cost SnS_x counter electrodes for dye-sensitized solar cells. *Chem. Commun.* 49: 5793–5795.

Chen, X., C. Jia, Z. Wan, and X. Yao. 2014b. Organic dyes with imidazole derivatives as auxiliary donors for dye-sensitized solar cells: Experimental and theoretical investigation. *Dyes Pigments* 104: 48–56.

Chen, Y., H. Zhang, Y. Chen, and J. Lin. 2012. Study on carbon nanocomposite counter electrode for dye-sensitized solar cells. *J. Nanomater.* doi:10.1155/2012/601736

Chiang, Y.-F., R.-T. Chen, A. Burke, U. Bach, P. Chen, and T.-F. Guo. 2013. Non-color distortion for visible light transmitted tandem solid state dye-sensitized solar cells. *Renew. Energy.* 59: 136–140.

Cho, W., J. W. Lee, Y.-S. Gal, M.-R. Kim, and S. H. Jin. 2014. Improved power conversion efficiency of dye-sensitized solar cells using side chain liquid crystal polymer embedded in polymer electrolytes. *Mater. Chem. Phys.* 143: 904–907.

Choi, H., J. K. Lee, K. Song, S. O. Kang, and J. Ko. 2007. Novel organic dyes containing bis-dimethylfluorenyl amino benzo[b]thiophene for highly efficient dye-sensitized solar cell. *Tetrahedron.* 63: 3115–3121.

Chung, I., B. Lee, J. He, R. P. H. Chang, and M. G. Kanatzidis. 2012. All-solid-state dye-sensitized solar cells with high efficiency. *Nature.* 485: 486–489.

Chung, J., J. Lee, and S. Lim. 2010. Annealing effects of ZnO nanorods on dye-sensitized solar cell efficiency. *Physica B.* 405: 2593–2598.

Dai, Q., D. B. Menzies, D. R. MacFarlane, S. R. Batten, S. Forsyth, L. Spiccia, et al. 2006. Dye-sensitized nanocrystalline solar cells incorporating ethylmethylimidazolium-based ionic liquid electrolytes. *C. R. Chim.* 9: 617–621.

Dare-Edwards, M. P., J. B. Goodenough, A. Hamnett, K. R. Seddon, and R. D. Wright. 1980. Sensitisation of semiconducting electrodes with ruthenium-based dyes. *Faraday Discuss. Chem. Soc.* 70: 285–298.

Desai, U. V., C. Xu, J. Wu, and D. Gao. 2013. Hybrid TiO_2_SnO_2 nanotube arrays for dye-sensitized solar cells. *J. Phys. Chem. C.* 117: 3232–3239.

Dkhissi, Y., F. Huang, Y.-B. Cheng, and R. A. Caruso. 2014. Quasi-solid-state dye-sensitized solar cells on plastic substrates. *J. Phys. Chem. C.* 118: 16366–16374.

Erten-Ela, S., S. Sogut, and K. Ocakoglu. 2014. Synthesis of novel ruthenium II phenanthroline complex and its application to TiO_2 and ZnO nanoparticles on the electrode of dye sensitized solar cells. *Mater. Sci. Semicon. Proc.* 23: 159–166.

Fan, J., L. Li, H.-S. Rao, Q.-L. Yang, J. Zhang, H.-Y. Chen, et al. 2014. A novel metal-organic gel based electrolyte for efficient quasi-solid-state dye-sensitized solar cells. *J. Mater. Chem. A.* 2: 15406–15413.

Fan, L., S. Kang, J. Wu, S. Hao, Z. Lan, and J. Lin. 2010. Quasi-solid state dye-sensitized solar cells based on polyvinylpyrrolidone with ionic liquid. *Energy Sources.* 32: 1556–1568.

Fang, X., M. Li, K. Guo, J. Li, M. Pan, L. Bai, M. Luoshan, and X. Zhao. 2014. Graphene quantum dots optimization of dye-sensitized solar cells. *Electrochim. Acta.* 137: 634–638.

Fukui, A., R. Komiya, R. Yamanaka, A. Islam, and L. Han. 2006. Effect of a redox electrolyte in mixed solvents on the photovoltaic performance of a dye-sensitized solar cell, *Solar Energy Mater. Solar Cells.* 90: 649–658.

Gerischer, H., M. E. Michel-Beyerle, F. Rebentrost, and H. Tributsch. 1968. Sensitization of charge injection into semiconductors with large band gap. *Electrochim. Acta.* 13: 1509–1515.

Gratzel, M. 2003. Dye-sensitized solar cells. *J. Photochem. Photobiol. C: Rev.* 4: 145–153.

Gratzel, M. 2005. Solar energy conversion by dye-sensitized photovoltaic cells. *Inorg. Chem.* 44: 6841–6851.

Guo, K., M. Li, X. Fang, M. Luoshan, L. Bai, and X. Zhao. 2014. Performance enhancement in dye-sensitized solar cells by utilization of a bifunctional layer consisting of core-shell β-NaYF$_4$:Er^{3+}/Yb^{3+}@SiO$_2$ submicron hexagonal prisms *J. Power Sources.* 249: 72–78.

Hao, S., J. Wu, Y. Huang, and J. Lin. 2006. Natural dyes as photosensitizers for dye-sensitized solar cell. *Solar Energy.* 80: 209–214.

Hara, K., M. Kurashige, Y. Dan-Oh, C. Kasada, A. Shinpo, S. Suga, K. Sayama, and H. Arakawa. 2003. Design of new coumarin dyes having thiophene moieties for highly efficient organic-dye-sensitized solar cells. *New J. Chem.* 27: 783–785.

Hara, K., K. Sayama, Y. Ohga, A. Shinpo, S. Suga, and H. Arakawa. 2001. A coumarin-derivative dye sensitized nanocrystalline TiO$_2$ solar cell having a high solar-energy conversion efficiency up to 5.6%. *Chem. Commun.* 6: 569–570.

Hara, K., Z.-S. Wang, T. Sato, A. Furube, R. Katoh, H. Sugihara, Y. Dan-oh, C. Kasada, A. Shinpo, and S. Suga. 2005. Oligothiophene-containing coumarin dyes for efficient dye-sensitized solar cells. *J. Phys. Chem. B.* 109: 15476–15482.

Horiuchi, T., H. Miura, K. Sumioka, and S. Uchida. 2004. High efficiency of dye-sensitized solar cells based on metal-free indoline dyes. *J. Am. Chem. Soc.* 126: 12218–12219.

Hou, Y., D. Wang, X. H. Yang, W. Q. Fang, B. Zhang, H. F. Wang, et al. 2013. Rational screening low-cost counter electrodes for dye-sensitized solar cells. *Nat. Commun.* 4: 1–8.

Hu, L., S. Dai, J. Weng, S. Xiao, Y. Sui, Y. Huang, et al. 2007. Microstructure design of nanoporous TiO$_2$ photoelectrodes for dye-sensitized solar cell modules. *J. Phys. Chem. B.* 111: 358–362.

Huo, J., Y. Hu, H. Jiang, W. Huang, and C. Li. 2014. SnO$_2$ nanorod@TiO$_2$ hybrid material for dye-sensitized solar cells. *J. Mater. Chem. A.* 2: 8266–8272.

Imahori, H., S. Hayashi, T. Umeyama, S. Eu, A. Oguro, S. Kang, Y. Matano, T. Shishido, S. Ngamsinlapasathian, and S. Yoshikawa. 2006. Comparison of electrode structures and photovoltaic properties of porphyrin-sensitized solar cells with TiO$_2$ and Nb, Ge, Zr-added TiO$_2$ composite electrodes. *Langmuir.* 22: 11405–11411.

Imahori, H., T. Umeyama, and S. Ito. 2009. Large π-aromatic molecules as potential sensitizers for highly efficient dye-sensitized solar cells. *Acc. Chem. Res.* 42: 1809–1818.

Islam, A., H. Sugihara, K. Hara, L.-P. Singh, R. Katoh, M. Yanagida, Y. Takahashi, S. Murata, and H. Arakawa. 2000. New platinum(II) polypyridyl photosensitizers for TiO$_2$ solar cells. *New J. Chem.* 24: 343–345.

Ito, S., N.-L. C. Ha, G. Rothenberger, P. Liska, P. Comte, S. M. Zakeeruddin, P. Péchy, M.-K. Nazeeruddin, and M. Grätzel. 2006. High-efficiency (7.2%) flexible dye-sensitized solar cells with Ti-metal substrate for nanocrystalline-TiO$_2$ photoanode. *Chem. Commun.* 38: 4004–4006.

Jayaweera, E. N., C. S. K. Ranasinghe, G. R. A. Kumara, W. M. N. M. B. Wanninayake, K. G. C. Senarathne, et al. 2015. Novel method to improve performance of dye-sensitized solar cells based on quasi-solid gel-polymer electrolytes. *Electrochim. Acta.* 152: 360–367.

Jayaweera, P. M., S. S. Palayangoda, and K. Tennakone. 2001. Nanoporous TiO$_2$ solar cells sensitized with iron(II) complexes of bromopyrogallol red ligand. *J. Photochem. Photobiol. A: Chem.* 140: 173–177.

Jiang, K.-J., J.-B. Xia, N. Masaki, S. Noda, and S. Yanagida. 2008. Efficient sensitization of nanocrystalline TiO$_2$ films with high molar extinction coefficient ruthenium complex. *Inorg. Chim. Acta.* 361: 783–785.

Jing, L., S. Ming-Xuan, Z. Xiao-Yan, and C. Xiao-Li. 2011. Counter electrodes for dye-sensitized solar cells. *Acta Phys. Chim. Sin.* 27: 2255–2268.

Joseph, J., K. M. Son, R. Vittal, W. Lee, and K.-J. Kim. 2006. Quasi-solid-state dye-sensitized solar cells with siloxane poly(ethylene glycol) hybrid gel electrolyte. *Semicond. Sci. Technol.* 21: 697.

Jovanovski, V., E. Stathatos, B. Orel, and P. Lianos. 2006. Dye-sensitized solar cells with electrolyte based on a trimethoxysilane-derivatized ionic liquid. *Thin Solid Films.* 511–512: 634–637.

Jun, H. K., M. A. Careem, and A. K. Arof. 2013. Quantum dot-sensitized solar cells-perspective and recent developments: A review of Cd chalcogenide quantum dots as sensitizers. *Renew. Sust. Energy Rev.* 22: 148–167.

Jung, H. S., J.-K. Lee, and M. Nastasi. 2005. Preparation of nanoporous MgO-coated TiO_2 nanoparticles and their application to the electrode of dye-sensitized solar cells. *Langmuir.* 21: 10332–10335.

Kalyanasundaram, K., and M. Grätzel. 1998. Application of functionalized transition metal complexes in photonic and optoelectronic devices. *Coord. Chem. Rev.* 177: 347–414.

Karthikeyan, C. S., and M. Thelakkat. 2008. Key aspects of individual layers in solid-state dye-sensitized solar cells and novel concepts to improve their performance. *Inorg. Chim. Acta.* 361: 635–655.

Kawano, R., H. Matsui, C. Matsuyama, A. Sato, M. A. B. H. Susan, N. Tanabe, and M. Watanabe. 2004. High performance dye-sensitized solar cells using ionic liquids as their electrolytes. *J. Photochem. Photobiol. A: Chem.* 164: 387–392.

Kay, A., and M. Grätzel. 1996. Low cost photovoltaic modules based on dye sensitized nanocrystalline titanium dioxide and carbon powder. *Solar Energy Mater. Solar Cells.* 44: 99–117.

Kohle, O., S. Ruile, and M. Grätzel. 1996. Ruthenium(II) charge-transfer sensitizers containing 4,4'-dicarboxy-2,2'-bipyridine. Synthesis, properties and bonding mode of coordinated thio- and selenocyanates. *Inorg. Chem.* 35: 4779–4787.

Komiya, R., L. Han, R. Yamanaka, A. Islam, and T. Mitate. 2004. Highly efficient quasi-solid-state dye-sensitized solar cell with ion conducting polymer electrolyte. *J. Photochem. Photobiol. A: Chem.* 164: 123–127.

Kong, E.-H., Y.-H. Yoon, Y.-J. Chang, and H. M. Jang. 2014. Hybrid photoelectrode by using vertically aligned rutile TiO_2 nanowires inlaid with anatase TiO_2 nanoparticles for dye-sensitized solar cells. *Mater. Chem. Phys.* 143: 1440–1445.

Kontos, A. G., T. Stergiopoulos, G. Tsiminis, Y. S. Raptis, and P. Falaras. 2008. In situ micro- and macro-Raman investigation of the redox couple behavior in DSSCs. *Inorg. Chim. Acta.* 361: 761–768.

Koo, H.-J., J. Park, B. Yoo, K. Yoo, K. Kim, and N.-G. Park. 2008. Size-dependent scattering efficiency in dye-sensitized solar cell. *Inorg. Chim. Acta.* 361: 677–683.

Krysova, H., J. Trckova-Barakova, J. Prochazka, A. Zukal, J. Maixner, and L. Kavan. 2014. Titania nanofiber photoanodes for dye-sensitized solar cells. *Catal. Today.* 230: 234–239.

Kuang, D., C. Klein, H. J. Snaith, R. Humphry-Baker, S. M. Zakeeruddin, and M. Grätzel. 2008. A new ion-coordinating ruthenium sensitizer for mesoscopic dye-sensitized solar cells. *Inorg. Chim. Acta.* 361: 699–706.

Kuang, D., C. Klein, Z. Zhang, S. Ito, J.-E. Moser, S.M. Zakeeruddin, and M. Grätzel. 2007. Stable, high-efficiency ionic-liquid-based mesoscopic dye-sensitized solar cells. *Small.* 3: 2094–2102.

Kubo, W., T. Kitamura, K. Hanabusa, Y. Wada, and S. Yanagida. 2002. Quasi-solid-state dye-sensitized solar cells using room temperature molten salts and a low molecular weight gelator. *Chem. Commun.* 4: 374–375.

Kumara, G. R. A., S. Kaneko, A. Konno, M. Okuya, K. Murakami, B. Onwona-agyeman, and K. Tennakone. 2006. Large area dye-sensitized solar cells: Material aspects of fabrication. *Prog. Photovoltaics: Res. Appl.* 14: 643–651.

Kumara, G. R. A., M. Okuya, K. Murakami, S. Kaneko, V. V. Jayaweera, and K. Tennakone. 2004. Dye-sensitized solid-state solar cells made from magnesium oxide-coated nanocrystalline titanium dioxide films: Enhancement of the efficiency. *J. Photochem. Photobiol. A: Chem.* 164: 183–185.

Lagref, J.-J., Md. K. Nazeeruddin, and M. Grätzel. 2008. Artificial photosynthesis based on dye-sensitized nanocrystalline TiO_2 solar cells. *Inorg. Chim. Acta.* 361: 735–745.

Lai, H., Y. Wang, G. Du, W. Li, and W. Han. 2014. Dual functional YVO4: Eu^{3+}, Bi^{3+}@SiO_2 submicron-sized core–shell particles for dye-sensitized solar cells: Light scattering and downconversion. *Ceram. Int.* 40: 6103–6108.

Lee, S., J. Y. Kim, K. S. Hong, H. S. Jung, J.-K. Lee, and H. Shin. 2006. Enhancement of the photoelectric performance of dye-sensitized solar cells by using a $CaCO_3$-coated TiO_2 nanoparticle film as an electrode. *Sol. Energy Mater. Solar Cells.* 90: 2405–2412.

Lee, W. J., E. Ramasamya, D. Y. Lee, and J. S. Song. 2008b. Performance variation of carbon counter electrode based dye-sensitized solar cell. *Sol. Energy Mater. Solar Cells.* 92: 814–818.

Lee, W., J. Y. Seng, and J.-I. Hong. 2013. Metal-free organic dyes with benzothiadiazole as an internal acceptor for dye-sensitized solar cells. *Tetrahedron.* 69: 9175–9182.

Lee, Y.-L., Y.-J. Shen, and Y.-M. Yang. 2008a. A hybrid PVDF-HFP/nanoparticle gel electrolyte for dye-sensitized solar cell applications. *Nanotechnology.* 19: 455201.

Li, B., G. Lü, L. Luo, and Y. Tang. 2010. TiO_2@MgO core-shell film: Fabrication and application to dye-sensitized solar cells. *Wuhan Univ. J. Nat. Sci.* 15: 325–329.

Li, P., Y. Duan, Q. Tang, B. He, and R. Li. 2015. An avenue of expanding triiodide reduction and shortening charge diffusion length in solid-state dye-sensitized solar cells. *J. Power Sources.* 273: 180–184.

Li, Q., Q. Tang, H. Chen, H. Xu, Y. Qin, B. He, et al. 2014. Quasi-solid-state dye-sensitized solar cells from hydrophobic poly(hydroxyethylmethacrylate/glycerin)/polyaniline gel electrolyte. *Mater. Chem. Phys.* 144: 287–292.

Li, S., Z. Chen, Z. Zhang, Y. Wang, B. Xu, T. Li, and W. Zhang. 2013. One-step electrophoretic deposition of magnesium oxide-coated tin oxide film and its application to flexible dye sensitized solar cells. *J. Electrochem. Soc.* 160: H513–H517.

Li, S.-L., K.-J. Jiang, K.-F. Shao, and L.-M. Yang. 2006. Novel organic dyes for efficient dye-sensitized solar cells. *Chem. Commun.* 2792–2794.

Li, Y., J. Hagen, W. Schaffrath, P. Otschik, and D. Haarer. 1999. Titanium dioxide films for photovoltaic cells derived from a sol-gel process. *Sol. Energy Mater. Solar Cells.* 56: 167–174.

Li. Z.-Y., M. S. Akhtar, J. H. Kuk, B.-S. Kong, and O.-B. Yang. 2012. Graphene application as a counter electrode material for dye-sensitized solar cell. *Mater. Lett.* 86: 96–99.

Lin, C.-C., H.-C. Chen, Y. L. Tsai, H.-V. Han, H.-S. Shih, Y.-A. Chang, et al. 2012. Highly efficient CdS-quantum-dot-sensitized GaAs solar cells. *Optics Express.* 20: A319–A326.

Lin, L.-Y., M.-H. Yeh, C.-P. Lee, J. Chang, A. Baheti, R. Vittal, et al. 2014. Insights into the co-sensitizer adsorption kinetics for complementary organic dye-sensitized solar cells. *J. Power Sources.* 247: 906–914.

Lindström, H., A. Holmberg, E. Magnusson, L. Malmqvist, and A. Hagfeldt. 2001. A new method to make dye-sensitized nanocrystalline solar cells at room temperature. *J. Photochem. Photobiol. A: Chem.* 145: 107–112.

Liu, Q.-P. 2014a. Analysis on dye-sensitized solar cells based on Fe-doped TiO_2 by intensity-modulated photocurrent spectroscopy and Mott–Schottky. *Chin. Chem. Lett.* 25: 953–956.

Liu, Q.-P. 2014b. Photovoltaic performance improvement of dye-sensitized solar cells based on Mg-doped TiO_2 thin films. *Electrochim. Acta.* 129: 459–462.

Lotey, G. S., and N. K. Verma. 2014. Synthesis and characterization of $BiFeO_3$ nanowires and their applications in dye-sensitized solar cells. *Mater. Sci. Semicond. Proc.* 21: 206–211.

Lu, L., R. Li, K. Fan, and T. Peng. 2010. Effects of annealing conditions on the photoelectrochemical properties of dye-sensitized solar cells made with ZnO nanoparticles. *Solar Energy.* 84: 844–853.

Luz, A., J. Conradt, M. Wolff, H. Kalt, and C. Feldmann. 2013. *p*-DSSCs with BiOCl and BiOBr semiconductor and polybromide electrolyte. *Solid State Sci.* 19: 172–177.

Ma, T., M. Akiyama, E. Abe, and A. Imai. 2005. High-efficiency dye-sensitized solar cell based on a nitrogen-doped nanostructured titania electrode. *Nano Lett.* 5: 2543–2547.

Manfredi, N., A. Bianchi, V. Causin, R. Ruffo, R. Simonutti, and A. Abbotto. 2014. Electrolytes for quasi solid-state dye-sensitized solar cells based on block copolymers. *J. Polym. Sci. A: Poly. Chem.* 52: 719–727.

Maruthamuthu, P., S. Fiechter, and H. Tributsch. 2006. Lipid (detergent)-based composite-dye solar cell. *C. R. Chimie.* 9: 684–690.

Massihi, N., M. R. Mohammadi, A. M. Bakhshayesh, and M. Abdi-Jalebi. 2013. Controlling electron injection and electron transport of dye-sensitized solar cells aided by incorporating CNTs into a Cr-doped TiO_2 photoanode. *Electrochim. Acta.* 111: 921–929.

Matsui, H., K. Okada, T. Kawashima, T. Ezure, N. Tanabe, R. Kawano, and M. Watanabe. 2004. Application of an ionic liquid-based electrolyte to a 100 mm × 100 mm sized dye-sensitized solar cell. *J. Photochem. Photobiol. A: Chem.* 164: 129–135.

Matsumura, M., S. Matsudaira, H. Tsubomura, M. Takata, and H. Yanagida, 1980. Dye sensitization and surface structures of semiconductor electrodes. *Ind. Eng. Chem. Prod. Res. Dev.* 19: 415–421.

Matteocci, F., S. Casaluci, S. Razza, A. Guidobaldi, T. M. Brown, A. Reale, et al. 2014. Solid state dye solar cell modules. *J. Power Sources.* 246: 361–364.

Mazille, F., Z. Fei, D. Kuang, D. Zhao, S. M. Zakeeruddin, M. Grätzel, and P. J. Dyson. 2006. Influence of ionic liquids bearing functional groups in dye-sensitized solar cells. *Inorg. Chem.* 45: 1585–1590.

Meng, Q.-B., K. Takahashi, X.-T. Zhang, I. Sutanto, T. N. Rao, O. Sato, et al. 2003. Fabrication of an efficient solid-state dye-sensitized solar cell. *Langmuir.* 19: 3572–3574.

Michaleviciute, A., M. Degbia, A. Tomkeviciene, B. Schmaltz, E. Gurskyte, J. V. Grazulevicius, J. Bouclé, and F. Tran-Van. 2014. Star-shaped carbazole derivative based efficient solid-state dye sensitized solar cell. *J. Power Sources.* 253: 230–238.

Mihalache, I., A. Radoi, M. Mihaila, C. Munteanu, A. Marin, M. Danila, et al. 2015. Charge and energy transfer interplay in hybrid sensitized solar cells mediated by graphene quantum dots. *Electrochim. Acta.* 153: 306–315.

Millington, K. R., K. W. Fincher, and A. L. King. 2007. Mordant dyes as sensitisers in dye-sensitised solar cells. *Sol. Energy Mater. Solar Cells.* 91: 1618–1630.

Miyasaka, T., Y. Kijitori, T. N. Murakami, M. Kimura, and S. Uegusa. 2002. Efficient nonsintering type dye-sensitized photocells based on electrophoretically deposited TiO_2 layers. *Chem. Lett.* 31: 1250–1251.

Mohmeyer, N., D. Kuang, P. Wang, H.-W. Schmidt, S. M. Zakeeruddin, and M. Grätzel. 2006. An efficient organogelator for ionic liquids to prepare stable quasi-solid-state dye-sensitized solar cells. *J. Mater. Chem.* 16: 2978–2983.

Moser, J. 1887. Note on reinforcement photoelectric currents by optical sensibilisirung. *Monatsh. Chem.* 8: 373.

Murakami, T. N., and M. Grätzel. 2008. Counter electrodes for DSC: Application of functional materials as catalysts. *Inorg. Chim. Acta.* 361: 572–580.

Nazeeruddin, M. K., F. D. Angelis, S. Fantacci, A. Selloni, G. Viscardi, P. Liska, et al. 2005. Combined experimental and DFT-TDDFT computational study of photoelectrochemical cell ruthenium sensitizers. *J. Am. Chem. Soc.* 127: 16835–16847.

Nazeeruddin, M., K. P. Péchy, and M. Grätzel. 1997. Efficient panchromatic sensitization of nanocrystalline TiO_2 films by a black dye based on atrithiocyanato-ruthenium complex. *Chem. Commun.* 1705–1706.

Nishikitani, Y., T. Kubo, and T. Asano. 2006. Modeling of photocurrent in dye-sensitized solar cells fabricated with PVDF-HFP-based gel-type polymeric solid electrolyte. *C. R. Chimie.* 9: 631–638.

Noh, Y., and O. Song. 2014. Properties of a Ru/Ti bilayered counter electrode in dye sensitized solar cells. *Electronic Mater. Lett.* 10: 271–273.

Olsen, E., G. Hagen, and S. E. Lindquist. 2000. Dissolution of platinum in methoxy propionitrile containing LiI/I_2. *Sol. Energy Mater. Solar Cells.* 63: 267–273.

O'Regan, B. C., and J. R. Durrant. 2009. Kinetic and energetic paradigms for dye-sensitized solar cells: Moving from the ideal to the real. *Acc. Chem. Res.* 42: 1799–1808.

O'Regan, B. C., and M. Grätzel. 1991. A low-cost, high-efficiency solar cell based on dye sensitized colloidal TiO_2 films. *Nature.* 353: 737–740.

Otaka, H., M. Kira, K. Yano, S. Ito, H. Mitekura, T. Kawata, and F. Matsui. 2004. Multi-colored dye-sensitized solar cells. *J. Photochem. Photobiol. A: Chem.* 164: 67–73.

Palomares, E., J. N. Clifford, S. A. Haque, T. Lutz, and J. R. Durrant. 2002. Slow charge recombination in dye-sensitised solar cells (DSSC) using Al_2O_3 coated nanoporous TiO_2 films. *Chem. Commun.* 14: 1464–1465.

Park, G. W., C. G. Hwang, J. W. Jung, and Y. M. Jung. 2012. Improvement in long-term stability and photovoltaic performance of UV cured resin polymer gel electrolyte for dye sensitized solar cell. *Bull. Korean Chem. Soc.* 33: 4093–4097.

Park, N.-G., J. V.-de Lagemaat, and A. J. Frank. 2000. Comparison of dye-sensitized rutile- and anatase-based TiO_2 solar cells. *J. Phys. Chem. B.* 104: 8989–8994.

Park, S. K., T. K. Yuna, J. Y. Baea, and Y. S. Wonb. 2013. Combined embedding of N-doping and $CaCO_3$ surface modification in the TiO_2 photoelectrodes for dye-sensitized solar cells. *App. Surf. Sci.* 285: 789–794.

Passoni, L., F. Ghods, P. Docampo, A. Abrusci, J. Martí-Rujas, M. Ghidelli, et al. 2013. Hyperbranched quasi-1D nanostructures for solid-state dye-sensitized solar cells. *ACS Nano.* 7: 10023–10031.

Paulsson, H., A. Hagfeldt, and L. Kloo. 2003. Molten and solid trialkylsulfonium iodides and their polyiodides as electrolytes in dye-sensitized nanocrystalline solar cells. *J. Phys. Chem. B* 107: 13665–13670.

Pazoki, M., P. W. Lohse, N. Taghavinia, A. Hagfeldt, and G. Boschloo. 2014. The effect of dye coverage on the performance of dye-sensitized solar cells with a cobalt-based electrolyte. *Phys. Chem. Chem. Phys.* 16: 8503–8508.

Pichot, F., J. R. Pitts, and B. A. Gregg. 2000. Low-temperature sintering of TiO_2 colloids: Application to flexible dye-sensitized solar cells. *Langmuir.* 16: 5626–5630.

Rai, P., R. Khan, K.-J. Ko, J.-H. Lee, and Y.-T. Yu. 2014. CeO2 quantum dot functionalized ZnO nanorods photoanode for DSSC applications. *J. Mater. Sci.: Mater. Electronics.* 25: 2872–2877.

Ramasamy, K., B. Tien, P. S. Archana, and A. Gupta. 2014. Copper antimony sulfide ($CuSbS_2$) mesocrystals: A potential counter electrode material for dye-sensitized solar cells. *Mater. Lett.* 124: 227–230.

Ranjusha, R. P., K. R. V. Lekha, Subramanian, V. Nair Shantikumar, and A. Balakrishnan. 2011. Photoanode activity of ZnO nanotube based dye-sensitized solar cells. *J. Mater. Sci. Technol.* 27: 961–966.

Rho, W.-Y., J.-W. Choi, H.-Y. Lee, S. Kyeong, S. H. Lee, H. S. Jung, et al. 2014. Dye-sensitized solar cells with silica-coated quantum dot-embedded nanoparticles used as a light-harvesting layer. *New J. Chem.* 38: 910–913.

Roh, D. K., W. S. Chi, H. Jeon, S. J. Kim, and J. H. Kim. 2014. High efficiency solid-state dye-sensitized solar cells assembled with hierarchical anatase pine tree-like TiO_2 nanotubes. *Adv. Funct. Mater.* 24: 379–386.

Rühle, S., M. Shalom, and A. Zaban. 2010. Quantum-dot-sensitized solar cells. *Chem. Phys. Chem.* 11: 2290–2304.

Saito, Y., W. Kubo, T. Kitamura, Y. Wada, and S. Yanagida. 2004. I^-/I_3^- redox reaction behavior on poly(3,4-ethylenedioxythiophene) counter electrode in dye-sensitized solar cells. *J. Photochem. Photobiol. A: Chem.* 164: 153–157.

Sakaguchi, S., H. Ueki, T. Kato, T. Kado, R. Shiratuchi, W. Takashima, K. Kaneto, and S. Hayase. 2004. Quasi-solid dye sensitized solar cells solidified with chemically cross-linked gelators: Control of TiO2/gel electrolytes and counter Pt/gel electrolytes interfaces. *J. Photochem. Photobiol. A: Chem.* 164: 117–122.

Sapp, S. A., C. M. Elliott, C. Contado, S. Caramori, and C. A. Bignozzi. 2002. Substituted polypyridine complexes of cobalt(II/III) as efficient electron-transfer mediators in dye-sensitized solar cells. *J. Am. Chem. Soc.* 124: 11215–11222.

Sarkar, S., A. Makhal, K. Lakshman, T. Bora, J. Dutta, and S. Kumar Pal. 2012. Dual-sensitization via electron and energy harvesting in CdTe quantum dots decorated ZnO nanorod-based dye-sensitized solar cells. *J. Phys. Chem. C.* 116: 14248–14256.

Sayama, K., H. Sugihara, and A. Hironori. 1998. Photoelectrochemical properties of a porous Nb_2O_5 electrode sensitized by a ruthenium dye. *Chem. Mater.* 10: 3825–3832.

Sayama, K., S. Tsukagoshi, K. Hara, Y. Ohga, A. Shinpou, Y. Abe, S. Suga, and H. Arakawa. 2002. Photoelectrochemical properties of J aggregates of benzothiazole merocyanine dyes on a nanostructured TiO_2 film. *J. Phys. Chem. B.* 106: 1363–1371.

Schmidt-Mende, L., U. Bach, R. Humphry-Baker, T. Horiuchi, H. Miura, S. Ito, S. Uchida, and M. Grätzel. 2005. Organic dye for highly efficient solid-state dye-sensitized solar cells. *Adv. Mater.* 17: 813–815.

Senevirathna, M. K. I., P. K. D. D. P. Pitigala, E. V. A. Premalal, K. Tennakone, G. R. A. Kumara, and A. Konno. 2007. Stability of the SnO_2/MgO dye-sensitized photoelectrochemical solar cell. *Sol. Energy Mater. Solar Cells.* 91: 544–547.

Shalan, A. E., and M. M. Rashad. 2013. Incorporation of Mn^{2+} and Co^{2+} to TiO_2 nanoparticles and the performance of dye-sensitized solar cells. *Appl. Surf. Sci.* 283: 975–981.

Shaterian, M., M. Barati, K. Ozaee, and M. Enhessari. 2014. Application of $MnTiO_3$ nanoparticles as coating layer of high performance TiO_2/$MnTiO_3$ dye-sensitized solar cell. *J. Ind. Eng. Chem.* 20: 3646–3648.

Shi, J., J. Chen, Y. Li, Y. Zhu, G. Xu, and J. Xu. 2015. Three-dimensional network electrolytes with highly efficient ion-transporting channels for quasi-solid-state dye-sensitized solar cells. *J. Power Sources.* 282: 51–57.

Shi, J., S. Peng, J. Pei, Y. Liang, F. Cheng, and J. Chen. 2009. Quasi-solid-state dye-sensitized solar cells with polymer gel electrolyte and triphenylamine-based organic dyes. *ACS Appl. Mater. Interfaces.* 1: 944–950.

Shinpuku, Y., F. Inui, M. Nakai, and Y. Nakabayashi. 2011. Synthesis and characterization of novel cyclometalatediridium(III) complexes for nanocrystalline TiO_2-based dye-sensitized solar cells. *J. Photochem. Photobiol. A: Chem.* 222: 203–209.

Smith, W. 1873. Effect of light on selenium during the passage of an electric current. *Nature.* 7: 303.

Sun, M., P. Song, J. Li, and X. Cui. 2013. Preparation, characterization and applications of novel carbon and nitrogen codoped TiO_2 nanoparticles from annealing TiN under CO atmosphere. *Mater. Res. Bull.* 48: 4271–4276.

Suzuki, K., M. Yamaguchi, S. Hotta, N. Tanabe, and S. Yanagida. 2004. A new alkyl-imidazole polymer prepared as an inonic polymer electrolyte by *in situ* polymerization of dye sensitized solar cells. *J. Photochem. Photobiol. A: Chem.* 164: 81–85.

Suzuki, K., M. Yamaguchi, M. Kumagai, N. Tanabe, and S. Yanagida. 2006. Dye-sensitized solar cells with ionic gel electrolytes prepared from imidazolium salts and agarose. *C. R. Chim.* 9: 611–616.

Suzuki, K., M. Yamaguchi, M. Kumagai, and S. Yanagida. 2003. Application of carbon nanotubes to counter electrodes of dye-sensitized solar cells. *Chem. Lett.* 32: 28–29.

Tamilavan, V., A.-Y. Kim, H.-B. Kim, M. Kang, and M. H. Hyun. 2014. Structural optimization of thiophene-(N-aryl)pyrrole-thiophene-based metal-free organic sensitizer for the enhanced dye-sensitized solar cell performance. *Tetrahedron.* 70: 371–379.

Tennakone, K., P. K. M. Bandaranayake, P. V. V. Jayaweera, A. Konno, and G. R. R. A. Kumara. 2002. Dye-sensitized composite semiconductor nanostructures. *Physica E.* 14: 190–196.

Tennakone, K., A. R. Kumarasinghe, G. R. R. A. Kumara, K. G. U. Wijayantha, and P. M. Sirimanne. 1997. Nanoporous TiO_2 photoanode sensitized with the flower pigment cyanidin. *J. Photochem. Photobiol. A: Chem.* 108: 193–195.

Thomas, S., T. G. Deepak, G. S. Anjusree, T. A. Arun, S. V. Nair, and A. S. Nair. 2014. A review on counter electrode materials in dye-sensitized solar cells. *J. Mater. Chem. A* 2: 4474–4490.

Tian, J., Q. Zhang, L. Zhang, R. Gao, L. Shen, S. Zhang, X. Qu, and G. Cao. 2013. ZnO/TiO$_2$ nanocable structured photoelectrodes for CdS/CdSe quantum dot co-sensitized solar cells. *Nanoscale.* 5: 936–943.

Tian, H., L. Hu, C. Zhang, W. Liu, Y. Huang, L. Mo, L. Guo, J. Sheng, and S. Dai. 2010. Retarded charge recombination in dye-sensitized nitrogen-doped TiO$_2$ solar cells. *J. Phys. Chem. C.* 114: 1627–1632.

Törngren, B., K. Akitsu, A. Ylinen, S. Sandén, H. Jiang, J. Ruokolainen, et al. 2014. Investigation of plasmonic gold–silica core–shell nanoparticle stability in dye-sensitized solar cell applications. *J. Colloid Interface Sci.* 427: 54–61.

Usui, H., H. Matsui, N. Tanabe, and S. Yanagida. 2004. Improved dye-sensitized solar cells using ionic nanocomposite gel electrolytes. *J. Photochem. Photobiol. A: Chem.* 164: 97–101.

Wang, C., F. Meng, T. Wang, T. Ma, and J. Qiu. 2014d. Monolithic coal-based carbon counter electrodes for highly efficient dye-sensitized solar cells. *Carbon.* 67: 465–474.

Wang, G., Z. Cai, F. Li, S. Tan, S. Xie, and J. Li. 2014a. 2% ZnO increases the conversion efficiency of TiO$_2$ based dye sensitized solar cells by 12%. *J. Alloy. Compd.* 583: 414–418.

Wang, H., W. Wei, and Y. H. Hu. 2013a. Efficient ZnO-based counter electrodes for dye-sensitized solar cells. *J. Mater. Chem. A.* 1: 6622–6628.

Wang, H.-T., D. K. Mishra, P. Chen, and J.-M. Ting. 2014b. p-Type dye-sensitized solar cell based on nickel oxide photocathode with or without Li doping. *J. Alloy. Compd.* 584: 142–147.

Wang, P., S. M. Zakeeruddin, P. Comte, I. Exnar, and M. Grätzel. 2003b. Gelation of ionic liquid-based electrolytes with silica nanoparticles for quasi-solid-state dye-sensitized solar cells. *J. Am. Chem. Soc.* 125: 1166–1167.

Wang, P., S. M. Zakeeruddin, I. Exnar, and M. Grätzel. 2002. High efficiency dye-sensitized nanocrystalline solar cells based on ionic liquid polymer gel electrolyte. *Chem. Commun.* 2972–2973.

Wang, P., S. M. Zakeeruddin, J.-E. Moser, and M. Grätzel. 2003a. A new ionic liquid electrolyte enhances the conversion efficiency of dye-sensitized solar cells. *J. Phys. Chem. B.* 107: 13280–13285.

Wang, P., S. M. Zakeeruddin, J.-E. Moser, R. Humphry-Baker, and M. Grätzel. 2004b. A solvent-free, SeCN$^-$/(SeCN)$_3^-$ based ionic liquid electrolyte for high-efficiency dye-sensitized nanocrystalline solar cells. *J. Am. Chem. Soc.* 126: 7164–7165.

Wang, Q., W. M. Campbell, E. E. Bonfantani, K. W. Jolley, D. L. Officer, P. J. Walsh, et al. 2005b. Efficient light harvesting by using green Zn-porphyrin-sensitized nanocrystalline TiO$_2$ films. *J. Phys. Chem. B.* 109: 15397–15409.

Wang, S. H., X. W. Zhou, X. R. Xiao, Y. Y. Fang, and Y. Lin. 2014c. An increase in conversion efficiency of dye-sensitized solar cells using bamboo-type TiO$_2$ nanotube arrays. *Electrochim. Acta.* 116: 26–30.

Wang, Y. F., J. M. Zhang, X. R. Cui, P. C. Yang, and J. H. Zeng. 2013b. A novel organic ionic plastic crystal electrolyte for solid-state dye-sensitized solar cells. *Electrochim. Acta* 112: 247–251.

Wang, Z. S., H. Kawauchi, T. Kashima, and H. Arakawa. 2004a. Significant influence of TiO$_2$ photoelectrode morphology on the energy conversion efficiency of N719 dye-sensitized solar cell. *Coord. Chem. Rev.* 248: 1381–1389.

Wang, Z.-S., K. Sayama, and H. Sugihara. 2005c. Efficient eosin Y dye-sensitized solar cell containing Br$^-$/Br$_3^-$ electrolyte. *J. Phys. Chem. B.* 109: 22449–22455.

Wang, Z.-S., T. Yamaguchi, H. Sugihara, and H. Arakawa. 2005a. Significant efficiency improvement of the black dye-sensitized solar cell through protonation of TiO$_2$ films. *Langmuir.* 21: 4272–4276.

Wolfbauer, G., A. M. Bond, J. C. Eklund, and D. R. MacFarlane. 2001. A channel flow cell system specifically designed to test the efficiency of redox shuttles in dye sensitized solar cells. *Sol. Energy Mater. Solar Cells.* 70: 85–101.

Wongcharee, K., V. Meeyoo, and S. Chavadej. 2007. Dye-sensitized solar cell using natural dyes extracted from rosella and blue pea flowers. *Sol. Energy Mater. Solar Cells.* 91: 566–571.

Wu, M., and T. Ma. 2014. Recent progress of counter electrode catalysts in dye-sensitized solar cells. *J. Phys. Chem. C.* 118: 16727–16742.

Wu, S., H. Han, Q. Tai, J. Zhang, S. Xu, C. Zhou, et al. 2008. Enhancement in dye-sensitized solar cells based on MgO-coated TiO$_2$ electrodes by reactive DC magnetron sputtering. *Nanotechnology.* 19: 215704.

Wu, W.-Q., Y.-F. Xu, H.-S. Rao, H.-L. Feng, C.-Y. Su, and D.-B. Kuang. 2014. Constructing 3D branched nanowire coated macroporous metal oxide electrodes with homogeneous or heterogeneous compositions for efficient solar cells. *Angew. Chem. Int. Ed.* 53: 4816–4821.

Xu, J., M. Li, L.Wu, Y. Sun, L. Zhu, S. Gu, et al. 2014. A flexible polypyrrole-coated fabric counter electrode for dye-sensitized solar cells. *J. Power Sources.* 257: 230–236.

Xu, P., W. Meng, F. X. Qing, Z. C. Neng, H. Z. Peng, and D. S. Yuan. 2013. Ionic liquid crystal-based electrolyte with enhanced charge transport for dye-sensitized solar cells. *Sci. China. Chem.* 56: 1463–1469.

Xue, Z., C. Jiang, L. Wang, W. Liu, and B. Liu. 2014. Fabrication of flexible plastic solid-state dye-sensitized solar cells using low temperature techniques. *J. Phys. Chem. C.* 118: 16352–16357.

Yang, J., C. Bao, K. Zhu, T. Yu, F. Li, J. Liu, et al. 2014. High catalytic activity and stability of nickel sulfide and cobalt sulfide hierarchical nanospheres on the counter electrodes for dye-sensitized solar cells. *Chem. Commun.* 50: 4824–4826.

Yodyingyong, S., Q. Zhang, K. Park, C. S. Dandeneau, X. Zhou, D. Triampo, et al. 2010. ZnO nanoparticles and nanowire array hybrid photoanodes for dye-sensitized solar cells. *Appl. Phys. Lett.* 96: 73115.

Yue, G.-T., J.-H. Wu, Y.-M. Xiao, J.-M. Lin, M.-L. Huang, L.-Q. Fan, et al. 2013. A dye-sensitized solar cell based on PEDOT: PSS counter electrode. *Chin. Sci. Bull.* 58: 559–566.

Yum, J.-H., P. Chen, M. Grätzel, and M.-K. Nazeeruddin. 2008. Recent developments in solid-state dye-sensitized solar cells. *ChemSusChem.* 1: 699–707.

Yum, J.-H., S.-S. Kim, D.-Y. Kim, and Y.-E. Sung. 2005. Electrophoretically deposited TiO_2 photo-electrodes for use in flexible dye-sensitized solar cells. *J. Photochem. Photobiol. A: Chem.* 173: 1–6.

Yum, J.-H., S. Nakade, D.-Y. Kim, and S. Yanagida. 2006. Improved performance in dye-sensitized solar cells employing TiO_2 photoelectrodes coated with metal hydroxides. *J. Phys. Chem. B.* 110: 3215–3219.

Yusoff, A., N. T. R. N. Kumara, A. Lim, P. Ekanayake, and K. U. Tennakoon, 2014. Impacts of temperature on the stability of tropical plants pigments as sensitizer for dye sensitized solar cells. *J. Biophys.* 2014: 739514.

Zhang, G., H. Bala, Y. Cheng, D. Shi, X. Lv, Q. Yu, and P. Wang. 2009. High efficiency and stable dye-sensitized solar cells with an organic chromophore featuring a binary π-conjugated spacer. *Chem. Commun.* 16: 2198–2200.

Zhang, D., S. M. Lanier, J. A. Downing, J. L. Avent, J. Lum, and J. L. McHale. 2008. Betalain pigments for dye-sensitized solar cells. *J. Photochem. Photobiol. A: Chem.* 195: 72–80.

Zhang, D. W., X. D. Lia, H. B. Lia, S. Chena, Z. Suna, X. J. Yinb, et al. 2011. Graphene-based counter electrode for dye-sensitized solar cells. *Carbon.* 49: 5382–5388.

Zhang, J., S. Najmaei, H. Lin, and J. Lou. 2014. MoS_2 atomic layers with artificial active edge sites as transparent counter electrodes for improved performance of dye-sensitized solar cells. *Nanoscale.* 6: 5279–5283.

Zheng, Y.-Z., H. Ding, X. Tao, and J.-F. Chen. 2014. Investigation of iodine dopant amount effects on dye-sensitized hierarchically structured ZnO solar cells. *Mater. Res. Bull.* 55: 182–189.

Zhou, H., L. Wu, Y. Gao, and T. Ma. 2011. Dye-sensitized solar cells using 20 natural dyes as sensitizers. *J. Photochem. Photobiol. A: Chem.* 219: 188–194.

7 Photogalvanic Cells

Yasmin, Abhilasha Jain, Pinki B. Punjabi, and Suresh C. Ameta

CONTENTS

7.1 Introduction 115
    7.1.1 Cell Efficiency 117
    7.1.2 Requirements for the Efficient Photogalvanic Cell 117
7.2 Dyes 119
7.3 Complexes 124
7.4 Miscellaneous 129
References 132

7.1 INTRODUCTION

The world is experiencing adverse consequences from using current commercial energy resources that are based on fossil and nuclear fuels. Attention needs to focus on research to find clean and renewable energy sources. One suitable source of energy in this context is photogalvanic cells. They are also called liquid-junction solar cells. They can generate electrical energy from solar energy as well as electrochemical energy and provide the basis for a system with an energy storage component.

Photogalvanic cells offer a promising area for exploration of the direct use of sunlight. Rideal and Williams (1925) first discovered the photogalvanic effect, but it was systematically investigated by Rabinowitch (1940a, 1940b) for an iron–thionine system. Weber and Matijević (1947) made an attempt to systematize the phenomena in terms of the *Becquerel effect*. Becquerel's photogalvanic effect has been thoroughly investigated on systems composed of different organic redox-dyes and organic acceptors, especially the speed of changes in potential and the influence of reducing and oxidizing agents.

Some photogalvanic cells using the iron–thionine system as the photosensitive fluid were built and tested to explore this suggestion. The observed maximum power conversion efficiency was $3 \times 10^{-4}\%$, depending on light absorbed. The principal reason for this low efficiency may be polarization of the polished platinum electrodes. Coating these electrodes with platinum black reduced polarization sufficiently; as a result, it was possible to achieve an efficiency of $6 \times 10^{-2}\%$, although this value was not actually observed. It may be possible to further increase efficiency by increasing electrode area and decreasing electrolyte resistance (Potter and Thaller 1959).

A photogalvanic device or cell is defined as a battery, where the cell solution absorbs light directly to generate photochemical species which, upon back-reaction through an external circuit in the presence of suitable electrodes, produce electrical power. It may store a significant amount of energy as chemical potential under open-circuit conditions and release it as electricity, when the external circuit is closed. This cell functions as a light recharged storage battery, if the chemical transformations occur in it without significant degradation over a number of cycles. Such photogalvanic cells based upon light-sensitive materials in solution are distinguished from photovoltaic cells, which are based on inorganic semiconductors (purely solid-state electronic devices). The photovoltaic devices depend on direct excitation of electrons and their separation from their geminate holes to produce electrical currents. They have a demerit that they lack inherent capacity for storage. All practical systems in use for direct conversion of sunlight into electricity utilize solid-state photovoltaic devices, particularly silicon p-n junctions.

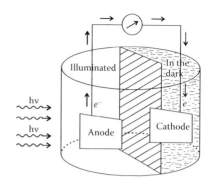

FIGURE 7.1 Photogalvanic cell.

In the conventional photogalvanic cell (Figure 7.1), a set of electrodes is immersed in an electrolyte solution, which contains some light-sensitive materials such as organic dye, complex, and so on. When the solution in the vicinity of one electrode in the photogalvanic cell is irradiated with light at a suitable wavelength, and the solution in the vicinity of the second electrode is kept in dark, then the different energy states of electrochemical equilibrium in the light and dark portions of the solution will develop a potential difference between these two electrodes. As a result, a shift in electrochemical equilibrium at the electrodes will generate an electrical current in the cell. Thus, the photogalvanic cell based on light-sensitive materials in electrolyte solution can be used indefinitely, because its function depends only on reversible shifts in electrical equilibrium in the cell solution in light and dark. In this cell, no material is consumed as fuel.

The following reactions take place in a photogalvanic cell at electrode and in solution:
In solution,

$$A + Z \underset{\text{Dark}}{\overset{h\nu}{\rightleftharpoons}} B + Y \tag{7.1}$$

At electrode,

$$B \rightleftharpoons A + e^- \tag{7.2}$$

$$Y + e^- \rightleftharpoons Z \tag{7.3}$$

Almost all photogalvanic cells are based upon these photochemical electron-transfer reactions, because these are relatively free from any irreversible side reactions and yield electrode-responsive products. The photoredox process involved is

$$A + Z \xrightarrow{h\nu} B + Y \, (\Delta G > 0) \tag{7.4}$$

The process may be initiated by absorption of light by A, Z, or a sensitizer. Then A/B and Z/Y are respective redox couples. The A is reduced to B by accepting an electron, while Z is oxidized. The reverse processes will take place in a reversible shift in the electrochemical equilibrium:

$$A + e^- \rightleftharpoons B \tag{7.5}$$

$$Y + e^- \rightleftharpoons Z \tag{7.6}$$

The standard redox potential (E°) of redox couple A/B must be less than that of the other redox couple Y/Z. The permissible electrode reactions are represented by Equations 7.5 and 7.6, although

Photogalvanic Cells

only certain combinations of these can produce electrical power. The cell reaction should ideally be the reverse of Equation 7.4:

$$B + Y \xrightarrow{K} A + Z (\Delta G > 0) \quad (7.7)$$

It should occur only at the electrode. This is a spontaneous process and occurs before B and Y can diffuse through solution and react at the electrodes. This homogeneous back-reaction may greatly reduce the efficiency of the cell. The photogalvanic cell will have good storage capacity, if the back-reaction is slow enough so that B and Y could be separated or it is so slow that B and Y could coexist in appreciable concentration in solution. In other words, light energy could be converted to chemical energy on a recharging process, and chemical energy to electrical energy on discharging process.

7.1.1 Cell Efficiency

It has been predicted theoretically that the power conversion efficiency of a photogalvanic cell could be appreciably as large as 18%, but the experimental conversion efficiencies are much lower than expected. The major reasons for this are back-electron transfer, low stability of dyes, and the aggregation of dye molecules around the electrode. It is very difficult to fulfill all the necessary conditions for better efficiency. It has been estimated that the maximum power conversion efficiency achieved from a photogalvanic cell lies somewhere between 5% and 9%.

Real systems will exhibit certain extrinsic, nonideal, and device-dependent losses, and as a result, such systems will exhibit comparatively lower efficiencies.

7.1.2 Requirements for the Efficient Photogalvanic Cell

The performance of photogalvanic cells for the direct conversion of solar energy to electrical energy depends upon several factors, for example, solution photochemistry, other homogeneous kinetics, mass transport, electrode kinetics, and load on the cell.

The electrode behavior should be selective, and the photogalvanic cells may be classified in two categories:

- Concentration cell, where the electrodes are identical, and both electrodes are reversible to one couple (*A/B*) and irreversible to the other (*Y/Z*). Here, the concentration ratio of B to A is different in two compartments (i.e., dark and illuminated chambers). Therefore, the cell yields power on irradiation. However, the maximum power point is quite close to the short-circuit condition, so it is not possible to draw significant currents from this type of concentration cell, and at the same time, it maintains very different concentrations at the two electrodes (Figure 7.2).
- Differential cell, where the illuminated electrode is reversible to *A/B* and irreversible to *Y/Z*, but the dark electrode is either reversible to *Y/Z* and irreversible to *A/B* or reversible to both couples. This type of cell gives much better performance in principle. If all the light is absorbed close to the illuminated electrode, then these two possibilities produce identical performances (Figure 7.3).

To be useful for conversion and storage of solar energy, there are some requirements for selecting a photochemical reaction:

- The reaction must be endergonic (energy storing).
- The reaction must be reversible (or cyclic). The backward reaction should be slow enough to make long-term storage possible under ambient conditions, but it should proceed under specific conditions to release the energy stored as and when needed.

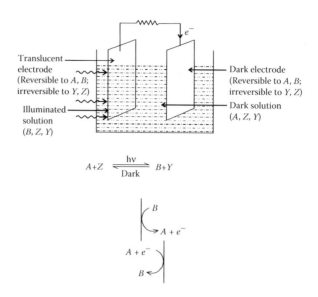

FIGURE 7.2 Concentration cell.

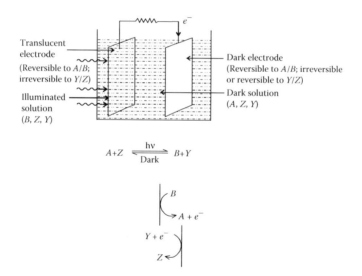

FIGURE 7.3 Differential cell.

- The side reactions must be absent, which may lead to the photochemical degradation (irreversible) of the reactants.
- Such photochemical reactions that proceed smoothly at any reasonable temperature higher than the ambient temperature should be investigated.
- It is not necessary for the system to absorb the light radiation directly. Under such circumstances, a sensitizer may be used with the absorption bands in the visible region and more preferably in the near-infrared region.
- The photoproducts of the reaction must be easily separable and should be convenient to store and/or transport.
- The reactants should be as low cost as possible, even though these are used again and again in a cyclic process with a forward photoreaction and a backward thermal reaction.

- The inorganic substrates are preferred over organic compounds, because the organic reaction is normally associated with some side reactions. The organic compounds are consumed during the progress of the reaction; thus, coordination compounds will have an edge over organic compounds in the future.
- The liquid and solid reactants and products are preferred over gaseous reactants because of size considerations; however, the gaseous products are not excluded.
- The reactants must be nontoxic, and the reactions should remain unaffected by the atmospheric oxygen.

Photosynthesis is the only photochemical storage system available in nature that can satisfy almost all of these conditions; despite efforts made by various scientists all over the world to mimic this process under laboratory conditions, this has not been accomplished. However, some, but not all, photovoltaic systems satisfy most of these requirements for direct conversion of solar energy into electricity.

A number of compounds have been investigated, and the search is still on, but no system could fulfill all of the desired conditions. The system fulfilling a majority of these conditions will be preferred over other systems for photochemical conversion and storage of solar energy. Different sensitizers, including dyes and complexes, have been used in photogalvanic cells by various workers.

7.2 DYES

Clark and Eckert (1975) reported a photogalvanic cell based on the iron–thionine system, where the power conversion efficiency for absorbed monochromatic light is 1.5%. Gomer (1975) carried out an analysis of photogalvanic cells and applied it to a specific system, iron–thionine, with only slight simplifications. A general treatment has been developed for calculating the change in potential at the point of zero current, when a photogalvanic cell containing two electrochemical redox couples (such as the iron–thionine system) was illuminated. Potential change on illumination from a thermodynamic dark potential to a mixed potential may be positive, negative, or zero. It depends on the balance between the electrode kinetics of the two couples in the solution. The rate of transport to the electrode and homogeneous back-reaction in solution also affects the magnitude of the potential shift (Albery and Archer 1976).

Sensitization of a totally illuminated thin-layer photogalvanic cell based on the iron-thiazine photoredox reaction has been demonstrated by both current action spectra and enhanced white light output. Photogalvanic output has been obtained upon illumination of a single solution containing two photoredox dyes and three sensitizers with monochromatic light in the wavelength range of ~375–700 nm. Sensitization with rhodamine 6 G considerably enhanced cell output under illumination with white light. However, white light output was inhibited with some sensitizers (Wildes et al. 1977).

Photoredox reactions and photogalvanic effects of gel systems containing thionine, Fe(II) salt, and gelatinizing agents were studied by using a thin-layer photocell, with SnO_2 and Pt electrodes. The photopotential and photocurrent of such a gel system depends on the characteristics of the gelatinizing agent. Their values were relatively large and did not decrease with illumination time, when the gelatinizing agent had intramacromolecular hydrogen bonds. The reason for these large values and good electron recycling inside the cell may be due to the hydrogen transfer–type electron-exchange reaction between Fe(III) and Fe(II). It was concluded that the primary structure of the gelatinizing agent was the most important factor determining the frequency of the electron-exchange reaction (Shigehara et al. 1978).

Tien et al. (1978) proposed and tested a novel photoelectrochemical cell based on a combined principle of photogalvanic and photovoltaic effects. The principal element of this cell consists of a pigmented membrane separating two aqueous solutions, one of which contains thionine dye and ferrous ions. It was observed that the photo-electromotive force (emf) generated across the cell was equal to the sum of the voltages derived from photogalvanic and photovoltaic processes. Hall et al.

(1977) reported a thin-layer photogalvanic cell based on the iron–thionine system with an efficiency of 0.03%. The functioning of this cell depends on the selectivity of the tin oxide anode, which responded preferentially to the thionine–leucothionine couple. The selective response of this electrode was explained in terms of electronic energy levels of semiconductor bands and redox species.

Albery et al. (1979a) used thionine-coated electrode and observed that up to 20 monolayers of thionine can be reversibly coated on platinum or tin oxide electrode. The kinetics of this thionine–leucothionine couple was reversible or quasi-reversible. It was shown that reduction of Fe(III) is much slower on a coated electrode, which means that thionine-coated electrode is one of the ideal electrodes for a cell.

Tsubomura et al. (1979) studied the photovoltages and photocurrents in various photogalvanic cells containing a dye (thionine, riboflavin, or proflavin) and a reducing agent (Fe^{2+}, hydroquinone, EDTA, or triethanolamine). The results indicate that those systems having EDTA or triethanolamine as a reducing agent gave much higher photocurrents than those having Fe^{2+} or hydroquinone. Such a difference may be due to irreversible reactions of EDTA or triethanolamine taking place after their photooxidation.

The theory of the operation of the ideal photogalvanic cell for solar energy conversion has been described, and the crucial kinetic characteristics that the system must possess were deduced for the homogeneous kinetics, the mass transfer, and the electrode kinetics (Albery and Foulds, 1979). Solar and Getoff (1979) investigated photophysical and chemical processes affecting the stability of the thiazine dye–iron system as a photogalvanic cell. Photoelectrochemical devices can be used under suitable conditions for the production of hydrogen.

Albery et al. (1979b) used the transparent disc electrode to investigate photogalvanic cells using three possible systems: the iron–thionine system and two iron–ruthenium systems. The modulated photocurrent was produced when the light source was modulated. Brokken-Zijp and De Groot (1980) analyzed the kinetics of four reactions of ferrous-thionine photogalvanic cell, which take place during the illumination. It was concluded that the pseudo-exponential description that was given in the literature is incorrect. Archer et al. (1980) reported the electrochemical behavior of the iron–thionine system in acid aqueous sulfate solution in dark. This study was carried out on the bare platinum electrode and the multilayer thionine-coated platinum. The ferric/ferrous couple has shown quasi-reversible electrode kinetics at the platinum covered with a submonolayer of thionine, which was reversibly adsorbed. It was found to be quite irreversible at a thionine-coated electrode.

Albery et al. (1980) synthesized and investigated six new modified thiazine dyes to meet the photochemical and kinetic requirements of the photogalvanic cell, because efficiency of the iron-thionine photogalvanic cell for solar energy conversion was severely limited due to the insolubility of thionine. They explained that for efficient energy conversion, it is also necessary to have a selective electrode and show that for the thionine-coated electrode it was suitable for the cell.

Fox and Singletary (1980) made use of chemically modified electrodes in dye-sensitized photogalvanic cells. They observed the photocurrents when arene-derivatized electrodes were used as anodes in an aqueous rhodamine B–hydroquinone photogalvanic cell. These are consistent with electron injection into the semiconductor from either an excited state or a reduced photoproduct. The attached molecules presumably function only as energy or electron relays.

Albery and Foulds (1981) studied the suitability of new methylene blue N,N as a dye for the photogalvanic cell. Low quantum efficiency of this type of cell was observed, which may be due to the dimerization in addition to the diffusion-controlled self-quenching by monomers and dimers. They concluded that this dye is unsuitable, and that self-quenching may be a serious problem for other Fe-thiazine systems. The open-circuit voltage of the ferrous-thionine photogalvanic cell has been reported (Brokken-Zijp et al. 1981). Here, Pt or Au electrode was used in the illuminated chamber. It was found that experimental open-circuit voltage versus light intensity curves were in agreement with the calculated values.

Albery et al. (1981) gave two reasons for the poor performance of the iron–thionine photogalvanic cell: (1) the low solubility of thionine, which means that the incident solar radiation was absorbed too far from the illuminated electrode; and (2) the formation of dimers, which do not undergo the desired photoredox reaction. They reported synthesis and photoelectrochemical characterization of several isomeric disulfonated thionines. All the isomers examined were sufficiently soluble to absorb solar radiation close to the illuminated electrode in a photogalvanic cell.

Tamilarasan and Natarajan (1981) observed that thionine dye oxidizes Fe^{2+} ions in solution on excitation by light, and the photoproducts undergo dark reactions to restore the starting materials. The photogalvanic (photovoltaic) potentials were generated by this cyclic process by a cell with two platinum electrodes, keeping one illuminated and the other in the dark chamber. Light was converted more effectively into electricity in the totally illuminated thin-layer cell with platinum and tin oxide electrodes, but the efficiency of the conversion was limited by energy-wasting back-reactions in solution and by a restriction on the concentration of the dye. They showed that these limitations can be overcome if thick films (~10 μm) of thionine condensed with macromolecules are coated onto inert electrodes.

Sharon et al. (1981) studied a thionine/Fe^{2+}/Fe^{3+} system and found it to be nonreversible after 30 min of exposure in solar radiation. A detailed study was made on a Fe^{2+}/Rh-B system. They reported that the cell is fully reversible in solar photons giving 40–35 mV potential. This cell can be charged in 29 min. The rate of discharge in the dark was slow. The mechanism of photoreaction of the role of rhodamine-B dye was discussed.

Murthy et al. (1982) determined the photovoltages and photocurrent in a photogalvanic cell containing flavin mononucleotide and a reducing agent, ethylenediaminetetraacetic acid. The efficiency of the cell has been estimated as 0.048%. Srivastava et al. (1982) studied the photogalvanic behavior of iron–thionine and iron–methylene blue systems in the presence of surfactants. Their data indicate that the surfactant micelles can be utilized for the storage of light energy. Murthy and Reddy (1983b) also studied the electrochemical behavior of the systems containing toluidine blue by cyclic voltammetry. The photovoltage and photocurrent of this cell containing dye toluidine blue and different reducing agents like EDTA, triethanol amine, triethylamine, and ferrous ions were measured, and it was found that photo-outputs with EDTA and amines were higher than with Fe(II).

Roy and Aditya (1983) set up a photoelectrochemical cell based on the photochemistry of anthraquinone-2-sulfonate (D). They reported that in the presence of formate at pH 11.0, D on illumination produces $D^-\bullet$ or D^{2-}. At the platinum electrode, the anodic reaction is $D^-\bullet \rightarrow D + e^-$, or $D^{2-} \rightarrow D + 2e^-$ in the absence of oxygen. At the dark electrode, the cathodic reaction is $O_2 + 2H_2O + 4e^- \rightarrow 4OH^-$. The open-circuit potential of the cell is 500 mV, and the short-circuit current is 180 μA. This cell has been recycled at least eight times. The efficiency of the cell increases with platinized platinum electrode in the dark chamber. The steady current under illumination is 65 μA with the same open-circuit voltage of 500 mV. The short-circuit current is 250 μA. The efficiency was even better with a CdS electrode in the illuminated chamber. Here, the open-circuit voltage is 560 mV. After charging the cell by illumination for 8 h, a steady current of 120 μA can be drawn from the cell, with illumination off, for 40 h. The short-circuit current of this cell was 450 μA. The maximum power output was found to be 4.2×10^{-6} W. The cell can be recycled at least four times without any loss in efficiency. Gray deposition on the CdS electrode indicates possible electrode decomposition.

The effect of temperature on phenosafranine-EDTA photogalvanic cell has been studied by Rohatgi-Mukherjee et al. (1983). They showed that the photovoltage and photocurrent increase linearly with increasing temperature, and the time to attain both the equilibrium values also diminished gradually. Photovoltage and photocurrent appear only in unstirred solutions. The power conversion efficiency was only 0.24 μW cm^{-2}, and the solar energy efficiency was 10^{-3}%. The current–potential curve indicates high activation overpotential, which is responsible for the low efficiency of the cell.

Pan et al. (1983) developed a photoelectrochemical cell involving thiazine photoreduction by EDTA as the electron donor together with photosynthetic electron transport in chloroplasts to generate electricity with a higher power conversion efficiency than that obtained for thiazine-EDTA alone. They reported that the thiazine photogalvanic cell can be combined with the chloroplast solar battery in order to increase the power conversion efficiency of the system.

The improved photo-output in the Fe-thionine photogalvanic cell with polypyrrole-coated selective electrodes was reported by Murthy and Reddy (1983a). It was found that these electrodes are sensitive to the ferrous/ferric couple only. Various dyes like thionine, oxonine, Nile blue A, and neutral red modified electrodes have been developed for the photogalvanic cell (Bauldreay and Archer 1983). It was found that the coating of these dyes forms more easily in acidic media than in aprotic solvents. Acridine yellow/EDTA/K_2PtCl_6 has proved to be an excellent system for the photoproduction of hydrogen (Bi and Tien 1984). Further enhancement can be achieved with the addition of Triton X-100 in this system. Naman and Karim (1984) studied photogalvanic cells using identical Pt electrodes for different dyes (thionine, methylene blue, eosin, lumiflavine, rhodamine 6G, and rhodamine B). They recorded the maximum sunlight engineering efficiency (SEE) for these cells and studied the possibility of a multidye photogalvanic cell.

A quantitative study of electrode selectivity in the performance of a Fe(II)/thionine cell has been carried out by Groenen et al. (1985) by using various carbon materials as illuminated electrodes. The open-circuit voltage of this photogalvanic cell depends strongly on the type of carbon electrode used. These electrodes generated high open-circuit voltage and were selective due to the suppression of the ferric reduction. The photogalvanic effect of phenosafranine (PSF) dye in an aqueous solution containing three types of surfactants—CTAB (cationic), SLS (anionic), and Triton X-100 (neutral)—has been investigated (Rohatgi-Mukherjee et al. 1985). It was found that only the PSF/Triton X-100 system showed a photogalvanic effect.

Aliwi et al. (1985) studied the photogalvanic effect in 13 different organic dyes and vanadium(III) tris(acetylacetonate) in acetonitrile aqueous solution in a photogalvanic cell. They observed the effect of acetonitrile concentration for these dyes (new methylene blue, thionine, methylene blue, eosin, safranine T, rhodamine B, eriochrome black T, Bengal rose, eriochrome blue black B, eriochrome red B, vitamin B_2, proflavine, and eriochrome blue black R). The sunlight engineering efficiencies (SEEs) were determined for each dye, and it was found that the best SEE was obtained when new methylene blue was used at concentration 1×10^{-5} M. The V(III) complex concentration was 1×10^{-4} M at pH 4, and the acetonitrile solution was 40%. They evaluate the reversibility of this system by monitoring the open-circuit current as a function of time.

Riefkohl et al. (1986) reported results on the dark electrochemistry and photoelectrochemistry of thionine and two of its disulfonated derivatives (2,6-DST and 4,6-DST) on the electrode (RODRE), which clearly indicate that coating the electrode with a thin layer of polymeric thionine is crucial to measuring sizable photoinduced ring currents. The parent thionine does not require the polymeric coat because, due to its adsorption and aggregation properties, a one- to two-monolayer film of aggregated molecules spontaneously coats the electrode. They reported that the measured efficiency of the parent iron–thionine system drops by a factor of 19 or more when semitransparent gold electrodes coated with thick polythionine layers were utilized in experimental photogalvanic cells. A phenosafranine-amine photoelectrochemical cell has been studied by Bhowmik and Roy (1987). The photovoltage generated in these systems was found to correlate with the ionization potential of the amine used.

The thionine-EDTA system has been used for solar energy conversion by Ameta et al. (1989a), where a photocurrent of 80 µA and photovoltage of 900 mV were obtained. The effects of various working parameters on the electric output of this cell have also been studied. The photogalvanic effect of the viologen in the presence of oxygen has been studied by Kaneko and Wohrle (1991). This viologen cation radical is formed electrochemically on the surface of indium tin oxide–coated electrode dipped in aqueous electrolyte. The mechanism for generation of photocurrent in the reaction of the photoexcited cation radical with oxygen was proposed.

A novel photoelectrochemical cell has been developed by Bhattacharya et al. (1991), which consists of dye photosafranine and EDTA in various surfactants solution, like CTAB (cationic), SLS (anionic), and Triton X-100 (neutral); this solution was separated from a saturated aqueous solution of iodine by Pyrex-sintered glass membrane.

Different redox couples have been used to construct a photogalvanic cell by Jana et al. (1993). In the construction of these types of cells, semiconductor (InO_3)/phenosafranine dye and EDTA aqueous solution were placed in one compartment of an H-shaped cell, and the other compartment contained Cu^+/Cu^{2+}, I^-/I_2, $Fe(CN)_6^{4-}/Fe(CN)_6^{3-}$, or Fe^{2+}/Fe^{3+} redox couples. These cells showed two to three times greater efficiency than the same cell with illuminated Pt electrode. A higher photovoltage was reported for a 10-layer cell arrangement of type SnO_2/10 mM phenosafranine + 10 mM brilliant cresyl blue + 0.5 EDTA/Pt. In a photogalvanic cell with tin oxide (working electrode) and SCE electrode, the photopotentials of aqueous solutions containing xanthenes dyes and electron donors were determined by Zhou et al. (1994). The hydrogen production was reported for the system containing an aqueous solution of erythrosine and triethanolamine. An aqueous solution of safranine T and triethanolamine was used in this cell, and it was found that this cell shows three to five times greater values of V_{oc} and 10 times greater values of I_{sc} compared to without safranine T.

The photoelectrochemical behavior of inclusion complexes of dye with β-cyclodextrin has been investigated by Raj and Ramaraj (1996). It was reported that the addition of β-cyclodextrin to the dye solution resulted in a twofold increase in the photogalvanic output for the dye-Fe^{2+} photoredox system. This can be attributed to the deaggregation of the photoinactive dimer dye into a photoactive monomer dye. The photoelectrochemical properties of thionine dye covalently attached to poly(acryloamidoglycolic acid) [P-(AGA)] were studied by Vishwanathan and Natarajan (1996). Here, the electron transfer processes are more reversible for [P-(AGA)]-TH^+ in homogeneous solution.

The photogalvanic effect of Fe(II)-β-diketonate/thionine systems in aqueous acetonitrile was investigated by Hamdi and Aliwi (1996). They studied the photogalvanic effect of ferrous bis-(acetylacetonate) $(Fe(II)(acac)_2)$ and ferrous bis-(trifluoroacetylacetonate) $(Fe(II)(tfac)_2)$ complexes in aqueous acetonitrile thionine dye solutions in a photogalvanic cell. They determined the theoretical sunlight engineering efficiency for both complexes and found that the best SEE was obtained when $Fe(II)(tfac)_2$ was used at a concentration of 1.5×10^{-4} mol/dm³ with a thionine concentration of 1×10^{-4} mol/dm³ at pH = 4 in 40% aqueous acetonitrile.

Cherepy et al. (1997) have constructed a photoelectrochemical cell by utilizing flavanoid anthocyanin dyes along with TiO_2 powder, which showed 0.56% efficiency under full sun. Jana and Bhowmik (1997) studied electrode kinetics of photoinduced redox reactions of phenazine dye-EDTA aqueous systems at different pH levels. They concluded that the photovoltage growth and decay curves follow the functional forms related to the relaxation times. The inverse of the relaxation time or the rate of electrode reaction was pH dependent under the experimental conditions.

Ghosh and Bhattacharya (2002) carried out spectral studies on the interaction of safranine T, phloxin, and fluorescein with inorganic ions and observed a photogalvanic effect. They reported that the system consisting of anionic dye fluorescein with inorganic ions in aqueous solution generates photovoltage in a photogalvanic cell. Fluorescein dye produces much greater photovoltage compared to other anionic dyes like phloxin (2,4,5,7-tetrabromo-3′,4′,5′,6′-tetrachlorofluorescein) and cationic dye safranine T (3,6-diamino-2,7-dimethyl-5-phenylphenazinium chloride). The open-circuit photovoltage, the short-circuit current, and the solar energy efficiencies of these systems were also determined.

The polarity of the microenvironments within a nafion (Nf) film was studied by electrochemical and photoelectrochemical techniques using a phenothiazine dye and thionine (TH) as a probe by Abraham and Ramaraj (2004). They observed that a cathodic photocurrent was observed when the Nf/TH film was exposed to visible light, whereas an anodic photocurrent was observed for the TBA-Nf/TH film (TBA, tetrabutylammonium). The observed difference in the polarity of the photocurrent reiterates the polar and less polar environments experienced by TH in Nf and TBA-Nf films, respectively. The polarity of the photocurrent changes from anodic to cathodic when the TBA-Nf/TH film is soaked in the photogalvanic cell solution for longer times.

Jana and Rajavenii (2004) reported that the phenozine dyes such as phenosafranine, safranine-O, and safranine-T form a 1:1 charge-transfer or electron donor–acceptor complex with Triton X-100. They observed that the photogalvanic and photoconductivity studies also support the above molecular interaction. Some anionic xanthene dyes show enhancement of fluorescence intensity with a red shift and develop photovoltage in a photoelectrochemical cell in the presence of Triton X-100 (Bhowmik and Ganguly 2005).

Ameta et al. (2006) studied the use of a bromophenol red-EDTA system for generation of electricity in a photogalvanic cell. The photopotential and photocurrent generated by this cell were 581.0 mV and 45 µA, respectively, while the use of mixed dyes toluidine blue and azur-B system and EDTA as a reductant in the photogalvanic cell for solar energy conversion and storage was investigated by Lal and Yadav (2007). The photocurrent and photopotential generated by this system were 40.0 µA and 802.0 mV, respectively. The observed conversion efficiency was 0.0708%, and maximum output of the cell was 32.08 µW. The cell can be used for 2 h in the dark. Genwa et al. (2009) studied the use of photosensitizers toluidine blue and malachite green in the presence of NaLS for photogalvanic solar energy conversion. They found that the conversion efficiency and storage capacity of the developed cells were 0.1448% and 123 min with the toluidine blue system, and were 0.059% and 32 min with the malachite green system.

Halls et al. (2012) studied wastewater as a photoelectrochemical fuel source. They developed a photogalvanic cell that employed 2,4-dicholorophenol as a fuel source, an N-substituted phenothiazine as the light harvester, and a sacrificial zinc anode. This cell has approximately 0.4% light to electrical power conversion efficiency in violet light. Mahmuoud et al. (2014) observed an improvement in a photogalvanic cell for solar energy conversion and storage using a rose Bengal-oxalic acid-tween 80 system and studied its commercial viability. The cell, as developed, can work for 230 min in the dark on irradiation for 72 min.

Azmat and Uddin (2012) observed photocurrent during the reduction of methylene blue with maltose into leucomethylene blue. The dye molecule presumably functions as an energy or electron relay. Absorption maxima, quantum yields, and lifetimes of the triplet transient species were reported. Results showed that transmitted light intensity depends upon the concentration of reducing sugar, solution acidity, and temperature.

Solid-state photogalvanic dye-sensitized solar cells were developed by Berhe et al. (2014). They examined thermal electron transfer from molecular sensitizers to nanostructured semiconductor electrodes composed of TiO_2 nanorods. They also studied electron-accepting molecular dyes along with an aryl amine as the electron donor. It was observed that the dyes operating by thermal injection into TiO_2 function work better in solid-state photoelectrochemical cells than in liquid-junction cells due to the kinetic advantage of solid-state cells with respect to photoinduced acceptor quenching to form the necessary radical anion sensitizers.

Genwa and Chouhan (2004) investigated the use of three heterocyclic dyes, azur A, azur B, or azur C, in photogalvanic cells for solar energy conversion and storage with a NaLS-ascorbic acid system. They reported the conversion efficiencies for azur A/azur B/azur C-NaLS-ascorbic acid systems to be 0.5461%, 0.9646%, and 0.4567% and storage capacities 110, 135, and 95 min, respectively.

On the basis of experimental observations, a general mechanism for a photogalvanic cell has been derived. The nature of the electroactive species was confirmed by the effect of diffusion length on i_{max} and i_{eq}. The reduced form of dye (semireduced or leuco form of dye) has been proposed as an electroactive species at the illuminated electrode (platinum electrode), while dye itself acted as an electroactive species in a dark chamber at the counterelectrode (SCE).

Many researchers have tried photogalvanic cells containing different combinations of dyes, reductants, and/or surfactants. Their results have been summarized in Table 7.1.

7.3 COMPLEXES

Hoffman and Lichtin (1979) examined the photochemical determinants of the efficiency of photogalvanic cell operation. They are (1) the absorption spectral characteristics of the cell solution, (2) the efficiency of the formation of separated charge carriers, and (3) the lifetimes of the carriers toward back-electron transfer. They achieved modulation of bulk solution dynamics by variation of the solution medium and discussed the photochemical determinants with particular reference to the use of thionine or $[Ru(bpy)_3]^{2+}$ as the light-absorbing species. Current–time and potential–time characteristics were measured, and the transient processes in the photogalvanic cell containing

TABLE 7.1
Some Photogalvanic Cells

System: Dye-Reductant-Surfactant	Photopotential (mV) Photocurrent (µA) Power (µW)	Conversion Efficiency (%) Fill Factor Performance in Dark (min)	Reference
Methylene blue-EDTA-NaLS	654 190 —	— — —	Ameta et al. (1989b)
Azur A-glucose-NaLS	811 1470 —	— — —	Khamesra et al. (1990)
Rhodamine 6G-oxalic acid-CTAB	414 90 25.22	0.55 0.45 —	Genwa and Mahaveer (2007)
Methyl orange-DTPA-brij-35	625 95 28.16	0.2707 0.40 94	Genwa and Khatri (2009)
Safranine O-EDTA-tween-80	785 300 101.6	0.9769 0.34 60	Gangotri and Gangotri (2009)
Rhodamine B-DTPA-NaLS	843 185 77.28	0.74 0.4955 85	Genwa and Kumar (2010)
Bromo cresol green-ascorbic acid-NaLS	834 350 83.52	0.80 0.23 140	Genwa and Singh (2013)
Rhodamine 6G-EDTA-NaLS	1162 510 131.6	1.26 — 168	Meena et al. (2013)
Biebrich scarlet-ascorbic acid-tween-60	919 210 93.15	0.8967 — 75	Genwa and Sagar (2013)
Rose bengal-oxalic acid-tween-80	666 119.5 22.30	0.258 0.328 230	Mahmoud et al. (2014)
Azur C-EDTA	879 75 —	— — —	Lodha et al. (1991)
Azur A-NTA	362 60 17.82	— — —	Dube et al. (1993)
Azur B-NTA	340 60 18.75	— — —	Dube et al. (1997)
Methylene blue-oxalic acid	312 110 12.6	0.211 0.28 35	Gangotri and Meena (2001)
Safranine T-I⁻	503 2.0 —	0.0031 — —	Ghosh and Bhattacharya (2002)

(Continued)

TABLE 7.1 (Continued)
Some Photogalvanic Cells

System: Dye-Reductant-Surfactant	Photopotential (mV) Photocurrent (µA) Power (µW)	Conversion Efficiency (%) Fill Factor Performance in Dark (min)	Reference
Phloxin-Br⁻	410	0.0027	
	2.0	—	
	—	—	
Fluorescein-Br⁻	1274	0.0059	
	4	—	
	—	—	
Azur B-NTA	996	0.1685	
	70	0.25	
	17.52	12	
Rose Bengal-EDTA	520	0.9706	Madhwani et al. (2007a)
	56	0.36	
	29.12	28	
Fuchsine basic-EDTA	650	—	Madhwani et al. (2007b)
	60	—	
	—	—	
Fluorescein-EDTA	418	—	Madhwani et al. (2007c)
	42	—	
	—	—	
Brilliant cresyl blue-fructose	—	—	Sharma et al. (2011)
	590	—	
	183.3	—	
Eosin-fructose	848	0.84	Gangotri and Bhimwal (2011)
	240	0.3404	
	87.52	55	
Fast Green FCF-fructose	1083	1.33	Koli (2014)
	431	—	
	138.6	70	
Azur-(mannitol + NTA)	347	—	Dube (1993)
	80	—	
	—	—	
(Toluidine blue + thionine)-EDTA	695	0.16	Lal and Gangotri (2011)
	105	—	
	72.9	42	
(Brilliant cresyl blue + toluidine)-ethylene glycol-NaLS	871	1.87	Gangotri and Mahawar (2012)
	630	—	
	548.73	70	
(Brilliant green + celestine blue)-EDTA	636	0.31	Yadav and Lal (2013)
	93	—	
	59.1	65	

Ru(bipy)$_3^{2+}$ and iron have been studied by Daul et al. (1980). A maximum was observed in the voltage-time curve when high light intensity and low to medium work load of the cell was employed.

A photogalvanic cell based on photolysis of rubidium anion in THF has been investigated by Goldstein et al. (1980). This photogalvanic cell was made up of two half-cells of type [Rb-THF]$_I$ and [Rb- THF, crown ether (or kryptand)]$_{II}$. Values of 30 nA photocurrent and 250 mV photovoltage were obtained by irradiation with the wavelength above 320 nm, whereas with the radiations with wavelength above 620 nm, only 6 nA photocurrent and 80 mV photovoltage were generated.

A model system for singlet state–driven photogalvanic cells has been reported involving photoreduction of 3,7-diaminophenoxazinylium chloride (oxonine) by iron(II) (Creed et al. 1981). Yamase and Ikawa (1980) observed the pH dependence of the alkylammonium molybdate system and production of hydrogen by a photogalvanic cell.

The alternating current photogalvanic cell has been developed by Daul et al. (1981). This cell has a $Ru(bipy)_3^{2+}/Fe^{3+}$ system and symmetrical electrode configuration. Here, the two electrodes were illuminated by light beams, alternatively, and resulted in almost rectangular current–time and voltage–time responses. This cell is stable and eliminates the disadvantages of slow diffusion processes, because mass transfer is not necessary.

Kawai and Yamamura (1982) have developed a thin-layer photocell, where the working electrode was made up of tin oxide modified with tris(2,2-bipyridine)ruthenium complex immobilized in a polyvinylcinnamate membrane. Platinum was used as the counterelectrode, and an aqueous solution containing the disodium salt of EDTA was used as an electrolyte solution. It was also reported that the addition of the tetracyanoquinodimethane salt (TCNQ) of poly-4-vinylpyridine and neutral tetracyanoquinodimethane enhanced the generation of photocurrent.

4-Phenylpyridine (as ancillary ligand) in Ru(II) polypyridyl complexes has been used by Garcia et al. (1998) for sensitization of the n-type TiO_2 electrode. Two novel molecular sensitizers, cis-$\{(dcbH_2)_2Ru(ppy)_2\}^{2+}$ and cis-$\{(dcbH_2)_2Ru(ppy)H_2O\}^{2+}$ were prepared, where $dcbH_2$ = 4,4'-$(CO_2H)_2$-2,2'-bipyridine and ppy = 4-phenylpyridine. They achieved a higher incident photon-to-current conversion efficiency value and broader spectral response in longer wavelengths using a derivative with one coordinated ppy, cis-$\{(dcbH_2)_2Ru(ppy)(H_2O^{2+})\}$. It was observed that the number of azine ligands coordinated to the nonattached side extends spectral sensitivity to the visible light.

The generation of photocurrent with the graphite electrodes coated with polymer films containing tris(2,2'-bipyridyl)Ru(II) and viologen was observed by Oyama et al. (1982). Hoffman and Sima (1983) examined a new technique for photochemical conversion of solar energy based on ligand photodissociation from metal complexes. The concept was illustrated with a photogalvanic cell, where voltages are generated by photodissociation of CO from carbonylferroheme, and with a cell, where the illuminated electrode was coated with an iron tetraphenylporphyrin. A singlet state–driven photogalvanic cell based on the photoreduction of 3,7-diaminophenoxazinylium chloride (oxonine) by iron(II) has been investigated by Fawcett et al. (1985).

The photoredox couple comprising vanadium(III) bis (2,2'-bipyridyl) chloride-Fe(III) also showed photogalvanic effect (Aliwi et al. 1986). It was reported that generation of photocurrent and photovoltage is directly proportional to the incident light intensity. Casado et al. (1990) investigated the photogalvanic behavior of $K_3Mn(CN)_6$ in aqueous cyanide solution. It was observed that irradiation of anodic solution containing Mn(II) complex with UV light resulted in increasing current. The stationary value of the photocurrent lasted for 6.5 h after putting off the light, when the current reached its maximum value.

Aliwi and Hanna (1987) determined new Z-values for 20 pure solvents and their aqueous mixtures at 25%, 50%, and 75% water by volume by using 1-methyl-4-carbomethoxy-pyridinium iodide at room temperature. They investigated the effect of pure solvents and their water mixtures on the VCl_3^--thionine photogalvanic system. They determined the voltage efficiency of these photogalvanic cells for each solution and correlated that with the Z-values. They reported that the highest voltage efficiency was obtained with acetonitrile and acetone and aqueous mixtures of these two solvents.

Operating characteristics of a photogalvanic cell using RS rubrum chromatophores have been studied by Erabi et al. (1987). They investigated operating characteristics of a photogalvanic cell using a Pt vertical chromatophore, m-PMS, ascorbate vertical SnO//z system. Open-circuit photovoltage and short-circuit photocurrent about 60 mV and 200–500 nA, respectively, were reported.

Photogalvanic generation of dihydrogen by water splitting using $[MoS_4]^{2-}$ as catalyst has been reported by Bhattacharyya et al. (1988). They achieved 4% solar energy efficiency for photogeneration of dihydrogen using $[MoS_4]^{2-}$ as catalyst, when the anode compartment of a photogalvanic cell was illuminated. The generated photocurrent rises slowly with time and reaches a limiting value. A secondary dark reaction sets in after 10 h, which produces H_2 and photocurrent, even when the light was switched off. They also suggested possible mechanisms in both of these cases.

The photogalvanic effect has been reported for alkylpyridinium salt in water or in water–methanol solution by Markov and Novkirishka (1991). This process was found to be reversible. Here, the value of the electrode potential depends on the type of salt and solvent used. It was observed that the absence of atmospheric oxygen in the irradiated solution affects the energy as well as potential of this photogalvanic cell.

Photogalvanic cells based on thionine-ferrous, $[Ru(bipy)_3]^{2+}/[PtCl_6]^{2-}$, $[Ru(5-Clphen)_3]^{2+}/Fe[(5-Clphen)_3]^{3+}$ systems were investigated by Titse et al. (1993). Various methods of electrode surface chemical modification were applied in order to transfer the photoelectrochemical processes in a solid phase. Gobi and Ramaraj (1993) studied the electron transfer reaction at Nafion and montmorillonite clay coated electrode containing $[Ru(bipy)_3]^{2+}$ in the presence of Fe^{3+} and $HClO_4$. These electrodes showed different photoelectrochemical properties depending on the coating materials. A novel photogalvanic cell has been constructed by using Pt/Nafion-$[Ru(bipy)_3]^{2+}$ and Pt/MM-$[Ru(bipy)_3]^{2+}$ as a photoanode and photocathode, respectively.

Gobi and Ramaraj (1994) also constructed a photogalvanic cell using Nafion and montmorillonite (MM) clay adsorbed $[Ru(bpy)_3]^{2+}$ coated electrodes in the presence of Fe^{3+} ions. They observed that the Nafion and MM clay adsorbed $[Ru(bpy)_3]^{2+}$ electrodes exhibited different photoelectrochemical behavior depending on the nature of the coating material. Anodic behavior was observed at a Nafion-$[Ru(bpy)_3]^{2+}$ electrode with reference to an inert electrode in a photogalvanic cell, while the MM clay-$[Ru(bpy)_3]^{2+}$ electrode showed cathodic polarity. A new photogalvanic cell was constructed by coating one electrode with Nafion-$[Ru(bpy)_3]^{2+}$ and the other electrode with MM clay-$[Ru(bpy)_3]^{2+}$. They concluded that this new photogalvanic cell showed an additive photogalvanic response on visible light irradiation.

Suresh et al. (1999) reported manganese-molybdenum-diethyldithiocarbamate complex as a potential system for solar energy conversion. They reported that the manganese-molybdenum-diethyldithiocarbamate $[MnMoO_2(Et_2dtc)_4(H_2O)]$ {Et_2dtc = diethyldithiocarbamate} complex exhibited reversible photogalvanic behavior in aqueous dimethylformamide medium in a Honda cell. The photogalvanic behavior was observed by varying the pH, temperature, and photosensitizers. UV, visible, and sunlight were used as the light sources. A potential of 345 mV was obtained at 80°C in visible light, and the system was found to be reversible for several cycles. They constructed a photoelectrochemical cell by coupling a charged nickel electrode with the complex electrode, which incorporates the experimental compound in acetylene black on a nickel substrate. They also reported a maximum potential of 1.08 V with 80 µA current, when irradiated with a tungsten lamp.

Suresh Raj et al. (2000) did photoelectrochemical studies on $[MnMoO_2(NCS)(Ox)_3(H_2O)_2]$ (Ox = 8-quinolinol) complex in aqueous dimethylformamide medium in a Honda cell. They reported that this system developed a maximum potential of 335 mV, when exposed to visible light at 30°C, and it was reversible. When a temperature difference was maintained between the illuminated and dark half-cells, this system generated 410 mV at 60°C. They also reported a solid-state galvanic cell, using the complex mixed with tetraethylammonium perchlorate (TEAP), which showed a maximum voltage of 25 mV. A sandwich galvanic cell, constructed from transparent tin oxide-glass/complex/platinum system was also developed with a maximum photovoltage of 88 mV, when irradiated with a tungsten halogen lamp.

Matsumoto et al. (2001) constructed a photogalvanic cell using a photosynthetic reaction center complex and cytochrome c_2 from *Rhodospirillum rubrum*. This photosynthetic reaction center complex was immobilized on a *p*-benzoquinonethiol-modified Au electrode. The potential of the reaction center–immobilized Au electrode in 0.1 M (M = mol/dm³) Tris-HCl buffer (pH 7.5) was shifted to negative direction upon illumination, and the value increased with increasing intensity of incident light. It was observed that the potential shift was enhanced by adding cytochrome c_2 to the solution. A photogalvanic cell was constructed using this electrode, cytochrome c_2, and an ITO

Photogalvanic Cells

electrode with hydrophilic surface. The cell performance was examined, where open-circuit photovoltage and short-circuit photocurrent were about 5 mV and 40 μA, respectively.

The photogalvanic behavior of $[Cr_2O_2S_2(1\text{-Pipdtc})_2(H_2O_2)]$ was investigated in a Honda cell using 100% DMF and different percentages of water–DMF systems (Pokhrel and Nagaraja 2009). The maximum potential of 200 mV with 18 μA current was generated in DMF. The system was found to be reversible when the irradiated solution was aerated immediately; the solution was found to be irreversible when it was kept in the dark for a long time (12 h) and then aerated. The Cr(V) is photoreduced to Cr(IV) with the light irradiation, and the unstable Cr(IV) reverts to Cr(V) by aerial oxidation.

Long-lived photoinduced charge separation in a $[Ru(Bpy)_3]^{2+}$/viologen system at the Nafion membrane-solution interface was observed by Yi et al. (2000). They examined the photoinduced electron transfer from the excited state of tris(2,2'-bipyridine)ruthenium(II) $[Ru(bpy)_3]^{2+}$ incorporated into Nafion membranes to propylviologen sulfonate (PVS) in the surrounding solution by photochemical and photoelectrochemical measurements. Here, N,N'-tetramethylene-2,2'-bipyridinium $[DQ]^{2+}$ entrapped in the Nafion membranes was used as an electron relay. An electrode was fabricated by coating $[Ru(bpy)_3]^{2+}$-$[DQ]^{2+}$-incorporated Nafion film on an ITO glass. The photoinduced voltage of this electrode was measured with a saturated calomel reference electrode in PVS solution to be ~350 mV, when the light intensity of the order ~60 mW cm^{-2} was used. This electrode was used as the illuminated electrode to construct a photogalvanic cell with a platinum electrode as the dark electrode. Irradiation of this electrode with visible light results in cathodic photocurrent. There is no net chemical change associated with the functioning of this cell, which converts light to electricity.

7.4 MISCELLANEOUS

The photoelectrolysis of water has been investigated using cells that consist of an illuminated n-TiO$_2$ (rutile) anode, an aqueous electrolyte, and a platinized-Pt cathode (Mavroides et al. 1975). It has been found that such cells operate either in the photogalvanic mode (no H$_2$ evolved) or in the photoelectrolytic mode (H$_2$ evolved at the cathode by decomposition of water). Maximum values of 80%–85% for the external quantum efficiency for current production in the photogalvanic mode have been measured at $h\nu \approx 4$ eV with single-crystal and polycrystalline TiO$_2$ anodes. Similar results were obtained in preliminary experiments with SrTiO$_3$ anodes. The internal quantum efficiencies, corrected for reflection and absorption losses, are close to 100%. This indicated that the band bending in TiO$_2$ under photogalvanic conditions was sufficient to separate the electron–hole pairs generated by photon absorption and also that the oxygen overvoltage for charge transfer at the semiconductor–electrolyte interface was negligible for illuminated anodes.

Fong and Winograd (1976) reported the photogalvanic effects arising from chlorophyll a-quinhydrone (Chl:H$_2$Q, 1:1) half-cell reactions using platinum electrodes. A photocurrent was developed in the presence of light as Chl a aggregates presumably undergo a charge-transfer interaction that results in the creation of a p-type semiconductor film. Photopotential at the Pt-Chl a electrode was positive. When the light was turned off, the half-cells regressed toward the preillumination conditions, and a reverse current was observed. The spectral response of the photogalvanic effect has also been determined. It was concluded that a distribution of Chl a-H$_2$O aggregates contributes to the observed photocurrent.

A general treatment has been developed for the current–voltage characteristics of photogalvanic cells containing two identical electrodes and operating by virtue of homogeneous reactions. This treatment involved the electrode and homogeneous kinetics, the relative magnitudes of the diffusion and reaction layers, and the position of the photostationary state. It was observed that power cannot be drawn from this cell if both the couples were either highly reversible or highly irreversible. The best performance was obtained if one couple is highly reversible, and the other is highly irreversible. If the homogeneous kinetics are rapid, an important contribution to the current delivered by the cell from catalytic currents due to the presence of Y and Z close to the electrode was observed (Albery and Archer 1977).

Zeichner et al. (1978) studied photogalvanic effects in the aqueous ferric bromide system. This system exhibited a considerable photogalvanic effect, with the illuminated half-cell normally positive in potential with respect to the nonilluminated one. They determined the photogalvanic open-circuit potential as a function of solution composition and incident light intensity. The results are in agreement with the kinetic mechanisms for the photolysis, subsequent reactions, and mixed potential measurements.

Mountz and Tien (1978) constructed a novel type of photoelectrochemical cell based on a unique combination of a photogalvanic and photovoltaic effect and demonstrated the feasibility of photo-galvano-voltaic cell for light conversion. Tien et al. (1979) also described a new type of electrochemical photocell. It was based on the combined principles of the photovoltaic (PV) and photogalvanic (PG) phenomena. This system had the advantages of both cells (PV and PG) and is, therefore, called the photogalvanovoltaic (PGV) cell. The key element of the cell responsible for the PV effect was a porphyrin-coated glassy carbon electrode. Pt or glassy carbon served as the counterelectrode.

Fox and Kabir-ud-Din (1979) developed a new carbanionic photogalvanic cell. Cyclooctatetraenyl dianion functioned as an electron source when excited with visible light at the surface of a TiO_2 semiconductor electrode in a liquid NH_3 photoelectrochemical cell. Since the conduction band of TiO_2 in ammonia is located below the oxidation potential of disodium or dipotassium salts of cyclooctatetraene at −0.55 eV versus Ag, the efficiency of the photoinduced electron transfer may depend on high-lying surface states, kinetic retardation of electron exchanges at the ammonia-semiconductor interface, or other relaxation phenomena.

The theory of transient processes in photogalvanic cells has been developed by Daul et al. (1979). The systems consisted of a photoactive species B, which can be oxidized (or reduced) to A in its excited state B^* by the one-electron acceptor Z, which itself is reduced (oxidized) to Y. It was based on an assumption that both redox couples undergo one-electron transfer reactions at the metal electrodes. The electrode kinetics for both redox couples was taken into account. A computer program was described, where data such as current, potential of the illuminated electrode, potential of the dark electrode, power, and the concentration profiles of A, B, Z, and Y near the electrodes can be obtained as functions of time, after illumination has started.

The relationship between open-circuit photovoltage and light intensity in photogalvanic cells has been studied by Quickenden and Yim (1979). All the expressions gave logarithmic relationships between photovoltage and light intensity when the concentration perturbations are small. Under certain other conditions, the logarithmic expressions approximate linear relationships.

Fox and Singletary (1980b) described the synthesis and the chemical and physical properties of some arene-derivatized tin oxide semiconductor electrodes. They employed several techniques for this attachment (esterification, silanation, silanation/amidation, and the use of cyanuric chloride as a linking agent). The benefits of each technique were evaluated. The relationship between the observed properties of the attached arenes and the potential utility of the derivatized electrodes as anodes in photogalvanic cells was discussed. Murthy et al. (1980) examined the photogalvanic cell, making use of the photoreduction of riboflavin with ethylenediaminetetraacetic acid. They observed an appreciable photoinduced voltage in this cell, but the photoinduced current was low. The estimated sunlight engineering efficiency has been estimated to be approximately 0.018%.

Haas (1980) described various processes in the photogalvanic cell and their possibilities in solar energy storage and also limitations of these cells. The photolysis of various hydrocarbon anions with the visible light leads to those photoreactions, which have applications in photochemical energy storage reactions and photogalvanic cells (Fox and Singletary 1980).

The photogalvanic effect has been observed by Katz et al. (1981) during the photooxidation of pheophytin a in dry acetone. It was observed that the photopotential can be increased by the addition of Mn^{2+} ions for this cell. Open-circuit photovoltage of 750 mV and short-circuit current of 100 µA cm^{-2} can be obtained with the photoanode on the base of pheophytin-sensitized oxidation of Mn^{2+} ions.

Stevenson and Erbelding (1981) reported a photogalvanic cell where visible light was converted into electricity via the photodissociation of iodine in aqueous and nonaqueous solutions. They explained a mechanism for the aqueous system that probably involves the formation of I_2^- or I radical, which is reduced at the transparent illuminated electrode. The efficiency was sensitive to formal concentration of I_2, and wavelength, and somewhat less on electrode material, with the highest efficiency observed (0.03%) in a cell containing indium tin oxide electrode at 4047 Å and 9F I_2, 3F NaI in acetonitrile.

Dixit and Mackay (1982) investigated the current and voltage responses of the totally illuminated thin-layer photogalvanic cell in microemulsions, micellar solution, and water. A significant enhancement of power output in anionic microemulsion was reported as compared to that in cationic microemulsion, micellar solution, and water. They also examined the power conversion efficiency of the photocell as a function of the various parameters, such as dye concentration, light intensity, pH, and distance between the electrodes. The highest percent solar engineering efficiency obtained was 0.33×10^{-3}.

Photogalvanic cells using heteropoly electrolytes have been developed by Papaconstantunou and Ioannidis (1983). These compounds undergo reduction on illumination with UV-vis light, whereas oxidation of organic species takes place. The potential difference has been developed between the photoreduced light half-cell and similar dark half-cell. Groenen et al. (1984) analyzed the effect of the nonionic amphiphile Triton X-100 micelles on the electrical performance of the ferrous/thionine photogalvanic cell. They reported that the addition of Triton X-100 micelles to the aqueous acidic cell solution leads to solubilization of the dye thionine in the outer poly(oxyethylene) spheres of the micelles. As a result, the solubility of thionine increases, and thermal back-reactions were suppressed, whereas the diffusion of species toward the electrodes slowed. An overall efficiency increase of fivefold has been reached relative to the photogalvanic cell, free of micelles.

Kamat (1985, 1986) studied photoelectrochemistry in colloidal systems using a photogalvanic cell based on TiO_2 semiconductor colloid. The feasibility of employing a colloidal semiconductor system in the operation of a photogalvanic cell was demonstrated. A photogalvanic power conversion efficiency of 0.002% was achieved with thionine and colloidal TiO_2 upon bandgap excitation. The use of TiO_2 colloid as a carrier for the deposition of a photoactive or an electroactive species on the electrode surface was discussed.

A riboflavin-diethanolamine system has also shown photogalvanic effects (Prajuntaboribal and Chaikum 1987). Photogalvanic properties of an Ag/Langmuir-Blodgett film/GaAs structure based on the fluorocarbon polymer have been studied by Znamensky et al. (1992). It was reported that the efficiency of such a type of structure with the thickness of 10–16 nm of Langmuir-Blodgett films was six times higher than simple Ag/GaAs diodes.

Markov et al. (1996) worked on photopotential and photocurrent induced by an aqueous solution of aliphatic alcohols. They reported the photogalvanic behavior of aliphatic alcohols in aqueous solution, where UV illumination caused the appearance of photopotential and photocurrent. The values of the electrode potential and energy of this photogalvanic cell depend on the type of alcohol used, its concentration, and the temperature and the pH of the irradiated solution. An important characteristic of the phenomenon observed was its reversibility. After discharge of a photogalvanic cell by application of an external load in the circuit, a subsequent irradiation again leads to the appearance of photoinduced electrode potential and photocurrent.

Mbindyo et al. (1997) prepared and evaluated the performance of a small laboratory-scale reactor employing a nanocrystalline titanium dioxide anode and a platinum black cathode for pollutant decomposition with simultaneous reduction of water to produce hydrogen. This reactor requires only light as an energy input, and it operates as a photogalvanic cell to produce electricity. They achieved oxidative photodegradation of some phenols: 4-chlorophenol, 2,4,5-trichlorophenol, and 4,4'-dichlorobiphenyl. It was observed that the solutions of 0.1 mM chlorophenols were decomposed in 2–3 h, with an average turnover of 5.4×10^{15} molecules cm^{-2} s^{-1}, whereas complete degradation of chlorophenols to carbon dioxide, water, and chloride ion was achieved in less than 6 h.

4,4′-Dichlorobiphenyl was poorly water soluble; hence, it was adsorbed onto soil and suspended in a pH 13 anode solution for its decomposition. Hydrogen gas was produced at the cathode at a rate 1.4 mL h^{-1} cm^{-2} using an anode solution of pH 13 and a cathode solution of pH 1. The average reactor potential under these conditions was 1.53 V, and the power output was 0.36 mW at a current density of 2 mA cm^{-2}.

Li et al. (2002) fabricated an electrode by transferring the Langmuir-Blodgett film on an ITO glass and used this electrode as the illuminated electrode in a photogalvanic cell with a platinum electrode as the dark electrode. They found that irradiation of this electrode with visible light results in anodic photocurrent, and there is no net chemical change associated with the function of the cell converting light to electricity.

Properties and applications of semiconductor-nanoparticle-donor nanochemical composites as energy sources of the future have been analyzed by Tsivadze et al. (2012). An idea of using nanofilm-nanoparticle-electron donor composites as the energy sources to substitute oil and gas has been developed. Electrons are knocked out from nanoparticles under the effect of visible light, and are absorbed by the nanofilm, which acts as a reservoir. Finally, these are attached to the donor, and then the cycle was repeated. They analyzed the electronic structure, properties, and interactions between all the nanocomposite components, as well as their interaction with the solvent and protective shell.

Halls and Wadhawan (2012) developed a lightweight, autonomous, and practical concept for an electrical power source. The quasi-biphasic, entirely new concept electrochemical cell, based an electron transfer rather than ion transfer, has been shown to act as a photogalvanic device, where violet light irradiation exhibited maximum light to electrical power conversion efficiency of ~2%. It has the additional ability to act as an electrically rechargeable electrochemical capacitor of voltage efficiency 85% and power efficiency ~80%.

Halls and Wadhawan (2013) also considered a mathematical model for a photosynthesis-inspired regenerative photogalvanic device. The performance of this system as a solar cell (for cells constructed from electrochemically reversible redox couples with fast photoinduced electron transfer reaction) is critically dependent on the concentration of the supersensitizer. Based on the numerical simulations, it has been suggested that this regenerative system has a solar-to-electrical power conversion efficiency of 4.5%, an attractive realistic single cell value.

Although the photogalvanic cells can be used in the dark for a relatively longer time (especially in the presence of surfactant in the form of micelles), the conversion efficiency of these systems is too low to make it commercially useful. The search is still on for newer photogalvanic cells with greater efficiency, but there are many more miles to go before such cells become economically viable and useful for such applications. However, such cells can be used for applications in electronic systems, where current at the microampere limit is required.

REFERENCES

Abraham, J. S., and R. Ramaraj. 2004. Microenvironment effects on the electrochemical and photoelectrochemical properties of thionine loaded Nafion films. *J. Electroanal. Chem.* 561: 199–126.

Albery, W. J., and M. D. Archer. 1976. Photogalvanic cells-I. The potential of zero current. *Electrochim. Acta.* 21: 1155–1163.

Albery, W. J., and M. D. Archer. 1977. Photogalvanic cells. 2. Current-voltage and power characterstics. *J. Electrochem. Soc.* 124: 688–697.

Albery, W. J., P. N. Bartlett, J. P. Davies, A. W. Foulds, A. R. Hillman, and F. S. Bachiller. 1980. New thiazine dyes for photogalvanic cells. *Faraday Discuss. Chem. Soc.* 70: 341–357.

Albery, W. J., P. N. Bartlett, A. W. Foulds, F. A. Souto-Bachiller, and R. Whiteside. 1981. Photogalvanic cells. Part 14. The synthesis and characterization of disulphonated thionines. *J. Chem. Soc., Perkin Trans.* 2: 794–800.

Albery, W. J., W. R. Bowen, F. S. Fisher, A. W. Foulds, K. J. Hall, A. R. Hillman, R.l G. Egdell, and A. F. Orchard. 1979a. Photogalvanic cells: Part XI. The thionine-coated electrode. *J. Electroanal. Chem. Interfacial Electrochem.* 107: 37–47.

Albery, W. J., W. R. Bowen, F. S. Fisher, and A. D. Turner. 1979b. Photogalvanic cells. Part IX. Investigations using the transparent rotating disc electrode. *J. Electroanal. Chem.* 107: 11–22.

Albery, W. J., and A. W. Foulds. 1979. Photogalvanic cells. *J. Photochem.* 10: 41–57.

Albery, W. J., and A. W. Foulds. 1981. Photogalvanic cells XIII: New methylene blue *NN*. *J. Photochem.* 15: 321–328.

Aliwi, S. M., and E. M. Hanna. 1987. New measured Z-values and their effects on the voltage efficiency of VCI_3^- thionine photogalvanic system. *Magallat buhut al-taqat al-Samsiyyat.* 5: 39–52.

Aliwi, S. M., S. A. Naman, and I. K. Al-Daghstani. 1985. Photogalvanic effect in organic dyes/vanadium (III) tris (acetylacetonate) in aqueous acetonitrile solution. *Magallat buhut al-taqat al-Samsiyyat.* 3: 49–61.

Aliwi, S. M., S. A. Naman, and I. K. Al-Daghstani. 1986. Photogalvanic effect in the vanadium(III) bis(2,2'-bipyridyl) chloride Fe(III) system. *Sol. Cells.* 18: 85–91.

Ameta, S. C., S. Khamesra, A. K. Chittora, and K. M. Gangotri. 1989b. Use of sodium lauryl sulphate in a photogalvanic cell for solar energy conversion and storage: Methylene blue-EDTA system. *Int. J. Energy Res.* 13: 643–647.

Ameta, S. C., S. Khamesra, S. Lodha, and R. Ameta. 1989a. Use of the thionine-EDTA system in photogalvanic cells for solar energy conversion. *J. Photochem. Photobiol. A: Chem.* 48: 81–86.

Ameta, S. C., P. B. Punjabi, J. Vardia, S. Madhwani, and S. Chaudhary. 2006. Use of bromophenol red–EDTA system for generation of electricity in a photogalvanic cell. *J. Power Sources.* 159: 747–751.

Archer, M. D., M. Isabel, C. Ferreira, W. J. Albery, and A. R. Hillman. 1980. Photogalvanic cells: Part XII. The dark electrochemistry of the iron-thionine system at platinum. *J. Electroanal. Chem. Interfacial Electrochem.* 111: 295–308.

Azmat, R., and F. Uddin. 2012. Photogalvanic effect of maltose/methylene blue system in aqueous methanol. *Asian J. Chem.* 24: 2833–2838.

Bauldreay, J. M., and M. D. Archer. 1983. Dye-modified electrodes for photogalvanic cells. *Electrochim. Acta.* 28: 1515–1522.

Berhe, S. A., H. B. Gobeze, S. D. Pokharel, E. Park, and W. J. Youngblood. 2014. Solid state photogalvanic dye sensitized solar cells. *ACS Appl. Mater. Interfaces.* 6: 10696–10705.

Bhattacharya, S., A. K. Jana, and B. B. Bhowmik. 1991. Storage solar cell consisting of phenosafranin-EDTA in surfactant solution. *J. Photochem. Photobiol. A: Chem.* 56: 81–87.

Bhattacharyya, R. G., D. P. Mandal, and K. K. Rohatgi Mukherjee. 1988. Photogalvanic generation of dihydrogen by water splitting using $[MoS_4]^{2-}$ as catalyst. *Bull. Mater. Sci.* 10: 373–379.

Bhowmik B. B., and P. Ganguly. 2005. Photophysics of xanthene dyes in surfactant solution. *Spectrochim. Acta A.* 61: 1997–2003.

Bhowmik, B. B., and S. Roy. 1987. A phenosafranin-amine photoelectrochemical cell. *Energy.* 12: 519–521.

Bi, Z.-C. and H. T. Tien. 1984. Photoproduction of hydrogen by dye-sensitized systems. *Int. J. Hydrogen Energy.* 9: 717–722.

Brokken-Zijp, J. C. M., and M. S. De Groot. 1980. The kinetics of chemical reactions in ferrous-thionine photogalvanic solutions. *Chem. Phys. Lett.* 76: 1–6.

Brokken-Zijp, J. C. M., M. S. De Groot, and P. A. J. M. Hendriks. 1981. The open-circuit voltage of the ferrous-thionine photogalvanic cell. *Chem. Phys. Lett.* 81: 129–135.

Casado, J., J. Peral, J. Balué, and X. Domenech. 1990. Photogalvanic behaviour of $K_3Mn(CN)_6$ in CN^- aqueous solutions. *Electrochim. Acta.* 35: 427–429.

Cherepy, N. J., G. P. Smestad, M. Grätzel, and J. Z. Zhang. 1997. Ultrafast electron injection: Implications for a photoelectrochemical cell utilizing an anthocyanin dye-sensitized TiO_2 nanocrystalline electrode. *J. Phys. Chem. B.* 101: 9342–9351.

Clark, W. D. K., and J. A. Eckert. 1975. Photogalvanic cells. *Sol. Energy.* 17: 147–150.

Creed, D., N. C. Fawcett, and R. L. Thompson. 1981. Photoreduction of 3,7-diaminophenoxazinylium chloride ('oxonine') by iron(II). A model system for singlet-state driven photogalvanic cells. *J. Chem. Soc., Chem. Commun.* 497–499.

Daul, C., O. Haas, A. Lottaz, A. V. Zelewsky, and H.-R. Zumbrunnen. 1980. Transient processes in photogalvanic cells: Part II. The Ru $(bipy)_3^{2+}/Fe^{3+}$ cell. *J. Electroanal. Chem. Interfacial Electrochem.* 112: 51–61.

Daul, C., O. Haas, and A. Von Zelewsky. 1979. Transient processes in photogalvanic cells. Part I. Fundamentals. *J. Electroanal. Chem.* 107: 49–58.

Daul, C., O. Haas, A. V. Zelewsky, and Hans-Rudolf Zumbrunnen. 1981. Transient processes in photogalvanic cells: Part III. The alternating current photogalvanic cell. *J. Electroanal. Chem. Interfacial Electrochem.* 125: 307–313.

Dixit, N. S., and R. A. Mackay. 1982. Microemulsions as photogalvanic cell fluids. The surfactant thionine-iron(II) system. *J. Phys. Chem.* 86: 4593–4598.

Dube, S. 1993. Simultaneous use of two reductants in a photogalvanic cell for solar-energy conversion and storage. *Int. J. Energy Res.* 17: 311–314.

Dube, S., A. Lodha, S. L. Sharma, and S. C. Ameta. 1993. Use of an Azur-A-NTA system in a photogalvanic cell for solar energy conversion. *Int. J. Energy Res.* 17: 359–363.

Dube, S., S. L. Sharma, and S. C. Ameta. 1997. Photogalvanic effect in azur B-NTA system. *Energy Conver. Manage.* 38: 101–106.

Erabi, T., T. Sengoku, K. Nishimura, and M. Wada. 1987. Operating characteristics of a photogalvanic cell using RS. Rubrum chromatophores. 1987. *Chem. Express.* 2: 455–458.

Fawcett, N. C., D. Creed, R. L. Thompson, and D. W. Presser. 1985. A singlet state- driven photogalvanic cell based on the photoreduction of 3,7-diaminophenoxazinylium chloride ('oxonine') by iron(II). *J. Chem. Soc., Chem. Commun.* 11: 719–720.

Fong, F. K., and N. Winograd. 1976. *In vitro* solar conversion after the primary light reaction in photosynthesis. Reversible photogalvanic effects of chlorophyll-quinhydrone half-cell reactions. *J. Am. Chem. Soc.* 98: 2287–2289.

Fox, M. A., and Kabir-ud-Din. 1979. A new carbanionic photogalvanic cell. *J. Phys. Chem.* 83: 1800–1801.

Fox, M. A., and N. J. Singletary. 1980. Solar energy utilization by carbanion photolysis. *Sol. Energy.* 25: 225–229.

Gangotri, K. M., and M. K. Bhimwal. 2011. The photochemical conversion of solar energy into electrical energy: Eosin-fructose system. *Environ. Prog. Sust. Energy.* 30: 493–499.

Gangotri, K. M., and A. K. Mahawar. 2012. Comparative study on effect of mixed photosensitizer system for solar energy conversion and storage: Brilliant cresyl blue + toluidine blue–ethylene glycol–NaLS system. *Environ. Prog. Sust. Energy.* 31: 474–480.

Gangotri, K. M., and R. C. Meena. 2001. Use of reductant and photosensitizer in photogalvanic cells for solar energy conversion and storage: Oxalic acid-methylene blue system. *J. Photochem. Photobiol. A: Chem.* 141: 175–177.

Gangotri, P., and K. M. Gangotri. 2009. Studies of the micellar effect on photogalvanics: Solar energy conversion and storage in EDTA-Safranine O-Tween-80 system. *Energy Fuels.* 23: 2767–2772.

Garcia, C. G., N. Y. Murakami Iha, R. Argazzi, and C. A. Bignozzi. 1998. 4-Phenylpyridine as ancillary ligand in ruthenium(II) polypyridyl complexes for sensitization of n-type TiO_2 electrodes. *J. Photochem. Photobiol. A: Chem.* 115: 239–242.

Genwa, K. R., and A. Chouhan. 2004. Studies of effect of heterocyclic dyes in photogalvanic cells for solar energy conversion and storage: NaLS-ascorbic acid system. *J. Chem. Sci.* 116: 339–345.

Genwa, K. R., and N. C. Khatri. 2009. Use of Brij-35-Methyl orange-DTPA system in photogalvanic cell for solar energy conversion and storage. *Indian J. Chem. Technol.* 16: 396–400.

Genwa, K. R., and A. Kumar. 2010. Role of Rhodamine B in photovoltage generation using anionic surfactant in liquid phase photoelectrochemical cell for solar energy conversion and storage. *J. Indian Chem. Soc.* 87: 933–939.

Genwa, K. R., A. Kumar, and A. Sonel. 2009. Photogalvanic solar energy conversion: Study with photosensitizers toluidine blue and malachite green in presence of NaLS. *Appl. Energy.* 86: 1431–1436.

Genwa, K. R., and Mahaveer. 2007. Role of surfactant in the studies of solar energy conversion and storage: CTAB-rhodamine 6G-oxalic acid system. *Indian J. Chem. Sec. A.* 46: 91–96.

Genwa, K. R., and C. P. Sagar. 2013. Energy efficiency, solar energy conversion and storage in photogalvanic cell. *Energy Conver. Manage.* 66: 121–126.

Genwa, K. R., and K. Singh. 2013. Use of bromocresol green-ascorbic acid-NaLS system in photogalvanic cell for solar energy conversion and storage. *J. Indian Chem. Soc.* 90: 813–819.

Getoff, N. 1990. Photoelectrochemical and photocatalytic methods of hydrogen production: A short review. *Int. J. Hydrogen Energy.* 15: 407–417.

Ghosh, J. K., and S. C. Bhattacharya. 2002. Spectral studies on the interaction of safranine T, phloxin, fluorescein with inorganic ions and their photogalvanic effect. *J. Indian Chem. Soc.* 79: 225–230.

Gobi, K. V., and R. Ramaraj. 1993. Electron transfer reactions at Nafion and clay adsorbed Ru $(bpy)^{2+}_3$ coated electrodes. *J. Mol. Catal.* 84: 187–192.

Gobi, K. V., and R. Ramaraj. 1994. Photoinduced electron transfer reactions of $[Ru(bpy)_3]^{2+}$ adsorbed onto Nafion® and clay coated electrodes in the presence of Fe^{3+} ions. *J. Electroanal. Chem.* 368: 75–85.

Goldstein, S., S. Jaenicke, and H. Levanon. 1980. Photogalvanic cell based on the photolysis of rubidium anions in THF. *Chem. Phys. Lett.* 71: 490–493.

Gomer, R. 1975. Photogalvanic cells. *Electrochim. Acta.* 20: 13–20.

Groenen, E. J. J., M. S. de Groot, and R. de Ruiter. 1985. Carbon electrodes in the ferrous/thionine photogalvanic cell: A quantitative study of electrode selectivity. *Electrochim. Acta.* 30: 1199–1204.

Groenen, E. J. J., M. S. De Groot, R. De Ruiter, and N. De Wit. 1984. Triton X-100 micelles in the ferrous/thionine photogalvanic cell. *J. Phys. Chem.* 88: 1449–1454.

Haas, O. 1980. Processes in photogalvanic cells and their possibilities in solar energy storage In J. Silverman (Ed.), *Energy Storage: A Vital Element in Mankind's Quest for Survival and Progress.* Oxford: Pergamon, pp. 433–435.

Hall, D. E., W. D. K. Clark, J. A. Eckert, N. N. Lichtin, and P. D. Wildes. 1977. Photogalvanic cell with semiconductor anode. *Am. Ceram. Soc. Bull.* 56: 408–411.

Halls, J. E., T. Johnson, A. A. Altalhi, and J. D. Wadhawan. 2012. Wastewater as a photoelectrochemical fuel source: Light-to-electrical energy conversion with organochloride remediation. *Electrochem. Commun.* 22: 4–7.

Halls, J. E., and J. D. Wadhawan. 2012. Photogalvanic cells based on lyotropic nanosystems: Towards the use of liquid nanotechnology for personalized energy sources. *Energy Environ. Sci.* 5: 6541–6551.

Halls, J. E. and J. D. Wadhawan. 2013. A model for efficient, semiconductor-free solar cells via supersensitized electron transfer cascades in photogalvanic devices. *Phys. Chem. Chem. Phys.* 15: 3218–3226.

Hamdi, S. T., and S. M. Aliwi. 1996. The photogalvanic effect of Fe(II)-β-diketonate/thionine systems in aqueous acetonitrile. *Monatsh. Chem.* 127: 339–346.

Hoffman, M. Z., and N. N. Lichtin. 1979. Photochemical determinants of the efficiency of photogalvanic conversion of solar energy. *Trans. J. Brit. Ceram. Soc.* 153–187.

Hoffman, B. M., and P. D. Sima. 1983. Solar energy conversion through ligand and photodissociation. *J. Am. Chem. Soc.* 105: 1776–1778.

Jana, A. K., and B. B. Bhowmik. 1997. Electrode kinetics of photoinduced redox reactions: Phenazine dye-EDTA aqueous systems at different pH. *J. Photochem. Photobiol. A: Chem.* 110: 41–46.

Jana, A. K., and S. Rajavenii. 2004. Studies on the molecular interaction of phenazine dyes with Triton X-100. *Spectrochim. Acta A.* 60: 2093–2097.

Jana, A. K., S. Roy, and B. B. Bhowmik. 1993. Photoelectrochemical cells consisting of phenosafranin-EDTA and different redox couples with illuminated semiconductor electrode. *Sol. Energy.* 51: 313–316.

Kamat, P. V. 1985. Photoelectrochemistry in colloidal systems. Part 2. A photogalvanic cell based on TiO_2 semiconductor colloid. *J. Chem. Soc., Faraday Trans. 1.* 81: 509–518.

Kamat, P. V. 1986. Erratum: Photoelectrochemistry in colloidal systems. Part 2. A photogalvanic cell based on TiO_2 semiconductor colloid. *J. Chem. Soc., Faraday Trans. 1.* 82: 1031.

Kaneko, M., and D. Wöhrle. 1991. Novel photogalvanic effect of the viologen cation radical formed electrochemically in the presence of oxygen. *J. Electroanal. Chem. Interfacial Electrochem.* 30: 209–215.

Katz, E. Y., Y. N. Kozlov, and B. A. Kiselev. 1981. Photoanode on the base of pheophytin-sensitized reactions. *Energy Convers. Manage.* 21: 171–174.

Kawai, W., and S. Yamamura. 1982. A thin layer photocell based on a tris(2,2'-bipyridine)ruthenium complex immobilized in a membrane containing an electronically conductive polymer. *J. Memb. Sci.* 12: 107–117.

Khamesra, S., R. Ameta, M. Bala, and S. C. Ameta. 1990. Use of micelles in photogalvanic cell for solar energy conversion and storage. Azur A-glucose system. *Int. J. Energy Res.* 14: 163–167.

Koli, P. 2014. Solar energy conversion and storage: Fast green FCF-fructose photogalvanic cell. *Appl. Energy.* 118: 231–237.

Lal, C., and K. M. Gangotri. 2011. Energy conversion and storage potential of photogalvanic cell based on mixed dyes system: Ethylene diaminetetraacetic acid-toluidine blue-thionine. *Environ. Prog. Sust. Energy.* 30: 754–761.

Lal, C., and S. Yadav. 2007. Use of mixed dyes in photogalvanic cell for solar energy conversion and storage: EDTA-toluidiene blue and azur-B system. *Asian J. Chem.* 19: 981–987.

Li, J. Y., M. L. Peng, L. Zhang, L. Zhu Wu, B. Wang, and C. Tung. 2002. Long-lived photoinduced charge separation in carbazole-pyrene-viologen system incorporated in Langmuir-Blodgett films of substituted diazacrown ethers. *J. Photochem. Photobiol.: A Chem.* 150: 101–108.

Lodha, S., S. Khamesra, B. Sharma, and S. C. Ameta. 1991. Use of an Azur C-EDTA system in a photogalvanic cell for solar energy conversion. *Int. J. Energy Res.* 15: 431–435.

Madhwani, S., R. Ameta, J. Vardia, P. B. Punjabi, and V. K. Sharma. 2007c. Fluoroscein-EDTA system in photogalvanic cell for solar energy conversion. *Energy Sources, Part A.* 29: 721–729.

Madhwani, S., S. Chaudhary, P. B. Punjabi, V. K. Sharma, and S. C. Ameta. 2007a. Use of rose bengal-EDTA system for the generation of electricity in a photogalvanic cell. *J. Indian Chem. Soc.* 84: 181–183.

Madhwani, S., J. Vardia, P. B. Punjabi, and V. K. Sharma. 2007b. Use of fuchsine basic: Ethylenediaminetetraacetic acid system in photogalvanic cell for solar energy conversion. *Proc. Inst. Mech. Engg. Part A: J. Power Energy.* 221: 33–39.

Mahmoud, S. A., B. S. Mohamed, A. S. El-Tabei, M. A. Hegazy, M. A. Betiha, H. M. Killa, E. K. Heikal, S. A. K. Halil, M. Dohium, and S. B. Hosney. 2014. Improvement of the photogalvanic cell for solar energy conversion and storage: Rose Bengal-oxalic acid-Tween 80 system. *Energy Procedia.* 46: 227–236.

Markov, P., and M. Novkirishka. 1991. On the photogalvanic properties of alkylpyridinium salts in solution. *Electrochim. Acta.* 36: 1287–1289.

Markov, P., M. Novkirishka, and K. Aljanapy. 1996. Photopotential and photocurrent induced by an aqueous solution of aliphatic alcohols. *J. Photochem. Photobiol. A: Chem.* 96: 161–165.

Matsumoto, K., S. Fujioka, Y. Mii, M. Wada, and T. Erabi. 2001. Construction of a photogalvanic cell using photosynthetic reaction center complex and cytochrome c_2 from *Rhodospirillum rubrum*. *Electrochem.* 69: 340–343.

Mavroides, J. G., D. I. Tchernev, J. A. Kafalas, and D. F. Kolesar. 1975. Photoelectrolysis of water in cells with TiO_2 anodes. *Mater. Res. Bull.* 10: 1023–1030.

Mbindyo, J. K. N., M. F. Ahmadi, and J. F. Rusling. 1997. Pollutant decomposition with simultaneous generation of hydrogen and electricity in a photogalvanic reactor. *J. Electrochem. Soc.* 144: 3153–3158.

Meena, A. S., Rishikesh, and R. C. Meena. 2013. Electrochemical studies of anionic and nonionic micelles with dyes and reductant in photogalvanic cell. *Int. J. Innov. Res. Sci. Eng. Technol.* 2: 6118–6123.

Mountz, J. M., and H. T. Tien. 1978. The photogalvanovoltaic cell. *Sol. Energy.* 21: 291–295.

Murthy, A. S. N., R. Bhargava, and K. S. Reddy. 1982. Flavin mononucleotide (FMN)-ethylenediaminetetra acetic acid (EDTA) photogalvanic cell. *Int. J. Energy Res.* 6: 389–395.

Murthy, A. S. N., H. C. Dak, and K. S. Reddy. 1980. Photogalvanic effect in riboflavin-ethylenediaminetetraacetic acid system. *Int. J. Energy Res.* 4: 339–343.

Murthy, A. S. N., and K. S. Reddy. 1983a. Polypyrrole coated selective electrodes for iron-thionine photogalvanic cell. *Electrochim. Acta.* 28: 473–476.

Murthy, A. S. N., and K. S. Reddy. 1983b. Studies on photogalvanic effect in systems containing toluidine blue. *Sol. Energy.* 30: 39–43.

Naman, S. A., and A. S. R. Karim. 1984. Efficiency of photogalvanic cells with various dyes. *J. Sol. Energy Res.* 2: 31–41.

Oyama, N., S. Yamaguchi, M. Kaneko, and A. Yamada. 1982. Photocurrent generation of graphite electrodes coated with polymer films confining tris(2,2′-bipyridyl) ruthenium (II) and viologen. *J. Electroanal. Chem. Interfacial Electrochem.* 139: 215–222.

Pan, R. L., R. Bhardwaj, and E. L. Gross. 1983. Photochemical energy conversion by a thiazine photosynthetic-photoelectrochemical cell. *J. Chem. Technol. Biotechnol. Chem. Technol.* 33 A: 39–48.

Papaconstantinou, E., and A. Ioannidis. 1983. Photogalvanic cells using heteropoly electrolytes. *Inorg. Chim. Acta.* 75: 235–236.

Pokhrel, S., and K. S. Nagaraja. 2009. Photogalvanic behaviour of $[Cr_2O_2S_2(1\text{-Pipdtc})_2(H_2O)_2]$ in aqueous DMF. *Sol. Energy Mater. Solar Cells.* 93: 244–248.

Potter, Jr., A. E., and L. H. Thaller. 1959. Efficiency of some iron-thionine photogalvanic cells. *Sol. Energy.* 3: 1–7.

Prajuntaboribal, K., and N. Chaikum. 1987. Photogalvanic effect in the riboflavin-diethanolamine system. *Sol. Energy.* 38: 149–153.

Quickenden, T. I., and G. K. Yim. 1979. The relationship between open circuit photovoltage and light intensity in photogalvanic cells—An extension of Albery and Archer's treatment. *Electrochim. Acta.* 24: 143–146.

Rabinowitch, E. 1940a. The photogalvanic effect I: The photochemical properties of the thionine-iron system. *J. Chem. Phys.* 8: 551–559.

Rabinowitch, E. 1940b. The photogalvanic effect II. The photogalvanic properties of the thionine-iron system. *J. Chem. Phys.* 8: 560–566.

Raj, C. R., and R. Ramaraj. 1996. Electrochemistry and photoelectrochemistry of phenothiazine dye-β-cyclodextrin inclusion complexes. *J. Electroanal. Chem.* 405: 141–147.

Rideal, E. K., and E. G. Williams. 1925. The action of light on the ferrous iodine iodide equilibrium, *J. Chem. Soc. Trans.* 127: 258–269.

Riefkohl, J., L. Rodriquez, L. Romero, and F. Souto. 1986. Photoelectrochemistry with the rotating optical disk ring electrode: II. The thionine-coated electrode is selective to the leuco forms of the parent thionine and the 2,6- and 4,6-disulfonated thionine derivatives. *J. Electrochem. Soc.* 133: 1828–1834.

Rohatgi-Mukherjee, K. K., R. Chaudhuri, and B. B. Bhowmik. 1985. Molecular interaction of phenosafranin with surfactants and its photogalvanic effect. *J. Colloid Interface Sci.* 106: 45–50.

Rohatgi-Mukherjee, K. K., M. Roy, and B. B. Bhowmik. 1983. Photovoltage generation in the solid phenosafranine-EDTA sandwich cell. *Solar Energy.* 31: 417–419.

Roy, A., and S. Aditya. 1983. A novel photogalvanic cell using anthraquinone-2-sulfonate. *Int. J. Hydrogen Energy.* 8: 91–96.

Sharma, U., P. Kohli, and K. M. Gangotri. 2011. Brilliant cresyl blue-fructose for enhancement of solar energy conversion and storage capacity of photogalvanic solar cells. *Fuel.* 90: 3336–3342.

Sharon, M., S. G. Sharan, A. Sinha, and B. M. Prasad. 1981. Saur vidyut kosh-solar photogalvanic cell-II. *J. Electrochem. Soc. India.* 30: 200–203.

Shigehara, K., M. Nishimura, and E. Tsuchida. 1978. Photogalvanic effect of thin-layer photocells composed of thionine/Fe(II) systems. *Electrochim. Acta.* 23: 855–860.

Solar, S., and N. Getoff. 1979. Photophysical and chemical processes affecting the stability of the thiazine dye-iron system. *Int. J. Hydrogen Energy.* 4: 403–410.

Srivastava, R. C., P. R. Marwadi, P. K. Latha, and S. B. Bhise. 1982. Solar energy storage using surfactant micelles. *Int. J. Energy Res.* 6: 247–251.

Stevenson, K. L., and W. F. Erbelding. 1981. A photogalvanic cell utilizing the photodissociation of iodine in solution. *Sol. Energy.* 27: 139–141.

Suresh Raj, A. M. E., J. Pragasam, F. P. Xavier, and K. S. Nagaraja. 2000. Photoelectrochemical studies on [$MnMoO_2(NCS)(Ox)_3(H_2O)_2$] (Ox = 8-quinolinol): A novel system for solar energy conversion. *Int. J. Energy Res.* 24: 1351–1358.

Suresh, E., J. Pragasam, F. P. Xavier, and K. S. Nagaraja. 1999. Investigation of manganese-molybdenum-diethyldithiocarbamate complex as a potential system for solar energy conversion. *Int. J. Energy Res.* 23: 229–233.

Tamilarasan, R., and P. Natarajan. 1981. Photovoltaic conversion by macromolecular thionine films. *Nature.* 292: 224–225.

Tien, H. T., J. Higgins, and J. Mountz. 1979. Photogalvanovoltaic cells and photovoltaic cells using glassy carbon electrodes. *Trans. J. Br. Ceram. Soc.* 203–235.

Tien, T. H., J. M. Mountz, and M. James. 1978. Photo-galvano-voltaic cell: A new approach to the use of solar energy. *Int. J. Energy Res.* 2: 197–200.

Titse, A. M., A. M. Timonov, and G. A. Shagisultanova. 1993. Photosensitive chemically modified electrodes for photogalvanic cells. *Coord. Chem. Rev.* 125: 43–52.

Tsivadze, A. Y., G. V. Ionova, V. K. Mikhalko, and I. S. Ionova. 2012. Semiconductor-nanoparticle-donor nanochemical composites: Properties and applications as energy sources of the future. *Prot. Metals Phys. Chem. Surfaces.* 48: 1–26.

Tsubomura H., Y. Shimoura, and S. Fujiwara. 1979. Chemical processes and electric power in photogalvanic cells containing reversible or irreversible reducing agents. *J. Phys. Chem.* 83: 2103–2106.

Viswanathan, K., and P. Natarajan. 1996. Studies on the photoelectrochemical properties of thionine dye covalently bound to poly (acrylamidoglycolic acid). *J. Photochem. Photobiol. A: Chem.* 95: 255–263.

Weber, K., and E. Matijević. 1947. Über photogalvanische Erscheinungen bei organischen Redoxsystemen. *Experientia.* 3: 280–281.

Wildes, P. D., D. R. Hobart, N. N. Lichtin, D. E. Hall, and J. A. Eckert. 1977. Sensitization of an iron-thiazine photogalvanic cell to the blue: An improved match to the insolation spectrum. *Sol. Energy.* 19: 567–570.

Yadav, S., and C. Lal. 2013. Optimization of performance characteristics of a mixed dye based photogalvanic cell for efficient solar energy conversion and storage. *Energy Conver. Manage.* 66: 271–276.

Yamase, T., and S. Ikawa. 1980. Production of hydrogen by photogalvanic cell. Part 2. pH dependence of alkylammonium molybdate system. *Inorg. Chim. Acta.* 45: L55–L57.

Yi, X.-Y., L.-Z. Wu, and C.-H. Tung. 2000. Long-lived photoinduced charge separation in [$Ru(Bpy)_3$]$^{2+}$/viologen system at nafion membrane-solution interface. *J. Phys. Chem. B.* 104: 9468–9474.

Zeichner, A., J. R. Goldstein, and G. Stein. 1978. Photogalvanic effect in ferric bromide solutions. *J. Phys. Chem.* 82: 1687–1692.

Zhou, R-L., Y. G. Yang, and Y. Y. Han. 1994. The study of photoelectrochemical cells based on chlorophyll a and safranine T. *J. Photochem. Photobiol. A: Chem.* 81: 59–63.

Znamensky, D. A., R. G. Yusupov, and B. V. Mislavsky. 1992. Langmuir-Blodgett mono- and multilayers of fluorocarbon amphiphilic polymers and their application in photogalvanic metal-insulator-semiconductor structures. *Thin Solid Films.* 219: 215–220.

8 Hydrogen
An Alternative Fuel

Neelu Chouhan, Rajesh Kumar Meena, and Ru-Shi Liu

CONTENTS

- 8.1 Introduction 140
- 8.2 Hydrogen 141
  - 8.2.1 Chemical Fuel 141
  - 8.2.2 Hydrogen as a Sustainable and Clean Energy Source 142
  - 8.2.3 Hydrogen Fuel: Is It Safe? 142
- 8.3 Hydrogen Production: Past, Present, and Future 142
- 8.4 Concept of Photochemical Hydrogen Generation 144
- 8.5 Technology for Hydrogen Generation 147
  - 8.5.1 Kvaerner Carbon Black and Hydrogen Process 148
  - 8.5.2 Biological Production 148
  - 8.5.3 Electrolysis of Water 148
  - 8.5.4 Concentrating Solar Thermal Power 149
  - 8.5.5 Thermal Electrolysis of Water 150
    - 8.5.5.1 Two-Step Cycles 150
    - 8.5.5.2 Three-Step Cycles: Main Cycles 151
  - 8.5.6 Photocatalytic and Photoelectrocatalytic Hydrogen Production 151
  - 8.5.7 Hydrogen as a By-Product of Other Chemical Processes 152
    - 8.5.7.1 Catalytic Reforming of Oil Refinery 153
    - 8.5.7.2 Chlor-Alkali Plants 153
    - 8.5.7.3 Hydrocarbon Waste of High-Temperature Fuel Cells 153
    - 8.5.7.4 Waste Biomass of Wine Industries 154
- 8.6 Photocatalytic Hydrogen Generation 155
- 8.7 Mechanisms of Photocatalytic Water Splitting 156
- 8.8 Modification of Photocatalysts 157
  - 8.8.1 Photosensitization: Dyes/Quantum Dots 157
  - 8.8.2 Reducing Bandgaps through Doping 159
  - 8.8.3 Engineered Solid Solutions 159
  - 8.8.4 Macromolecular Systems for Water Splitting 159
  - 8.8.5 Plasmonic Nanostructures with Surface Plasmon Resonance 161
  - 8.8.6 Nanostructuring of the Photocatalyst 162
  - 8.8.7 Composite Photocatalyst 163
  - 8.8.8 Photoelectrochemical Water Splitting 163
- 8.9 Electrochemical Aspect 164
- 8.10 Hydrogen as a Key Solution 167
- References 169

8.1 INTRODUCTION

Hydrogen is the most widely occurring element of the universe. It is available on Earth in multiple chemical forms such as hydrocarbon, hydrides, prebiotic organic compounds, water, and so on. All the major common conventional energy sources, such as oil, gas, and coal, are not renewable. Moreover, their amounts on Earth are being exhausted rapidly. Therefore, the trend is now shifting toward renewable energy sources. In the last few decades, renewable energy has become an apparent choice over conventional energy. The prime driving forces behind this trend are future carbon emissions constraints, energy security, downfall of subsisting energy sources (raising prices), resulting changes in climate and environment, escalating energy demands of industrial and economic development, inextricable link between nuclear weapons and nuclear power, abundant and free renewable energy sources, consumer awareness, financial risk mitigation, flexibility, and resilience. Renewable energy will no longer be a minor player as it surpasses nuclear energy and other traditional energy sources. This will occur only on the condition that one happens to crack the code of sustainable economic growth by reducing energy demands. A futuristic energy scenario is the alternative image of how the future might unfold with an appropriate tool to analyze how driving forces may influence future outcomes and also to assess the associated uncertainties.

Hydrogen is a renewable and clean fuel that offers us the greatest promise of improving our energy problems. The only combustion product of hydrogen is water vapor without any trace of noxious carbon emissions in the form of carbon dioxide, carbon monoxide, or unburned hydrocarbons. Now the question arises, how can one get this hydrogen? Because 90% of the material available on Earth contains hydrogen, it is very tempting to use hydrogen as an energy source (fuel) and energy carrier. To utilize this resource, pure hydrogen must be obtained in a cost-effective, benign, and reliable way. Due to its qualities, hydrogen has attracted much attention from scientists as the fuel of the future. Currently, the cleavage of natural gas methane is the major source of hydrogen production, but this requires comparatively harsh conditions of high temperature and pressure and results in the production of large amounts of greenhouse gas emissions, which have already been linked to ozone depletion and global warming. Hence, this method is not considered environmentally friendly.

The eco-friendly production of hydrogen in its purest form is really a difficult task. Environmentally benign hydrogen can be produced by the cleavage of water by using renewable energy sources such as solar energy. Photocatalytic or photoelectrochemical water-splitting techniques are used for hydrogen production. In this context, one must contend with the difficulty of, the low density of the energy when considering the use of solar energy. A large area is required in order to harvest a reasonable amount of solar energy. Although there are many methods of splitting water—thermochemically, biophotolytically, mechanocatalytically, plasmolytically, magnetolytically, and radiolytically—photocatalytic water splitting will be advantageous for the large-scale application of solar hydrogen production because of its simplicity. It is the most promising route, and it is

$$H_2O + h\nu + \text{Photocatalyst} \rightarrow \frac{1}{2}O_2 + H_2 \qquad (8.1)$$

Nowadays, hydrogen is primarily used as a fuel in spaceships and space shuttles to lift them into space and supply electricity to all the systems during the flight. Speedy development in related technologies, such as hydrogen fuel cells, made it possible to use this gas not only in space shuttles but also in ordinary road vehicles, cars and buses, making them much cleaner without posing any threats to the environment. Hydrogen is also considered as a potential energy carrier, just like electricity, which is the primary energy carrier of the world and is utilized for transporting persons and goods from one place to another. On mixing hydrogen with oxygen in fuel cells, electricity is generated, with water being the only by-product and without any trace of harmful carbon emission. It represents an absolutely green form of energy that can be directly used as an energy source like oil, gas, or coal.

This reaction is extraordinarily slow at ambient temperature and moderately accelerated in the presence of a catalyst, such as platinum, or an electric spark. The production of hydrogen will allow

Hydrogen

for energy independence. The benefits of hydrogen vehicles are clear and more realistic in fifth-generation vehicles that feature new design possibilities (stack of 200 cells size equal to a home PC), quieter operation (no moving parts in the fuel cell stack), eco-friendly operation (zero carbon emission), possibility of interchangeable bodies (skateboard-like chassis), no engine compartment, larger cabin 82 hp engine 0–62 mph quick start (in 16 s), and high speed. The Honda FCX concept car, with a maximum speed of 93 mph, 80 hp, and 201 lb-ft torque (Burns et al. 2002), is about to come on the market, and it is hoped that the dream to drive a hydrogen car will be fulfilled by 2020.

The development and use of hydrogen fuel cells mean a decreased need for Middle Eastern oil or low imports of hydrogen by the United States or a non-oil-producing country. It can also lead to possible economic collapse in main oil-exporting nations. Hydrogen fuel cells can be used to power third world countries. In underdeveloped countries, the governments would not have to import oil. Solar or wind power collectors could produce energy, which would make hydrogen. Hydrogen fuel cells will drastically change our cars and the ways we heat and power our homes and businesses. This will lead to positive and negative changes in the lifestyles of laymen and political scenarios of the Middle East and third world countries.

8.2 HYDROGEN

8.2.1 Chemical Fuel

Hydrogen is the simplest element in the universe with one proton and one electron. It is the most abundant element on Earth. Fortunately or unfortunately, it cannot remain alone. At normal temperature, it shows low reactivity unless it has been activated by an appropriate catalyst, but at high temperatures, it becomes highly reactive. Atomic hydrogen is treated as a powerful reductive agent at room temperature and produces hydrogen peroxide, H_2O_2, with oxygen. It reacts with the oxides and chlorides of metals, like silver, copper, lead, bismuth, mercury, and so on, to produce free metals. It reduces some salts to their metallic state, like nitrates, nitrites, and sodium and potassium cyanide. It reacts with a number of elements, metals, and nonmetals, to produce hydrides, like NAH, KH, H_2S, and PH_3. Similarly, it reacts with organic compounds to form a complex mixture of products (e.g., with ethylene, C_2H_4, the products are ethane, C_2H_6, and butane, C_4H_{10}).

Hydrogen makes up almost 95% of the visible matter (mass) on the Earth. It is an odorless, invisible, nontoxic and nonpoisonous, smoke free, and highly buoyant (lighter than air) gas that rises and diffuses when leaked. The H_2 gas can be most easily identified by the thermal wave, which produces low radiant heat. With the widest range of flammability (H_2: 4%–75%, gasoline: 1%–7.6%), low autoignition temperature (400°C), and high octane rating (130; gasoline: 87–93), hydrogen has the highest heating value among all available fuels at nearly 52,000 Btu lb^{-1} (British thermal units per pound mass). Its flammability range allows for lean mixtures with better fuel economy and lower combustion temperatures than gasoline; hence, hydrogen engines perform more efficiently than gasoline engines. It possesses a compression ratio higher than gasoline and lower than diesel, which makes it beneficial when used as a fuel for internal combustion engines (ICEs), because lean mixtures increase efficiency (diesel engines > hydrogen engines > gasoline [petrol] engines), especially at low power and at engine idle, and lower combustion temperatures can help to suppress the amount of emissions of nitrogen oxides, which can still occur in ICEs fueled by hydrogen. Moreover, in terms of efficiency, 57 million metric tons of hydrogen is equal to about 170 million tons of oil equivalent. The growth rate of hydrogen production is around 10% per year. These superb properties of hydrogen strongly support hydrogen being considered as a green and clean fuel. Some other well-known alternative fuels are biodiesel, bioalcohol (methanol, ethanol, butanol), chemically stored electricity (batteries and fuel cells), nonfossil methane, nonfossil natural gas, vegetable oil, propane, and other biomass sources. The main purpose of any fuel is to store energy, which should be in a stable form and can be easily transported to the place of production. Almost all fuels are chemical fuels. The user employs this fuel to generate heat or perform mechanical work, such as powering an engine. It may also be used to generate electricity, which is then used for heating, lighting, or other purposes. Furthermore, high diffusivity, low viscosity, unique chemical nature, combustibility, and electrochemical properties are the characteristics that also make the hydrogen a different or better fuel than other gases.

8.2.2 Hydrogen as a Sustainable and Clean Energy Source

Renewable energy sources for electricity constitute a diverse group by including the wind, solar, tidal, and wave energy to hydro, geothermal, and biomass-based power generation. Apart from hydro power in the few places where it is plentiful, none of these is suitable, intrinsically or economically, for large-scale power generation, where a continuous and reliable supply of energy is needed. Growing use will, however, be made of renewable energy sources in the years ahead, although their role is limited by their intermittent nature. Another issue associated with renewable energy source use is economic feasibility, which becomes crucial when an intermittent supply of electricity is demanded at a small scale. In the Organisation for Economic Co-operation and Development (OECD) countries, about 2% of electricity is harnessed from renewable sources other than hydro, and this is expected to increase to 4% by 2015.

Hydrogen production is a large and growing industry and is useful as a compact energy source in fuel cells and batteries. Hydrogen can be produced using renewable energy sources like wind, solar, waves, hydro, and biomass. The modest form of renewable energy can be stored in water as hydrogen, produced by splitting water by means of electrolysis, and used for electricity. Relying only on renewable energy may solve the threats of oil depletion and pollution in the present energy system. This also makes it possible for everybody to produce their own energy, creating more political stability and benefits to all of us. After the production, hydrogen functions as an energy carrier that can be used to supply energy wherever it is needed.

Hydrogen and fuel cells are comparatively new and different energy systems that either use energy directly as electricity or store it in the form of hydrogen for use in transportation or store for the time when the sun is not shining or the wind is not blowing. The fuel cells are used to convert the hydrogen into energy, where hydrogen and oxygen (air) react and produce water as the only emission. The reaction creates electricity and heat that can be used in various applications. Fuel cells are versatile in their application and can be used in many devices that need energy, including mobile phones, cars, buses, and even heat and power plants. Fuel cell technology is the next innovative development that will bring progress and prosperity to society and will remarkably impact our lives, as significantly as did the steam engine and the combustion engine.

8.2.3 Hydrogen Fuel: Is It Safe?

The superiority of hydrogen fuel over other fuels is an established fact. All fuels are hazardous, but hydrogen is comparably or less so because of its nonpoisonous and highly buoyant nature. Spills present very little danger as it rapidly evaporates and disperse. However, it can still displace oxygen in confined spaces. There is a need to handle it with care, because it can burn with a clear flame and without smell, cannot sear from a distance, burns first with no smoke, does not puddle, and is hard to make explode in free air (22 times less explosive power). Beyond the Hindenburg myth (1937) (McAlister 1999), until now nobody was killed by hydrogen fire, which is completely unrelated to hydrogen bombs. For a vehicle, it can be stored onboard in either pressurized gas or liquid form in composite tanks that utilize carbon fiber for additional safety.

Hence, one can say that hydrogen is a safe energy source (fuel), if one can handle it carefully.

8.3 HYDROGEN PRODUCTION: PAST, PRESENT, AND FUTURE

There are some inherent advantages of hydrocarbons (natural gas and petroleum), including feedstock availability, cost competitiveness, convenience of storage and distribution, purity of the product, and relatively high H/C ratio. Scientists have been highly engaged in research in the field of hydrogen production. The current world's major resources for hydrogen generation are coal (18%), natural gas

Hydrogen

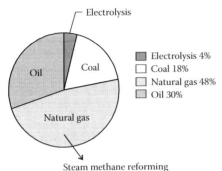

FIGURE 8.1 Sources of hydrogen generation from natural gases, oil, coal, and electrolysis of water.

(40%), petroleum (38%), and electrolysis of water (4%) that consume approximately 50 kW h^{-1} of electricity for per kilogram of hydrogen production (Hairston 1996) (Figure 8.1). Most of the industrial hydrogen consumers are chemical process industries (73.9%), metallurgical industries (2.7%), food (3.6%), electronics (8.1%), petroleum, pharmaceuticals (11.7%), and other industries (Hairston 1996).

Scientists have found dramatic growth in the demand for renewable energy over the past few decades, which reflects the "take-off" phase of the renewables market along with energy supply security, autonomy, resilience, jobs, industrial development, financial profit, portfolio risk mitigation, price risks of fossil fuels, rural energy access, climate change, environmental sustainability, and reduction in nuclear accidents/wastes. Improvement in the economy for large-scale manufacturing, radical improvements in technology, advances in performance, and reductions in the costs of renewable energy (wind, hydropower, water, geothermal, and biomass), categorize hydrogen as the most important source of energy. Onshore wind power is closest to commercial maturity. Moreover, offshore wind power is more expensive than onshore, but it has a large (although uncertain because of fluctuating density) potential for cost reductions, not just for turbines, but also for logistics, long-term operations, and maintenance costs. Remarkable reduction in the cost of solar photovoltaics (PV) has been seen in recent years. Solar energy input in a year can replace the total world storage of conventional fuel and nuclear fuel. Therefore, at present, annual global investment in power generation is flowing to renewable energy instead of fossil fuels and nuclear. The growth of renewable energy worldwide began in the 1990s and has greatly accelerated in the 2000s.

Many have credited this growth to the proliferation of supportive government policies, rising costs of conventional energy, dramatic reductions in renewable energy technology costs, and economies of scale in manufacturing. For these major reasons, dynamic growth has accelerated over the past decades, and past projections about renewable energy have fallen short. The International Energy Agency (IEA) in 2000 projected 34 GW (gigawatts) of wind power globally by 2010, while the actual level reached was 200 GW. In 1996, the World Bank projected 9 GW of wind power and 0.5 GW of solar PV in China by 2020, while the actual levels reached in 2011 were 62 GW of wind power and 3 GW of solar PV (Hairston 1996). The history of energy scenarios is full of similar projections for renewable energy that proved to be too low by a factor of 10, or were achieved a decade earlier than expected.

Projections show continued cost reductions, many possible technology advances, and competitiveness with retail electricity prices without subsidies occurring in many jurisdictions soon and in many more places around the world by 2020. Concentrated solar power (CSP) has a large cost reduction potential, with future opportunities for bulk power supply, for dedicated applications such as industrial heat supply and desalination, and power grid balancing using multihour and multiday embedded heat storage. Another promising advance in these series is the *first generation of biofuels*. Many projections show large future markets for advanced biofuels from agricultural and forestry wastes, and from crops grown on unproductive lands. A wide variety of new approaches for use of biomass are also projected, such as growing international commodity markets for wood pellets and bio-heating oil, greater use of biogas in a variety of applications, new types of biorefineries in agriculture and forestry, and greater use of biomass in heat supply.

At least 30 countries already have shares of renewable energy more than 20%. Some 120 countries have various types of policy targets for long-term shares of renewable energy (e.g., 20% target for the European Union by 2020). Some countries have long-term policy targets that will put them in the high renewables domain by 2030 or 2050, such as Denmark (100%) and Germany (60%). A diverse group of 20 countries outside of Europe target 10%–50% energy shares in the 2020–2030 time frame (Ogden 1999).

Renewable markets will become even broader based in a larger number of countries beyond 2020, as developing countries increasingly share leadership. Unique opportunities for renewable energy exist in future development that includes new electric power infrastructure, diesel generator replacement, new settlements, new power-market rules, regional cooperation frameworks, local manufacturing, and rural (off-grid) energy services. Furthermore, the predicted decrease in conventional sources as well as the interrupted imported oil supply due to political instability in the Middle East, will result in increased petroleum prices.

Hydrogen fuel is rich in energy and is basically proposed as an alternative fuel that has the potential to dramatically decrease the energy consumption of vehicles. On using it in a fuel cell, electricity and heat are produced by burning H_2 and O_2. The by-products in this process are H_2O and, of course, small amounts of nitrogen oxide. Because the consumption of hydrogen as a fuel produces nonpollutant flues, many experts believe that hydrogen fuel could be an eco-friendly alternative to our current energy sources, which can be commercialized in the future on a large scale. Hydrogen was first commercially used as a fuel in a fuel cell in the 1970s for launching NASA's space shuttles and other space rockets. Since then, hydrogen fuel cells have also been used in the production of electricity for spacecrafts while in orbit. Astonishingly, the water produced from burning H_2 and O_2 during long travels in space can even be used for drinking purposes for the crew, which proves that hydrogen fuel is indeed an eco-friendly energy source.

Fuel cells are often compared to batteries. Both produce energy by conversion of a chemical reaction into usable electric power. However, the fuel cell will produce electricity as long as fuel (hydrogen) is supplied, never losing its charge. A fuel cell is a promising technology for use as a source of heat and electricity for buildings and electric motors propelling vehicles. Fuel cells operate best on pure hydrogen. Fuels like natural gas, methanol, or even gasoline can be reformed to produce the hydrogen required for fuel cells. Some fuel cells can be fueled directly with methanol, without using a reformer.

Outstanding fuel efficiency of hydrogen is always welcome, but associated technological challenges owing to its lightness (i.e., safety in production, storage, transportation, distribution, and end use) are making its path difficult (Ogden 1999; Szklo and Schaffer 2007; Scholz 1993). Because hydrogen is a low-density gas, imperfect packing and careless refilling into a cylinder may cause its leakage into the atmosphere. It was estimated by Caltech Research that atmospheric emission will grow from 60 to 120 trillion (four to eight times more hydrogen than the normal human action release into air) a year by replacing fossil fuel with hydrogen globally. Although the figures remain uncertain until scientists gain more understanding of the effects of using hydrogen as an energy source, some approximations have been made of a 10% decrease in ozone concentration in the planet's atmosphere caused by the inevitable leaks. That was attributed to moisture buildup from hydrogen combining with atmospheric oxygen. As a consequence, the upper atmosphere cools, causing indirect destruction of the ozone.

Hydrogen has the potential to be used as an energy carrier in electricity production. It can be considered as an energy carrier, which moves and may deliver energy in a usable form to consumers as and when required. Other renewable energy sources, such as the Sun and wind, cannot produce energy all of the time. But they could produce electric energy and hydrogen, which can be stored for future needs. Hydrogen can also be transported (like electricity) to locations where it is needed.

8.4 CONCEPT OF PHOTOCHEMICAL HYDROGEN GENERATION

Among the variety of available hydrogen-containing compounds on Earth, water is a carbon-free abundant source of hydrogen. It is transparent by appearance and categorized as almost unbreakable (stable) under normal circumstances. Moreover, due to the low autoionization ($K_w = 1.0 \times 10^{-14}$), pure water falls into the category of insulator at room temperature that poorly conducts current, 0.055 μS·cm^{-1}. Therefore, very large potential is required to increase the autoionization of water. Furthermore, kinetic and thermodynamic decomposition of pure water into hydrogen and oxygen is not suitable at standard temperature and pressure due to large positive Gibbs free energy ($\Delta G = 237$ kJ mol^{-1}); thus, the backward reaction (recombination of hydrogen and oxygen into water) easily proceeds (Chouhan and Liu 2011). Therefore, activation energy, ion mobility (diffusion) and concentration, wire resistance, surface hindrance including bubble formation (causes electrode area blockage), and entropy require a greater applied potential to overcome these factors. The following oxidation and reduction reactions are involved in water splitting (Equations 8.2 and 8.3) at the anode and cathode:

$$\text{Anode: } 2H_2O(l) \rightarrow O_2(g) + 4H^+(aq) + 4e^- \quad E_{o,ox} = -1.23 \text{ V} \quad (E_{o,red} = 1.23) \tag{8.2}$$

$$\text{Cathode: } 2H^+(aq) + 2e^- \rightarrow H_2(g) \quad E_{o,red} = 0.00 \text{ V} \tag{8.3}$$

Thus, the standard potential of the water electrolysis cell is –1.23 V at 25°C at pH = 0 (H$^+$ = 1.0 M) and pH = 7 (H$^+$ = 1.0×10^{-7} M), based on the Nernst equation.

A well-known example of photocatalysis occurring in nature is photosynthesis by plants, where chlorophyll serves as the photocatalyst. Photocatalytic splitting of water mimics the photosynthesis process (Figure 8.2) as both work on the same principle—that is, under irradiation of light an electron is excited from the valence band into the conduction band to result in the formation of an electron (e^-)–hole (h^+) pair in a photocatalyst. These e^- and h^+ reduce and oxidize any substrate on the surface of the photocatalyst, respectively; if they recombine, then there will be no net chemical reaction. The original structure (or chemical composition) of the photocatalyst remains unchanged if an equal number of e^- and h^+ are consumed for chemical reaction and/or recombination.

Because of its abundance and free availability, solar energy has become the most obvious and preferred option for water breaking, which needs photons of the light wavelength shorter than 1008 nm, which corresponds to 1.23 eV. Fortunately, water is transparent; hence, automatically, it is unable to break down in ordinary conditions. Theoretically, if the energy gap of samples is kept at 1.23 V, then the maximum value of solar energy conversion efficiency will be about 32%. But practically, the conversion efficiency should be in the range of 10%–15%, because of the various overpotentials of the system. The bandgap is kept between 1.8 and 2.5 eV to avoid recombination (2.46 eV). Therefore, the material issue became vital for the water-splitting process, but the nature of electrolytes used and fabrication technology are also important concerns for their industrial applications.

Several terms have been adopted to describe the efficiency for converting solar energy, namely, the incident photon-to-current efficiency (IPCE), absorption photon-to-current efficiency (APCE), solar-to-hydrogen (STH) conversion efficiency, and quantum efficiency (QE) (Murphy et al. 2006).

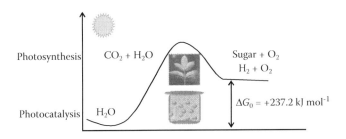

FIGURE 8.2 Analogy of photosynthesis with photocatalytic water splitting.

Percentage IPCE is calculated using Equation 8.4, which includes photocurrent density, wavelength of light, and irradiance:

$$\text{IPCE}(\%) = \frac{(1240 \times \text{Photocurrent density})}{(\lambda \times \text{Irradiance})} \quad (8.4)$$

The APCE is obtained by dividing the IPCE by the fraction of incident photons absorbed at each wavelength by light-harvesting efficiency (LHE), calculated through the absorbance A (Equations 8.5 and 8.6). APCE is usually used to characterize the photoresponse efficiency of a photoelectrode material under an applied voltage:

$$\text{APCE} = \frac{\text{IPCE}}{\text{LHE}(\lambda)} \quad (8.5)$$

$$\text{LHE}(\lambda) = \text{TTCO}(\lambda) \cdot \left(\frac{\alpha_{\text{dye}}}{\alpha_{\text{film}}}\right) \times (1 - e^{\alpha}) \quad (8.6)$$

where:
TTCO is the transmittance of the transparent conductive oxide
α_{film} is the absorption coefficient of the entire film
α_{dye} is the absorption coefficient due to the dye molecules

This is the simplest approach for calculating APCE, where second-order reflectance terms are not considered. The LHE is estimated by using the information about injection and collection efficiencies (Equation 8.6). The solar-to-hydrogen conversion efficiency STH of the water-splitting reaction can be determined using Equation 8.7:

$$\text{STH}(\%) = j_p (V_{\text{WS}} - V_{\text{Bias}}) \times \frac{100}{eE_s} \quad (8.7)$$

where:
j_p is the photocurrent density (mA cm^{-2}) produced per unit irradiated area
V_{WS} (1.23 eV) is the water-splitting potential per electron
V_{Bias} is the bias voltage applied between the working and the counterelectrodes
E_S is the photon flux
e is the electronic charge

According to Equation 8.7, every electron contributing to the current produces half an H_2 molecule. The photoconversion efficiency (η) is a percentage conversion of light energy into chemical energy in the presence of applied external voltage, and it can be calculated using Equations 8.8 and 8.9:

$$\eta(\%) = (\text{Total power output} - \text{Electrical power output}) \times 100 \quad (8.8)$$

$$= j_p \left[\frac{(E_{0,\text{rev}} - |E_{\text{app}}|)}{I_0}\right] \times 100 \quad (8.9)$$

where:
$E_{\text{app}} = E_{\text{meas}} - E_{\text{aoc}}$
$E_{0\text{rev}}$ is the standard reversible redox potential of the water and is equal to the value 1.23 versus NHE

E_{meas} is the electrode potential (vs. Ag/AgCl or saturated calomel electrode [SCE]) of the working electrode at which photocurrent was measured under light illumination

E_{aoc} is the electrode potential (vs. Ag/AgCl) of the same working electrode at open-circuit condition under the same illumination and in the same electrolyte solution

I_0 is the intensity of incident light (mW cm^{-2})

Quantum efficiency (QE) represents the characteristic photon conversion of photoactive films. It is defined as the percentage of generated electrons and incident photons, while the photoactive films are irradiated under a specific wavelength (Equation 8.10). It is noted that QE neglects the energy loss of solar irradiance and the chemical conversion efficiency. Therefore, it is used to qualify the photoactive films but not to represent the water-splitting reaction conversion efficiency. The apparent quantum efficiency was measured under the xenon arc lamp or four 420 nm-LEDs (3 W), used to trigger the photocatalytic reaction:

$$\text{QE}(\%) = \left(\frac{\text{Number of evolved H}_2 \text{ molecules} \times 2}{\text{Number of incident photons}} \right) \times 100 \tag{8.10}$$

8.5 TECHNOLOGY FOR HYDROGEN GENERATION

Progress of the human race revolved around the axis of fulfillment of our energy demand, which can facilitate the development of technologies and the productivity that is necessary for a higher standard of living. Therefore, energy is as vital as raw materials, capital, and labor, for growth. Fossil fuels are the most commonly available form of the energy and transportation fuel, but they are not ideal fuels because of the following:

- The combustion of fossil fuels, such as natural gas, coal, and petroleum, emits carbon dioxide and other pollutants, which is one of the major components of greenhouse gases that cause notorious climate changes.
- The supply of hydrocarbon resources in the world is limited, and the demand for hydrocarbon fuels is increasing day by day, particularly in China, India, and other developing countries. Various biological and nonbiological processes have taken place for millions of years, resulting in the storage of energy in mineral organic compounds or fossil fuels such as coal, petroleum, and natural gas. If their rates of production do not match the rate of consumption, that will lead to them being quickly exhausted.
- The supply of fossil fuels is powered by a limited number of nations. As a result, a significant amount of time and money will be spent for the relocation and distribution of these fuels. It was also seen that centralized power will cause the gasoline resources war among the nations.
- World transportation is primarily fueled by petroleum fuel (hydrocarbon economy). Therefore, it will become necessary for us to develop a sustainable energy source that can be easily produced at low cost and, of course, in an environmentally friendly way. Several alternative renewable energies have been developed to replace or reduce the use of fossil fuels. Alternative energies have lower carbon emissions as compared to conventional energy sources. These energy sources include wind, hydropower, solar, geothermal, and so on, which can be utilized to generate clean and efficient fuel (i.e., hydrogen, which is abundantly available in nature in various forms). The Earth's crust contains only 0.15% natural H_2 and 1.0% in atmosphere as free hydrogen. Hydrogen is a major constituent (0.5 ppm) of the most abundantly available resource in substituted form (98%)—that is, water (70%) and biomass (14%). Presently, hydrogen is mainly produced from fossil sources. This stimulated great interest in alternative, cheaper means of hydrogen production. Proponents of a world-scale hydrogen economy argue that hydrogen can be an environmentally cleaner source of energy to end users, particularly in transportation applications,

without the release of pollutants (such as particulate matter) or carbon dioxide at the point of end use. An analysis asserted that most of the hydrogen supply chain pathways would release significantly less carbon dioxide into the atmosphere than would gasoline used in hybrid electric vehicles. As a result, a significant reduction in carbon dioxide emissions would be possible, if carbon capture or carbon sequestration methods were utilized at the site of energy or hydrogen production.

Centralized production of hydrogen requires a large capital investment for the construction of the infrastructure and distribution to filling stations that may put a fleet of light-duty fuel cell vehicles on the road. Furthermore, the technological challenge of providing safe, energy-dense storage of the hydrogen on board a vehicle must be overcome to provide sufficient distance between fill-ups. Some chemical/technical methods involved in hydrogen production are discussed here.

8.5.1 Kvaerner Carbon Black and Hydrogen Process

A Norwegian company (Kvaerner) developed an endothermic method for the production of hydrogen from hydrocarbons (C_nH_m), such as methane, natural gas, and biogas, in the 1980s. However, it was first commercially exploited in 1999. It is an efficient mode of complete transformation of the hydrocarbon/methane to pure carbon (40%), heat (2%), and hydrogen (48%), using a plasma converter at around 1600°C in comparison to other reformation methods such as steam reforming and partial oxidation the natural gas (Naess et al. 2009):

$$C_nH_m \rightarrow nC + \frac{m}{2H_2} + \text{Heat} \tag{8.11}$$

8.5.2 Biological Production

Biohydrogen is manifested by a diverse group of living objects, such as bacteria, plants, multi-enzyme systems on organic substrate, and so on, involving steps similar to anaerobic conversion through photosynthesis, fermentation, and microbial electrolysis of cells. It was discovered in the 1990s that if an algae is deprived of sulfur, it will switch from the production of oxygen (normal photosynthesis) to the production of hydrogen. Biological hydrogen production provides a possibility of being renewable and carbon neutral as a method of H_2 production, and it can also be produced in an algae bioreactor, using sulfur-deprived algae.

The process involves bacteria feeding on hydrocarbons and excreting hydrogen and CO_2. The CO_2 can be sequestered successfully by several methods that result in the production of pure hydrogen gas. A prototype hydrogen bioreactor using waste as a feedstock is already in operation. Biological hydrogen can also be generated by the fermentation process in dark and light. In dark, fermentation reactions do not require light energy, but the photofermentation proceeds in the presence of light, and it can be employed in the conversion of small molecular fatty acids into hydrogen by using *Rhodobacter sphaeroides* SH2C (Tao et al. 2007). They are capable of constant production of hydrogen from organic compounds throughout the day and night.

Another important means of biological hydrogen production is biocatalyzed electrolysis (i.e., electrolysis using microbes). Hydrogen is generated after running through the microbial fuel cell with biocatalyzed electrolysis and a variety of aquatic plants, such as reed sweet grass, cord grass, rice, tomatoes, lupines, and algae (Rosenbaum et al. 2010).

8.5.3 Electrolysis of Water

At ambient conditions, pure water cannot dissociate due to a very poor degree of self-ionization ($K_w = 1.0 \times 10^{-14}$). Furthermore, the decomposition of pure water into hydrogen and oxygen is

Hydrogen

thermodynamically not favorable at standard temperature and pressure (Gibb's free energy $\Delta G_0 = 237$ kJ mol^{-1}; 2.46 eV per molecule) (Equation 8.12). Therefore, it is necessary to associate a large potential to induce electrolysis in water. This additional potential is known as overpotential. When this reaction is about to complete after applying potential, it is required to overcome certain factors (i.e., activation energy, ion mobility [diffusion] and concentration, wire resistance, surface hindrance including bubble formation [causes electrode area blockage], entropy, etc.):

$$\text{Overall: } 2H_2O(l) + 4e^- \rightarrow O_2(g) + 2H_2(g) + 4e^- \quad E = -1.23 \text{ V at } 25°C; \text{ pH} = 0 \text{ and } 7 \quad (8.12)$$

Thus, the standard potential of the water electrolysis cell is -1.23 V at 25°C, based on the Nernst equation. Water-soluble electrolytes or electrocatalysts are added to improve the conductivity of the water along with the applied potential. Salts of cations having electrode potential lower than H$^+$, such as Li$^+$, K$^+$, Na$^+$, Rb$^+$, Cs$^+$, Mg^{2+}, Sr^{2+}, Ca^{2+}, and Ba^{2+}, are used as the electrolyte (Pauling 1970). A solid polymer electrolyte such as Nafion (1.5 V) can also be utilized. The electrolyte disassociates into cations and anions in water. Afterward, these cations and anions will rush toward the cathode and anode to neutralize the negatively charged OH$^-$ and positively charged H$^+$ ions, at respective electrodes. This phenomenon allows the continued flow of electricity (Pauling 1970). The electrocatalysts, such as platinum alloys, molybdenum sulfide (Kibsgaard et al. 2014), graphene quantum dots (Fei et al. 2014), carbon nanotubes, and perovskite (Luo et al. 2014), are suitable for anode material utilized for the oxidation. Similarly, a two-electron reaction to produce hydrogen at the cathode can be electrocatalyzed with almost no overpotential by platinum and less overpotential with other materials.

Deiman and van Troostwijk were the first ones, who used gold electrodes for electricity generation from water. Later on, Volta invented the Voltaic pile for electrolysis of water. In this series, Gramme invented the Gramme machine for hydrogen production through the electrolysis of water, while Lachinov developed a method for industrial hydrogen production through the electrolysis of water. With the appropriate electrodes and electrolyte, oxygen will collect at the positively charged electrode (anode), and hydrogen will collect at the negatively charged electrode (cathode).

The efficiency of modern hydrogen generators is measured by power consumed per standard volume of hydrogen (MJ m^{-3}), assuming standard temperature and pressure of the H$_2$. A 100% efficient electrolyzer would consume 11.7 MJ m^{-3}. Lower is the actual power used, higher is the efficiency. There are two main technologies available in the market, cheaper alkaline (nickel catalysts, 60%–75%) and more expensive proton exchange membrane (PEM) (platinum group metal catalysts, 65%–90%) electrolyzers. Then 1 kg hydrogen (which has a specific energy of 143 MJ kg^{-1}, about 40 kWh kg^{-1}) generation through water electrolysis requires 50–79 kWh. It is 3–10 times the price of hydrogen produced from steam reformation of natural gas (NREL 2009). The price difference is due to the efficiency of direct conversion of fossil fuels to produce hydrogen, rather than burning fuel to produce electricity. Hydrogen from natural gas, used to replace, for example, gasoline, emits more CO$_2$ than the gasoline it would replace, and so this process is of no help in reducing greenhouse gases (Crabtree et al. 2004). Hydrogen can be made via high-pressure electrolysis or low-pressure electrolysis of water. The current best processes have an efficiency of 50%–80% (Zittel and Wurster 1996).

8.5.4 Concentrating Solar Thermal Power

CSP systems generate solar power by using solar concentrators (namely, parabolic trough, enclosed trough, dish Stirlings, concentrated linear Fresnel reflector, solar power tower, etc.) that concentrate a large area of sunlight onto a small area. The concentrated light is converted into heat, which drives a heat engine (usually a steam turbine) connected to an electrical power generator. It generates electricity, which is used to break water (Figure 8.3).

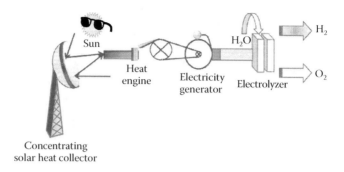

FIGURE 8.3 Solar thermal hydrogen production system.

CSP is an emerging technology in the field of electricity generation through thermal power, which turns sunlight either directly into electricity or first thermally breaks water into H_2 and O_2 and then into electricity. CSP requires clear skies and strong sunlight, which has virtually unlimited power. The suitable sites for CSP technologies are found between 15° and 40° parallels and occasionally at higher latitudes.

This is unconventional route for cost-effective hydrogen production from solar energy without emitting carbon dioxide. Solarothermal electrolysis of water was triggered by solar-to-thermal conversion followed by two steps: (1) thermal to electrical and (2) electrical to chemical conversions. Very high temperatures (2500 K) are required to dissociate water into hydrogen and oxygen.

New CSP stations present a healthy and economic competition with fossil fuels. This technology works efficiently under the high solar radiation zone. It was indicated that the metal nitrates (calcium, potassium, sodium)-based thermal storage systems will make the CSP plants highly effective.

8.5.5 Thermal Electrolysis of Water

Solar energy could be used for cost-effective hydrogen production via water splitting without emitting carbon dioxide. A very high temperature (2500 K) was required for dissociation of the water into hydrogen and oxygen. Plenty of catalysts are available to be used in these processes at comparatively feasible temperatures. There are more than 352 thermochemical cycles that can be used for water splitting, and about a dozen of these cycles, such as the iron oxide cycle, cerium (IV) oxide–cerium (III) oxide cycle, zinc–zinc oxide cycle, sulfur–iodine cycle, copper–chlorine cycle, and hybrid sulfur cycle, are actively being researched or in trail phases to produce hydrogen and oxygen from water and heat without using electricity (Kolb et al. 2007; Chen et al. 2007; Perret et al. 2007).

Abanades and Flamant (2006) have studied a single-step thermal decomposition of methane without using catalyst that produces hydrogen-rich gas and high-grade carbon black from concentrated solar energy and methane. A solar concentrator (2 m diameter) with 1500 K mean temperature at nozzle has shown a thermal-to-chemical conversion efficiency in the range of 2.6%–98%). Charvin et al. (2008) made a process analysis of ZnO/Zn, Fe_2O_3/FeO, and Fe_2O_3/Fe_3O_4 thermocycles and found these to be potentially efficient (two- or three-step routes) for large-scaled hydrogen production by concentrated solar thermal energy that operated at a temperature up to 2000 K with real efficiency reported as 25.2%, 28.4%, and 22.6%, respectively, for water splitting. Perkins and Weimer (2004) studied the solar-to-thermal water splitting efficiency of 16%–21% using a three-step Mn_2O_3/MnO cycle.

8.5.5.1 Two-Step Cycles

$$\frac{Fe_3O_4}{FeO}\left(\eta=17.4\%, \text{cost}=7.86-14.75\$ \text{ kg}^{-1}\text{ H}_2\right)$$

Hydrogen

$$\text{(i) } 3\text{FeO} + \text{H}_2\text{O}(g) \rightarrow \text{Fe}_3\text{O}_4 + \text{H}_2 \rightarrow 850\text{ K} \tag{8.13}$$

$$\text{(ii) } \text{Fe}_3\text{O}_4 \rightarrow 3\text{FeO} + \frac{1}{2}\text{O}_2(g) \rightarrow 1900\text{ K} \tag{8.14}$$

$$\frac{\text{Cd}}{\text{CdO}} (\eta = 48.3\%, \text{cost} = 4.5\$ \text{ kg}^{-1}\text{ H}_2)$$

$$\text{(i) } \text{Cd}(s, l) + \text{H}_2\text{O} \rightarrow \text{CdO}(s) + \text{H}_2(g) \rightarrow 298-573\text{ K} \tag{8.15}$$

$$\text{(ii) } \text{CdO}(l) \rightarrow \text{Cd}(g) + \frac{1}{2}\text{O}_2(g) \rightarrow 1773\text{ K} \tag{8.16}$$

$$\frac{\text{CeO}_2}{\text{Ce}_2\text{O}_3} (\eta = \text{NA cost} = \text{NA})$$

$$\text{(i) } 2\text{CeO}_2(s) \rightarrow \text{Ce}_2\text{O}_3(s) + \frac{1}{2}\text{O}_2(g) \rightarrow 2300\text{ K} \tag{8.17}$$

$$\text{(ii) } \text{Ce}_2\text{O}_3(s) + \text{H}_2\text{O}(g) \rightarrow 2\text{CeO}_2(s) + \text{H}_2(g) \rightarrow 700\text{ K} \tag{8.18}$$

8.5.5.2 Three-Step Cycles: Main Cycles

$$\frac{\text{Fe}_2\text{O}_3}{\text{Fe}_3\text{O}_4} (\eta = 18.6\%, \text{cost} = 8.4\$ \text{ kg}^{-1}\text{ H}_2)$$

$$\text{(i) } 3\text{Fe}_2\text{O}_3 \rightarrow 2\text{Fe}_3\text{O}_4 + \frac{1}{2}\text{O}_2 \rightarrow 1600\text{ K} \tag{8.19}$$

$$\text{(ii) } 2\text{Fe}_3\text{O}_4 + 6\text{KOH} \rightarrow 6\text{KFeO}_2 + \text{H}_2 + 2\text{H}_2\text{O} \rightarrow 673\text{ K} \tag{8.20}$$

$$\text{(iii) } 6\text{KFeO}_2 + 3\text{H}_2\text{O} \rightarrow 3\text{Fe}_2\text{O}_3 + 6\text{KOH} \rightarrow 373\text{ K} \tag{8.21}$$

$$\frac{\text{Mn}_2\text{O}_3}{\text{MnO}} (\eta = 16-21\%, \text{cost} = \text{NA})$$

$$\text{(i) } \text{Mn}_2\text{O}_3 \rightarrow 2\text{MnO} + \frac{1}{2}\text{O}_2 \rightarrow 1873\text{ K} \tag{8.22}$$

$$\text{(ii) } 2\text{MnO} + 2\text{NaOH} \rightarrow \text{H}_2 + 2\text{NaMnO}_2 \rightarrow 900\text{ K} \tag{8.23}$$

$$\text{(iii) } 2\text{NaMnO}_2 + \text{H}_2\text{O} \rightarrow \text{Mn}_2\text{O}_3 + 2\text{NaOH} \rightarrow 373\text{ K} \tag{8.24}$$

These processes can be more efficient than high-temperature electrolysis, typical in the range from 35% to 49% LHV efficiency. Although several thermochemical methods of hydrogen generation have been demonstrated in laboratories, none has been utilized at production levels because of high energy requirements and production costs. Chemical-solar routes of hydrogen production with different combinations are represented in Figure 8.4.

8.5.6 Photocatalytic and Photoelectrocatalytic Hydrogen Production

Hydrogen and electricity play a crucial role as energy carriers and fuel in setting our future energy endeavors or economics. It is known that the theoretical potential for the water-splitting process is 1.23 eV per H_2O molecule. This energy corresponds to the wavelength 1008 nm that makes almost 70% of the solar irradiated photon eligible for water splitting. But practically, our needs are for

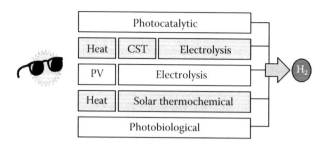

FIGURE 8.4 Technical routes for solar to hydrogen conversion through photocatalytic, concentrating solar thermal (CST) and photovoltaic and solar thermochemical (STCH) methods.

higher energy (1.8–2.6 eV) than the minimum energy 1.23 eV, which is required for compensating for intrinsic energy losses associated with the redox reactions, proceeding on the surface of the photocatalyst (Dincer and Joshi 2013). Photocatalytic materials require a minimum of sunlight exposure to generate the photoexcited electrons and holes on the surface of the catalyst that would allow the photoelectrocatalytic synthesis of hydrogen and oxygen, by reduction and oxidation of water, respectively. Varieties of materials like oxides, sulfides, selenides, tellurides, nitrides, oxynitrides, oxy sulfides, and so on, of metals, are considered good materials for overall water-splitting purposes.

The hydrogen produced would be the fuel to store the Sun's energy, and it may be converted back into electricity in a fuel cell, using the reverse reaction to water electrolysis. It is generally recognized that one of the key fundamental restrictions in making the photoelectrochemical generation of fuels a viable technology is the poor kinetics of the anode reaction (i.e., the water oxidation to oxygen). Although sunlight-driven water splitting shows a potential route to sustainable hydrogen fuel production, their widespread implementation has been hampered by the expense of photovoltaic and photoelectrochemical apparatus and also by the stability of the catalytic material.

A search is on for a highly efficient and low-cost water-splitting cell combining a state-of-the-art solution, in the form of the new materials in this field, such as bifunctional Earth-abundant catalyst NiFe-layered double hydroxide in perovskite tandem solar cell (12.3%), gold electrode covered with an iron-sulfur complex with doped layers of indium phosphide (InP) nanoparticles (60%), $[Mo_3S_{13}]_{22}$ clusters, Fe-doped graphene oxide QDs (2.5%), bimetal oxynitride (5.2%), metallo-oxysulfides, composite nanomaterials, etc. (Kibsgaard et al. 2014; Zeng and Zhang 2010; Luo et al. 2014). The catalyst electrode exhibits high activity toward both the oxygen and hydrogen evolution reactions in electrolyte on submerging in water and irradiating with light under the small electric current.

8.5.7 Hydrogen as a By-Product of Other Chemical Processes

Technological developments, economic conditions, and governmental policy set the trends that aligned to create a significant market opportunity for waste-to-energy (WTE) plants, which utilize the considerable amount of hydrogen as a by-product of industries, municipal solid waste (MSW), and wastewater treatment plants for the production of electricity and heat. It can be generated as the industrial by-product of diverse process industries (such as electrolysis in vinyl industries, the unsaturation process of hydrocarbons, distillation or dissociation of long-chain hydrocarbons in petroleum industries, chlorine industries, electrolysis of water, etc.) that can generate a sizable amount of hydrogen by-product. In 2011, the United States developed 40,000 waste treatment facilities that could be modified to generate hydrogen. A connection between energy source, type, technology, and production method of hydrogen has been demonstrated (Table 8.1).

In addition to the above processes, there are a few more industrial processes that are used to produce hydrogen as a by-product, as discussed below.

TABLE 8.1
Hydrogen Production Using Diverse Forms of Technologies and Energy Sources

Energy Source	Energy Type	Energy Technology	Production Method
Renewable	Power	Wind	Electrolysis
		Hydro	
		Waves	Photovoltaic
		Sun	
	Fuel	Biomass	Reforming
			Biochemical
Fossil	Power	Nuclear	Electrolysis
			Thermo
	Fuel	Natural	Reforming
		Oil	
		Gas	Gasification
		Coal	

8.5.7.1 Catalytic Reforming of Oil Refinery

Dehydrogenation occurs during catalytic or thermal cracking of the petroleum. The residence time of naphtha feedstock in traditional cracker furnaces is very low (less than milliseconds) to avoid the formation of undesirable heavy hydrocarbon or coke formation, because the coke layer developed inside the furnace reduces the heat transfer rate. The furnace is taken out of service, and decoking is carried out with air and steam. Thus, the cracker furnace operates in a cyclic manner between the cracking and decoking operation. Usually, steam is introduced with the feed to remove the coke layer on the tube surface by converting coke into carbon monoxide and hydrogen by gas reaction (Equation 8.25):

$$C + H_2O \text{ (steam)} \rightarrow CO + H_2 \tag{8.25}$$

8.5.7.2 Chlor-Alkali Plants

Chlor-alkali plants are based on the electrolysis of sodium chloride (brine water) for industrial production of chlorine and sodium hydroxide (caustic soda). The H_2 is created as a by-product in manufacturing chemicals like chlorine and NaOH. A basic membrane cell was used in the electrolysis of brine. At the anode, chloride (Cl^-) ions were oxidized to chlorine. The ion-selective membrane allows the counter-Na^+ ion to freely flow across the membrane but prevents anions such as hydroxide (OH^-) and chloride from diffusing across. At the cathode, water is reduced to hydroxide and hydrogen gas. The net process is the electrolysis of an aqueous solution of NaCl, which produced industrially useful products sodium hydroxide (NaOH) (1.1 tons) and chlorine gas (1 ton) (Equation 8.26). This overall reaction for every ton of chlorine produced during the electrolytic process of brine, 28 kg hydrogen as a by-product is also generated. Four percent of total worldwide hydrogen production is created by this process, using seawater (Figure 8.5 and Equation 8.26) (Chouhan and Liu 2011):

$$\text{Overall reaction: } 2NaCl + 2H_2O \rightarrow Cl_2 + H_2 + 2NaOH \tag{8.26}$$

8.5.7.3 Hydrocarbon Waste of High-Temperature Fuel Cells

High-temperature fuel cells based on molten carbonate (MCFC) or solid oxide (SOFC) technology operate at sufficiently high temperatures and run directly on methane (Chouhan and Liu 2011). This is sometimes called "internal reforming." Thus, MCFC and SOFC systems do not need a

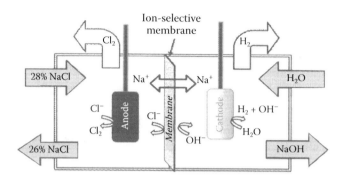

FIGURE 8.5 Ion-selective membrane cell for electrolysis of brine water for electrolysis of water. (Adapted from Boettcher, S. W., E. L. Warren, M. C. Putnam, E. A. Santori, D. Turner-Evans, M. D. Kelzenberg, M. G. Walter, J. R. McKone, B. S. Brunschwig, H. A. Atwater, and N. S. Lewis, *J. Am. Chem. Soc.*, 133: 1216–1219, 2011. With permission.)

pure or relatively pure hydrogen stream as do proton exchange membrane (PEM) and phosphoric acid (PAFC) systems, but these systems can run directly on natural gas, biogas, or landfill gas. Furthermore, such systems can be designed to produce additional purified hydrogen as a by-product, by feeding additional fuel and then purifying the hydrogen-rich "anode tail gas" from the fuel cell into purified hydrogen.

Some cleanup of the methane stream may be required depending on the source of the methane. However, projects such as the Ford Motors plant in Canada-FCE's and Sierra Nevada Brewery in Chico, California, have successfully demonstrated MCFC systems running on a blend of natural gas and brewery wastewater treatment digester gas. These and other wastewater treatment and landfill gas systems are attractive opportunities for renewably powered fuel cell systems.

8.5.7.4 Waste Biomass of Wine Industries

A typical winery in California can generate up to 10–12 million gallons of wastewater per year. The Napa Wine Company of Oakville, California, produced hydrogen by microbial electrolysis. It was the first out-of-laboratory renewable method for hydrogen production from wastewater using a microbial electrolysis cell (MEC) system. The refrigerator-sized hydrogen generator takes winery wastewater and specific bacteria along with a small amount of electrical energy to convert the organic material into hydrogen. MEC is an electrolysis cell, where exo-electrogenic bacteria oxidize biodegradable substrates and produce electrons and protons at the cathode and anode, respectively. Hydrogen gas is produced at the cathode through a recombination of electrons with protons, assisted by an additional voltage supplied by an external power source.

Experiments have determined that one group of bacteria turns unused sugar and unwanted vinegar from improper fermentation into electricity at a working potential of around 0.3 V, which is insufficient for the electrolysis of water. Theoretically, only an additional 0.11 V was needed to produce hydrogen, but in practice, an additional 0.25 V is required due to overpotential at the cathode (Logan and Regan 2006). This means that approximately half the energy needed for electrolysis is possible with MECs, but overall system efficiency is to be explored. The complete system includes maintaining MEC operating conditions, delivering feedstocks to the production facility, replacing other expendable materials, and requiring gas separation and cleanup. One of the biggest problems encountered with this project is maintenance of the effective concentration of bacteria at a certain level that maintains the production rate. Another major issue related to this technique is greater production of methane than hydrogen because hydrogen is being consumed by "methanogenic" microbes before leaving the solution. This methane could then be "reformed" into hydrogen by SMR, but the direct production would be far preferable from an overall energy use standpoint.

8.6 PHOTOCATALYTIC HYDROGEN GENERATION

It would be not overstated to call the twenty-first century the Age of Light (Fujishima and Tyrk 2012). Researchers and industrialists are looking for light-induced chemical and electrochemical reactions that can be performed in an eco-benign way by reducing their energy demands. In this context, annual sunlight exposure on the Earth's surface contained an ample amount of energy—that is, 3×10^{24} J, which is almost 10^4 times greater than the worldwide energy requirement or consumption. Efforts are on to develop efficient materials that can convert solar energy into some useful form of energy (electricity, chemical energy, etc.). This is one of the holy grails of future research and technology development that can lead to the path of clean energy. The following criteria must be satisfied, simultaneously, in designing or producing that kind of substance:

- Absorption of sunlight by some chemical substance that leads to the generation of electrons and holes
- Effective separation of these photogenerated electron–hole pairs at little energetic loss
- Efficient charge transfer on the surface of the material that can promote oxidation and reduction sites on the same surface
- Durability of material so that it should be stable enough at required conditions
- Apt energetics of the material that means their band edges must straddle between the redox potentials of the substance (H_2O, H_2S, pollutant, dye, $CO_2 + H_2O$, NH_3, etc.) to be broken
- Complementary metal-oxide semiconductor (CMOS) and biocompatibility
- Affordable cost of materials
- Nontoxic and easy-to-handle materials and sensitizers

Moreover, nature provides us a big clue about the enigmatic material in the form of the photosynthesis process in living plants or in some bacteria. It is a remarkable example of solar energy conversion into chemically stored energy. Lipids in chloroplast of green plants oxidized water into O_2 and reduced CO_2 into useful organic chemicals in the presence of sunlight. Therefore, photosynthesis fuels life on Earth by providing not only the carbohydrates (food) and oxygen (respiration) but also the fossil fuels.

In looking to the potential of hydrogen fuel, there is a growing interest in storing chemical energy or energy carriers as hydrogen. But fortunately or unfortunately, molecular hydrogen is not freely available on Earth in convenient natural reservoirs. Most of the hydrogen on Earth is bonded to oxygen in the form of water. Currently, elemental hydrogen is manufactured by the consumption of a conventional steam reforming or decomposition of long-chain hydrocarbon present in fossil fuel. In the former case, decomposition of water requires electrical or heat input, generated from some primary energy source (fossil fuel, nuclear power, or renewable energy) at an industrial level. The latter consumes the fossil resource and produces carbon dioxide, but often it requires no further energy input beyond the fossil fuel.

We need first to invest energy to extract the energy that is contained in the hydrogen fuel. A foremost method of hydrogen generation from fossil fuel, steam reforming, leads to carbon dioxide emissions, the same as a car engine produces. One needs to think and act on the line of renewables for hydrogen generation to substitute for fossil fuels, because we need to develop a method with less dependence on fast depleting fossil fuel sources, stability in most regions of crude oil–possessing countries, geopolitical independence, attributes to combat global warming. Therefore, producing hydrogen in an eco-friendly manner is welcome. Moreover, nature provides us with the most abundant resource of hydrogen in the form of water and free sunlight that can split water into hydrogen and oxygen without releasing carbon content into the environment. This process can be named *artificial photosynthesis*. Worldwide, a very small portion of the hydrogen (4% in 2006) is produced by electrolysis of water using electricity that consumes approximately 50 kWh of electricity per kilogram of hydrogen production at the cost of the environmental degradation. Hence, the use of solar energy to generate hydrogen and oxygen out of water, is an attractive choice to meet the growing demand for developing clean energy systems, without disturbing the eco-balance. Now the question arises about the material, which can efficiently split the water. A number of series of semiconductors have been tested on the criteria of cost, stability, and efficiency.

In the early 1970s, Fujishima and Honda (1972) used titanium dioxide (TiO_2) as a first photocatalytic material for water splitting by achieving 1% efficiency. Since then, the titanium dioxide serves as a model material for the mechanistic studies of photocatalysis, but it has a serious limitation as a photocatalyst for water splitting. Its bandgap is too large (3.2 eV) for absorbing visible photons in sunlight (46% Sun radiation); therefore, it is necessary to synthesize or modify the photocatalysts workable with sunlight. Though the UV-based photocatalysts will perform better per photon efficiency than visible light–based photocatalysts due to the higher photon energy, far more visible light than UV light reaches the Earth's surface. Thus, a less efficient photocatalyst that absorbs visible light may ultimately be more useful than a more efficient photocatalyst absorbing solely light with smaller wavelengths. In a single-step photocatalysis, the H_2 evolution rate and STH energy conversion efficiency are treated as a function of the bandgap energy of the photocatalysts with a 100% quantum yield.

It is assumed that photons with wavelengths shorter than 1008 nm convert water to $½H_2$. If the photocatalysts utilize visible light up to 600–700 nm with a quantum yield of 50%, the STH efficiency reaches ~10%, which is similar to some of the commercial photovoltaic systems. But the photocatalytic water-splitting devices are simpler than photovoltaic systems, because vacuum processing is not required in the preparation of devices.

Copious materials of different categories have been studied, and research is still ongoing to determine the visible light active proper material (efficiency more than 10%) for photocatalytic water splitting. Usually, photocatalytic materials along with their highest oxidation states, containing metal cations with d^0 and d^{10} electronic configurations, are suitable for overall water splitting. Heterogeneous photocatalyst materials include metal oxides, nitrides, and sulfides, metal (oxy) sulfides, and metal (oxy) nitrides. These are rigorously tested on the criteria for better efficiency (Kudo and Miseki 2009).

8.7 MECHANISMS OF PHOTOCATALYTIC WATER SPLITTING

The phenomena of water splitting followed four main steps (Figure 8.6):

- Sunlight with photonic energy more than 1008 nm is absorbed on the photocatalytic surface.
- Charge carriers (electron and holes pairs) are created in conduction band (CB) and valance band (VB), respectively.
- The grooves on the photocatalytic surface functioned as an anode (oxidation site) that produced O_2 using holes.

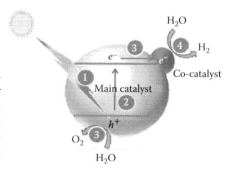

FIGURE 8.6 Four steps involved in photocatalytic water splitting for hydrogen production: (1) receiving photons of energy more than 1008 nm: (2) Formation of photocarriers (i.e., photoelectron and holes): (3) Separation of carriers by movement of e^- (reduction site co-catalyst) and hole toward oxidation site (grooves on catalyst surface): (4) Reduction and oxidation of water at reduction and oxidation site, respectively.

Hydrogen

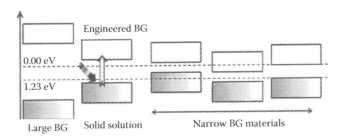

FIGURE 8.7 Bandgap engineering in solid solution photocatalysts, where conduction band and valence band are represented by white and gray boxes, respectively.

- Finally, electrons move toward the co-catalyst (anodic site) for reduction of water to produce hydrogen.

The reduction and oxidation reactions are the basic mechanisms of photocatalytic water splitting. The CB level should be more negative than the hydrogen production level $E_{H_2/H_2O} = 0.00$ eV for hydrogen production, while the VB should be more positive than the water oxidation level $E_{O_2/H_2O} = 1.23$ eV for efficient oxygen production from water by photocatalysis (Figure 8.7). Theoretically, all types of semiconductors that satisfy these requirements can be used as photocatalysts for hydrogen and oxygen production through water splitting.

Although solar water splitting does not suffer from electricity storage problems, unlike the photovoltaics, it can provide two important feedstocks: hydrogen and oxygen. But their efficiency of energy conversion has remained much lower than that of photovoltaics (Fujishima and Tyrk 2012). The upgradation in efficiency based on the emergence of new materials with efficient solar energy harvesting can be possible by lowering bandgap with the assistance of band position engineering, nanostructuring, sensitization with dyes or quantum dots, and addition of proper dopant/alloy and active co-catalyst to main photocatalyst (Bao 2014). The introduction of new systems like composite materials, solid solutions, or plasmonic materials with their fine-tuned surface plasmon resonance and supramolecular assemblies, as well as introduction of the active co-catalysts for hydrogen and oxygen evolutions may enhance the efficiency of water splitting. Recently, one of the most important advancements reported is in the form of the functionalized graphene oxides that can perform overall water splitting without co-catalysts and a sacrificial reagent. Some of the representative examples along with their water-splitting rates (WSR) and quantum yield (η) are as follows.

8.8 MODIFICATION OF PHOTOCATALYSTS

8.8.1 PHOTOSENSITIZATION: DYES/QUANTUM DOTS

Sensitization is a widely used technique to utilize visible light for energy conversion. Some dyes having redox property and visible light sensitivity can be used in solar cells as well as photocatalytic systems (Choi et al. 1994; Maeda and Domen 2010). Under visible light illumination, the excited dyes/quantum dots (QDs) can inject electrons to the CB of the photocatalyst to initiate the catalytic reactions (Figure 8.8). Photoactivity of the whole system depends on the efficient absorption of the visible light and efficient transfer of the electrons from excited dyes/QDs to the CB of TiO_2. Afterward, the CB electrons can be transferred to the co-catalyst (noble metal such as Pt) loaded on the surface to reduce the water. Redox systems or sacrificial agents, such as the I^{3-}/I^- pair and EDTA, can be added to rejuvenate dyes that sustain the reaction cycle.

The selection of QDs/dye-sensitized systems benefits the reaction by improving electron–hole separation or suppression of charge recombination centers, thus improving the solar visible light–harvesting power of the photocatalyst, and the selectivity or yield of a particular product. It has been shown that the inorganic sensitizers, organic dyes, and coordination metal complexes can be used as the effective

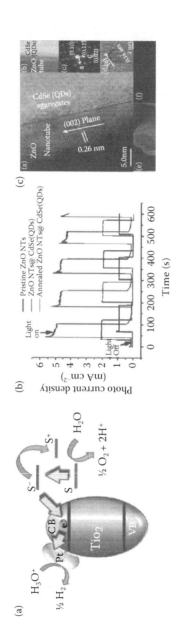

FIGURE 8.8 (a) Dye sensitization of the photocatalyst for overall water splitting. (b) Efficient and sustainable QD CdSe sensitized ZnO nanotubes used for water splitting. (c) High-resolution transmission electron microscopy (HRTEM) image, showing the lattice fringes of CdSe(QDs) loaded on ZnO NTs; transmission electron microscopy (TEM) image of aggregates of CdSe QDs loaded on 002 plane of ZnO nanotubes; SAED pattern of pristine ZnO NTs and HRTEM lattice image of single CdSe QD. (Adapted from Chouhan, N., C. L. Yeh, S. F. Hu, R. S. Liu, W. S. Chang, and K. H. Chen, *Chem. Commun.*, 2011, 47, 3493–3495. With permission.)

Hydrogen

photosensitizers (Gurunathan et al. 1997; Sauve et al. 2000; Jing and Guo 2007). QD sensitization with their significant qualities like quantum confinement, efficient charge transfer, and separation are attracting the attention of researchers. Photosensitization by dyes is common in solar cells due to their prominent visible light absorption properties, ease of change in their structures, and low cost.

8.8.2 Reducing Bandgaps Through Doping

Noble metal, transition metals, or nonmetals (B, C, N, F, P, S) alloying/doping to the main photocatalytic system is also a popular mode of reducing bandgaps of the photocatalysts by adjusting Fermi levels (Ni et al. 2007; Zhang et al. 2013) (Figure 8.9).

8.8.3 Engineered Solid Solutions

Solid solutions are usually produced by the combination of large-bandgap materials with low-bandgap solids at high temperature (Figure 8.7). These solutions contain four or more elements. The $(Ga_{1-x}Zn_x)(N_{1-x}O_x)$ and Ta-based semiconductors are well-known examples of such semiconductor alloys. UV-based photocatalyst $NaTaO_3$:La (WSR = 9.7 mmol h^{-1} and η = 56%) (Kato et al. 2003) and $K_3Ta_3B_2O_{12}$ (WSR = 1.21 mmol h^{-1}, and η = 6.5%) without co-catalyst assistance and use of sacrificial reagents (Kurihara et al. 2006), yields pretty high efficiency for water splitting. Visible light active $Ga_{0.82}Zn_{0.18}N_{0.82}O_{0.18}$ (WSR = 0.4 mmol h^{-1} and η = 5.9%) (Maeda et al. 2008), loaded with co-catalyst $Rh_{2-y}Cr_yO_3$ and without utilizing the sacrificial reagents, has the highest quantum yield along with a water-splitting rate in the presence of visible light. These catalysts can be tuned for high efficiency by calcination that reduced the amount of surface defects, which can act as recombination sites.

8.8.4 Macromolecular Systems for Water Splitting

Thermodynamics of the solar water splitting involves an energetically uphill chemical reaction that required 1.23 V. Water reduction is energetically favorable via a multielectron pathway (1.23 V vs. NHE) compared to via a single-electron pathway (5 V vs. NHE) that makes the design of a complete water-splitting system quite complicated. Therefore, an in-depth understanding of either oxidation or reduction by replacing the other half of the reaction with a sacrificial electron acceptor (EA) and an electron donor (ED), is required. Many macromolecular systems are emerging as the photocatalytic active substances that include organized zeolites-based photoredox systems. A synthetic strategy for the construction of zeolite-entrapped organized molecular assemblies has to be developed. This can be achieved by interacting adjacent cage dyads composed of two polypyridine complexes of Ru(II)

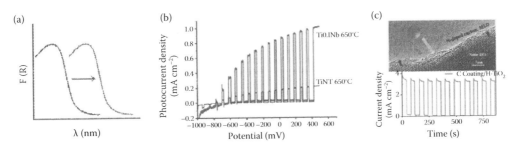

FIGURE 8.9 (a) Modification of UV photocatalysts by cation/anion doping. (b) Nb-doped TiO_2 nanotube versus Ti–Nb alloys, showing strongly increased and stable photoelectrochemical water splitting. (c) Enhanced water-splitting performance of TiO_2 nanotube arrays coated with an ultrathin nitrogen-doped carbon film. (Adapted from Das, C., P. Roy, M. Yang, H. Jha, and P. Schmuki, *Nanoscale*, 3, 3094–3096, 2011, and Tong, X., P. Yang, Y. Wang, Y. Qin, and X. Guo, *Nanoscale*, 6, 6692–6700, 2014. With permission.)

or from heterosupramolecular chemistry, metal clusters, dendrimers, and so on (Sykora et al. 1998; Arachchige et al. 2009). Among them, the most common photocatalysts with normal structure and diverse function solar hydrogen–producing capacity, are the systems that include Ru, Pt, Ru–Rh, and Rh–Rh supramolecular architecture.

In this context, supramolecular chemistry provides an ideal solution to study, design, and develop a solar hydrogen photocatalyst for overall water splitting. Overall photocatalytic activities of the supramolecules involve some individual component, which performs a specific act of either oxidation or reduction. Consequently, the assembly of supramolecules can be varied in a fine-tunable fashion, along with maintenance of the overall water-splitting process. At least two electrons are collected at the central site of the supramolecule after optical excitation by means of a photoinitiated electron collection (PEC) process. For an efficient PEC, supramolecules can be designed to have a TL/ED-LA-BL-EC-BL-LA-ED/TL structural motif, where the term TL is a terminal ligand (e.g., two bpy, two dpp, two dpb, two phen, or tpy and Cl ligands; bpy = 2,2′-bipyridine, phen = 1,10-phenanthroline, tpy = 2,2′:6′,2″-terpyridine, dpp = 2,3-bis(2-pyridyl)pyrazine, dpb = 2,3-bis(2-pyridyl) benzoquinoxaline), LA is light absorber, BL is bridging ligand (dpp, whereas the Ir(III) system uses the dpb ligand), and EC is electron collector ($M'X_2$; M' = Rh(III) or Ir(III) and X = Cl or Br).

The complexes possess intense MLCT transitions in the visible region of the spectrum, making them efficient LAs. Light absorption is dictated by the TL-LA-BL subunit. The presence of two LAs connected to a single EC through bridging ligands allows electron collection with multiple photoexcitation steps and intense metal ligand charge-transfer transitions in the visible region of the spectrum, making them efficient LAs. Light absorption is dictated by the TL-LA-BL subunit. Thus, the supramolecular motifs have great potential to exploit the light radiation for breaking H_2O for hydrogen generation. The first functioning PEC material was $[(bpy)_2Ru(dpb)_2IrCl_2][PF_6]_5$, which collects two electrons on the dpb(π^*) orbital upon sequential excitation and electron transfer events. The three most active photocatalysts, $[\{(bpy)_2Ru(dpp)\}_2RhCl_2](PF_6)_5$, $[\{(bpy)_2Ru(dpp)\}_2RhBr_2](PF_6)_5$, and $[\{(phen)_2Ru(dpp)\}_2RhCl_2](PF_6)_5$ were used for the higher hydrogen yield using the DMA electron donor (Balzani et al. 1997). The results of the screening of these new photocatalysts reveal that the designs are functional for supramolecular complexes as photocatalysts in general or complexes that can be predictable on the basis of their basic chemical properties. For systems that do function, optimization of the reaction conditions, by way of component modi-cation, requires identifying new electron donors and their optimized experimental conditions.

Dedicated efforts of the researchers in this direction are underway to gear the better PEC in the supramolecular architecture for achieving higher hydrogen generation efficiency. The first rhodium-centered trimetallic supramolecular $[\{(bpy)_2Ru(dpp)\}_2RhCl_2]^{5+}$ was reported that collects multiple electrons and is capable of photocatalyzing hydrogen production from water. This supramolecular system incorporated two $[(bpy)_2Ru(dpp)]^{2+}$ LAs and a single Rh EC, which allowed for the multiple electron collection necessary for the multielectron reduction of the substrate. The monodentate Rh-bound ligands can be lost after photoreduction of the substrate, which provides a site for substrate binding for photoreactivity to the supramolecular complex.

The redox properties of $[(bpy)_2Ru(dpp)_2RhCl_2]^{5+}$ demonstrated that the Ru LA are largely electronically isolated important PEC. Its multielectron photochemical component was modified by changing the terminal ligand (TL), TA metal, and Rh-bound halides. Subsequently, it afforded a series of additional photoinitiated ECs having a LA-BL-Rh_2X-BL-LA structure motif (LA = Ru″ or Os″ polyazine X = Cl^-, Br^-, I^-, BL-dpp), and these cells are capable of catalyzing solar hydrogen production. Studies have established $[(bpy)_2Ru(dpp)_2 RhCl_2]^{5+}$ and $[[(phen)_2Ru (dpp)]_2RhX_2]^{5+}$ X = Cl^-, Br^-, as photochemical molecular devices for PEC hydrogen production photocatalysis with a hydrogen yield ϕ = 0.01 (Dincer and Zamfirescu 2012). Ozawa et al. (2006, 2007) investigated a $[Pt^{II}(bpy)_2Cl_2]$ component linked to a ruthenium LA through an amide linkage to ruthenium capable of photochemically producing hydrogen from water ($\phi \approx 0.01$ and 5 turnovers in 10 h) in the presence of ethylenediaminetetraacetic acid. Rau et al. (2006) reported a Ru-Pd bimetallic system that photochemically produces hydrogen in the presence of an electron donor, TEA, affording 56 turnovers in ~30 h.

Recent studies on platinum (Du et al. 2008) and palladium (Lei et al. 2008) systems suggested that hydrogen production can be attributed to the decomplexation of the reactive metal to form colloidal metal, which can function as the hydrogen generation site. Elvington and Brewer (2006) reported the photoinitiated electron collection by a metal center in $[\{(bpy)_2Ru(dpp)\}_2RhCl_2](PF6)_5$. When excited with visible light, this complex is reduced, converting Rh(III) to Rh(I), which undergoes chloride loss to form $[\{(bpy)_2Ru(dpp)\}_2RhI]^{5+}$ (Elvington et al. 2007). This complex photochemically reduces water to hydrogen using visible light excitation with $\phi \approx 0.01$ in the presence of a sacrificial electron donor DMA. The ability of the system to undergo photoinitiated electron collection on the rhodium center with the molecular architecture remaining intact is unprecedented and allows its use in multielectron photochemistry.

The design considered for mixed-metal supramolecular complexes capable of photoinitiated electron collection provides an insight into general molecular architectures for photocatalysts for solar energy–driven hydrogen generation from water. Some of the polymetallic complexes that couple ruthenium or osmium polyazine LAs to rhodium or iridium metal centers through BLs have been reported with many interesting light-activated DNA photocleavage agents (Elvington et al. 2007; Arachchige et al. 2008). Existence of the supramolecular catalyst in the photocatalytic world is prominent, but it is a major challenge to maintain photoelectrons for a sufficiently long duration at the supramolecular level.

8.8.5 Plasmonic Nanostructures with Surface Plasmon Resonance

Nanoparticles of pure metals show active absorption in the visible region because of the existence of the unique feature known as surface plasmon resonance (SPR). Plasmonic photocatalyst has two prime parts (1) nanoparticles of noble metals with LSPR and (2) support/carrier (semiconductor/polar material) with Schottky junction. Nanoparticles of metals contribute strong absorption of visible light and excitation of active charge carriers, whereas the latter induces the charge separation and transfer. Free electrons of the noble metals will integrate with the photon energy that produces subwave and conducting electrons in oscillating mode and offers localized surface plasmonic resonance (Ritchie 1957).

Resonance frequency of the plasmonic material can be tuned by varying the nanoparticle size, shape, material, and proximity to other nanoparticles. Due to LSPR, NPs of noble metals can act as the thermal redox reaction-active centers on catalyst that can trap, scatter, and concentrate light. Furthermore, integration of the plasmonic metals with the semiconductor includes plasmon resonance energy transfer (PRET), photon scattering effect, hot electron transfer, plasmon-induced heating, reflection reduction, reduced e/h diffusion length, enhanced local electric effect and molecular polarization effect, quantum tunneling effect, and high catalytic effect, at the metal/semiconductor interface (Barnes et al. 2003; Maier and Atwater 2005; Maier 2007; Lal et al. 2007). The typical examples of exhibiting these properties are plasmonic silver or gold nano particle–assisted oxide semiconductor $N-TiO_2$ and $CdS@SiO_2//Au@SiO_2$ (Figures 8.10 and 8.11).

In typical PEC water splitting with plasmonic metal integrated metal oxide, hot electron of noble metal is injected to conduction bands of metal oxides, which increases the power absorbed in this region and enhances the efficiency of the device. Few of the plasmonic photocatalytic systems for enhancement of the renewable hydrogen evolution through photocatalytic water splitting are Au/graphene/TiO_2 (12 mmol/2.4 mmol h^{-1}) (Singh et al. 2014), Au/ZnO nanorods (11.2 and 4.4 µmol h^{-1}) (Chen et al. 2012), $CdS@SiO_2//Au@SiO_2$ (10 mmol h^{-1} in 300 min exposure) (Torimoto et al. 2011), and many more. By integrating plasmonic materials with hydrogen and oxygen co-catalyst on the main catalyst, overall water splitting will be further enhanced. Because the scattering of photons by plasmonic nanoparticles promotes PRET phenomena that allow efficient electron transfer and lead selective formation and separation of the electron–hole pairs in the near-surface region of the

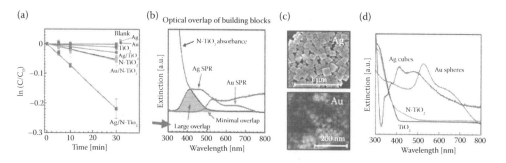

FIGURE 8.10 (a) Aqueous methylene blue decomposition over different photocatalysts: Ag cubes only, Au spheres only, P25 TiO_2 only, N-TiO_2 only, Ag/TiO_2 composite, Au/N-TiO_2 composite, and Ag/N-TiO_2. (b) Overlap between the illumination source spectrum, semiconductor absorbance spectrum, and metal nanoparticle surface plasmon resonance spectrum. (c) Scanning electron micrograph (SEM) of the Ag nanocube (edge length = 118, 25 nm) and spherical gold particles (diameter 25.4, 4.5 nm). (d) Diffuse reflectance UV-visible spectra of TiO_2, N-TiO_2, Ag nanocubes, Au spheres. TiO_2 and N-TiO_2 source. (Adapted from Ingram, D. B., P. Christopher, J. L. Bauer, and S. Linic, *ACS Catal.*, 1, 1441–1447, 2011. With permission.)

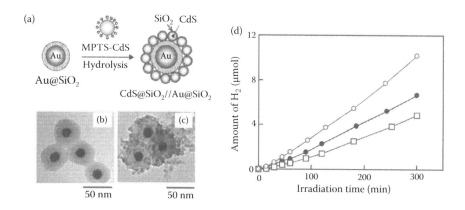

FIGURE 8.11 (a) Immobilization of CdS@SiO_2 on Au@SiO_2 particles. (b) TEM images of Au@SiO_2 prepared with Au particles of 19 nm in diameter as a core. (c) CdS@SiO_2-deposited Au@SiO_2 particles CdS@SiO_2//Au@SiO_2. (d) Time courses of hydrogen evolution by photocatalysts of CdS@SiO_2 (solid circles) and CdS@SiO_2//Au@SiO_2 particles with SiO_2 layer thicknesses of 17 nm (open circles) and 2.8 nm (open squares) on Au cores. The experiments were performed by irradiation light from a 300 W Xe lamp ($\lambda > 350$ nm). (Adapted from Torimoto, T., H. Horibe, T. Kameyama, K. Okazaki, S. Ikeda, M. Matsumura, A. Ishikawa, and H. Ishihara, *J. Phys. Chem. Lett.*, 2, 2057–2062, 2011. With permission.)

semiconductor. In totality, one can say that water splitting via SPR is a complex multifaceted process, and enhancement in its efficiency requires the thoughtful attention to various parameters such as morphology, stability, structural defects, bandgap and band-edge position of the metal oxides, and size, morphology, plasmonic particle areal density, and location of plasmonic metal nanoparticles.

8.8.6 Nanostructuring of the Photocatalyst

Nanoparticles continue to exhibit much higher activity than their bulk counterparts due to the high surface-to-volume ratio. These two-dimensional nanomaterials have the potential to be excellent catalysts, as they can harvest solar energy and generate photoelectrons and holes, along with provide paths for the separation and diffusion of photoexcited carriers. Liao et al. (2014) studied the cobalt (II) oxide (CoO) nanoparticles that can carry out overall water splitting with a solar-to-hydrogen

Hydrogen

efficiency of around 5%. Bismuth-based systems have also been demonstrated to have an efficiency of 5% with the advantage of a simple and cheap catalyst (Abdi et al. 2013). Furthermore, the p-type conductivity results in the formation of an accumulation layer at the graphene (GO)/water interface, which is favorable for water reduction to hydrogen. Modification of a graphene sheet to exhibit both p- and n-type conductivities may produce a photocatalytic medium, which is effective for overall water splitting into H_2 and O_2. Here, effective exciton separation and charge transfer are essential factors responsible for overall water splitting. Reducing the size of the GO sheets may lower the recombination probability of the photogenerated charges (Yeh et al. 2014).

8.8.7 Composite Photocatalyst

Oxide photocatalysts are the most commonly used photocatalysts in various applications of optoelectronic materials due to their potential thermal and chemical stability, low cost, and environmentally friendly aspects, but these are UV light–active materials due to their wide bandgap. Therefore, some suitable nonoxide/oxide with a proper band-positioned photocatalyst is used to couple with them to make them visible light active. The whole assembly of these systems is known as a composite material. These were found to be suitable for water splitting because of their excellent photocorrosion resistance, which is ascribed to a considerable decrease of surface defect density on the photocatalytic surface and the reduction of holes. A few of their examples are $CdS-TiO_2$, $CdS-ZnO$, $CdS-AgI$ (Yamada et al. 2005), $ZnO/Ag/CdS$ ($\eta = 3.5$ mL h^{-1}) (Zhang et al. 2014), $ZnO/Ag/AgCl$ (Pirhashemi and Yangjeh 2014), and so on. The electron transfer mechanism for water splitting has been demonstrated in Figure 8.12.

Material with more than 10% efficiency would be a major breakthrough for hydrogen production by using solar water splitting (Tachibana et al. 2012). Recent progress toward developing artificial photosynthetic devices, together with their analogies to biological photosynthesis, including technologies that focus on the development of visible light active heterostructures still require an understanding of the underlying interfacial carrier dynamics. Of course, it has definitely proposed a vision for a futuristic development of the novel efficient materials along with the improved designs of the rational device, which will be based on theory and experimental simulations.

8.8.8 Photoelectrochemical Water Splitting

Electricity produced by photovoltaic systems offers one of the cleanest ways to produce hydrogen. We can use photonic energy from sunlight to subsidize its electricity demand. These reactions are known as photoelectrochemical (PEC) reactions. When water is broken into hydrogen and oxygen by electrolysis under light exposure in a PEC device, the process is named *artificial photosynthesis*. These PEC devices are made of three electrodes: photoanodes; photocathodes, mainly platinum

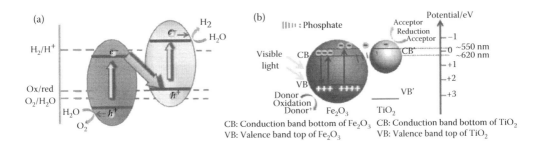

FIGURE 8.12 (a) Water splitting in a composite system. (b) Improved photoactivity of TiO_2–Fe_2O_3 nanocomposites for visible light water splitting after phosphate bridging and its mechanism. (Adapted from Luan, P., M. Xie, X. Fu, Y. Qu, X. Sun, and L. Jing, *Phys. Chem. Chem. Phys.*, 17, 5043–5050, 2015. With permission.)

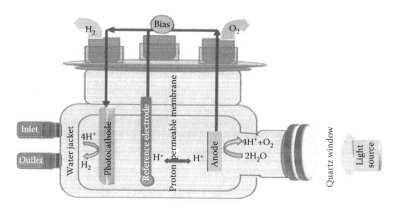

FIGURE 8.13 Photoelectrochemical cell utilized for water splitting.

wires; and reference electrodes (calomel or Ag/AgCl electrode) dipped in aqueous electrolyte (Figure 8.13). Photoelectrochemical solar devices must accomplish several tasks during water splitting. The steps involved in water splitting are

1. The photocatalyst absorbs short-wavelength, high-energy photons on a catalytic surface and transmits longer-wavelength photons.
2. The mobile photocharge carriers (i.e., photoelectrons and holes) are generated.
3. The generated charge carriers are transported toward catalytic reduction and oxidation sites.
4. The H_2 and O_2 are produced at the cathode and anode, respectively.

Electrochemical mass transport can be categorized according to three driving forces: convection, migration, and diffusion. Mechanical movement of the bulk solution generates convective transport, which is suppressed by structured photoelectrodes. Migration refers to the motion of charged species under the influence of electric fields, the operation in either highly acidic or highly alkaline media, and the excess of the supporting electrolyte, which effectively suppresses ion migration forces. Ultimately, the diffusion is likely to dominate mass transport in the vicinity of structured photoelectrodes, as used for water-splitting purposes.

8.9 ELECTROCHEMICAL ASPECT

When a nonilluminated semiconductor electrode is in contact with the solution of a redox couple electrolyte, then equilibrium taking place at the interface and the Fermi level of the semiconductor can be adjusted with the same redox couple electrolyte. During this process, two layers are formed. The first is the space charge layer/depletion layer ($1-10^3$ mm, depending upon the charge density and dielectric constant) formed inside the semiconductor of the thin region close to its surface. Mobile charge carriers have diffused away, or have been forced away by an electric field in the space charge layer. The only elements left in the depletion region are ionized donor or acceptor impurities. The second layer is the interfacial layer between the electrode surface and the electrolyte solution. Bending (upward for n-type) or (downward for p-type) changing the direction of electronic bands (valence and conduction bands) occurs at the space charge layer. The size of the bending depends on the relationship between the Fermi level of the semiconductor and the redox couple solution, as depicted by Figure 8.14.

If the semiconductor is illuminated by the light having photonic energy greater than the bandgap of the semiconductor, then the electron–hole pair is generated and separated in the space charge

Hydrogen

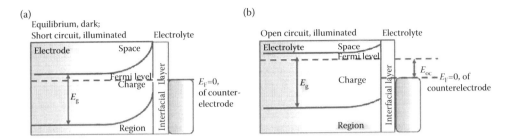

FIGURE 8.14 Energy level of generic n-type semiconductor in a representative PEC: (a) equilibrium, dark and short circuit, illuminated condition; (b) open circuit, illuminated condition.

layer. In an n-type semiconductor, the electric field exists across the space charge layer, it drives holes to move toward the interfacial layer (electrolyte/solution), and the electron moves to the interior of the electrode and from there to the external circuit. In an open circuit, electrons flow toward the bulk semiconductor and begin to accumulate, which will raise the Fermi level. This level will continue to increase depending on the light intensity and recombination rate, until a steady state is reached. There is no net change in the number of oxidized and reduced forms of the redox couple because the same number of oxidized forms is reduced at the semiconductor surface as the number of reduced forms produced by the oxidized redox couple.

At illuminated semiconductor–liquid junctions, the open-circuit photovoltage, V_{oc} represents a fundamental figure of the merit for photoelectrochemical water splitting (Equation 8.27). As the flow of current is a kinetic phenomenon, V_{oc} is a kinetic parameter, not a thermodynamic parameter (Boettcher et al. 2010):

$$V_{oc} = n \frac{K_B T}{q} \ln\left(\frac{J_{ph} N_{D:A} L_{min}}{q D_{min} n_i^2}\right) \tag{8.27}$$

where:
- n is the diode ideality factor
- J_{ph} is the short-circuit photocurrent density (photocurrent per unit area) under illumination
- N_D (N_A) is the donor (acceptor) density
- L_{min} is the minority carrier diffusion length
- D_{min} is the minority carrier diffusion coefficient
- n_i is the intrinsic carrier density

The term saturation current density J_s, is equal to $-q D_{min} n_i^2 / N_{D:A} L_{min}$, as demonstrated by Equation 8.28 (Boettcher et al. 2011):

$$V_{oc} = n \frac{K_B T}{q} \ln\left(\frac{J_{ph}}{J_s}\right) \tag{8.28}$$

The dimensionless figure γ is introduced in Equation 8.28 for structured materials used in bulk photoelectrochemical water splitting, to reconcile the incongruity between the two areas (i.e., J_{ph} and J_s) (Equation 8.29) (Stempel et al. 2008):

$$\gamma = \frac{\text{Geometric surface area}}{\text{Projected surface area}} \tag{8.29}$$

Incorporation of term γ into Equation 8.29 enables a straightforward comparison of the currents and current densities (Equation 8.30):

$$V_{oc} = n\frac{K_B T}{q} \ln\left(\frac{J_{ph}}{\gamma J_s}\right) \quad (8.30)$$

Equation 8.30 was introduced for the small contact area heterojunctions, optical concentrators, or a nanoemitter-styled photoelectrochemical cell (Yamada et al. 2005; Zhang et al. 2014; Pirhashemi and Yangjeh 2014).

Ayers (1983) of Energy Conversion Devices (United States), has demonstrated and patented the first multijunction high-efficiency photoelectrochemical system for direct splitting of water. This group demonstrated wireless solar water splitting with low-cost thin-film amorphous silicon multijunction sheets with metal substrate at back and sandwiched Nafion membrane (provided a path for ion transport), merged in water. The H_2 was evolved on the front catalyst–decorated amorphous silicon surface, while oxygen was evolved off the back of the metal substrate. Research continues toward developing high-efficiency PEC technology at academic and industrial levels. Few ambassador PEC electrode materials are mentioned here to cover the whole range of the photocatalytic semiconductors. In particular, doped and undoped oxide photocatalyst ZnO (Ahn et al. 2007, 2008; Yang et al. 2009) are used for water splitting because of their unmatched piezo-electric properties. Anodically fabricated TiO_2 nanotube arrays have proved to be robust and cost-effective functional materials. These have been widely investigated in many applications, especially those related to energy conversion, such as photoelectrochemical water splitting and solar cells (Murphy 2008; Grimes and Mor 2009; Allam and Grimes 2009; Hamedani et al. 2011; Sharmoukh and Allam 2012).

Simultaneously, the compositional doping of TiO_2 with different elements was considered as an approach for bandgap engineering (Sharmoukh and Allam 2012). The $(Ga_{1-x}Zn_x)(N_{1-x}O_x)$ modified with $Rh_{2-y}Cr_yO_3$ holds the highest AQY for overall water splitting using a single photocatalyst under visible light (5.1% at 410 nm) (Maeda and Domen 2010). However, solid solution $(Ga_{1-x}Zn_x)(N_{1-x}O_x)$ did not exhibit good activity beyond 440 nm for water-splitting devices that can typically harvest visible light and have a low solar-to-hydrogen efficiency of around 0.1%. Co-based photocatalysts are emerging materials for hydrogen evolution with a solar-to-hydrogen efficiency of 5%. They add volume to the developing potential of Co-based photocatalysts (Liao et al. 2014). Additionally, some chalcogenides containing CuI ions, such as $CuGaSe_2$, Cu_2ZnSnS_4, and $Cu(Ga,In)(S,Se)_2$, function as p-type semiconductors and can also be employed as photovoltaic cells (Ma et al. 2011; Yokoyama 2010; Moriya et al. 2013). IrO_2-loaded $LaTiO_2N$ electrodes have been used with a Ta contact layer between the $LaTiO_2N$ particles and the Ti conductor layer in the electrolyte of aqueous 1 M Na_2SO_4 solution with pH = 13.5 (adjusted by the addition of NaOH) (Minegishi et al. 2013).

CoO_x was also found to be effective for nitride and oxysulfide photocatalysts such as Ta_3N_5 (Ma et al. 2013) and $Sm_2Ti_2S_2O_5$ (Ma et al. 2012). CdS is an n-type semiconductor and can form a good p-n heterojunction with $CuGaSe_2$. The photoelectrode surface was modified with Pt to catalyze the hydrogen evolution reaction. The band bending initiated in CdS on $CuGaSe_2$ is the key to enhance the photocurrent (Figure 8.15a) (Moriya et al. 2013). The highest AQY of 6.3% at 420 nm was achieved with a Z-scheme system based on Pt-loaded $ZrO_2/TaON$ as a hydrogen evolution photocatalyst, Pt-loaded WO_3 as an oxygen evolution photocatalyst, and IO_3^-/I^- as a redox couple (Figure 8.15b) (Maeda et al. 2010), which has shown almost double activity.

Scientists are also interested in graphene species, a promising material for synthesizing metal-free, cost-effective, and environmentally friendly catalysts for overall water splitting under solar illumination. Although the water-splitting activity of the nitrogen-doped graphene oxide quantum

Hydrogen

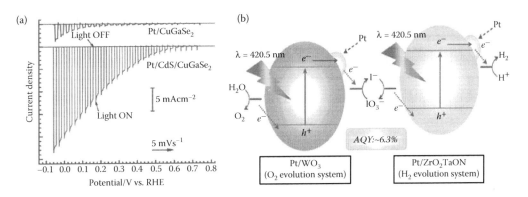

FIGURE 8.15 Photoelectrochemical: (a) current density versus voltage curve for Pt/CuGaSe$_2$ and Pt/CdS/CuGaSe$_2$. (Adapted from Moriya, M., T. Minegishi, H. Kumagai, M. Katayama, J. Kubota, and K. Domen, *J. Am. Chem. Soc.*, 135, 3733–3735. With permission.) (b) Z-scheme based on Pt-loaded ZrO$_2$/TaON as a hydrogen evolution photocatalyst, Pt-loaded WO$_3$ as an oxygen evolution photocatalyst, and IO$_3^-$/I$^-$ as a redox couple. (Adapted from Maeda, K., M. Higashi, D. Lu, R. Abe, and K. Domen, *J. Am. Chem. Soc.*, 132, 5858–5868. With permission.)

dots was reported to be almost half that of the Rh$_{2-y}$Cr$_y$O$_3$/GaN:ZnO, it has a lot of possibilities for the future (Maeda et al 2010; Yeh et al. 2014). The photocurrent densities of the gold nanorod–TiO$_2$/platinum electrode correspond to the hydrogen and oxygen production under the full sunlight irradiation. This system produced 5 × 10^{13} H$_2$ molecules cm^{-2} s^{-1} under 1 sun illumination (AM 1.5 G and 100 mW cm^{-2}), with unprecedented long-term operational stability. This is high among the reported plasmonic systems in terms of solar-to-hydrogen efficiency (≈0.1%) but low enough for practical use (Mubeen et al. 2013).

8.10 HYDROGEN AS A KEY SOLUTION

In looking to the recent developments in hydrogen production, storage, and potential as a fuel and energy carrier, one can easily say that hydrogen will be a key solution to the current energy challenges faced by the different facets of economy/society. Hydrogen economy as a primary energy source and feedstock produced from sources other than coal, oil, and natural gas, would cut down the release of the greenhouse gases into the atmosphere, which is the primary feature associated with the combustion of fossil energy resources. The major benefit of the hydrogen economy would be that energy generation and its use could be decoupled. Therefore, the primary energy source would not have to travel with the vehicle, as currently required for hydrocarbon fuels. Instead of tailpipe-dispersed energy emissions, the energy could be generated from point sources, which are otherwise unfeasible for mobile applications.

Centralized energy plants promise higher hydrogen production efficiency, but suffer from difficulties in high volume and long-range hydrogen transportation (due to factors such as hydrogen damage and the ease of hydrogen diffusion through solid materials), which makes electrical energy distribution attractive within the hydrogen economy. Under the distributed energy generation schemes, the small-scale renewable energy sources could be associated with hydrogen stations. The proper balance between hydrogen distribution and long-distance electrical distribution is one of the primary questions raised about the hydrogen economy. Again, the dilemma of production sources and transportation of hydrogen can now be overcome using on-site (home, business, or fuel station) generation of hydrogen from off-grid renewable sources.

Distributed electrolysis would bypass the problems of distributing hydrogen by distributing electricity instead of hydrogen. Generated electricity can be transported to the end user by using an existing electrical network or on-site electrolyzer located at filling stations. Hydrogen has been explored as an excellent fuel for passenger vehicles, where it can be used in fuel cells to power electric motors or burned in internal combustion engines (ICEs). In both ways, it is an environmentally friendly fuel that has the potential to dramatically reduce our dependence on imported oil. The alternative fuel hydrogen falls into a category of nonconventional or advanced fuels.

The combination of the fuel cell and electric motor is two to three times more efficient than an ICE. However, the high capital cost of fuel cells is one of the major obstacles in its development that has to be overcome in the future. Until then, the fuel cells are technically but not economically more efficient than an ICE (Tester et al. 2005). Other technical obstacles include hydrogen storage issues and the purity requirement of hydrogen used in fuel cells. With current technology, an operating fuel cell requires the purity of hydrogen to be as high as 99.999%. Hydrogen engine conversion technology is more economical than fuel cells. Hydrogen produces no air pollutants or greenhouse gases when used in fuel cells; however, it produces only nitrogen oxides (NO_x), when burned in ICEs. There are several significant challenges (price, hydrogen storage issues, and the purity of hydrogen for fuel cells) and until solutions are available for these problems, it cannot be widely used. In the event that fuel cells become price competitive with ICEs and turbines, large gas-fired power plants could adopt this technology. Hydrogen can also be utilized as the raw material in many chemical industries, cooking gas, fuel for house warming, fuel cells, power production, high-temperature welding, and vehicles. Therefore, hydrogen is a potential candidate for providing a clean and green energy solution to our present and future energy needs.

To sum up the story on hydrogen energy, it is worthy to summarize the benefits and drawbacks of this alternative source of energy as a fuel and energy carrier. Furthermore, a hydrogen economy should be duly established and its importance in present and futuristic points of view is to be stressed. Hydrogen is popular for its high energy density-by-weight ratio that supported its candidacy as a clean fuel. The obvious advantage of hydrogen is its environmentally favorable profile and the availability of multiple sources to produce hydrogen, such as water, methane, gasoline or coal, and so on. Out of these sources, water seems to be an ultimate source of hydrogen because its decomposition produced hydrogen and oxygen without releasing noxious carbon content in the air.

Several popular methods have been discussed for water splitting including the solar cleavage of water using photocatalytic and photoelectrocatalytic techniques accompanied with their related conceptual details and representative examples. But these techniques still need improvements to meet required efficiency. Research into the development of an appropriate material continues for large-scale hydrogen production with high efficiency. It is hoped that we will hear good news from the photocatalytic world of materials. The efficiency of hydrogen production can be increased by improving material engineering and systems design. In order to meet the ultimate target of achieving benchmark STH efficiency (i.e., more than 10% for the solar water splitting), one has to make it commercially competitive and resolve the hurdles in the path of economical, green, and clean production of hydrogen. Unlike past practices, where materials are discovered through trial and error, we are going to possess more novel materials and rational device designs based on theory and experimental simulations.

The hydrogen economy is proposed to suppress the negative effects of conventional fuels, which release the load of the carbon into the atmosphere, and hydrogen is promoted as an environmentally cleaner source of energy to end users (especially in transportation applications), to cut down the release of air pollutants including carbon dioxide into the atmosphere at the point of end use. Other technical obstacles include the issue of storage and the generation of highly pure (99.999%) hydrogen that is required for fuel cells. Although hydrogen engine conversion technology is more economical than fuel cells, the most significant drawback is its high production cost and difficulties in storage during transport in comparison to traditional sources of energy. Future innovations in technology and chemistry will help us to solve this issue. The moral of the story is that hydrogen can be considered as a fuel of the future that still needs advancement in available technologies for trouble-free storage and ample supply at low cost.

REFERENCES

Abanades, S., and G. Flamant. 2006. Thermochemical hydrogen production from two step solar driven water splitting cycle based on cerium oxides. *Solar Energy.* 80: 1611–1623.

Abdi, F. F., L. Han, A. H. M. Smets, M. Zeman, B. Dam, and R. Van de Krol. 2013. Efficient solar water splitting by enhanced charge separation in a bismuth vanadate-silicon tandem photoelectrode. *Nat. Commun.* 4: 2195. doi:10.1038/ncomms3195

Ahn, K. S., Y. Yan, S. Shet, T. Deutsch, J. Turner, and M. Al-Jassim. 2007. Enhanced photoelectrochemical responses of ZnO films through Ga and N codoping. *Appl. Phys. Lett.* 91: 231909.

Ahn, K. S., Y. Yan, S. Shet, K. Jones, T. Deutsch, J. Turner, and M. Al-Jassim. 2008. ZnO nanocoral structures for photoelectrochemical cells. *Appl. Phys. Lett.* 93: 163117.

Allam, N. K., and C. A. Grimes. 2009. Effect of rapid infrared annealing on the photoelectrochemical properties of anodically fabricated TiO_2 nanotube arrays. *J. Phys. Chem. C.* 113: 7996–7999.

Arachchige, S. M., J. Brown, and K. J. Brewer. 2008. Photochemical hydrogen production from water using the new photocatalyst $[\{(bpy)_2Ru(dpp)_2RhBr_2\}(PF_6)_5$ *J. Photochem. Photobiol. A: Chem.* 197: 13–17.

Arachchige, S. M., J. R. Brown, E. Chang, A. Jain, D. F. Zigler, et al. 2009. Design considerations for a system for photocatalytic hydrogen production from water employing mixed-metal photochemical molecular devices for photoinitiated electron collection. *Inorg. Chem.* 48: 1989–2000.

Ayers, W. 1983. US Patent 4,466,869 Photolytic production of hydrogen. November.

Balzani, V., A. Credi, and M. Venturi. 1997. Photoprocesses. *Curr. Opin. Chem. Biol.* 1: 506–513.

Bao, J. 2014. Recent developments in photocatalytic solar water splitting, *Mater. Today.* 17: 208–209.

Barnes, W. L., A. Dereux, and T. W. Ebbesen. 2003. Surface plasmon subwavelength optics. *Nature.* 424: 824–830.

Boettcher, S. W., J. M. Spurgeon, M. C. Putnam, E. L. Warren, D. B. Turner-Evans, M. D. Kelzenberg, et al. 2010. Energy-conversion properties of vapor-liquid-solid-grown silicon wire-array photocathodes. *Science.* 327: 185–187.

Boettcher, S. W., E. L. Warren, M. C. Putnam, E. A. Santori, D. Turner-Evans, M. D. Kelzenberg, et al. 2011. Photoelectrochemical hydrogen evolution using Si microwire arrays. *J. Am. Chem. Soc.* 133: 1216–1219.

Burns, L. D., J. B. McCormick, and E. E. Borroni-Bird. 2002. Vehicle of change. *Sci. Am.* 287: 64–73.

California Energy Commission. 2002. Fuel cells: Cost. California distributed energy resource guide. January 18.

Charvin, P., S. Abanades, F. Lemort, and G. Flamant. 2008. Analysis of solar chemical processes for hydrogen production from water splitting thermochemical cycles. *Energy Convers. Manage.* 49: 1547–1556.

Chen, H. M., C. K. Chen, C.-J. Chen, L.-C. Cheng, P. C. Wu, B. H. Cheng, et al. 2012. Plasmon inducing effects for enhanced photoelectrochemical water splitting: X-ray absorption approach to electronic structures. *ACS Nano.* 6: 7362–7372.

Chen, H., Y. Chen, H.-T. Hsieh, and N. Siegel. 2007. CFD Modeling of gas particle flow within a solid particle solar receiver. *ASME J. Solar Energy Eng.* 129: 160–170.

Choi, W., A. Termin, and M. R. Hoffmann. 1994. The role of metal ion dopants in quantum-sized TiO_2: Correlation between photoreactivity and charge carrier recombination dynamics. *J. Phys. Chem.* 98: 13669–13679.

Chouhan, N., and R.-S. Liu. 2011. Electrochemical technologies for energy storage and conversion. In: *Electrochemical Technologies for Energy Storage and Conversion*, Vol. 1, J. Zhang, L. Zhang, H. Liu, A. Sun, and R.-S. Liu (Eds.), Wiley-VCH, Weinheim, p. 27.

Chouhan, N., C. L. Yeh, S. F. Hu, R. S. Liu, W. S. Chang, and K. H. Chen. 2011. Photocatalytic CdSe QDs-decorated ZnO nanotubes: An effective photoelectrode for splitting water. *Chem. Commun.* 47: 3493–3495.

Crabtree, G. W., M. S. Dresselhaus, and M. V. Buchanan. 2004. The hydrogen economy. *Phys. Today.* 57: 39.

Das, C., P. Roy, M. Yang, H. Jha, and P. Schmuki. 2011. Nb-doped TiO_2 nanotube for enhanced photoelectrochemical water-splitting. *Nanoscale.* 3: 3094–3096.

Dincer, I., and A. S. Joshi. 2013. *Solar Based Hydrogen Production System.* Springer, New York, pp. 27–69.

Dincer, I., and C. Zamfirescu. 2012. *Sustainable Energy Systems and Application.* New York: Springer, pp. 519–632.

Du, P., J. Schneider, F. Li, W. Zhao, U. Patel, F. N. Castellano, and R. Eisenberg. 2008. Bi- and terpyridyl platinum(II) chloro complexes: Molecular catalysts for the photogeneration of hydrogen from water or simply precursors for colloidal platinum? *J. Am. Chem. Soc.* 130: 5056–5058.

Elvington, M., and K. J. Brewer. 2006. Photoinitiated electron collection at a metal in a rhodium-centered mixed-metal supramolecular complex. *Inorg. Chem.* 45: 5242–5244.

Elvington, M., J. Brown, S. M. Arachchige, and K. J. Brewer. 2007. Photocatalytic hydrogen production from water employing a Ru, Rh, Ru molecular device for photoinitiated electron collection. *J. Am. Chem. Soc.* 129: 10644–10645.

Fei, H., R. Ye, G. Ye, Y. Gong, Z. Peng, Z. Fan, et al. 2014. Boron- and nitrogen-doped graphene quantum dots/graphene hybrid nanoplatelets as efficient electrocatalysts for oxygen reduction. *ACS Nano.* 8: 10837–10843.

Fujishima, A., and K. Honda. 1972. Electrochemical photolysis of water at a semiconductor electrode. *Nature.* 238: 37–38.

Fujishima, A., and D. A. Tryk. 2012. Energy carriers and conversion systems: Vol. 1. Photochemical and photoelectrochemical water splitting. In: *Encyclopedia of Life Support Systems*, T. Ohta and T. N. Vezirozeu (Eds.). ELOSS & UNESCO.

Grimes, C. A., and G. K. Mor. 2009. *TiO_2 Nanotube Arrays: Synthesis, Properties, and Applications.* Springer, New York.

Gurunathan, K., P. Maruthamuthu, and M. V. C. Sastri. 1997. Photocatalytic hydrogen production by dye-sensitized Pt/SnO_2 and $Pt/SnO_2/RuO_2$ in aqueous methyl viologen solution. *Int. J. Hydrogen Energy.* 22: 57–62.

Hairston, D. 1996. Hail hydrogen. *Chem. Eng.* 103: 59–62.

Hamedani, H. A., N. K. Allam, H. Garmestani, and M. A. El-Sayed. 2011. Electrochemical fabrication of strontium-doped TiO_2 nanotube array electrodes and investigation of their photoelectrochemical properties. *J. Phys. Chem. C.* 115: 13480–13486.

Ingram, D. B., P. Christopher, J. L. Bauer, and S. Linic. 2011. Predictive model for the design of plasmonic metal/semiconductor composite photocatalysts. *ACS Catal.* 1: 1441–1447.

Jing, D., and L. Guo. 2007. WS2 sensitized mesoporous TiO_2 for efficient photocatalytic hydrogen production from water under visible light irradiation. *Catal. Commun.* 8: 795–799.

Kato, H., K. Asakura, and A. Kudo. 2003. Highly efficient water splitting into H and O over lanthanum-doped NaTaO photocatalysts with high crystallinity and surface nanostructure. *J. Am. Chem. Soc.* 125: 3082.

Kibsgaard, J., T. F. Jaramillo, and F. Besenbacher. 2014. Building an appropriate active-site motif into a hydrogen-evolution catalyst with thiomolybdate $[Mo_3S_{13}]_2$ clusters. *Nat. Chem.* 6: 248–253.

Kolb, G. J., N. P. Siegel, and R. B. Diver. 2007. Central-station solar hydrogen power plant, *ASME J. Solar Energy Eng.* 129: 179–183.

Kudo, A., and Y. Miseki. 2009. Heterogeneous photocatalyst materials for water splitting. *Chem. Soc. Rev.* 38: 253–278.

Kurihara, T., H. Okutomi, Y. Miseki, H. Kato, and A. Kudo. 2006. Highly efficient water splitting over $K_3Ta_3B_2O_{12}$ photocatalyst without loading cocatalyst. *Chem. Lett.* 35: 274–275.

Lal, S., S. Link, and N. Halas. 2007. Nano-optics from sensing to waveguiding, *Nat. Photonics.* 1: 641–648.

Lei, P., M. Hedlund, R. Lomoth, H. Rensmo, O. Johansson, and L. Hammarström. 2008. The role of colloid formation in the photoinduced H_2 production with a RuII–PdII supramolecular complex: A study by GC, XPS, and TEM. *J. Am. Chem. Soc.* 130: 26–27.

Liao, L., Q. Zhang, Z. Su, Z. Zhao, Y. Wang, Y. Li, et al. 2014. Efficient solar water-splitting using a nanocrystalline CoO photocatalyst. *Nat. Nanotechnol.* 9: 69–73.

Logan, B. E., and J. M. Regan. 2006. Electricity-producing bacterial communities in microbial fuel cells. *Trend. Microbiol.* 14: 512–518.

Luan, P., M. Xie, X. Fu, Y. Qu, X. Sun, and L. Jing. 2015. Improved photoactivity of TiO_2 Fe_2O_3 nanocomposites for visible-light water splitting after phosphate bridging and its mechanism. *Phys. Chem. Chem. Phys.* 17: 5043–5050.

Luo, J., J.-H. Im, M. T. Mayer, M. Schreier, M. K. Nazeeruddin, N.-G. Park, et al. 2014. Water photolysis at 12.3% efficiency via perovskite photovoltaics and Earth-abundant catalysts. *Science.* 345: 1593–1596.

Ma, S. S. K., T. Hisatomi, K. Maeda, Y. Moriya, and K. Domen. 2012. Enhanced water oxidation on Ta3N5 photocatalysts by modification with alkaline metal Salts. *J. Am. Chem. Soc.* 134: 19993–19996.

Ma, S. S. K., K. Maeda, T. Hisatomi, M. Tabata, A. Kudo, and K. Domen. 2013. A redox-mediator-eree solar-driven Z-scheme water-splitting system consisting of modified Ta_3N_5 as an oxygen-evolution photocatalyst. *Chem. Eur. J.* 19: 7480–7486.

Ma, G., T. Minegishi, D. Yokoyama, J. Kubota, and K. Domen. 2011. Photoelectrochemical hydrogen production on Cu_2ZnSnS_4/Mo-mesh thin-film electrodes prepared by electroplating. *Chem. Phys. Lett.* 501: 619–622.

Maeda, K., and K. Domen. 2010. Photocatalytic water splitting: Recent progress and future challenges. *J. Phys. Chem. Lett.* 1: 2655–2661.

Maeda, K., M. Higashi, D. Lu, R. Abe, and K. Domen. 2010. Efficient nonsacrificial water splitting through two-step photoexcitation by visible light using a modified oxynitride as a hydrogen evolution photocatalyst. *J. Am. Chem. Soc.* 132: 5858–5868.

Maeda, K., K. Teramura, and K. Domen. 2008. Effect of post-calcination on photocatalytic activity of $(Ga_{1-x}Zn_x)(N_{1-x}O_x)$ solid solution for overall water splitting under visible light. *J. Catal.* 254: 198–204.

Maier, S. A. 2007. *Plasmonics Fundamentals and Applications*. Springer, New York, pp. 21–34, 65–88.

Maier, S. A., and H. A. Atwater. 2005. Plasmonics: Localization and guiding of electromagnetic energy in metal/dielectric structures. *J. Appl. Phys.* 98: 011101.

McAlister, R. 1999. What causes the Hindenberg disaster? *Hydrogen Today.* 10: 6–8.

Minegishi, T., N. Nishimura, J. Kubota, and K. Domen. 2013. Photoelectrochemical properties of $LaTiO_2N$ electrodes prepared by particle transfer for sunlight-driven water splitting. *Chem. Sci.* 4: 1120–1124.

Moriya, M., T. Minegishi, H. Kumagai, M. Katayama, J. Kubota, and K. Domen. 2013. Stable hydrogen evolution from CdS-modified $CuGaSe_2$ photoelectrode under visible-light irradiation. *J. Am. Chem. Soc.* 135: 3733–3735.

Mubeen, S., J. Lee, N. Singh, S. Krämer, G. D. Stucky, and M. Moskovits. 2013. An autonomous photosynthetic device in which all charge carriers derive from surface plasmons. *Nat. Nanotechnol.* 8: 247–251.

Murphy, A. B. 2008. Does carbon doping of TiO_2 allow water splitting in visible light? Comments on nanotube enhanced photoresponse of carbon modified (CM)-n-TiO_2 for efficient water splitting. *Sol. Energy Mater. Solar Cells.* 92: 363–367.

Murphy, A. B., P. R. F. Barnes, L. K. Randeniya, I. C. Plumb, I. E. Grey, M. D. Horne, et al. 2006. Efficiency of solar water splitting using semiconductor electrodes. *Int. J. Hydrogen. Energy.* 31: 1999–2017.

Naess, S. N., A. Elgsaeter, G. Helgesen, and K. D. Knudsen. 2009. Carbon nanocones: Wall structure and morphology. *Sci. Technol. Adv. Mater.* 10: 065002.

Ni, M., M. K. H. Leung, D. Y. C. Leung, and K. Sumathy. 2007. A review and recent developments in photocatalytic water-splitting using TiO_2 for hydrogen production. *Renew. Sust. Energy Rev.* 11: 401–425.

NREL. 2009. Current state-of-the-art hydrogen production cost estimate using water electrolysis. NREL/BK-6A1-46676 (http://1.usa.gov/VFOj30), pp. 23–24.

Ogden, J. M. 1999. Prospects for building a hydrogen energy infrastructure. *Ann. Res. Energy Environ.* 24: 227–279.

Ozawa, H., M. Haga, and K. Sakai. 2006. A photo-hydrogen-evolving molecular device driving visible-light-induced EDTA-reduction of water into molecular hydrogen. *J. Am. Chem. Soc.* 128: 4926–4927.

Ozawa, H., Y. Yokoyama, M. Haga, and K. Sakai. 2007. Syntheses, characterization, and photo-hydrogen-evolving properties of tris(2,2′-bipyridine)ruthenium(II) derivatives tethered to a cis-Pt(II)Cl_2 unit: Insights into the structure-activity relationship. *Dalton Trans.* 1197–1206.

Pauling, L. 1970. *General Chemistry*, Section 15-2. San Francisco, CA, Dover.

Perkins, C., and A. W. Weimer. 2004. Likely near-term solar-thermal water splitting technologies. *Int. J. Hydrogen Energy.* 29: 1587–1599

Perret, R., W. Alan. G. Besenbruch, R. Diver, M. Lewis, Y. Chen, and K. Roth. 2007. Development of solar-powered thermochemical production of hydrogen from water, *DOE Hydrogen Program FY Annual Progress Report*, pp. 128–135.

Pirhashemi, M., and A. H.-Yangjeh. 2014. Preparation of AgCl–ZnO nanocomposites as highly efficient visible-light photocatalysts in water by one-pot refluxing method. *J. Alloys Compds.* 601: 1–8.

Rau, S., B. Schäfer, D. Gleich, E. Anders, M. Rudolph, M. Friedrich, et al. 2006. A supramolecular photocatalyst for the production of hydrogen and the selective hydrogenation of tolane. *Angew. Chem. Int. Ed.* 45: 6215–6218.

Ritchie, R. H. 1957. Plasma losses by fast electrons in thin films. *Phys. Rev.* 106: 874–881.

Rosenbaum, M., Z. He, and L. T. Angenent. 2010. Light energy to bioelectricity: Photosynthetic microbial fuel cells. *Curr. Opin. Biotechnol.* 21: 259–264.

Sauve, G., M. E. Cass, G. Coia, S. J. Doig, I. Lauermann, K. E. Pomykal, N. S. Lewis, et al. 2000. Dye sensitization of nanocrystalline titanium dioxide with osmium and ruthenium polypyridyl complexes. *J. Phys. Chem. B.* 104: 6821–6836.

Scholz, W. 1993. Processes for industrial production of hydrogen and associated environmental effect. *Gas Sep. Purif.* 7: 131.

Sharmoukh, W., and N. K. Allam. 2012. TiO_2 Nanotube-based dye-sensitized solar cell using new photosensitizer with enhanced open-circuit voltage and fill factor. *ACS Appl. Mater. Interfaces.* 4: 4413–4418.

Singh, G. P., K. M. Shrestha, A. Nepal, K. J. Klabunde, and C. M. Sorensen. 2014. Graphene supported plasmonic photocatalyst for hydrogen evolution in photocatalytic water splitting, *Nanotechnology.* 25: 265701.

Stempel, T., M. Aggour, K. Skorupska, A. Munoz, and H. J. Lewerenz. 2008. Efficient photoelectrochemical nanoemitter solar cell. *Electrochem. Commun.* 10: 1184–1186.

Sykora, M., K. Maruszewski, S. M. Treffert-Ziemelis, and J. R. Kincaid. 1998. A synthetic strategy for the construction of zeolite-entrapped organized molecular assemblies. Preparation and photophysical characterization of interacting adjacent cage dyads comprised of two polypyridine complexes of Ru(II). *J. Am. Chem. Soc.* 120: 3490–3498.

Szklo, A., and R. Schaeffer. 2007. Fuel specification, energy consumption and CO_2 emission in oil refineries. *Energy.* 32: 1075–1092.

Tachibana, Y., L. Vayssieres, and J. R. Durrant. 2012. Artificial photosynthesis for solar water-splitting. *J. Nat. Photonics.* 6: 511–518.

Tao, Y., Y. Chen, Y. Wu, Y. He, and Z. Zhou. 2007. High hydrogen yield from a two-step process of dark-and photo-fermentation of sucrose, *Int. J. Hydrogen Energy.* 32: 200–206.

Tester, J. W., E. M. Drake, M. J. Driscoll, M. W. Golay, and W. A. Peters. 2005. *Sustainable Energy: Choosing among Options.* Cambridge, MA: MIT Press.

Torimoto, T., H. Horibe, T. Kameyama, K. Okazaki, S. Ikeda, M. Matsumura, et al. 2011. Plasmon-enhanced photocatalytic activity of cadmium sulfide nanoparticle immobilized on silica-coated gold particles. *J. Phys. Chem. Lett.* 2: 2057–2062.

Yamada, S., A. Y. Nosaka, and Y. Nosaka. 2005. Fabrication of US photoelectrodes coated with titania nanosheets for water splitting with visible light. *J. Electroanal. Chem.* 585: 105–112.

Yang, X. Y., A. Wolcott, G. M. Wang, A. Sobo, R. C. Fitzmorris, F. Qian, et al. 2009. Nitrogen-doped ZnO nanowire arrays for photoelectrochemical water splitting. *Nano Lett.* 9: 2331–2336.

Yeh, T. F., C. Y. Teng, S. J. Chen, and H. Teng. 2014. Nitrogen-doped graphene oxide quantum dots as photo-catalysts for overall water-splitting under visible light illumination. *Adv. Mater.* 26: 3297–3303.

Yokoyama, D., T. Minegishi, K. Jimbo, T. Hisatomi, G. Ma, M. Katayama, et al. 2010. H_2 Evolution from water on modified Cu_2ZnSnS_4 photoelectrode under solar light. *Appl. Phys. Express.* 3: 101202.

Zeng, K., and D. Zhang. 2010. Recent progress in alkaline water electrolysis for hydrogen production and applications. *Prog. Energy Combust. Sci.* 36: 307–326.

Zhang, C., Y. Z. Jia, Y. Jing, Y. Yao, J. Maa, and J. Sun. 2013. Effect of non-metal elements (B, C, N, F, P, S) mono-doping as anions on electronic structure of $SrTiO_3$. *Comp. Mater. Sci.* 79: 69–74.

Zhang, X., Y. Li, J. Zhao, S. Wang, Y. Li, H. Dai, and X. Sun. 2014. Advanced three-component ZnO/Ag/CdS nanocomposite photoanode for photocatalytic water splitting. *J. Power Sources.* 269: 466–472.

Zittel, W., and R. Wurster. 1996. Production of hydrogen. In *Part 4: Production from electricity by means of electrolysis. HyWeb: Knowledge—Hydrogen in the Energy Sector.* Ludwig-Bölkow-Systemtechnik GmbH.

9 Photocatalytic Reduction of Carbon Dioxide

Guoqing Guan, Xiaogang Hao, and Abuliti Abudula

CONTENTS

9.1 Introduction 173
9.2 Characteristics of CO_2 Molecule 174
9.3 Mechanism of Photocatalytic Reduction of CO_2 174
9.4 Photocatalysts for Reduction of CO_2 177
    9.4.1 TiO_2 and Related Titanium-Containing Solids 177
    9.4.2 Other Metal Oxide Photocatalysts 179
    9.4.3 Nonoxide Semiconductor Photocatalysts 179
9.5 Operation Conditions 181
    9.5.1 Pressure Effect 181
    9.5.2 Reductant Effect 181
    9.5.3 Temperature Effect 181
    9.5.4 Reactor-Type Effect 182
9.6 Conclusions and Outlook 182
References 183

9.1 INTRODUCTION

Carbon dioxide (CO_2) is the chief greenhouse gas that results from human activities, which causes global warming and climate change. Based on the atmospheric CO_2 measurement data at Mauna Loa Observatory, Hawaii, the concentration of CO_2 in the atmosphere is increasing each decade at an accelerating rate (ESRL 2014). The upper safety limit for atmospheric CO_2 is 350 ppm; however, atmospheric CO_2 levels have remained higher than 350 ppm since early 1988, now reaching levels of 399 ppm. There are three possible strategies for reducing atmospheric CO_2:

- Reduction of the produced amount of CO_2 from power generation stations and industrial sites
- Usage of CO_2 as a feedstock of nearly zero cost for conversion to fuels and chemicals
- Storage of it using carbon sequestration and storage (CCS) technologies (Centi and Parathoner 2009; Mikkelsen et al. 2010; Richter et al. 2013; Hu et al. 2013)

The conversion of CO_2 to fuels, chemicals, and materials has attracted much attention in recent years (Ma et al. 2009; Morris et al. 2009; Jensen et al. 2011; Windle and Perutz 2012; Liu et al. 2012; Inglis et al. 2012; Tran et al. 2012; Dhakshinamoorthy et al. 2012; Mori et al. 2012; Fan et al. 2013; Izumi 2013; Handoko et al. 2013; Tahir and Amin 2013; Stechel and Miller 2013; Mao et al. 2013; Kondratenko et al. 2013; Das and Daud 2014). However, CO_2 is a kinetically and thermodynamically stable compound, which is difficult to oxidize or reduce to other fuels and/or chemicals at low temperatures. Reduction of CO_2 needs high energy input, cofeeding of a high-energy reactant such as H_2, and an excellent catalyst capable of driving its selective conversion to targeted chemicals. CO_2 can be theoretically reduced to hydrocarbons and/or alcohols at high temperature and high pressure. However, the process is quite complex, and a large amount of energy is essential. It would be attractive to capture CO_2 from the atmosphere or the exhaust of power stations or factories, and convert it to fuel and chemicals by using photocatalysts and a sustainable source of energy such as sunlight. In combination with photocatalytic H_2 production from water, the solar-driven CO_2 reduction to fuels and chemicals could be a very promising process for utilization of CO_2. Here, the mechanism of photocatalytic reduction of CO_2, the catalytic activity of the developed catalysts for the CO_2 reduction, and its advantages and challenges are introduced and discussed.

9.2 CHARACTERISTICS OF CO_2 MOLECULE

The CO_2 molecular geometry is linear in its ground state with two polar C=O bonds, in which oxygen atoms show a weak Lewis basicity, while the carbon atom is electrophilic (Centi and Parathoner 2009; Mikkelsen et al. 2010). Because it is centrosymmetric, the molecule has no electrical dipole, and only two vibrational bands are observed in the infrared (IR) spectrum: an antisymmetric stretching mode at 2349 cm^{-1} and a bending mode near 666 cm^{-1}. There is also a symmetric stretching mode at 1388 cm^{-1} in its Raman spectrum. The CO_2 is soluble in water, where it reversibly forms weak acid H_2CO_3, and easily reacts with basic compounds. The reactions of CO_2 are dominated by nucleophilic attack at the carbon atom, and it reacts with electron-donating reagents. Carbon dioxide has low reactivity, but it can be activated from nucleophilic centers on the surfaces of solids or in organometallic complexes. However, it is necessary to overcome a thermodynamic barrier by providing the energy for the reaction. In other words, thermochemical reduction of CO_2 is always an endothermic reaction, and it has to be realized at high temperatures even in the presence of catalysts. For instance, the reduction of CO_2 by H_2O to form methanol and methane requires 726 and 802 kJ mol^{-1} thermal energy, respectively.

9.3 MECHANISM OF PHOTOCATALYTIC REDUCTION OF CO_2

Photocatalytic reduction of CO_2 with a reducing agent such as H_2 and H_2O into useful fuels and chemicals such as CO, HCOOH, HCHO, CH_4, CH_3OH, and C_2H_5OH has been studied for decades (Centi and Parathoner 2009; Ma et al. 2009; Morris et al. 2009; Mikkelsen et al. 2010; Jensen et al. 2011; Windle and Perutz 2012; Liu et al. 2012; Inglis et al. 2012; Tran et al. 2012; Dhakshinamoorthy et al. 2012; Mori et al. 2012; Richter et al. 2013; Hu et al. 2013; Fan et al. 2013; Izumi 2013; Handoko et al. 2013; Tahir and Amin 2013; Stechel and Miller 2013; Mao et al. 2013; Kondratenko et al. 2013; Das and Daud 2014). Direct photolysis of CO_2 to CO and O_2 in the absence of a reducing agent is also possible (Corma and Garcia 2013). Carbon dioxide can only absorb the photons below 200 nm, which belongs to the deep UV spectral region:

$$2CO_2 + h\nu(<200 \text{ nm}) \rightarrow 2CO + O_2 \quad \Delta H = +257 \text{ kJ mol}^{-1} \quad (9.1)$$

In order to get the light in this region, pure quartz or metal halides and vacuum are required. However, the key issues of the direct photolysis of CO_2 are the efficiency and energy consumption per mole of CO_2 conversion. It is estimated that the energy consumption value is about 0.3 GJ mol^{-1} CO_2 for this process. This energy consumption is high if the efficiency of the electricity to deep UV light conversion for mercury lamps is considered. As a sustainable source of energy, only a small amount of deep UV light (100–280 nm) is present in the UV light that reaches Earth. How to use a large amount of other photons included in the solar insolation for the CO_2 reduction should be one of the main challenges in the field of photocatalytic reduction of CO_2.

The reduction of CO_2 by one electron to form a carbon dioxide anion radical ($CO_2^{-\bullet}$) is highly unfavorable, requiring a highly negative potential of −1.9 V versus NHE (Figure 9.1) (Morris et al.

Photocatalytic Reduction of Carbon Dioxide

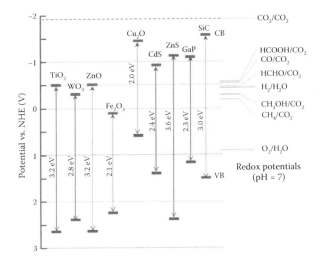

FIGURE 9.1 Band-edge positions of typical semiconductors and redox potentials at pH = 7 for CO_2 reduction.

2009; Corma and Garcia 2013). Rapid reduction requires an overpotential of at least 0.6 V due to the kinetic restrictions imposed by the structural difference between linear CO_2 and bent $CO_2^{-\bullet}$. In the photocatalysis process, electrons are provided by a semiconductor exposed to light (Inoue et al. 1979). The photocatalysis over a semiconductor is initiated by the adsorption of photons in the light with equal or higher energy than the bandgap energy of the semiconductor. The excitation of the electron from the valence band (VB) to the conduction band (CB) results in electron vacancy called a *hole* in the VB, and the electron–hole pair is generated in the semiconductor (Equation 9.2) (Figure 9.2).

Thereafter, the separation and migration of these photogenerated electrons and holes to the surface sites for the oxidation of H_2O (donor) by holes to O_2 (Equation 9.3) and the reduction of H_2O or CO_2 (acceptor) by electrons to H_2 or CO and other organic compounds become important (Equations 9.4 through 9.10). Otherwise, the photogenerated electron–hole pairs will recombine. The recombination time of the electron–hole pairs is of the order of 10^{-9} s, whereas the chemical interaction with the adsorbed species has a longer time of 10^{-8} to 10^{-3} s. Therefore, it is fundamentally important to

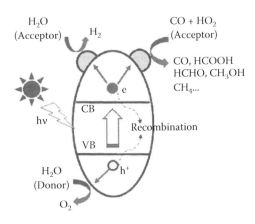

FIGURE 9.2 The photocatalytic reduction of CO_2 on the catalyst surface.

avoid the recombination by improving their separation and migration to the surface sites during the photocatalysis process:

$$\text{Photocatalyst} + h\nu \;(\geq \text{Band gap}) \rightarrow e^- + h^+ \quad (9.2)$$

$$2H_2O + 4h \rightarrow O_2 + 4H^+ \rightarrow E^0 \text{ versus NHE} = +0.81 \text{ V} \quad (9.3)$$

$$2H^+ + 2e^- \rightarrow H_2 \rightarrow E^0 \text{ versus NHE} = -0.42 \text{ V} \quad (9.4)$$

$$CO_2 + e^- \rightarrow CO_2^{-\bullet} \rightarrow E^0 \text{ versus NHE} = -1.90 \text{ V} \quad (9.5)$$

$$CO_2 + 2H^+ + 2e^- \rightarrow CO + H_2O \rightarrow E^0 \text{ versus NHE} = -0.53 \text{ V} \quad (9.6)$$

$$CO_2 + 2H^+ + 2e^- \rightarrow HCOOH \rightarrow E^0 \text{ versus NHE} = -0.61 \text{ V} \quad (9.7)$$

$$CO_2 + 4H^+ + 4e^- \rightarrow HCHO + H_2O \rightarrow E^0 \text{ versus NHE} = -0.48 \text{ V} \quad (9.8)$$

$$CO_2 + 6H^+ + 6e^- \rightarrow CH_3OH + H_2O \rightarrow E^0 \text{ versus NHE} = -0.38 \text{ V} \quad (9.9)$$

$$CO_2 + 8H^+ + 8e^- \rightarrow CH_4 + 2H_2O \rightarrow E^0 \text{ versus NHE} = -0.24 \text{ V} \quad (9.10)$$

Thermodynamically, proton-coupled electrochemical reactions are possible at much less negative potentials. Thus, photocatalytic reduction of CO_2 to either CH_3OH or CH_4 seems to be more feasible than reduction to $CO_2^{-\bullet}$. However, the lower oxidation states of carbon in both are achieved through kinetically unfavorable routes, because they require the transfer of six and eight electrons to each carbon center, respectively, along with complex chemical steps involving proton transfers (Inglis et al. 2012). It should be noted that the standard reduction potential of H_2O to produce H_2 is much lower than that of CO_2 to generate $CO_2^{-\bullet}$. Thus, in the case of the photocatalytic reduction of CO_2 with H_2O, the photocatalytic reduction of H_2O to generate molecular hydrogen at a much lower reduction potential will compete with the reduction of CO_2. Therefore, this thermodynamic and kinetic contradiction should be considered by creating spatially separated centers for electrons and protons to avoid their collapse before being transferred to CO_2.

The photocatalytic reduction of CO_2 using water as the reducing agent always results in a small amount of organic products (Tahir and Amin 2013). Several factors can be considered:

- The solubility of CO_2 at pH = 7 is very low so that it is more difficult to contact the catalyst surface than H_2O. Photocatalytic reduction of CO_2 can be performed in alkaline solution to solve this problem. The formation of CO_3^{2-} and HCO_3^- in the alkaline solution can accelerate the CO_2 reduction. Furthermore, the OH^- in the alkaline solution can serve as a strong hole scavenger to promote the photocatalytic activity (Mao et al. 2013).
- H_2O is a poor electron donor. In this case, it is possible to improve the efficiency by employing other sacrificial electron donors such as isopropyl alcohol, dimethylformamide, and phosphoric acid (Mikkelsen et al. 2010, Dhakshinamoorthy et al. 2012).
- Photocatalytic activity of the developed semiconductors is low; hence, in addition to photoreduction mechanisms, other factors should be considered for a practical process.

9.4 PHOTOCATALYSTS FOR REDUCTION OF CO_2

Many traditional semiconductors such as TiO_2, CdS, Fe_2O_3, WO_3, ZnO, ZrO_2, and their various combinations have been tested for the photoreduction of CO_2 (Centi and Parathoner 2009; Ma et al. 2009; Morris et al. 2009; Mikkelsen et al. 2010; Jensen et al. 2011; Windle and Perutz 2012; Liu et al. 2012; Inglis et al. 2012; Tran et al. 2012; Dhakshinamoorthy et al. 2012; Mori et al. 2012; de Richter et al. 2013; Hu et al. 2013; Fan et al. 2013; Izumi 2013; Handoko et al. 2013; Tahir and Amin 2013; Stechel and Miller 2013; Mao et al. 2013; Kondratenko et al. 2013; Das and Daud 2014). Recently, some new semiconductors such as copper chalcopyrite compounds, $CuInSe_2$, $CuInS_2$, and $CuInGaSe_2$, and carbon nitride (g-C_3N_4) were also considered as the photocatalysts for the photoreduction of CO_2 (Tran et al. 2012; Kondratenko et al. 2013). In general, as a good photocatalyst, these should have high catalytic activity with high stability and durability, with a low tendency to photocorrosion.

9.4.1 TiO_2 AND RELATED TITANIUM-CONTAINING SOLIDS

Inoue et al. (1979) first reported photocatalytic reduction of CO_2 to organic compounds including HCOOH, CH_3OH, and/or HCHO by irradiation of aqueous saturated CO_2 using a Xe lamp in the presence of various semiconductors such as TiO_2, ZnO, and WO_3. Since then, the study on artificial photosynthesis boomed with typical focus on TiO_2. Comparing to other semiconductors, TiO_2 is relatively low cost, nontoxic, and fairly stable. Three phases of TiO_2 (i.e., rutile, anatase, and brookite) exist naturally. Anatase and rutile phases reveal definite bandgap energy at about 3.2 and 3.0 eV, respectively. The best photoactivity can be achieved by combining anatase with a slight amount of rutile (Das and Daud 2014). Pure TiO_2 in its anatase form shows good catalytic performance when it is irradiated with photons of wavelength shorter than 380 nm. Generally, P25, a kind of TiO_2 made by Degussa with a particle size of about 30 nm, containing rutile, is considered as the reference material for the photocatalytic applications of TiO_2 and other developing photocatalysts. The photocatalytic activity of TiO_2 can be enhanced by

- Control of its particle morphology by decreasing its size to nanolevel, synthesizing titania nanotube or nanorod, and supporting aluminosilicates
- Doping with metallic or nonmetallic elements
- Photosensitization

Photoreduction of CO_2 by using TiO_2 nanoparticles with size in the range of 4.5–29 nm were investigated by Kočí et al. (2009). As the nanoparticle size decreased, higher yields of CH_3OH and CH_4 were obtained. The optimum particle size corresponding to the highest yields of both products (CH_3OH: 1.15 μmol/g-catalyst; CH_4: 9.5 μmol/g-catalyst) was found to be 14 nm. Theoretically, the bandgap between the CB and VB becomes larger as the TiO_2 particle size decreases, making it suitable and applicable for the reduction of CO_2 (Mori et al. 2012). However, further decreasing the particle size resulted in a decrease in the catalytic activity due to the competing effects of specific surface area, charge–carrier dynamics, and light absorption efficiency.

The TiO_2 in the form of a nanotube or a nanorod also exhibits higher photocatalytic activity as compared to P25. The electron transport along the long axes of the crystallites of nanotube and nanorod is very fast, and it is beneficial for the charge separation. Furthermore, the light can be trapped inside the nanotube or nanorod, not only by the light scattering but also by the *crystal photonic* effect that appears when the wavelength of the photon matches the distances between the TiO_2 nanotubes or nanorods (Dhakshinamoorthy et al. 2012).

Titanium oxide moieties implanted and isolated in the silica matrices of zeolite and mesoporous silica materials at the atomic level, which have been called *single-site photocatalysts*, have also been tested for the photoreduction of CO_2 (Mori et al. 2012). In this case, the isolated Ti atoms can be

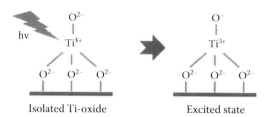

FIGURE 9.3 The electronic state change in isolated Ti-oxide molecular species.

substituted with Si atoms in the silica matrices (Si/Ti > 30) and coordinated tetrahedrally with oxygen atoms. If it is excited by UV irradiation, an electron motivation from the oxygen (O^{2-}) to Ti^{4+} ions will occur, and pairs of trapped hole centers (O^-) and electron centers (Ti^{3+}) are formed (Figure 9.3). Such excited electron–hole pairs locate quite near each other compared to the electron–hole pairs in the bulk TiO_2, and unique photocatalytic activity could be obtained for the photoreduction of CO_2. The results obtained by using micro- and mesoporous titanosilicates such as TS-1, Ti-MCM-41, Ti-MCM-48, Pt-Ti-MCM-48, titanium beta-zeolites (Ti-β) of hydrophilic Ti-β-OH and hydrophobic Ti-β-F, and Ti-FSM-16 indicated that Ti-MCM-48 exhibits the highest reactivity and selectivity for the formation of CH_3OH due to the combined contribution of a high dispersion of titanium oxide moieties, large pore size, and tridirectional channel structure. Pt doping in Ti-MCM-48 results in greater yield of CH_4 and a decrease in the CH_3OH yield (Dhakshinamoorthy et al. 2012, Mori et al. 2012).

Ti-β-OH shows five times higher reactivity compared to Ti-β-F for the formation of CH_4, but Ti-β-F shows higher selectivity for the formation of CH_3OH due to the different abilities of zeolite pores on the H_2O affinity.

TiO_2 with a wide bandgap mainly absorbs ultraviolet photons with wavelengths shorter than 380 nm. However, solar light contains about 5% only of ultraviolet photons, and a large amount of light in nature is visible light. Many efforts have been made to extend the spectral response of pure TiO_2 to visible light (Fujishima et al. 2008; Dhakshinamoorthy et al. 2012; Mori et al. 2012):

- Doping metal to replace Ti atoms in the lattice for introducing empty orbitals at energies below the CB of pure TiO_2 (Figure 9.4a)
- Coupling TiO_2 with narrow-bandgap semiconductors
- Preparing oxygen-deficient TiO_2
- Doping nonmetal atoms such as N, C, S, B, P, and F in the lattice to replace O atoms for introducing occupied orbitals above the energy of the VB of pure TiO_2 (Figure 9.4b)

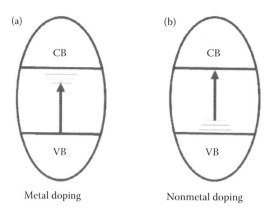

FIGURE 9.4 Change in bandgap structure by (a) replacing the Ti atom with another metal element in the TiO_2 lattice or (b) by replacing some O atoms in the TiO_2 lattice by other nonmetallic species.

However, it is found that metal doping always results in photocorrosion phenomenon during long-term operation and affects the stability and durability of the photocatalyst. In contrast, nonmetal doping renders more durable and stable photocatalysts working in visible light. Among nonmetal doping, N-doped TiO_2 is the most effective in narrowing the bandgap.

Another method of metal doping is doping the metal on the surface of TiO_2. Cu doping was found to be the most efficient way to enhance the selectivity toward CH_3OH production and formation of formic acid derivatives (Tseng et al. 2002, 2004; Wu et al. 2005). It was found that Cu^+ species on TiO_2 plays a significant role in the photoelectrochemical production of CH_3OH from CO_2 and H_2. Cu metal can serve as an electron trapper, and it prohibits the recombination of electron and hole, thus, significantly increasing the photoefficiency. In contrast, Pt or Pd doping favors selectivity toward CH_4 and CO (Dhakshinamoorthy et al. 2012). It should be noted that the doping amount has a great effect on catalytic activity. A third effect in the case of metal nanoparticle doping, especially for Au doping, is that a strong surface plasmon band in the visible light region can be observed. In this case, electrons expelled from the Au nanoparticle will have enough energy to move to CB in TiO_2, even when irradiated by visible light (Dhakshinamoorthy et al. 2012; Mori et al. 2012).

9.4.2 Other Metal Oxide Photocatalysts

In addition to TiO_2, other metal oxide semiconductors can be used as photocatalysts for the reduction of CO_2. Different metal oxides possess different bandgap energies and different CB and VB positions versus NHE (Figure 9.1). It was found that the yield of methanol increases as the CB is located more negative with respect to the redox potential of CH_3OH/CO_2. For instance, methanol cannot be produced over WO_3 catalyst, as its conduction band is more positive than the redox potential of CH_3OH/CO_2. However, methane can be generated from it, while it has low bandgap energy (2.8 eV) with respect to visible light irradiation (Figure 9.1). In order to improve the catalytic activity, various synthesis methods such as sol-gel, hydrothermal, and solid-state reaction methods are developed to get high-performance metal oxide semiconductors with some special structures for the photoreduction of CO_2. For instance, $W_{18}O_{49}$ nanowires, each with a diameter of about 0.9 nm and with a large number of oxygen vacancies, were synthesized, and these show potential for the photoreduction of CO_2 in the visible light region (Xi et al. 2012). Hydrothermally prepared monoclinic $BiVO_4$ catalysts show higher ethanol yields (110 μmol/h/g-catalyst) than tetragonal $BiVO_4$, as the anchored CO_3^{2-} to the Bi^{3+} sites on the external surface can receive the photogenerated electrons effectively via a weak Bi-O bond from V 3d-block bands of $BiVO_3$ (Liu et al. 2012). Table 9.1 summarizes some typical metal oxide photocatalysts for the reduction of CO_2 with relatively high yields.

9.4.3 Nonoxide Semiconductor Photocatalysts

Nonoxide semiconductor photocatalysts always have low bandgap energy so that they can respond to the visible region and show high photocatalytic activity (Das and Daud 2014). The main photocatalysts of this type are

- Metal sulfide semiconductors such as CdS and ZnS
- Metal phosphide semiconductors such as GaP and InP
- Others such as AgCl, AgBr, GaAs, and p-Si

Metal sulfides have relatively high CB states and are more appropriate for enhancing solar responses than metal oxide semiconductors, which are facilitated by the higher VB states consisting of S 3p orbitals. Especially, CdS with a narrow bandgap of 2.63 eV shows higher catalytic activity for CO_2

TABLE 9.1
Typical Metal Oxide Semiconductors for Photocatalytic Reduction of CO_2 with Relatively High Yields

Catalyst	Bandgap Energy (eV)	Major Product/Yield (μmol/h/g-catalyst)	Reference
TiO_2 (JRC-TIO)	3.47	CH_4/0.03	Anpo et al. (1995)
TiO_2	3.0	HCHO/16	Inoue et al. (1979)
ZnO	3.2	HCHO/17.14	Inoue et al. (1979)
$W_{18}O_{49}$	2.7	CH_4/666 ppm/h/g-catalyst	Xi et al. (2012)
HNb_3O_8	3.66	CH_4/3.58	Li et al. (2012)
$BiVO_4$	2.24	C_2H_5OH/110	Liu et al. (2009)
Montmorillonite-modified TiO_2	3.07	CH_4/441.5	Tahir and Amin (2013)
23.2% $AgBr/TiO_2$	2.9	CH_4/25.72; C_2H_5OH/15.57	Asi et al. (2011)
1% CuO/TiO_2	3.2	$HCOOCH_3$/1602	Qin et al. (2011)
3% CuO/TiO_2	2.88	CH_3OH/442.5	Slamet et al. (2005)
5% ZnO/C	3.2	CO/200000	Gokon et al. (2003)
1.7% Cu-0.9% Pt/TiO_2	—	CH_4/33	Zhai et al. (2013)
$Pt-KNbO_3$	3.1	CH_4/70 ppm/h/g-catalyst	Shi and Zou (2012)
$N-InTaO_4$/Ni-NiO	2.28	CH_3OH/165	Tsai et al. (2011)
Sensitized 45% $Cu-TiO_2$ on SBA-15	3.2	CH_3OH/475	Yang et al. (2009)
$Ag-BaLa_4Ti_4O_{15}$	—	CO/73; HCOOH/2.3	Iizuka et al. (2011)

reduction as compared to other metal sulfides. When it is modified with Bi_2S_3, the activity can be improved to some extent (Li et al. 2011). Bi_2S_3 has a higher photochemical activity and visible light response than CdS.

The product yields when nonoxide semiconductor photocatalysts are used for CO_2 reduction are still far lower than expected (Table 9.2).

Theoretically, the wide-bandgap semiconductors could be better photocatalysts for CO_2 reduction in aqueous solutions because photogenerated electrons in the bottom of the CB can have sufficient negative redox potential to drive CO_2 reduction, while the photogenerated holes in the VB can

TABLE 9.2
Typical Nonoxide Semiconductors for Photocatalytic Reduction of CO_2 with a Relatively High Yield

Catalyst	Bandgap Energy (eV)	Major Product/Yield (μmol/h/g-catalyst)	Reference
CdS	2.4	HCHO/29; CH_3OH/17	Inoue et al. (1979)
SiC	3.0	HCHO/14; CH_3OH/76	Inoue et al. (1979)
GaP	2.3	HCHO/14; CH_3OH/16	Inoue et al. (1979)
15% Bi_2S_3-CdS	1.28	CH_3OH/120	Li et al. (2011)
p-GaAs	—	HCOOH/170 μmol	Aurian-Blajeni et al. (1983)
RuO_2 or $Rh_{1.32}Cr_{0.66}O_3$-loaded $Cu_xAg_yIn_zZn_kS_m$	—	CH_3OH/118.5	Liu et al. (2011)
Ag@AgX (X = Cl, Br)	—	CH_3OH/45.3; C_2H_5OH/65.8	Jiang et al. (2010)

be sufficiently energetic to act as acceptors and oxidize water to O_2 (Jiang et al. 2010). However, the wide-bandgap semiconductors cannot use visible light efficiently. Although many nonoxide semiconductors can absorb visible light, only a few of them are catalytically active because the energy levels of neither CBs nor VBs are suitable for CO_2 reduction and/or water oxidation. Furthermore, many of these nonoxide semiconductors suffer from photocorrosion.

9.5 OPERATION CONDITIONS

9.5.1 Pressure Effect

Numerous photocatalysts have been applied for the reduction of CO_2. A green chemical approach is application of a stable photocatalyst, CO_2, and water for the production of fuels and chemicals. In this green approach, the conversion efficiency of CO_2 is greatly dependent on the molar ratio of H_2O to CO_2. The optimized H_2O/CO_2 molar ratio is near 5.0 for TiO_2 (Anpo et al. 1995). Thus, how to dissolve more CO_2 in H_2O becomes an important issue in this process. Increasing CO_2 pressure can increase its solubility in H_2O. It was found that an increase in CO_2 pressure accelerates CO_2 reduction in the water, and some products such as ethylene, which are not produced in ambient CO_2 pressure, can be traced under high CO_2 pressure. The methanol yield was increased from 175 to 230 μmol g-catalyst when CO_2 pressure was increased from 110 to 250 kPa, but decreased to 85 μmol g-catalyst with further increase of the pressure to 136 kPa (Tseng et al. 2002). Therefore, an optimum CO_2 pressure should exist for different photocatalytic reduction systems.

9.5.2 Reductant Effect

Photoreduction of CO_2 competes with H_2 formation in the reaction process, when only H_2O was used as the electron donor. In order to solve this problem, it is possible to use other reductants (sacrificial electron donors) instead of H_2O for CO_2 photoreduction. Triethylamine, triethanolamine, dimethylformamide, acetonitrile, propanol, and dichloromethane with various dielectric constants are considered as the good sacrificial electron donors (Mikkelsen et al. 2010; Liu et al. 2012). It was found that the amount of formate increases, while CO decreases with an increase in the dielectric constant of the reductants. Adding NaOH or other alkaline chemicals in the solutions can dissolve more CO_2 and simultaneously, OH^-, in aqueous solutions acts as a strong hole scavenger, enhancing CO_2 conversion. Furthermore, the photoreduction of CO_2 can also be performed in the gas phase using reductants such as CH_3OH, CH_4, H_2, and H_2S (Liu et al. 2012).

9.5.3 Temperature Effect

In general, photocatalytic reduction of CO_2 is always performed at room temperature. However, it was found that CO_2 reduction can be promoted by increasing the reaction temperature (Richter et al. 2013). Thermal energy could promote the activity of photocatalyst already activated by light irradiation. The desorption rate of the products produced on the photocatalyst surface could be improved by heating. However, CO_2 and reductant adsorptions on the photocatalyst surface should also be controlled by heating. Therefore, an optimum temperature might also exist in a photoreduction system. Thermal reduction of CO_2 will occur if the temperature is too high. In our previous studies (Guan et al. 2003a, 2003b), photoreduction of CO_2 with H_2O was performed under concentrated sunlight using photocatalysts combined with Fe- and Cu/ZnO-based thermal CO_2 reduction catalysts. The reaction temperature reached over 600 K by concentrating the solar irradiation, and CH_3OH and C_2H_5OH were generated together in the presence of Fe-based CO_2 reduction catalyst while only CH_3OH was detected in the presence of Cu/ZnO-based thermal CO_2 reduction catalyst.

9.5.4 Reactor-Type Effect

The physical structure of a photoreactor is very important to ensure photogenerated photos are effectively collected by the photocatalyst. The design of the irradiation device including mirror, reflector, window materials, and shape is a key issue for a photoreactor. Before designing a useful photoreactor, besides making sure that the reactants contacted with the photocatalyst surface well, two points should be considered:

- How to suspend or support the photocatalysts for utilizing maximum solar spectrum
- How to concentrate sunlight by a reflecting surface to maximize the light harvesting

The most common types of photoreactors for CO_2 reduction are slurry reactor, fixed-bed reactor, bubble-flow reactor, surface-coated reactor, and optical fiber and monolith reactor (Liu et al. 2012; Tahir and Amin 2013). Among photoreactors, optical fiber and monolith reactors are considered to be more prominent due to their higher illuminated surface area and efficient utilization of photon energy.

9.6 CONCLUSIONS AND OUTLOOK

Carbon dioxide is a zero-cost carbon source for the synthesis of carbon fuels and value-added chemicals. The catalyst plays a key role in CO_2 conversion. Numerous methods and catalysts have been developed to realize the high efficient conversion of CO_2. The photocatalytic reduction process is one of the best ways to solve the greenhouse gas emission and fuel crisis using the inexhaustible solar energy source. However, to date, the conversion efficiency and selectivity are still low. The yields of the main products are below 500 μmol/h/g-catalyst, and the product distributions are also complicated so that the separation of the products from an aqueous system is poor, which limits the application of this process. To solve these problems, the following issues should be considered in future studies:

- The mechanisms of photoreduction of CO_2, which include multielectron reductions, activation of two very stable molecules, CO_2 and H_2O, the link between the semiconductor structure and the product selectivity, the adsorption/desorption of CO_2, the formation of intermediate products, and the role of adsorbed H_2O, should be clarified for the design of novel photoreduction systems in order to improve the CO_2 conversion efficiency and the selectivity of some target products, such as CH_3OH.
- Novel photocatalysts should have the desired bandgap energy for visible light response and can withstand photocorrosion, long-term stability, and high selectivity. Many wide-bandgap semiconductors exhibit relatively long-term stability, but only a small amount of solar energy can be used. *Band engineering* to improve the visible light absorption ability of the wide-bandgap semiconductor via nonmetal doping, coupling with narrow-bandgap semiconductor, photosensitization by inorganic and/or organic dyes, noble metal with surface plasmon band, and quantum dot sensitization have been applied, but the results are still unsatisfactory. Further study of this strategy is necessary. The other way is to develop nanocomposite photocatalysts of semiconductor-cocatalyst, semiconductor-carbon materials such as CNT and graphene, and semiconductor-semiconductor to replace single-component semiconductors, which have been identified to improve the CO_2 conversion significantly. But the mechanisms of their function are still unclear, and it is necessary to improve their activity and selectivity.
- One should pay more attention to operating conditions and photoreactor design. Carbon dioxide pressure, a dissolved amount of CO_2 in water, the pH of the solution, the type of reductant and its amount, the reaction temperature, and the photoreactor type have a great effect on the product yield as well as the selectivity. The novel photoreduction system should be simple, convenient, and efficient for the effective utilization of solar energy.

- It is possible to combine photoreduction of CO_2 with other reduction processes, such as thermal reduction of CO_2 and electrocatalytic reduction of CO_2, in order to reach the maximum efficiency for the conversion of CO_2 to fuels and/or chemicals. Solar energy can provide photons and thermal energy simultaneously. Concentrated sunlight can also produce high temperatures for the thermal reduction of CO_2. If these two processes can be combined, total energy efficiency could be improved to a great extent. Furthermore, solar energy can be transferred to electricity, and the efficiency of the electrocatalytic reduction of CO_2 is generally much higher than that of the photoreduction process. Therefore, development of a photoelectrocatalytic combined process to replace the simple photoreduction process should be more attractive.

REFERENCES

Anpo, M., H. Yamashita, Y. Ichihashi, and S. Ehara. 1995. Photocatalytic reduction of CO_2 with H_2O on various titanium oxide catalysts. *J. Electroanal. Chem.* 396: 21–26.

Asi, M. A., C. He, M. Su, D. Xia, L. Lin, H. Deng, Y. Xiong, R. Qiu, and X. Li. 2011. Photocatalytic reduction of CO_2 to hydrocarbons using $AgBr/TiO_2$ nanocomposites under visible light. *Catal. Today.* 175: 256–263.

Aurian-Blajeni, B., M. Halmann, and J. Manassen. 1983. Electrochemical measurement on the photoelectrochemical reduction of aqueous carbon dioxide on p-Gallium phosphide and p-Gallium arsenide semiconductor electrodes. *Solar Energy Mater.* 8: 425–440.

Centi, G., and S. Perathoner. 2009. Opportunities and prospects in the chemical recycling of carbon dioxide to fuels. *Catal. Today.* 148: 191–205.

Corma, A., and H. Garcia. 2013. Photocatalytic reduction of CO_2 for fuel production: Possibilities and challenges. *J. Catal.* 308: 168–175.

Das, S., and W. M. A. W. Daud. 2014. A review on advances in photocatalysts towards CO_2 conversion. *RSC Adv.* 4: 20856–20893.

Dhakshinamoorthy, A., S. Navalon, A. Corma, and H. Garcia. 2012. Photocatalytic CO_2 reduction by TiO_2 and related titanium containing solids. *Energy Environ. Sci.* 5: 9217–9233.

Earth System Research Laboratory. 2014. Global monitoring division. http://www.esrl.noaa.gov/gmd.

Fan, W., Q. Zhang, and Y. Wang. 2013. Semiconductor-based nanocomposites for photocatalytic H_2 production and CO_2 conversion. *Phy. Chem. Chem. Phy.* 15: 2631–2649.

Fujishima, A., X. Zhang, and D. A. Tryk. 2008. TiO_2 photocatalysis and related surface phenomena. *Surf. Sci. Rep.* 63: 515–582.

Gokon, N., N. Hasegawa, H. Kaneko, H. Aoki, Y. Tamaura, and M. Kitamura. 2003. Photocatalytic effect of ZnO on carbon gasification with CO_2 for high temperature solar thermochemistry. *Sol. Energy Mater. Sol. Cells.* 80: 335–341.

Guan, G., T. Kida, T. Harada, M. Isayama, and A. Yoshida. 2003a. Photoreduction of carbon dioxide with water over $K_2Ti_6O_{13}$ photocatalyst combined with Cu/ZnO catalyst under concentrated sunlight. *Appl. Catal. A: Gen.* 249: 11–18.

Guan, G., T. Kida, and A. Yoshida. 2003b. Reduction of carbon dioxide with water under concentrated sunlight using photocatalyst combined with Fe-based catalyst. *Appl. Catal. B: Environ.* 41: 387–396.

Handoko, A. D., K. Li, and J. Tang. 2013. Recent progress in artificial photosynthesis: CO_2 photoreduction to valuable chemicals in a heterogeneous system. *Curr. Opin. Chem. Eng.* 2: 200–206.

Hu, B., C. Guild, and S. L. Suib. 2013. Thermal, electrochemical, and photochemical conversion of CO_2 to fuels and value-added products. *J. CO_2 Utilization* 1: 18–27.

Iizuka, K., T. Wato, Y. Miseki, K. Saito, and A. Kudo. 2011. Photocatalytic reduction of carbon dioxide over Ag cocatalyst-loaded A $La_4Ti_4O_{15}$ (A = Ca, Sr, and Ba) using water as a reducing reagent. *J. Am. Chem. Soc.* 133: 20863–20868.

Inglis, J. L., B. J. MacLean, M. T. Pryce, and J. G. Vos. 2012. Electrocatalytic pathways towards sustainable fuel production from water and CO_2. *Coord. Chem. Rev.* 256: 2571–2600.

Inoue, T., A. Fujishima, S. Konishi, and K. Honda. 1979. Photoelectocatalytic reduction of carbon dioxide in aqueous suspensions of semiconductor powders. *Nature.* 277: 637–638.

Izumi, Y. 2013. Recent advances in the photocatalytic conversion of carbon dioxide to fuels with water and/or hydrogen using solar energy and beyond. *Coord. Chem. Rev.* 257: 171–186.

Jensen, J., M. Mikkelsen, and F. C. Krebs. 2011. Flexible substrates as basis for photocatalytic reduction of carbon dioxide. *Sol. Energy Mater. Solar Cells.* 95: 2949–2958.

Jiang, Z., T. Xiao, V. L. Kuznetsov, and P. P. Edwards. 2010. Turning carbon dioxide into fuel. *Phil. Trans. R. Soc. A.* 368: 3343–3364.

Kočí, K., L. Obalová, L. Matějová, D. Plachá, Z. Lacny, J. Jirkovsky, and O. Šolcová. 2009. Effect of TiO_2 particle size on the photocatalytic reduction of CO_2. *Appl. Catal. B: Environ.* 89: 494–502.

Kondratenko, E. V., G. Mul, J. Baltrusaitis, G. O. Larrazabal, and J. Perez-Ramirez. 2013. Status and perspectives of CO_2 conversion into fuels and chemicals by catalytic, photocatalytic and electrocatalytic processes. *Energy Environ. Sci.* 6: 3112–3135.

Li, X., J. Chen, H. Li, J. Li, Y. Xu, Y. Liu, and J. Zhou. 2011. Photoreduction of CO_2 to methanol over Bi_2S_3/CdS photocatalyst under visible light irradiation. *J. Nat. Gas Chem.* 20: 413–417.

Li, X., H. Pan, W. Li, and Z. Zhuang. 2012. Photocatalytic reduction of CO_2 to methane over HNb_3O_8 nanibelts. *Appl. Catal. A.*, 413–414: 103–108.

Liu, G., N. Hoivik, K. Wang, and H. Jakobsen. 2012. Engineering TiO_2 nanomaterials for CO_2 conversion/solar fuels. *Sol. Energy Mater. Solar Cells.* 105: 53–68.

Liu, J., B. Garg, and Y. Ling. 2011. $Cu_xAg_yIn_zZn_kS_m$ solid solutions customized with RuO_2 or $Rh_{1.32}Cr_{0.66}O_3$ co-catalyst display visible light-driven catalytic activity for CO_2 reduction to CH_3OH. *Green Chem.* 13: 2029–2031.

Liu, Y., B. Huang, Y. Dai, X. Zhang, X. Qin, M. Jiang, and M-H. Whangbo. 2009. Selective ethanol formation from photocatalytic reduction of carbon dioxide in water with $BiVO_4$ photocatalysts. *Catal. Commun.* 11: 210–213.

Ma, J., N. Sun, X. Zhang, N. Zhao, F. Xiao, W. Wei, and Y. Sun. 2009. A short review of catalysis for CO_2 conversion. *Catal. Today.* 148: 221–231.

Mao, J., K. Li, and T. Peng. 2013. Recent advances in the photocatalytic CO_2 reduction over semiconductors. *Catal. Sci. Technol.* 3: 2481–2498.

Mikkelsen, M., M. Jorgensen, and F. C. Krebs. 2010. The teraton challenge. A review of fixation and transformation of carbon dioxide. *Energy Environ. Sci.* 3: 43–81.

Mori, K., H. Yamashita, and M. Anpo. 2012. Photocatalytic reduction of CO_2 with H_2O on various titanium oxide photocatalysts. *RSC Adv.* 2: 3165–7231.

Morris, A. J., G. J. Meyer, and E. Fujita. 2009. Molecular approaches to the photocatalytic reduction of carbon dioxide for solar fuels. *Acc. Chem. Res.* 42: 1983–1994.

Qin, S., F. Xin, Y. Liu, X. Yin, and W. Ma. 2011. Photocatalytic reduction of CO_2 in methanol to methyl formate over $CuO-TiO_2$ composite catalysts. *J. Colloid Interface Sci.*, 356: 257–261.

Richter, R. K. de, T. Ming, and S. Caillol. 2013. Fighting global warming by photocatalytic reduction of CO_2 using giant photocatalytic reactors. *Renew Sust. Energy Rev.* 19: 82–106.

Shi, H., and Z. Zou. 2012. Photophysical and photocatalytic properties of $ANbO_3$ (A = Na, K) photocatalysts. *J. Phys. Chem. Solids.* 73: 788–792.

Slamet, H. W. Nasution, E. Purnama, S. Kosela, and J. Gunlazuardi. 2005. Photocatalytic reduction of CO_2 on copper-doped titania catalysts prepared by improved-impregnation method. *Catal. Commun.* 66: 91–97.

Stechel, E. B., and J. E. Miller. 2013. Re-energizing CO_2 to fuels with the sun: Issues of efficiency, scale, and economics. *J. CO_2 Utilization.* 1: 28–36.

Tahir, M., and N. S. Amin. 2013. Advances in visible light responsive titanium oxide-based photocatalysts for CO_2 conversion to hydrocarbon fuels. *Energy Conv. Manage.* 76: 194–214.

Tahir, M., and N. S. Amin. 2013. Photocatalytic reduction of carbon dioxide with water vapors over montmorillonite modified TiO_2 nancomposites. *Appl. Catal. B: Environ.* 142–143: 512–522.

Tran, P. D., L. H. Wong, J. Barber, and J. S. C. Loo. 2012. Recent advances in hybrid photocatalysts for solar fuel production. *Energy Environ. Sci.* 5: 5902–5918.

Tsai, C.-W., H. M. Chen, R.-S. Liu, K. Asakura, and T.-S. Chan. 2011. Ni@NiO core-shell structure-modified nitrogen-doped $InTaO_4$ for solar-driven highly efficient CO_2 reduction to methanol. *J. Phys. Chem. C.* 115: 10180–10186.

Tseng, I.-H., J. C. S. Wu, and H.-Y. Chou. 2004. Effects of sol-gel procedures on the photocatalysis of Cu/TiO_2 in CO_2 photoreduction. *J. Catal.* 221: 432–440.

Tseng, I.-H., W.-C. Chang, and J. C. S. Wu. 2002. Photoreduction of CO_2 using sol-gel derived titania and titania-supported copper catalysts. *Appl. Catal. B: Environ.* 37: 37–48.

Windle, C. D., and R. N. Perutz. 2012. Advances in molecular photocatalytic and electrocatalytic CO_2 reduction. *Coord. Chem. Rev.* 256: 2562–2570.

Wu, J. C. S., H.-M. Lin, and C-L. Lai. 2005. Photo reduction of CO_2 to methanol using optical-fiber photoreactor. *Appl. Catal. A: Gen.* 296: 194–200.

Xi, G., S. Ouyang, P. Li, J. Ye, Q. Ma, N. Su, H. Bai, and C. Wang. 2012. Ultrathin $W_{18}O_{49}$ nanowires with diameters below 1 nm: Synthesis, near-infrared absorption, photoluminescence, and photochemical reduction of carbon dioxide. *Angew. Chem. Int. Ed.* 51: 2395–2399.

Yang, H.-C., H.-Y. Lin, Y.-S. Chien, J. C.-S. Wu, and H.-H. Wu. 2009. Mesoporous TiO_2/SBA-15, and Cu/TiO_2/SBA-15 composite photocatalysts for photoreduction of CO_2 to methanol. *Catal. Lett.* 131: 381–387.

Zhai, Q., S. Xie, W. Fan, Q. Zhang, Y. Wang, W. Deng, and Y. Wang. 2013. Photocatalytic conversion of carbon dioxide with water into methane: Platinum and copper(I) oxide co-catalysts with core-shell structure. *Angew. Chem.* 125: 5888–5891.

10 Artificial Photosynthesis

Neelam Kunwar, Sanyogita Sharma, Surbhi Benjamin, and Dmitry Polyansky

CONTENTS

10.1 Introduction 187
10.2 Natural Photosynthesis to Artificial Photosynthesis 188
10.3 Approaches in Artificial Photosynthesis 190
    10.3.1 Homogeneous Artificial Photosynthesis Systems 191
    10.3.2 Heterogeneous Artificial Photosynthesis Systems 195
10.4 Light Absorption 196
10.5 Charge Separation by Molecular Donor–Acceptor Systems 197
10.6 Light-Driven Catalytic Water Splitting 199
10.7 Catalytic Water Oxidation 199
    10.7.1 Ruthenium-Based Catalysts 200
    10.7.2 Iridium-Based Catalyst 202
    10.7.3 Mn-, Co-, and Fe-Based Catalysts 203
10.8 Catalytic Proton Reduction and Hydrogen Evolution 204
    10.8.1 Proton Reduction Catalyst by Hydrogenases 204
    10.8.2 Light-Driven Catalysis for Hydrogen Production 206
10.9 Catalysis for CO_2 Reduction and Fuel Production 208
References 213

10.1 INTRODUCTION

Energy is one of the key components sustaining life on Earth, including human existence. The technological progress of human civilization resulted in a steep increase in anthropogenic energy consumption, leading to rapid exploration of natural resources of fossil fuels. As on today, approximately 150,000 TW·h of energy is consumed globally per year with over 85% of that still coming from nonrenewable sources; demand is projected to increase up to 30% in the next 20 years (IEA 2014). The extensive reliance on fossil fuels is associated with several major risk factors, which highlight the necessity of finding alternative energy sources. One of these factors is the limited resource base, which when combined with the rapidly increasing demand for energy would result in the depletion of available natural resources. The uneven distribution of global fossil fuel deposits creates another challenge associated with the security of energy supplies, which is greatly dependent on geopolitical climate. And probably one of the most immediate problems resulting from the extensive exploration of nonrenewable energy sources is environmental impact, manifested in pollution and global climate change. The latter has drawn significant political attention in many industrial countries resulting in a wave of new policies, which may ultimately limit the use of fossil fuels in the future. Altogether, recent trends in the global energy landscape urge extensive exploration of alternative renewable energy sources (Figure 10.1).

Among many renewable energy sources that are currently being explored, solar energy presents the most abundant resource with over 5 million terawatt hours (TW·h) per year available for

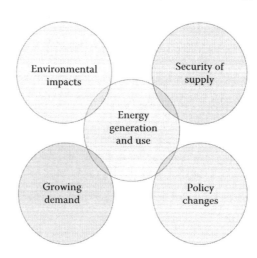

FIGURE 10.1 Major factors affecting global energy use and production.

practical use. However, one of the major challenges in efficient utilization of solar energy is energy storage. This is dictated by intermittency of solar flux, day/night cycles, and uneven energy demand, which significantly restricts the direct use of electricity generated from, for example, photovoltaic devices. One of the most efficient ways to store energy is in the form of chemical fuels. This provides the highest energy density (amount of energy stored per total weight of fuel and storage medium) and allows the use of energy in the existing infrastructure, such as transportation. The types of fuels that can be produced using the energy of solar light are hydrogen and fuels resulting from carbon reduction, such as alcohols or methane. In addition to these fuels, which can be used by a consumer directly, other products derived from the light-driven reduction of CO_2, such as carbon monoxide or formic acid, can be used in further industrial processes, such as syngas feedstock (H_2/CO mixture). The use of hydrogen is very attractive, as it is environmentally clean when produced using solar energy to split water. Water is the only product that is formed when hydrogen is burned or consumed in a fuel cell.

The production of chemical fuels using energy of light can be achieved using the artificial photosynthesis approach. This approach derives its inspiration from natural photosynthesis, where the energy of sunlight is used to convert water and carbon dioxide to carbohydrates, which are used later to build the plant's structure and sustain its functions. Artificial photosynthesis uses the same functional principles as its natural counterpart; however, components responsible for various functionalities are usually different. In this regard, artificial photosynthesis is a bioinspired approach (Graetzel 1983, Collings and Critchley 2005), where the important structural elements and reaction features of natural photosynthesis are used in conceptually simpler systems to achieve the results of natural photosynthesis. However, the solar energy conversion efficiency of natural photosynthesis is only a few percentages (0.02%–1%), creating additional challenges in creating artificial photosynthetic systems with efficiencies surpassing those of natural systems. If properly implemented, artificial photosynthesis should be capable of using solar energy to produce clean fuels at attractively high efficiencies (Cogdell et al. 2010).

10.2 NATURAL PHOTOSYNTHESIS TO ARTIFICIAL PHOTOSYNTHESIS

In the course of photosynthesis, plants harvest energy from the Sun and use it to split water and reduce carbon dioxide in order to store this energy in the form of chemical bonds. Photosynthesis reactions occur in two distinct photosynthetic reaction centers: photosystem I (PS I) and photosystem II (PS II). Light absorption by chlorophyll followed by a series of electron-transfer reactions

Artificial Photosynthesis

constitute the initial stage of the light-induced set of reactions. This is followed by oxidation of water to molecular oxygen which takes place at the oxygen-evolving complex (OEC) of PS II. The OEC is the natural catalytic center that facilitates the reaction of water oxidation, and it consists of a cluster of four manganese atoms and one calcium atom linked by oxo-bridges (McEvoy et al. 2005; McConnell et al. 2010).

In PS II, chlorophyll *a* (i.e., P680) absorbs photon energy, which excites electrons and transfers them to plastoquinone (Pq), an electron acceptor (Herek et al. 2002). Meanwhile, the photoexcited P680 (P680*) generates a driving force for extracting electrons from the water molecule at the calcium-manganese center, which acts as a water-oxidation catalyst (Umena et al. 2011). Umena et al. (2011) reported that the OEC is a Mn_4CaO_5 cluster. Several crystal structures of PS II have been resolved at various resolutions from 3.8 to 1.9 Å. It was observed that five oxygen atoms served as oxo-bridges linking five metal atoms, and that at least four water molecules were bound to the Mn_4CaO_5 cluster; some of them may, therefore, serve as substrates for dioxygen formation. More than 1300 water molecules were identified in each photosystem II monomer, out of which some formed extensive hydrogen-bonding networks that may serve as channels for protons or water or oxygen molecules.

The electron transport chain, which consists of Pq, cytochrome complex, and plastocyanine (Pc), carries electrons from PS II to PS I (Figure 10.2). The electrons are taken through a series of uphill and downhill steps to generate energy-rich intermediates in PS I reactions. In PS II, electrons excited by P700 are transferred to ferredoxin (Fd) and used to reduce $NADP^+$ to NADPH through Fd-NADP reductase. The newly formed NADPH is then consumed to fix CO_2 through the biocatalytic cascade of the Calvin cycle (Hohmann-Marriott and Blankenship 2011) when a series of light-independent (dark) reactions occur, where the products of the light-induced reactions (ATP and NADPH) are used to form C–C covalent bonds of carbohydrates, where the carbon source is CO_2, which is obtained from the atmosphere (Balzani et al. 2008).

Natural photosystems give an impetus to the work on artificial photosynthesis. In artificial photosynthesis, synthetic chromophores or even semiconductor materials can be used instead of chlorophyll, which can absorb solar light and produce charges with potential enough to, for example, drive the oxidation of water to produce O_2 and H^+. The artificial photosynthetic systems are not bound to the production of carbohydrates like plants but can be tuned to produce various products (e.g., H_2 through proton reduction) (Sayama et al. 2002), or they can include reduction of CO_2 to carbon-based fuels (methanol, methane, or formate).

Synthetic materials used in artificial photosynthesis have to efficiently realize various functions of components in natural photosynthesis. Light-absorbing components should efficiently capture

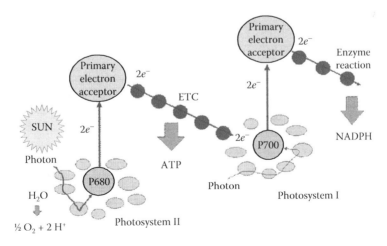

FIGURE 10.2 Electron transfer in the photosystem.

and transform photons of light into charges and transport them to catalytic centers, where synthetic catalysts should efficiently oxidize water to protons and O_2 on the oxidative side and reduce protons to H_2 or CO_2 to alcohols or methane on the reductive side. As an example of artificial design, an "artificial leaf" was introduced by Nocera and co-workers. The construction of an artificial leaf incorporates earth-abundant elements by interfacing a triple-junction, amorphous silicon photovoltaic with hydrogen and oxygen and evolving catalysts made from a ternary alloy (NiMoZn) and a cobalt-phosphate cluster (Co-Pi). It captures the structural and functional attributes of the combined PS II and PS I. The Co-Pi catalyst oxidizes H_2O by a proton-coupled electron transfer mechanism, where oxidation states Co-Pi are increased by four hole equivalents, similar to the S-states pumping of the Kok cycle of PS II (Nocera 2012).

10.3 APPROACHES IN ARTIFICIAL PHOTOSYNTHESIS

Understanding the biological processes of photosynthesis and utilizing this knowledge to design some novel synthetic molecular systems to provide good light to chemical energy conversion efficiency is the developing field of artificial photosynthesis (Barber 2009; Benniston and Harriman 2008; Gust et al. 2008). Different approaches have to be based on relationships, which are similar to those present in a photosynthetic apparatus in living organisms. The main characteristics of an artificial photosynthetic system must be obtained, taking into account nonvalence interactions between components of the corresponding model system. The cornerstone of successful design of functional artificial photosynthesis systems is in the development of efficient, robust, and synthetically accessible components for light absorption, charge separation, and catalysis, and the provision of efficient electronic and mass transport between these components. In general, one of the artificial photosynthetic architectures is envisioned in the form of a photoelectrochemical cell, where reductive and oxidative half-reactions take place in separate compartments separated by a proton-permeable membrane. Electrodes on each side contain catalysts for water oxidation and, for example, reduction of H^+ to hydrogen, respectively. The electrolysis is driven by the energy of solar radiation absorbed by chromophores or semiconductor materials and converted to electrical charges. The light-absorbing material can be located either outside the electrolysis cell (PV-to-electrolysis architecture) or deposited directly on the electrode.

The basic components of artificial photosynthetic systems include (1) light absorption, (2) excited-state electron transfer, (3) separation of electron transfer–generated oxidative and reductive equivalents by free energy gradients, (4) electron transfer activation of multielectron catalysts, (5) proton-electron coupled solar fuel half reactions, and (6) separation and collection of the products (Figure 10.3).

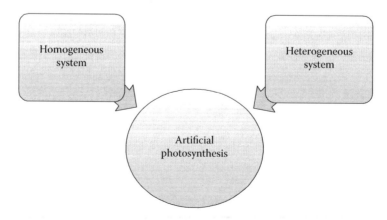

FIGURE 10.3 Systems of artificial photosynthesis.

Artificial Photosynthesis

Recently, many promising models have appeared by applying the basic concepts of this technology. Two major systems emerging from this are

- Homogeneous artificial photosynthesis systems
- Heterogeneous artificial photosynthesis systems

10.3.1 Homogeneous Artificial Photosynthesis Systems

Homogeneous artificial photosynthesis is a multielectron, multiproton process requiring the use of catalysts. The design of a homogeneous catalyst is more controllable than that of heterogeneous systems, and as a result their use can provide better insight into the reaction mechanism. The development of homogeneous catalysts is facilitated by access to a wide variety of spectroscopic, kinetic, and theoretical methods and the ability to relay obtained information to the functioning of the natural photosynthetic system. Some necessary components of an artificial photosynthesis system are antenna for light harnessing, a reaction center (RC) for charge separation, oxidation and reduction catalysts, and membrane to separate the generated products.

Allakhverdiev et al. (2009) examined the main pathways of H_2 photoproduction by using photosynthetic organisms and biomimetic photosynthetic systems. The systematic production of hydrogen and oxygen is shown in Figure 10.4.

Alstrum-Acevedo et al. (2005) reported light absorption and excited-state electron transfer to create oxidative and reductive equivalents for driving relevant fuel-forming half reactions such as the oxidation of water to O_2 and its reduction to H_2. They prepared an "integrated modular assembly," with separate components for light absorption, energy transfer, and long-range electron transfer by use of free energy gradients, which are integrated with oxidative and reductive catalysts into single molecular assemblies or on separate electrodes in photoelectrochemical cells. Derivatives of porphyrins and metalloporphyrins and metal polypyridyl complexes have been most commonly used for these assemblies.

Sun et al. (1997) prepared a series of compounds for developing models of PS II in green plants. In these compounds, a photosensitizer, ruthenium(II) tris(bipyridyl) complex (to mimic the function of P_{680} in PS II), was covalently linked to a manganese(II) ion through different bridging ligands. The structures of the compounds were characterized by electron paramagnetic resonance (EPR) measurements and electrospray ionization mass spectrometry (ESI-MS). The interaction between the ruthenium and manganese moieties within the complex was probed by steady-state and time-resolved emission measurements. They observed that on exposing the dinuclear complexes to flash photolysis in the presence of an electron acceptor such as methyl viologen (MV^{2+}), the initial

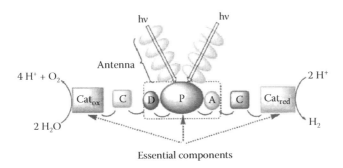

FIGURE 10.4 The artificial photosynthetic system for production of oxygen and hydrogen. (P, photosensitizer; A, electron acceptor; D, electron donor; C, electron carrier; Cat_{ox}, catalyst for oxidation of water; Cat_{red}, catalyst for reduction of H^+.) (Adapted from Allakhverdiev, J., *Photochem. Photobiol. B*, 104, 1–8, 2011. With permission.)

electron transfer from the excited state of Ru(II) in this compound formed Ru(III) and $MV^{+\bullet}$. An intramolecular electron transfer from coordinated Mn(II) to the photogenerated Ru(III) occurred with a first-order rate constant of 1.8×10^5 s^{-1}, regenerating Ru(II). Thus, a supramolecular system with a manganese complex has been used as an electron donor to a photooxidized photosensitizer. Balzani et al. (1997) suggested that the model systems can be developed, which are capable of mimicking the two fundamental steps of photosynthesis, namely, light harvesting and photoinduced charge separation.

The theoretical maxima of solar energy conversion efficiencies and productivities in oxygenic photosynthesis in a variety of photosynthetic organisms, including green microalgae, cyanobacteria, and C_4 and C_3 plants were evaluated by Melis (2009). He also reported that photosynthetic solar energy conversion efficiency and productivity can be improved by minimizing, or truncating, the chlorophyll antenna size of the photosystems up to threefold. The generation of truncated light-harvesting chlorophyll antenna size (tla) strains, in all classes of photosynthetic organisms, would help to reduce excess absorption of sunlight and the ensuing wasteful dissipation of excitation energy. It may also maximize solar-to-product energy conversion efficiency and photosynthetic productivity in high-density mass cultivations. Such a model may find application in the commercial exploitation of microalgae and plants for the generation of biomass, biofuels, and chemical feedstocks, as well as in the field of nutraceuticals and pharmaceuticals.

Absorption of sunlight by a chromophore is followed by the transfer of excitation energy to a reaction center (C). However, light absorption by a single molecule of a chromophore is limited and requires either the use of antenna arrays or multilayer structures for harnessing the light efficiently. Excitation of a chromophore molecule is then followed by photoinduced electron transfer to an electron acceptor (A), and a secondary thermal electron transfer from a donor component (D) to an oxidized chromophore (C*) (Figure 10.5). If the antenna array and reaction center (C) are not properly coupled in the dimensions of time, energy, and space, homogeneous artificial photosynthetic systems may not function properly.

Ort et al. (2011) discussed the theoretical and physiological factors that need to be considered in optimizing chlorophyll content for improved photosynthetic efficiency and maximum carbon gain by crops and microalgae. It was concluded that on optimizing the size of antenna (by reducing its size), the photosynthetic efficacy was also maximized. Multiporphyrin arrays, polypyridine complexes, and polymetallic complexes were among the molecules that have been widely investigated as antenna molecules. Balzani et al. (2008) reviewed a brief description of the mechanism at the basis of natural photosynthesis and its relevance to the field of photochemical conversion of solar energy. Solar energy, whether converted to wind, rain, biomass, or fossil fuels, was found to be the primary energy source for human-engineered energy transduction systems.

Hambourger et al. (2009) reviewed matches between some biological and technological energy systems. Herrero et al. (2008) described various aspects of the oxidizing enzyme or designing artificial catalysts, which have the ability to oxidize water with a low overpotential and could also greatly improve the efficiency of water electrolysis and photolysis. They provided an overview of the production of a biomimetic photocatalyst for water oxidation by focusing on the essential cofactors involved in the light-driven oxidation of water.

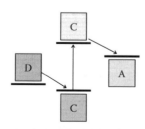

FIGURE 10.5 Excitation of chromophore.

An understanding of natural photosynthesis at the molecular level has been supported further by the creation of artificial photosynthetic model systems, such as donor–acceptor assemblies. The effort has been directed toward the development of artificial systems composed of molecular and supramolecular architectures. In molecular light-harvesting systems, the energy of many photons is collected and transferred to a final acceptor, resulting in charge separation. Light energy is used to move an electron from donor to acceptor, creating this charge-separated state. The resulting oxidizing and reducing equivalents are moved away from the site of charge separation to catalytic sites through subsequent electron transfer steps, where they are used to oxidize water to oxygen and carry out reductive interaction that can form fuels in oxygenic photosynthesis. Photoinduced electron transfer is also characteristic of these artificial reaction centers. Such systems have been developed to mimic the antenna effect.

Dyad and triad models based on the association of a photosensitizer to an electron donor, an electron acceptor, or both components, have also been designed to reproduce light-induced spatial charge separation. These samples show that transformation of light energy into chemical energy in the form of a charge-separated state can be achieved. However, the next step toward water decomposition would require the coupling of the relevant reduction and oxidation catalysts to a triad. The basic requirements for such a particular working photocatalytic system are

- Directionality of the electron transfer process for each individual charge separation step
- Tuning of the electronic coupling between the donor/acceptor and the photosensitizer in order to maximize electron transfer efficiency
- Long-lived charge-separated state
- Catalysts (donor/acceptor) capable of storing multiple oxidizing/reducing equivalents

Gust et al. (2001) reported that artificial light-harvesting antenna can be synthesized and linked to artificial reaction centers that convert excitation energy to chemical potential in the form of long-lived charge separation. These artificial reaction centers can form the basis for molecular-level optoelectronic devices. Fukuzumi (2008) presented a major strategy to reach long-lived charge separation states, which is an important requirement in view of the application of photogenerated electrons in catalysis.

Aratani et al. (2009) reviewed the synthesis of covalently and noncovalently linked discrete cyclic porphyrin arrays as models of the photosynthetic light-harvesting antenna complexes. They also prepared a series of extremely long but discrete meso-meso-linked porphyrin arrays and covalently linked large porphyrin rings on the basis of the silver(I)-promoted oxidative coupling strategy. The photophysical properties of these molecules were examined using steady-state absorption, fluorescence, fluorescence lifetime, fluorescence anisotropy decay, and transient absorption measurements. The exciton-exciton annihilation time and the polarization anisotropy rise time within these structures were described in terms of the Forster-type incoherent energy hopping model.

Meso-pyridine-appended zinc(II) porphyrins and their meso-meso-linked dimers spontaneously assemble in noncoordinating solvents such as $CHCl_3$ to form tetrameric porphyrin squares and porphyrin boxes, respectively. The rigorous homochiral self-sorting process and efficient excitation energy transfer (EET) have been demonstrated along these cyclic porphyrin arrays. The meso-cinchomeronimide appended zinc(II) porphyrin forms a cyclic trimer. It has also been shown that the corresponding meso-meso-linked diporphyrins undergo high-fidelity self-sorting assembling to form discrete cyclic trimer, tetramer, and pentamer with large association constants through perfect discrimination of enantiomeric and conformational differences of the meso-cinchomeronimide substituents (Naoki et al. 2009).

Wenger (2009) reported the long-range electron transfer with synthetic systems using d^6 or d^8 metal complexes as photosensitizers, which includes charge transfer over distances greater than 10 Å in both covalent and noncovalent donor–bridge–acceptor systems. The coordination compounds studied contained metals such as Ru(II), Os(II), Re(I), Ir(III), Ir(I), and Pt(II). Albinsson and Martensson (2008) reported that the property of a donor-bridge-acceptor (D-B-A) system is not a simple linear combination of properties of the individual components, but it depends on the specific building blocks and the style in which they are assembled. An important example is the ability of the bridge to support the intended transfer process. The mediation of the transfer is characterized by an attenuation factor (β) often viewed as a bridge-specific constant, but it also depends on the donor and the acceptor (i.e., the same bridge can either be poorly or strongly conducting depending on the nature of the donor and the acceptor). An account of the experimental exploration of the attenuation factor β in a series of bis(porphyrin) systems covalently linked by bridges of the oligo(phenyleneethynylene) (OPE) type has been given.

Attenuation factors for electron transfer (ET) as well as for both singlet and triplet excitation energy transfer (EET) are deliberated. A report is also given on the dependence of the transfer efficiency on the energy gap between the donor and bridge states relevant for the specific transfer process. The experimental variation of β with varying donor and acceptor components is shown for a range of conjugated bridges by representative examples. The theoretical rationalization for the observed variation is briefly discussed. Based on the Gamow tunneling model, the observed variations in β-values with varying donors and acceptors for the same bridges are simulated, and the observed energy gap dependence is modeled by Bo and Martensson (2008).

Harriman and Sauvage (1996) described a new approach for assembling the photosynthetic models with porphyrin-containing modules assembled around transition metals. Straight et al. (2008) prepared a molecular pentad consisting of two light-gathering antennas, a porphyrin electron donor, a fullerene electron acceptor, and a photochromic control moiety. This molecular system mimics the nonphotochemical quenching mechanism for photoprotection found in plants where the light conversion efficiency is altered in response to the incident photon flux. Specifically, this molecule assembly achieves photoinduced electron transfer with a quantum yield of 82% when the light intensity is low. As the light intensity was increased, the photochrome photoisomerizes, leading to the quenching of the porphyrin-excited state and reducing the quantum yield to as low as 27% (Figure 10.6).

The transition-metal coordination compounds are often used as photocatalysts for the reduction of carbon dioxide due to their ability to absorb a significant portion of the solar spectrum and promote activation of small molecules as reported by Morris et al. (2009). The review discussed four classes of transition-metal catalysts: (1) metal tetraaza-macrocyclic compounds; (2) supramolecular complexes; (3) metalloporphyrins and related metallomacrocycles; and (4) $Re(CO)_3(bpy)X$-based compounds (where bpy is 2,2'-bipyridine). Carbon monoxide and formate are the primary CO_2 reduction products, in addition to the formation of hydrogen and bicarbonate/carbonate.

Yin et al. (2010) reported the complex $[Co_4(H_2O)_2(PW_9O_{34})_2]^{10-}$, which includes a Co_4O_4 core and is stabilized by oxidatively resistant polytungstate ligands. It was described as a molecular fully inorganic water oxidation catalyst, which is hydrolytically and oxidatively stable and self-assembles in water from salts of earth-abundant elements (Co, W, and P). Substantial catalytic turnover

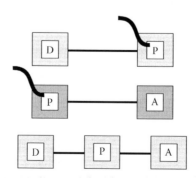

FIGURE 10.6 Dyad and triad models mimicking the charge separation state.

frequencies (TOFs) for O_2 production greater than or equal to 5 s^{-1} at pH = 8 were observed with $[Ru(bpy)_3]^{3+}$ (bpy is 2,2'-bipyridine) as the photogenerated oxidant. The sensitivity of the catalytic rate to the pH change reflects the pH dependence of the rate-limiting step in the overall four-electron H_2O oxidation. Extensive spectroscopic, electrochemical, and inhibition studies indicated that $[Co_4(H_2O)_2(PW_9O_{34})_2]^{10-}$ is a stable complex under catalytic turnover conditions.

10.3.2 Heterogeneous Artificial Photosynthesis Systems

The use of heterogeneous catalysts is ultimately more attractive for commercial-scale solar fuel production due to their higher stability toward oxidative degradation and more facile and economical fabrication. Moreover, heterogeneous catalysts can be more easily integrated into devices that are capable of combining processes of the water oxidation and proton reduction, making the complete artificial system functional. The term *artificial photosynthesis*, was initially applied to molecular systems, but now it is also applied for water-splitting and fuel-forming reactions at the semiconductor–electrolyte interface. Upon irradiation of a semiconductor-based photocatalyst with light having energy greater than the semiconductor's bandgap, electrons are promoted to the conduction band (CB), while holes are left in the valence band (VB). However, hole–electron pairs are still bound by Coulombic forces forming an exciton. These light-generated electron–hole pairs can be considered as localized versions of reductive and oxidative equivalents. Once formed, these electron–hole pairs are separated by internal electric fields and directed to spatially separate catalytic sites on the semiconductor surface. The photogenerated electrons and holes may undergo

- Recombination in bulk
- Recombination at the surface
- Reduction of suitable electron acceptor or oxidation of an electron donor

Electron–hole recombination is promoted by defects in the semiconductor materials and decreases the efficiency of photocatalytic activity. If an electron donor molecule (D) is present at the surface, then the photogenerated hole can react with this molecule to generate oxidizing products, D^+. Similarly, if an electron acceptor molecule (A) is present at the surface, then the photogenerated electrons can react with them to generate a reduced product, A^- (Prashant et al. 2003):

$$A + D \xrightarrow[\text{Semiconductor}]{h\nu} A^- + D^+ \tag{10.1}$$

In order to increase the light collection efficiency of wide-band semiconductors, their surfaces can be sensitized with molecular chromophores or dyes. In addition to extension of light-absorbing properties to the visible part of the solar spectrum, which increases the amount of photons collected by the absorber, dye molecules improve the efficiency of charge separation, due to ultrafast electron injection from the excited states of chromophores into the valence bands of semiconductor materials. Such dye/semiconductor interfaces make the basis of a dye-sensitized solar cell (DSSC). DSSC can also be considered as an example of a heterogeneous artificial photosynthesis system, which mimics the electron transfer processes in natural photosynthesis. While light-activated electron transfer processes in the natural photosystem produce charge gradient in the form of proton-motive force (pmf), the action of DSSC results in the production of electromotive force (emf), or in other words, electricity.

Abe et al. (2000) developed a stable dye-sensitized photocatalyst system. They reported that eosin Y attached to Pt-TiO_2 (EY-TiO_2) exhibited steady H_2 production from aqueous triethanolamine solution (TEOA aq.) under visible light irradiation for an extended time. The turnover number (TON) of the light-driven hydrogen production reached more than 10,000, and the quantum yield of the EY-TiO_2 at 520 nm was determined to be about 10%.

However, regardless of the mechanism of the generation of positive and negative charges, all current heterogeneous photosynthetic systems can be described in terms of the photoelectrochemical cell (PEC) concept, where electrons are transferred from a photoanode by oxidizing water to a cathode to reduce protons or CO_2 in the fuel-forming reactions. Bak et al. (2002) discussed the materials-related issues in the development of high-efficiency PECs. The property requirements for photoelectrodes, in terms of light absorption and electrochemical properties and their impact on the performance of PECs, were defined. Different types of PECs were overviewed, and the impacts of the PEC structure and materials selection on the conversion efficiency of solar energy were considered. They also reported that the performance of PECs is to be considered in terms of

- Excitation of electron–hole pair in photoelectrodes
- Charge separation in photoelectrodes
- Electrode processes and related charge transfer within PECs
- Generation of the PEC voltage required for water oxidation and reductive reactions

10.4 LIGHT ABSORPTION

Absorption of solar light is the first step in the sequence of reactions leading to the successful operation of artificial as well as natural photosystems. Nature has highly complex light-harvesting systems to utilize a substantial part of the solar spectrum by using chlorophyll dyes. The goal of artificial systems is to incorporate enhanced light-harvesting capacity and efficiency to mimic natural systems. In an artificial system, sufficient light harvesting can be achieved by using a single RC chromophore or by the excitation of a light-absorbing antenna array followed by the energy transfer sensitization of a reaction center (RC) (Alstrum-Acevedo et al. 2005). Antenna systems are capable of collecting light and transferring energy in an efficient and direct way. A large number of antenna chromophores surrounding the RC absorb incident photons if a RC is coupled to an antenna array. The resulting excited states transfer electronic energy to the RC before undergoing radiative or nonradiative deactivation. Highly branched tree-like dendrimers have been widely employed as antenna systems. Melis (2009) reported that the incorporation of antenna chromophores in artificial systems may not be beneficial, if the antenna array and the RC are not properly coupled in the dimensions of time, energy, and space. It has been shown that the photosynthetic solar energy conversion efficiency in natural systems will improve by reducing the antenna size.

Among various light-harvesting chromophores, porphyrins and phthalocyanines are closely related to chlorophyll derivatives. Röger et al. (2006) prepared a bichromophoric assembly consisting of blue naphthalene bisimide (NBI) dye at the periphery of aggregated zinc chlorines. Metal coordination compounds, exhibiting metal-to-ligand charge-transfer (MLCT) transitions at relatively low energy, have been widely employed as photosensitizers (Huynh et al. 2005) (Figure 10.7). It was reported that unlike chlorophyll a, the chromophore can be easily functionalized and incorporated into a wide variety of biomimetic electron donor–acceptor systems.

Meyer (1990) reported that polypyridyl complexes of Ru(II), Os(II), and Re(1) absorb visible light through MLCT transitions. They provided insight into the reactivity of MLCT excited states through the intramolecular electron and energy transfer processes. Lukas et al. (2002) prepared a green chromophore, 1,7-bis(pyrrolidin-1-yl)-3,4:9,10-perylene-bis(dicarboximide) (5PDI), that exhibits photophysical and redox properties similar to those of chlorophyll a (Chl a), but unlike Chl a, it can be easily functionalized and incorporated into a wide variety of biomimetic electron donor–acceptor systems. The N,N'-dicyclohexyl derivative (5PDI) absorbs strongly ($\varepsilon = 46,000$ M L cm^{-1}) at 686 nm in toluene and fluoresces at 721 nm with a 35% quantum yield. Additionally, 5PDI can be oxidized as well as reduced in CH_2Cl_2 at 0.57 V and –0.76 V versus SCE (saturated calomel electrode), respectively, making it a facile electron donor or acceptor. Rodlike covalent electron donor–acceptor molecules 5PDI-PI, 5PDI-NI, and 5PDI-PDI were prepared by linking the imide

Artificial Photosynthesis

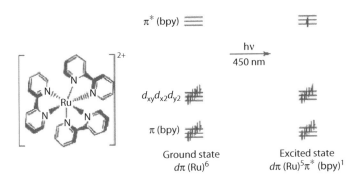

FIGURE 10.7 Energy-level diagram showing the ground and lowest MLCT excited state(s) of [Ru(bpy)3]2+. (Adapted from Huynh, M. H. V., D. M. Dattelbaum, and T. J. Meyer, *Coord. Chem. Rev.*, 249, 457–483, 2005. With permission.)

group of the 5PDI donor to pyromellitimide (PI), 1,8:4,5-naphthalene-bis(dicarboximide) (NI), and 1,7-bis(3,5-di-*tert*-butylphenoxy)-3,4:9,10-perylene-bis(dicarboximide) (PDI) acceptors via an N-N bond. The formation and decay of their excited and radical ion pair states were monitored directly by transient absorption spectroscopy following femtosecond laser excitation of 5PDI, 5PDI-PI, 5PDI-NI, and 5PDI-PDI in toluene and 2-methyltetrahydrofuran. The supramolecular light-harvesting arrays can be constructed by self-assembling chromophore building blocks in solution and on surfaces (Ahrens et al. 2004; Kelley et al. 2008).

10.5 CHARGE SEPARATION BY MOLECULAR DONOR–ACCEPTOR SYSTEMS

Charge separation processes play an essential role in the functionality of natural and artificial photosystems. Dinner and Babcock (1996) and Danielsson et al. (2006) reported the electron transfer sequence of light-driven water oxidation in PS II:

- P680 is excited by absorption of photons or energy transfer from light-harvesting antenna chlorophylls, and forms P680*, which transfers an electron to PS I via pheophytin (Phe), quinones A and B (QA and QB), resulting in a highly oxidizing P680$^{\cdot+}$.
- P680$^{\cdot+}$ oxidizes tyrosine Z (Tyrz) to a milder Tyrz radical species.
- Tyrz radical extracts an electron from the nearby OEC.
- Water oxidation occurs when four electrons are transferred.

Following the initial event of light absorption, the energy of photons is stored in metastable excited states, where the photoexcited electron is promoted from the highest occupied molecular orbital (HOMO) of the chromophore to a higher-energy molecular orbital of the same molecule. This produces a tremendous energy gradient within the photoexcited molecule, which forces an excited electron to return to its original state and release the excess energy in the form of heat (internal conversion) or light (luminescence). Photoexcited molecules can become better electron donors or acceptors, due to the charge separation character of the excited states. In order to avoid detrimental recombination processes and maximize the generation of charges, the charge separation character of the excited state should be enhanced, which can be achieved, for example, by incorporating donor–acceptor (D–A) assemblies in the light-absorbing units. These donor–acceptor assemblies facilitate the electron transfer, charge separation, and storage of redox equivalents for multielectron reactions by providing coupling between the donor or acceptor and the chromophore. Some of the donor–acceptor assemblies are designed as molecular dyads, such as the porphyrin quinone (P–Q) system. In the P–Q system, the covalently linked porphyrin and quinone function as the chromophore/electron donor and electron acceptor, respectively. Gust and Moore (1989) have described features of some dyad models, which mimic the photosynthetic electron transfer and singlet or triplet energy transfer. They also provided insight into how this type of dyad can be used for elucidating basic photochemical principles of artificial photosynthetic systems. Some limitations of these types of dyads in the area of temporal stabilization of electronic charge separation were also presented. This limitation inspired the development of multicomponent molecular devices with much more complex structures.

The more complex D–A architectures have also been explored. Moore et al. (1984) reported a molecular triad constructed by using a porphyrin (P) photosensitizer linked to a carotenoid (Car) electron donor on one side and a quinone (Q) electron acceptor on the other side. In this Car–P–Q triad, photoexcitation of the porphyrin yields the first excited singlet state, Car–$^1P^*$–Q. This excited state decays via a sequential, two-step, electron transfer process, leading to the formation of a Car•+–P–Q•– charge-separated state with a lifetime on the microsecond (µs) timescale. The charge recombination is significantly slowed because of the interposition of a neutral porphyrin between the widely separated ions. This triad approach mimics the strategy used in natural RCs, where multistep electron transfers occur through a series of donors and acceptors. The ability to optimize charge separation over charge recombination in any system, thereby creating long-lived charge-separated states, is essential for the development of efficient artificial photosynthetic systems. The electron transfer rate is a function of donor–acceptor orientation, the solvent, the intervening linkage or other medium, and the temperature in donor–acceptor assemblies.

Wiberg et al. (2007) observed that the donor–acceptor distances and donor bridge energy gaps influence the rates of charge separation and charge recombination differently in a donor-bridge-acceptor model system. It was reported that the exponential distance dependence is slightly higher for charge recombination in comparison with that for charge separation. They also have shown that the effect of the tunneling barrier height is different for charge separation and charge recombination, and that this difference is highly dependent on the electron acceptor. Sun et al. (2001) studied some supramolecular donor–acceptor model systems, which are capable of mimicking the light reactions on the donor side of photosystem II (PS II). They also reported that manganese complexes and tyrosine play the role of electron donors in the model system, similar to the manganese cluster of OEC and tyrosine in PS II. The donors have been covalently linked to a photosensitizer, a ruthenium(II) *tris*-bipyridyl complex, that plays the role of the P_{680} chlorophylls in PS II. It was observed that the incorporation of an intervening redox active link (such as tyrosine) might be crucial to the multistep electron transfer from the Mn cluster to the Ru center, as in PS II. The light-harvesting $[Ru(bpy)_3]^{2+}$ moiety was used as the replacement for P680; Mn complexes, such as a Mn(II,II) dimer were incorporated in the assemblies to mimic the OEC (Abrahamsson et al. 2002).

Light-induced accumulative electron transfer from the Mn(II,II) dimer to the photooxidized Ru center resulted in the formation of a Mn(III,IV) complex (Hammarström and Styring 2008). Huang et al. (2002) reported that the water-oxidizing complex in photosystem II, a dimeric manganese(II,II) complex, was linked to a ruthenium(II)*tris*-bipyridine ($Ru^{II}(bpy)_3$) complex via a substituted L-tyrosine, to form the trinuclear complex. The photolysis of the dimeric complex and $Ru^{II}(bpy)_3$ in aqueous solution, in the presence of an electron acceptor, resulted in the stepwise extraction of three electrons by $Ru^{III}(bpy)_3$ from the $Mn_2^{II,II}$ dimer, which then attained the $Mn_2^{III,IV}$ oxidation state. It indicates that oxidation from the $Mn_2^{II,II}$ state proceeds stepwise via intermediate formation of $Mn_2^{II,III}$ and $Mn_2^{III,III}$. In the presence of water, cyclic voltammetry showed an additional anodic peak beyond $Mn_2^{II,III/III,III}$ oxidation, which was significantly lower than in neat acetonitrile. Assuming that this peak is due to oxidation to $Mn_2^{III,IV}$ suggests that water is essential for the formation of the $Mn_2^{III,IV}$ oxidation state. A trinuclear complex is a structural mimic of the water-oxidizing complex, where it links a Mn complex via a tyrosine to a highly oxidizing photosensitizer. This complex also mimics mechanistic aspects of photosystem II, where the electron transfer to the photosensitizer is fast and results in several electron extractions from the Mn moiety.

Artificial Photosynthesis

10.6 LIGHT-DRIVEN CATALYTIC WATER SPLITTING

The overall reaction of the water splitting represents the transformation of two water molecules into one oxygen molecule and two hydrogen molecules. This reaction is endoergic and requires energy input, which can be provided by the energy of solar light:

$$2H_2O = O_2 + 2H_2 \quad \Delta G° = 474 \text{ kJ mol}^{-1} \quad (10.2)$$

The reaction of water splitting is typically separated into two of its half-reactions: catalytic water oxidation and catalytic proton reduction. Water oxidation is the most energy demanding, as compared with the latter. The difficulty of water oxidation is due to the complexity of the reaction, which involves multiple proton-coupled electron transfer (PCET) processes and the O–O bond formation:

$$2H_2O = O_2 + 4e^- + 4H^+ \quad \Delta G° = 474 \text{ kJ mol}^{-1} \quad (10.3)$$

$$2e^- + 2H^+ = H_2 \quad \Delta G° = 0 \text{ kJ mol}^{-1} \quad (10.4)$$

In electrochemical terms, the equilibrium potential required for achieving water oxidation is 1.23 V/NHE at pH = 0 and falls 59 mV per every unit of pH increase. In practice, a potential in excess of equilibrium potential is required to overcome the formation of high-energy intermediates and achieve reasonable reaction rates. As a result, a water oxidation catalyst (WOC) is required to minimize the overpotential and increase the rate of water oxidation. A qualified WOC has to fulfill the following criteria for large-scale application:

- Long-term durability
- Low overpotential
- High activity
- Low cost
- Low toxicity

10.7 CATALYTIC WATER OXIDATION

$$2H_2O \xrightarrow[\text{Photosensitizer}]{h\nu, \text{WOC}} O_2 + 4e^- + 4H^+ \quad (10.5)$$

This reaction is an essential part of water splitting. The net half-reaction is water oxidation to produce electrons and protons, with oxygen being a "by-product." A simple prototype architecture can be designed to study light-driven water oxidation by combining a water oxidation catalyst with a light-absorbing chromophore in a single assembly. A series of oligoproline-based light-harvesting chromophore water oxidation catalyst assemblies were presented by Ryan et al. (2014). This approach combines solid-phase peptide synthesis and also includes copper(I)-catalyzed azide-alkyne cycloaddition (CuAAC) as an orthogonal approach to install the chromophore and assemble the water oxidation catalyst. Alibabaei et al. (2013) presented a strategy for solar water splitting based on a dye-sensitized photoelectrolysis cell. It uses a derivatized, core–shell-nanostructured photoanode with the high surface area conductive indium tin oxide or antimony tin oxide core coated with a thin outer TiO_2 shell formed by atomic layer deposition. A *chromophore-catalyst assembly* $[(PO_3H_2)_2bpy)_2Ru(4\text{-Mebpy-4-bimpy}) Rub(tpy)(OH_2)]^{4+}$, which combines a light absorber and water oxidation catalyst in a single molecule, was attached to the TiO_2 shell. Visible light photolysis of the resulting core–shell assembly containing Pt cathode resulted in water splitting into hydrogen and oxygen. The strong electron-donating oxo and carboxylate ligands in the OEC play an essential role in stabilizing and lowering redox potentials of the OEC. This was an inspiration for designing a few families of synthetic WOCs based on transition metals and strong electron-donating ligands. A variety of homogeneous water oxidation catalysts have been developed which contain transition metals including Ru, Mn, Ir, Co, and Fe.

10.7.1 Ruthenium-Based Catalysts

Gersten et al. (1982) reported the molecular WOC cis,cis-[Ru(bpy)$_2$(H$_2$O)]$_2$(μ-O)]$^{4+}$ "blue dimer" with a TON of about 13 and a TOF of 0.004 s^{-1}. This was the first report of molecular water oxidation catalysts based on transition-metal complex. Thirty years of development on Ru WOCs has led to a dramatic improvement in the catalytic efficiency. Now, Ru-based WOCs are the most extensively studied systems, not only for the purpose of the mechanistic investigation of the water oxidation reaction but also for providing new insights into the structure-activity relationship.

Romain et al. (2009) reported that Ru–aqua complexes are capable of reaching high oxidation states as a result of the sequential or simultaneous loss of protons and electrons (Figure 10.8).

A solvent water molecule may or may not participate in the formation of the O–O bond. Accordingly, the two main pathways are

1. Solvent water nucleophilic attack (WNA) on the Ru=O group.
2. Interaction of two M–O units (I2M). Most of the complexes described belong to the WNA class, including a variety of mononuclear and polynuclear complexes containing one or several Ru–O units.

A common feature of these complexes is the generation of formal oxidation states as high as Ru(V) and Ru(VI), which render the oxygen atom of the Ru–O group highly electrophilic. On the other hand, only one symmetric dinuclear complex that undergoes an intramolecular O–O bond formation step has been described for the I2M class. It has a formal oxidation state of Ru(IV). It was also observed that Ru–OH$_2$ complexes contain redox active ligands (such as the chelating quinone). These ligands are capable of undergoing reversible redox processes, and thus generate a complex but fascinating electron transfer process between the metal and the ligand.

A general reactivity toward water oxidation in a class of Ru polypyridine complexes was investigated by tuning the properties of molecular catalysts by systematic synthetic variations of ligand manifolds (Concepcion et al. 2009). These molecules catalyze water oxidation driven either

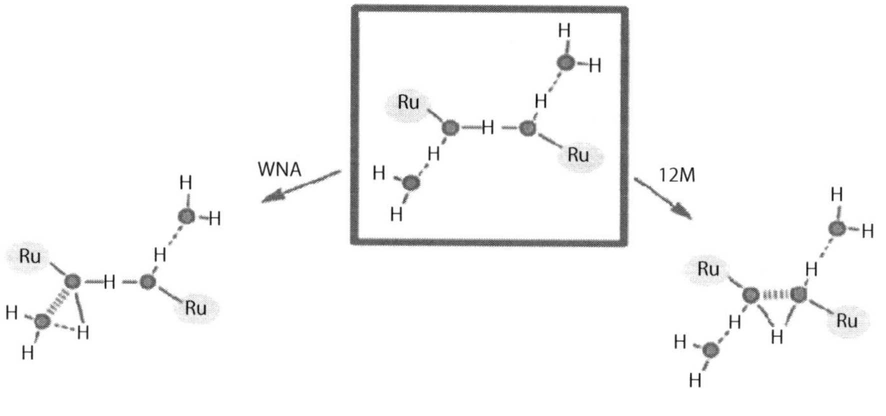

FIGURE 10.8 Ruthenium aqua complex. (Adapted from Romain, S., L. Vigara, and A. Llobet, *Acc. Chem. Res.*, 42, 1944–1953, 2009. With permission.)

electrochemically or by Ce(IV). The first two in the series were [Ru(tpy)(bpm)(OH$_2$)]$^{2+}$ and [Ru(tpy)(bpz)(OH$_2$)]$^{2+}$ (bpm is 2,2′-bipyrimidine; tpy is 2,2′:6′,2″-terpyridine), which undergo hundreds of turnovers without decomposition, with Ce(IV) as an oxidant. Detailed mechanistic studies and DFT calculations have revealed a stepwise mechanism: initial $2e^-/2\ H^+$ oxidation, to Ru^{IV}=O^{2+}, $1e-$ oxidation to Ru^V=O^{3+}, nucleophilic H$_2$O attack resulting in Ru^{III}–OOH^{2+}, further oxidation to Ru^{IV}(O$_2$)$^{2+}$, and, finally, oxygen loss, which is in competition with further oxidation of [Ru^{IV}(O$_2$)]$^{2+}$ to [Ru^V(O$_2$)]$^{3+}$, which loses O$_2$ more rapidly.

An extended family of 10–15 catalysts based on Mebimpy (Mebimpy is 2,6-bis(1-methyl-benzimidazol-2-yl)pyridine), tpy, and heterocyclic carbene ligands appears to share a common mechanism. The osmium complex [Os(tpy)(bpy)(OH$_2$)]$^{2+}$ also functions as a water oxidation catalyst. Mechanistic experiments have revealed additional pathways for water oxidation: one involving Cl− assisted catalysis, and another rate enhancement of O–O bond formation by concerted atom proton transfer (APT).

Sens et al. (2004) synthesized three new dinuclear ruthenium complexes. These are [Ru$_2^{II}$(bpp)(trpy)$_2$(μ-L)]$^{2+}$ (L = Cl or AcO) and [Ru$_2^{II}$(bpp)(trpy)$_2$(H$_2$O)$_2$]$^{3+}$. These three complexes have been characterized through the usual spectroscopic and electrochemical techniques. In aqueous acidic solution, the acetato bridge of the second complex is replaced by aqua ligands, generating the bis(aqua) complex, which is oxidized to its Ru^{IV}Ru^{IV} state. It has been shown to catalytically oxidize water to a molecular oxygen. The measured pseudo-first-order rate constant for the O$_2$ evolution process is 1.4×10^{-2} s^{-1}, which is more than three times larger than that reported earlier for Ru-O-Ru–type catalysts. This new water-splitting catalyst also has some improved stability.

Wada et al. (2000) synthesized the bridging ligand 1,8-bis(2,2′:6′,2″-terpyridyl) anthracene (btpyan). The resulting dinuclear complex [Ru$_2$(II)(OH)$_2$(3,6-(t)Bu$_2$sq)$_2$(btpyan)]^0 undergoes ligand-localized oxidation at $E_{1/2} = +0.40$ V (vs. Ag/AgCl) to give [Ru$_2$(II)(OH)$_2$(3,6-(t)Bu$_2$(qui)$_2$(btpyan)]$^{2+}$ in MeOH solution. Furthermore, metal-localized oxidation of [Ru$_2$(II)(OH)$_2$(3,6-(t)Bu$_2$(qui)$_2$(btpyan)]$^{2+}$ at E-p = + 1.2 V in CF$_3$CH$_2$OH/ether or water gives [Ru$_2$(III)(OH)$_2$(3,6-(t)Bu$_2$(qui)$_2$(btpyan)]$^{4+}$, which catalyzes water oxidation. Controlled potential electrolysis of [Ru$_2$(II)(OH)$_2$(3,6-(t)Bu$_2$qui)$_2$(btpyan)]$^{2+}$ (SbF$_6$)$_2$ at +1.70 V in the presence of H$_2$O in CF$_3$CH$_2$OH evolves dioxygen with a current efficiency of 91% (21 turnovers). The turnover number of O$_2$ evolution increases to 33,500 when the electrolysis is conducted in water (pH 4.0) by using a [Ru$_2$(II)(OH)$_2$(3,6-(t)Bu$_2$qui)$_2$(btpyan)]$^{2+}$ (SbF$_6$)$_2$-modified ITO electrode (Wada et. al. 2001).

A dinuclear ruthenium complex has been synthesized by Xu et al. (2010) (Figure 10.9).

It was employed to catalyze the homogeneous water oxidation. An exceptionally high TON was observed in the cases of chemical (Ce^{IV} as the oxidant) and light-driven ([Ru(bpy)$_3$]$^{2+}$-type photosensitizers) water oxidation. Cyclometalated dimeric ruthenium(II) complexes [Ru$_2$(cppd)(pic)$_6$]$^+$

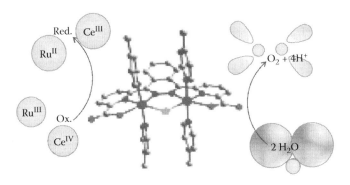

FIGURE 10.9 Homogeneous water oxidation. (Adapted from Xu, Y., A. Fischer, L. Duan, L. Tong, E. Gabrielsson, B. Åkermark, and L. Sun, *Angew. Chem. Int. Ed.*, 49, 8934–8937, 2010. With permission.)

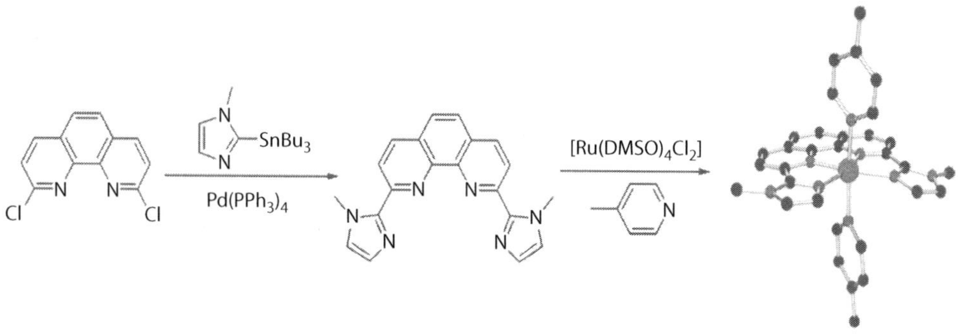

FIGURE 10.10 Ru(II) complex of tetradentate ligand. (Adapted from Zhang, G., R. Zong, H. W. Tseng, and R. P. Thummel, *Inorg. Chem, Angew.*, 47, 990–998, 2008. With permission.)

(H_3cppd = 3,6-bis-(6′-carboxypyrid-2′-yl)-pyridazine) and [Ru_2(cpph)(pic)$_4$(μ-Cl)]$^+$ (H_2cpph = 1,4-bis(6′-carboxypyrid-2′-yl)phthalazine) were also prepared by Xu et al. (2009).

The first family of mononuclear Ru–aqua WOCs trans-[Ru(pbn)(4-R-py)$_2$(OH$_2$)]$^{2+}$ (pbn = 2,2′-(4-(*tert*-butyl)pyridine-2,6-diyl)bis(1,8-naphthyridine); py = pyridine; R = Me, CF$_3$, and NMe$_2$) was reported by Zong and Thummel (2005) for water oxidation. Zhang et al. (2008) (Figure 10.10) and Tseng et al. (2008) have also reported a series of non-aqua Ru complexes that catalyze water oxidation effectively. One class of those complexes, [Ru(dpp)(4-R-py)$_2$]$^{2+}$ (dpp = 2,9-dipyrid-2′-yl-1,10-phenanthroline; R = Me, NMe$_2$, and CF$_3$), contain phenanthroline-based tetradentate ligands, which readily bind ruthenium(II) in an equatorial tetradentate fashion and monodentate pyridyl ligands at axial positions.

10.7.2 Iridium-Based Catalyst

Iridium oxide has long been recognized as a heterogeneous WOC with a low overpotential and long-term stability (Youngblood et al. 2009). McDaniel et al. (2008) published the first family of Ir WOCs, analogues of [Ir(ppy)$_2$(OH$_2$)$_2$]$^+$ (ppy = 2-phenylpyridine). Their TONs reported were impressively high (e.g., 2760). However, catalytic rates were relatively low; thus, the reaction needed over 1 week to reach completion. Hull et al. (2009) changed one of the ppy ligands to Cp* (C$_5$Me$_5$) and synthesized several [IrCp*(C^N)Cl] complexes (Figure 10.11).

As a step forward in developing Ir WOCs, they observed that the introduction of Cp* could dramatically increase the catalytic rates of Ir-based WOCs. Unfortunately, these [IrCp*(C^N)Cl] complexes showed lower TONs (1500 for [IrCp*(ppy)Cl]) than [Ir(ppy)$_2$(OH$_2$)$_2$]$^+$ and its analogues. Then [IrCp*(C^N)(OH$_2$)]$^+$ was proposed as the real WOC with WNA on the Ir=O species responsible for the O–O bond formation (Blakemore et al. 2010). Further, Lalrempuia et al. (2010) modified this complex and introduced an unusual pyridinium-carbene ligand instead of the ppy ligand. A new complex was obtained, which displayed a remarkably high TON of ~10,000. They suggested that the

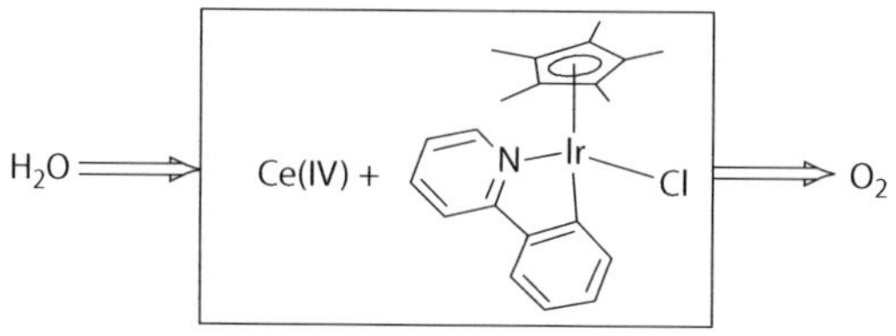

FIGURE 10.11 Water splitting by Ir complex. (Adapted from Hull, J. F., D. Balcells, J. D. Blakemore, C. D. Incarvito, O. Eisenstein, G. W. Brudvig, and R. H. Crabtree, *J. Am. Chem. Soc.*, 131, 8730–8731, 2009. With permission.)

Artificial Photosynthesis

high activity was due to the high electronic flexibility of the mesoionic pyridinium carbine ligand which stabilizes the low oxidation state of the Ir complex with its neutral carbene-type resonance form and the high oxidation state of the Ir complex with its charge-separated form.

10.7.3 Mn-, Co-, and Fe-Based Catalysts

Although nature has chosen Mn as the main building block of the catalytic OEC in the PS II, only a handful of synthetic Mn complexes have been reported, which are capable of catalyzing water oxidation. Thus, Mn(III,IV) dimer is capable of oxidizing water, when activated with a primary oxidant. It is thought to be biomimetic in nature, because some of its structural and mechanistic features are similar to those of the OEC. The catalytic mechanism of water oxidation based on the Mn(III,IV) dimer was thought to involve the formation of a high-valent $Mn^V=O$ or $Mn^{IV}-O^\bullet$ species (Mullins and Pecoraro 2008; Tagore et al. 2008). The oxyl radical is susceptible to nucleophilic attack by a water molecule, forming a hydroperoxo intermediate that rapidly decomposes into O_2 upon deprotonation. Besides Mn-based WOCs, some Co- and Fe-based WOCs were also discovered. Interestingly, these catalysts contain only earth-abundant elements and are relatively stable under catalytic conditions. Wasylenko et al. (2011) reported a Py_5-Co complex $[Co(Py_5)(OH_2)]^{2+}$ (Py_5 = 2,6-(bis(bis-2′-pyridyl)methoxy-methane)-pyridine, which efficiently mediates the oxidation of water electrochemically with a reaction rate of ~79 s^{-1}. The nucleophilic attack on the Co(IV)-hydroxy/oxo species by an incoming water/hydroxide substrate was suggested to be responsible for the formation of the O–O bond (Figure 10.12).

Dogutan et al. (2011) discovered a mononuclear cobalt hangman corrole complex, which is capable of catalyzing water oxidation electrochemically. When immobilized in Nafion films, the TOF for water oxidation at the single cobalt center of the hangman platform reached 0.81 s^{-1} driven at 1.4 V versus Ag/AgCl. The pendant –COOH group appears to benefit the O–O bond formation by preorganizing the incoming water in close proximity to the cobalt oxo center. Nakamura and Frei (2006) demonstrated that nanosized crystals of Co_3O_4 impregnated on mesoporous silica work efficiently as oxygen-evolving catalysts. A wet-impregnation procedure was used to grow Co clusters within the mesoporous Si as the template. The yield for clusters of cobalt oxide (Co_3O_4) nanosized crystals was about 1600 times higher than for micron-sized particles, and the TOF was about 1140 oxygen molecules per second per cluster.

The first family of Fe(III)-tetraamido macrocyclic ligand WOCs (Fe(III)-TAMLs) was discovered by Ellis et al. (2010). They reported that the oxygen evolution was rapid in the first 20 s and then it became very slow. Only 16 turnovers were achieved for the best catalyst with $TOF_{initial}$ = 1.3 s^{-1}. Both oxidative and hydrolytic decomposition pathways were proposed to limit the catalytic performance of Fe(III)-TAMLs. Because iron is one of the most earth-abundant metals and is environmentally friendly, this work will open up new avenues for the development of large-scale affordable WOCs.

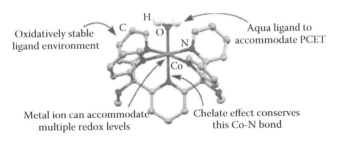

FIGURE 10.12 Catalytic water oxidation mediated by a high-valent cobalt complex. (Adapted from Wasylenko, D. J., C. Ganesamoorthy, J. Borau-Garcia, and C. P. Berlinguette, *Chem. Commun.*, 47, 4249–4251, 2011. With permission.)

The realization of artificial systems that perform water splitting requires catalysts that produce oxygen from water without the need for any excessive driving potential. Kanan and Nocera (2008) reported an amorphous cobalt–phosphate catalyst that oxidizes water to O_2. They reported that this catalyst forms upon the oxidative polarization of an inert indium tin oxide electrode in phosphate-buffered water containing cobalt(II) ions. The presence of phosphate was indicated in an approximate 1:2 ratio with cobalt in this material using various analytical techniques. The pH dependence of the catalytic activity also implicates the hydrogen phosphate ion as the proton acceptor in the oxygen-producing reaction. This catalyst not only forms *in situ* from earth-abundant materials but also operates in neutral water under ambient conditions (Kanan et al. 2009). The use of earth-abundant materials, operation in water at neutral pH, and formation of the catalyst *in situ* capture functional elements of the OEC of photosystem II.

10.8 CATALYTIC PROTON REDUCTION AND HYDROGEN EVOLUTION

$$2H^+ + 2e^- \xrightarrow[\text{Photosensitizer}]{h\nu, \text{PRC}} H_2 \tag{10.6}$$

The protons and electrons generated from catalytic water oxidation are combined to produce hydrogen. The process is assisted by proton reduction catalysts (PRCs).

10.8.1 Proton Reduction Catalyst by Hydrogenases

Hydrogenases are enzymes capable of catalyzing the oxidation of molecular hydrogen or its production from protons and electrons according to the reversible reaction:

$$H_2 \rightleftharpoons 2H^+ + 2e^- \tag{10.7}$$

Most of these enzymes fall into two major classes: NiFe and Fe-only hydrogenases (Leonard et al. 1998). Extensive spectroscopic, electrochemical, and structural studies have been carried out to explore the catalytic mechanism of hydrogenases. Although evolutionarily they are not related, the NiFe and Fe-hydrogenases share a common, unusual feature, and that is an active-site low-spin Fe center with CO and CN coordination.

Iron-iron hydrogenases (Nicolet et al. 2002; Peters 1999; Corr and Murphy 2011) are generally found capable of reducing protons, although some of them are used for hydrogen oxidation and bidirectional transformations. There are many different sources for molecular structures of Fe-Fe hydrogenases. Many structural similarities were found among them, although their structures are slightly different from each other under different crystallization states. Fe-Fe hydrogenases are well known for their abilities to reduce protons to hydrogen, at nearly Nernstian potentials, where the TOFs of Fe-Fe hydrogenase enzymes can reach a value of around 6000 mol of H_2/mol per hydrogenase enzyme per second. Ni-Fe hydrogenases are primarily used for hydrogen uptake (Ogata et al. 2002; Casalot and Rousset 2001).

Hydrogenases are very old redox enzymes and frequently present in microorganisms belonging to the Archaea and Bacteria domains of life; however, a few of them are found in Eukarya as well (Brown and Doolittle 1997). They display remarkable performance on the reversible interconversion between protons and hydrogen, since the purpose of hydrogenase enzymes is to facilitate a charge separation or combination.

Ogata et al. (2002) reported the characterization of [NiFe] hydrogenase using *Desulfovibrio vulgaris*. Characterization methods included x-ray crystallography and absorption and resonance Raman spectroscopy. After the addition of CO, it was found to be bound to the Ni atom at the Ni-Fe active site. The CO was not replaced with H_2 in the dark at 100 K, but it was found to be liberated by illumination with a strong white light. The Ni-C distances and Ni-C-O angles were about 1.77 Å and 160°, respectively, except for one case (1.72 Å and 135°), where an additional electron density peak between the CO and S_{gamma} (Cys546) was recognized. Distinct changes were observed in the electron density distribution of the Ni and S_{gamma} (Cys546) atoms between the CO-bound and CO-liberated structures for all the crystals tested. The novel structural features found near the Ni and S_{gamma} (Cys546) atoms suggest that these two atoms at the Ni-Fe active site play a role during the initial H_2-binding process. Anaerobic addition of CO to dithionite-reduced [NiFe] hydrogenase led to a new absorption band at about 470 nm (approximately 3000 $M^{-1} cm^{-1}$). Resonance Raman spectra (excitation at 476.5 nm) of the CO complex revealed CO-isotope-sensitive bands at 375/393 and 430 cm^{-1} (368 and 413 cm^{-1} for $^{13}C^{18}O$). The frequencies and relative intensities of the CO-related Raman bands indicated that the exogenous CO is bound to the Ni atom with a bent Ni-C-O structure in solution.

Casalot and Rousset (2001) observed the high degree of similarity that exists between all the [NiFe] hydrogenase operons. Near universality of hydrogen metabolism among microorganisms suggests that the microbial ability to metabolize hydrogen is of great importance. The large number of genes present in these operons are mostly involved in the maturation of the structural subunit, which is indicative of the complexity of the hydrogenase molecular structure. Two main groups of maturation genes can be differentiated based on the resulting phenotypes, when mutated:

- *Cis*-genes, encoding narrow specificity proteins. These are mainly located on the same transcription unit as the structural genes.
- *Trans*-genes, encoding broad specificity proteins. These are located on different operons. The maturation of the large subunit starts with the formation of a complex with the chaperone HypC, which remains bound to the amino terminus throughout processing. The ligands CN and CO, which are derived from carbamoyl phosphate, are then inserted via HypF and probably other accessory proteins. HypB is responsible for nickel atom delivery in a GTP-hydrolysis-dependent reaction. The last identified step in the large subunit maturation process is proteolytic cleavage at the carboxyl terminus.

The primary hydrogen binding site of Fe-hydrogenase has been identified by Nicolet et al. (2002). An extensive genome sequencing effort has shown that eukaryotic organisms contain putatively gene coding sequences that display significant homology to Fe-hydrogenases. A structural comparison was also carried out between Fe-hydrogenases and related proteins of unknown metal content from yeast, plant, worm, insect, and mammals.

Peters et al. (1999) gave a clarification of the structures of iron-only hydrogenases from the microorganisms *Clostridium pasteurianum* and *Desulfovibrio desulfuricans* and revealed that the presumed site of reversible hydrogen oxidation exists as a unique, protein-associated organometallic prosthetic group. The hydrogenase structures also provide insight into the chemical mechanism of this highly evolved catalyst.

Corr and Murphy (2011) reported that hydrogenases catalyze redox reactions with molecular hydrogen, either as substrate or product. The enzymes harness hydrogen as a reductant using metals that are abundant and economical, namely, nickel and iron.

Interestingly, the [Fe]-hydrogenase is used by certain microorganisms in the pathway that reduces carbon dioxide to methane. [Fe]-hydrogenase has consistently provided structural and mechanistic surprises since its discovery two decades ago, often requiring complete reevaluation of its mechanism of action.

Consequently, many synthetic catalysts have been inspired by the reactivity of hydrogenase enzymes. Therefore, many variations of electrochemical and photochemical hydrogen production/uptake systems based on Fe-complexes have been studied both experimentally and theoretically.

Inspired by the catalytic action of hydrogenase enzymes, a variety of synthetic molecular catalysts for hydrogen electro- and photochemical production have been developed. These catalysts are mainly based on transition-metal centers incorporated to organic ligand frameworks. A substantial number of studies have been conducted on cobalt (Krishnan and Sutin 1981; Hawecker et al. 1983; Hu et al. 2005, 2007; Razavet et al. 2005; Du et al. 2008; Khnayzer et al. 2014; Sun et al. 2011), nickel (Helm et al. 2011; Small et al. 2011), iron (Gloaguen et al. 2001; DuBois and DuBois 2008; Kaur-Ghumaan et al. 2010), and molybdenum (Karunadasa et al. 2010; Merki and Hu 2011) complexes, which have been shown to catalyze proton reduction in acidic acetonitrile or water to produce H_2. Extensive investigations of the proton reduction using molecular catalysts provided deep insights into the mechanism of this reaction. It was found that the proton reduction catalysis mainly proceeds through the sequence of reduction-protonation steps, through the formation of metastable metal–hydride intermediate, resulting in hydrogen evolution. Several interesting synthetic models of hydrogenase have been developed by the group of DuBois (Helm et al. 2011). These Ni coordination complexes contained pendant base groups in the second coordination sphere of the metal center, which resulted in tremendous enhancement of catalytic rates, which have reached an impressive 100,000 s^{-1} in wet acetonitrile. These complexes are considered as the closest synthetic models of hydrogenase and one of the best electrocatalysts for the reduction of protons.

10.8.2 Light-Driven Catalysis for Hydrogen Production

Photochemical production of hydrogen from water can be considered to consist of two subsystems:

- Photochemical or electrochemical component, where the required oxidizing or reducing equivalents are generated.
- Suitable redox catalysts assisting the formation of hydrogen and oxidation of water. Most up-to-date efforts revolve around the second component, identifying suitable redox catalysts. Noble metals such as Pt are known as good catalysts for H_2 evolution for artificial systems but not for water oxidation. Electrolysis of water is best achieved using a Pt electrode as cathode and a metal oxide such as RuO_2 or IrO_2 as the anode (Anthony and Critchley 2005; Graetzel 1983).

Kiwi et al. (1982) discussed some basic features of light-induced electron transfer reactions in solutions. Particular emphasis was placed on charge separation and kinetic control of the events by organized molecular assemblies such as micelles and vesicles. They also defined some fundamental aspects of redox catalysis with colloidal dispersions of noble metals and their oxides. The application of these concepts to the problem of hydrogen and oxygen production from water by visible light was illustrated. Methods to obtain highly active and selective catalysts were analyzed in detail, and the performance of bifunctional redox catalysts in cyclic water cleavage systems by visible light has been described.

Borgarello et al. (1981a) reported that zinc tetramethylpyridylporphyrin [(ZnTMPyP]$^{4+}$ in acidic aqueous solution sensitizes oxygen generation efficiently by visible light in the presence of acceptors such as Fe^{3+} and Ag^+ ions and colloidal RuO_2/TiO_2 redox catalyst. Hydrogen and oxygen are generated simultaneously under visible light illumination of [ZnTMPyP]$^{4+}$ solutions, when a bifunctional catalyst (Pt and RuO_2 codeposited onto TiO_2) was employed. Borgarello et al. (1981b) observed that the conversion of light into chemical fuels in photochemical devices equipped with nonoxide semiconductor electrodes (e.g., n-CdS) is associated with a serious problem of photocorrosion, because holes produced in the valence band of the semiconductor upon irradiation migrate to the surface of the semiconductor, where photocorrosion occurs. It is remedied by a thin layer of RuO_2 in microheterogenous CdS systems. A CdS sol prepared in the presence of maleic anhydride/styrene copolymer has been loaded with RuO_2 and Pt. These CdS microelectrodes are surprisingly active catalysts for cleavage of H_2O and H_2S.

The latter materials show very low overvoltage for water oxidation to molecular O_2. A popular and commonly used procedure in photochemical and electrochemical studies of H_2 evolution is to use a one-electron redox reagent like methyl viologen (4,4'-dimethyl-bipyridinium chloride, MV^{2+}) as an electron shuttle. MV^{2+} is very soluble in water and has a redox potential slightly more negative than that of normal potential of hydrogen ($E° = -0.44$ V). This reagent is colorless in its oxidized form but turns deep blue in singly reduced form (MV^+). In the presence of suitable redox catalysts, the reduced form is reoxidized readily with concomitant evolution of H_2 from water as

$$MV^{2+} + e^- \rightarrow MV^+ \tag{10.8}$$

$$2MV^+ + 2H^+ \rightarrow 2MV^{2+} + H_2 \tag{10.9}$$

Photolysis of a $[Ru(bpy)_3]^{2+}$ complex with visible light in the presence of MV^{2+} (e^- acceptor), a sacrificial donor like EDTA disodium salt, and a redox catalyst leads to sustained evolution of H_2 from water. Enzymes such as hydrogenases can also be used in biomimetic systems because the photolysis can be carried out in neutral aqueous solutions.

Kohl et al. (2009) described a solution-phase reaction scheme that leads to the stoichiometric liberation of dihydrogen and dioxygen in consecutive thermal-driven and light-driven steps mediated by mononuclear ruthenium complexes. The initial reaction of water at 25°C with a dearomatized ruthenium(II) pincer complex yields a monomeric aromatic Ru(II) hydrido–hydroxo complex, which on further reaction with water at 100°C, releases H_2 and forms a cis-dihydroxo complex. Irradiation of this complex in the 320–420 nm range liberates oxygen and regenerates the starting hydrido–hydroxo Ru(II) complex, most probably by elimination of hydrogen peroxide, which rapidly disproportionates. Isotopic labeling experiments with $H_2^{17}O$ and $H_2^{18}O$ clearly show that the process of an oxygen–oxygen bond formation is intramolecular, thus establishing a previously elusive fundamental step toward dioxygen-generating homogeneous catalysis.

Many groups have shown that nanocrystals of catalytic metal oxides such as Ir-oxide or Nb-oxide can be used efficiently as catalysts for the water splitting. Maeda et al. (2008) have demonstrated that potassium hexaniobate nanoscrolls (NS-$K_4Nb_6O_{17}$) formed by exfoliation of lamellar $K_4Nb_6O_{17}$ can be used as redox catalysts for visible light–driven H_2 production ($\lambda > 420$ nm) from water when $[Ru(bpy)_3]^{2+}$ is photolyzed in the presence of EDTA as a sacrificial electron donor. After exposure to sunlight, the photosensitizer $[Ru(bpy)_3]^{2+}$ reaches its excited state $[Ru(bpy)_3]^{2+*}$ and is reduced to $[Ru(bpy)_3]^+$ quickly by sacrificial electron donors like ascorbate. It returns back to $[Ru(bpy)_3]^{2+}$ after transferring electrons to proton reduction catalysts, and the electrons transferred to the catalysts are used to reduce protons to produce hydrogen.

Hoertz et al. (2007) reported that dicarboxylic acid ligands such as malonate, succinate, and butylmalonate stabilize IrO_2 particles (2 nm diameter) synthesized by hydrolyzing aqueous $[IrCl_6]^{2-}$ solutions. It was observed that monodentate (acetate) and tridentate (citrate) carboxylate ligands, as well as phosphonate and diphosphonate ligands, are comparatively less effective as stabilizers and lead to different degrees of nanoparticle aggregation, as confirmed by transmission electron microscopy. Succinate stabilized 2 nm IrO_2 particles are good catalysts for water photooxidation in persulfate/sensitizer solutions. Ruthenium tris(2,2'-bipyridyl) sensitizers containing malonate and succinate groups in the 4,4'-positions are also good stabilizers of 2 nm–diameter IrO_2 colloids. The excited-state emission of these surface-bound succinate-terminated sensitizer molecules is efficiently quenched on a timescale of ~30 ns, most likely by electron transfer to Ir(IV). The excited state of the bound sensitizer is quenched oxidatively on the timescale of ~9 ns in 1 M persulfate solutions in pH 5.8 Na_2SiF_6/$NaHCO_3$ buffer solutions. Electron transfer from Ir(IV) to Ru(III) occurs with a first-order rate constant of $8 \times 10^2 \, s^{-1}$ evolving oxygen. The TON for oxygen evolution under these conditions was ~150. Thus, the sensitizer–IrO_2 diad is a functional catalyst for photooxidation of water, and it provides a useful building block for overall visible light water-splitting systems.

10.9 CATALYSIS FOR CO_2 REDUCTION AND FUEL PRODUCTION

Development of practical systems for converting CO_2 through photocatalytic reduction processes to useful chemicals using solar light is considered as one of the solutions to alleviate the twin problem of global climate change and renewable energy utilization. Two approaches for photocatalytic CO_2 reduction were reported: homogeneous reaction systems (mainly using transition-metal complexes) and heterogeneous systems (mainly using inorganic semiconductor as light absorber). The reduction of CO_2 is one of the important reactions related to photosynthesis. There is an urgent need to find a means of reducing CO_2 gas to other C-1 products such as alcohols or methane. Some other reduction products such as formic acid (HCOOH) and carbon monoxide (CO) can also be utilized (Yui et al. 2011).

While CO_2-derived fuels such as methanol or methane can be used directly (e.g., in combustion systems), some products require special use. For example, formic acid can be used inside a special fuel cell to produce electricity. The side products of the formic acid fuel operation are carbon dioxide and water. A formic acid fuel cell may be a direct alternative to a hydrogen fuel cell. Formic acid is advantageous as a renewable fuel because it is a liquid; therefore, it can be stored and treated much easier than pressurized or liquid hydrogen. However, current methods of formic acid production do not involve the use of renewable energy and require precious metals as catalysts as well as harsh reaction conditions (i.e., high CO_2/H_2 pressure and high temperature). For example, the CO_2 reduction to methane with hydrogen gas takes place in the presence of a transition-metal catalyst and at high temperature and high pressure:

$$CO_2 + 4H_2 \rightarrow CH_4 + 2H_2O \qquad \Delta G = -27 \text{ kcal/mol} \qquad (10.10)$$

This reverse water gas shift (RWGS) reaction has become of commercial interest for the manufacturing of natural gas from the products of coal gasification. Its inverse reaction is called *steam reforming*, which is useful for hydrogen production at the industrial level. Researchers have been trying to find different methods of catalytic hydrogenation of carbon dioxide to methane at milder conditions. There have been a number of efforts to reduce CO_2 photocatalytically using Ru-bpy, $Re(CO)_3bpy$, metalloporphyrin complexes with or without additional metal/metal oxide catalysts.

This prompted extensive research on the development of new catalytic systems for photodriven CO_2 reduction.

Lehn and Ziesel (1982) reported that when the solutions of $Ru(2,2'-bipyridine)_3^{2+}$, cobalt(II) chloride, and carbon dioxide in acetonitrile/water/triethylamine were irradiated by visible light, carbon monoxide and hydrogen were generated simultaneously. This reaction involves photoinduced reduction of CO_2 and H_2O. They also reported that triethylamine served as electron donor in the $Ru(2,2'-bipyridine)_3^{2+}/Co^{2+}$ system. The amount of gas produced and the selectivity ratio of CO/H_2 depend markedly on the composition of the system. When bipyridine was added, a decrease in CO generation and increase in H_2 production were observed. With the addition of different tertiary amines, NR_3, both the quantity $(CO + H_2)$ and the ratio of CO/H_2 increased along the sequence R = methyl, ethyl, propyl. On using triethanolamine instead of triethylamine, higher selectivity for CO_2 reduction to CO occurred in preference to water reduction. Co^{2+} was found to be the most efficient mediator for both CO and H_2 generation and specifically promotes CO formation, whereas salts of other cations yield only H_2. The mechanism of the reaction may involve intermediate formation of Co(I) species. These processes may be viewed as an abiotic photosynthetic system allowing simultaneous generation of CO and H_2 and regulation of the CO/H_2 ratio. The results obtained were also of significance for solar energy conversion with consumption of a pollutant CO_2.

Kalyanasundaram (1986) suggested that the excited state of tricarbonylchloro(polypyridyl) rhenium(I) complexes, $[Re(CO)_3(LL)(Cl)]$ (LL= 2,2-bipyridine, 4,4'-dimethyl-2,2'-bipyridine, 1,10-phenanthroline, 5-chloro-1,10-phenanthroline, and 2,2'-bipyrazine), is emissive from the metal-to-ligand charge-transfer (MLCT) state in solution at room temperature, and it undergoes facile electron transfer reactions with a variety of electron donor and acceptor molecules. Three aspects of the excited-state photophysics and photoredox chemistry were also presented on the basis of laser photolysis studies: (1) sensitivity of the room temperature absorption and emission to variations in the nature of the polypyridyl ligand and solvent; (2) excited-state absorption spectral features; and (3) reversible and irreversible "reductive" quenching (using various amines as electron donors) and their relevance to the photocatalytic reduction of CO_2 to CO.

Kobayashi et al. (2014) observed the selective formation of dialkyl formamides through photochemical CO_2 reduction. Photochemical CO_2 reduction catalyzed by a $[Ru(bpy)_2(CO)_2]^{2+}/[Ru(bpy)_3]^{2+}/Me_2NH/Me_2NH_2^+$ system in CH_3CN selectively produced dimethylformamide (DMF). A ruthenium carbamoyl complex ($[Ru(bpy)_2(CO)\text{-}(CONMe_2)]^+$) formed by the nucleophilic attack of Me_2NH on $[Ru(bpy)_2(CO)_2]^{2+}$ in this process worked as the precursor to DMF. Thus, Me_2NH acted as both the sacrificial electron donor and the substrate, while $Me_2NH_2^+$ functioned as the proton source. Similar photochemical CO_2 reductions using R_2NH and $R_2NH_2^+$ (R = Et, nPr, or nBu) also afforded the corresponding dialkyl formamides (R_2NCHO) together with HCOOH as a by-product. It was reported that the main product from the CO_2 reduction varies from R_2NCHO to HCOOH with AN increase in the alkyl chain length of the R_2NH. The selectivity between R_2NCHO and HCOOH was found to depend on the rate of $[Ru(bpy)_2(CO)(CONR_2)]^+$ formation.

The Re complexes were also explored in detail due to their advanced photophysical properties, making these complexes valuable not only as catalysts, but also in applications such as sensors, optical switches, light-emitting devices, nonlinear optical materials, and radiopharmaceuticals (Kumar et al. 2010; Coleman et al. 2008; Vlcek and Busby 2006). A major disadvantage associated with these Re-based photocatalysts is that UV light is required for excitation due to their low absorbance in the visible region. Therefore, solar-driven applications are not possible with these complexes until their absorption maxima are shifted to visible range by some modifications.

Supramolecular systems for photocatalytic reduction of CO_2 were developed to allow light absorption in the visible part of the spectrum and more efficient electron transfer. Heteromultinuclear Ru–Re complexes for the photocatalytic reduction of CO_2 were used by Gholamkhass et al. (2005).

Reithmeier et al. (2014) prepared mononuclear iridium (III) complexes $[Ir(mppy)(tpy)X]$ (mppy = 4-methyl-2-phenylpyridine, X = Cl, I) and binuclear analogues with various bis(2-phenyl-pyridin-4-yl) bridging ligands. Kinetic measurements of the photocatalytic two-electron reduction of CO_2 to CO were investigated, and the influence of intermolecular interactions between two active centers was observed. A comparison between the monometallic and the bimetallic complexes was made, which indicated an enhanced lifetime (TON) of the covalently linked complexes, causing an increased overall conversion of CO_2.

It was observed that the proton-assisted, multielectron methods for CO_2 reduction require much less energy than the one-electron process to CO_2/CO_2^- ($E° = -1.9$ V). It stimulated the use of these multi-electron-transfer routes for CO_2 reduction using transition-metal complexes. The most common electrochemical reactions leading to the reduction of CO_2 together with the competing reaction of hydrogen evolution follow. The electrochemical potentials of these redox processes are calculated at pH 7 and presented versus NHE:

$$2H^+ + 2e^- \rightarrow H_2 \qquad E^{o\prime} = -0.41 \text{ V} \tag{10.11}$$

$$CO_2 + 1e^- \rightarrow CO_2^- \qquad E^{o\prime} = -1.9 \text{ V} \tag{10.12}$$

$$CO_2 + 2H^+ + 2e^- \rightarrow CO + H_2O \qquad E^{o\prime} = -0.52 \text{ V} \tag{10.13}$$

$$CO_2 + 2H^+ + 2e^- \rightarrow HCOOH \qquad E^{o\prime} = -0.61 \text{ V} \tag{10.14}$$

$$CO_2 + 6H^+ + 6e^- \rightarrow CH_3OH + H_2O \qquad E^{o\prime} = -0.38 \text{ V} \tag{10.15}$$

$$CO_2 + 8H^+ + 8e^- \rightarrow CH_4 + 2H_2O \qquad E^{o\prime} = -0.24 \text{ V} \qquad (10.16)$$

Ishida et al. (1990) carried out an electrochemical reduction of CO_2 catalyzed by $[RuL^1(L^2)(CO)_2]^{2+}$ $[L^1,L^2 = (bipy)_2, (bipy)(dmbipy), (dmbipy)_2,$ or $(phen)_2]$, $[Ru(phen)_2(CO)Cl]^+$ (phen = 1,10-phenanthroline), and $[RuL(CO)_2Cl_2]$ [L = 2,2′-bipyridine (bipy) or 4,4′-dimethyl-2,2′-bipyridine (dmbipy)] by controlled potential electrolysis at −1.30 V versus saturated calomel electrode in acetonitrile–water (4:1, v/v), MeOH, or MeCN–MeOH (4:1, v/v). They observed that there was no difference in activities between various catalysts in acetonitrile–water (4:1, v/v). On introduction of the dmbipy ligand in MeOH, the amount of produced carbon monoxide became larger than that of HCO_2^-. This was attributed to the equilibrium constants among the reaction intermediates $[RuL^1(L^2)(CO)_2]^{2+}$, $[RuL^1(L^2)(CO)\text{-}\{C(O)OH\}]^+$, and $[RuL^1(L^2)(CO)(CO_2^-)]^+$, which become smaller on substitution of bipy by dmbipy, because of the donor property of the CH_3 group.

Photochemical CO_2 reduction is normally carried out in aqueous solutions or organic solvents under 1 atm CO_2 at room temperature. The concentration of dissolved CO_2 is substantially higher in organic solvents (0.28 M in acetonitrile) as compared to water (0.03 M). Many catalytic systems produce formate and CO as products; however, it was found that the formate-to-CO ratio varied from system to system. Metallocarboxylates (also metallocarboxylic acids) were postulated as intermediates in photochemical and electrochemical CO_2 reduction and the water gas shift reaction. Photolysis of particulate dispersions of TiO_2 loaded with Ru catalyst in aqueous solutions in the presence of CO_2 leads to selective formation of methane at ambient temperatures (Yamashita et al. 1998). However, reduction of water to H_2 gas takes place in the absence of CO_2. Photolysis of aqueous dispersions of titania loaded with Cu catalyst has been found to yield methanol as a major reduction product (Yui et al. 2011; Wu et al. 2005):

$$CO_2 + 2H_2O \xrightarrow[\text{Cu-TiO}_2]{h\nu} CH_3OH + \frac{3}{2}O_2 \qquad (10.17)$$

Morris et al. (2009) have described CO_2 reduction mediated by four classes of transition-metal catalysts:

- Metal tetraaza-macrocyclic compounds
- Supramolecular complexes
- Metalloporphyrins and related metallomacrocycles
- $Re(CO)_3(bpy)X$-based compounds

Carbon monoxide and formate were the primary CO_2 reduction products. Bicarbonate/carbonate production was also proposed.

Sutin et al. (1997) reported the transition-metal-based systems that generate hydrogen and/or reduce carbon dioxide upon irradiation with visible light. Most of the systems involve polypyridine complexes of the d^6 centers cobalt(III), rhodium(III), iridium(III), ruthenium(II), and rhenium(I). Complexes with diimine ligands serve as photosensitizers and/or catalyst precursors. The corresponding d^8 metal centers and d^6 hydrides are important intermediates: bimolecular reactions of the hydrides or their reactions with H_2O/H_3O^+ are responsible for the formation of hydrogen. When carbon dioxide is present, it may insert into the metal–hydride bond to yield formate. Mechanistic schemes for some dual-acting photoconversion systems that generate both hydrogen and carbon monoxide or formate were considered.

Kyle and Clifford (2014) presented the recent developments in the use of rhenium and manganese bipyridine carbonyl catalysts for the electrochemical reduction of CO_2. They described that 4,4′-*tert*-butyl-substituted complexes fac-$Re(bpy$-$tBu)(CO)_3X$ have been found to be more active than the parent 2,2′-bipyridine complexes. It was observed that the presence of Bronsted acids increases the activity of these catalysts, with stronger acids leading to more rapid catalysis.

Therrien et al. (2014) used a series of pyridine- and lutidine-linked bis-NHC palladium pincer complexes for electrocatalytic CO_2 reduction. Lutidine-linked complexes have shown catalytic activity at potentials as low as −1.6 V versus $Ag/AgNO_3$. According to DFT studies, the redox activity of the ligand contributed to the stability of the reduced species, potentially addressing a major deactivation pathway of previous palladium pincer CO_2 reduction electrocatalysts. Raebiger et al. (2006) described a bimetallic palladium complex, $\{m\text{-}C_6H_4(triphos)_2[Pd(CH_3CN)]_2\}(BF_4)_4$, for CO_2 reduction. They observed that this catalyst exhibits high catalytic rates and larger turnover numbers.

La Porte et al. (2014) prepared a homogeneous, integrated chromophore/two-catalyst system, which is capable of storing the light via photochemically driven reverse water gas shift reaction, where the reducing equivalents are provided by renewable H_2. The system consists of the chromophore zinc tetraphenylporphyrin (ZnTPP), H_2 oxidation catalysts of the form $[Cp(R)Cr(CO)_3]^-$, and CO_2 reduction catalysts of the type $[Re(bpy\text{-}4,4'\text{-}R_2)(CO)_3]Cl$.

Yuhas et al. (2011) presented a series of fully integrated porous materials containing Fe_4S_4 clusters, dubbed "biomimetic chalcogels." They examined the effect of third metal cations on the electrochemical and electrocatalytic properties of the chalcogels and found that ternary biomimetic chalcogels containing Ni or Co show increased activity in transformation of carbon dioxide. These can be thought of as solid-state analogues of NiFe or NiFeS reaction centers in enzymes.

Chiericato et al. (2000) reported that transition-metal complexes of Fe, Co, and Ni incorporating terdentate polyimine ligands derived from substituted DAPA (2,6-bis-[1-(phenylimino)ethyl]pyridine) are effective electrocatalysts for the reduction of carbon dioxide. All the prepared complexes exhibited some degree of electrocatalytic activity; however, this activity was strongly dependent on the nature of the ligand, the metal center, and the nature of the redox process, whether it was predominantly metal or ligand localized. $[Ni(v\text{-}DAPA)_2]^{2+}$ appeared to be the most active complex. This material was able to reduce carbon dioxide electrocatalytically at ca. −1.0 V, which represents a dramatic reduction of about 1 V in the overpotential.

Neri et al. (2015) prepared a low-cost nickel–cyclam complex covalently anchored to a metal oxide surface. The role of the surface immobilization on enhancing the rate of photoelectron transfer was confirmed using transient spectroscopy. $[Ni(1,4,8,11\text{-tetraazacyclo-tetradecane-6-carboxylic acid})]^{2+}$ complex was shown to be an active electrocatalyst in solution for CO_2 reduction.

A viable option for recycling carbon dioxide is through the sunlight-powered photocatalytic conversion of CO_2 and water vapor into hydrocarbon fuels over highly active nanocatalysts. Many different approaches have been developed for the heterogeneous photocatalytic reduction of CO_2 on TiO_2, ZnO, and various other metal oxides. Rani et al. (2014) prepared (Cu, Pt)-sensitized TiO_2 nanoparticle wafers for the photocatalytic conversion of CO_2 and water vapor to hydrocarbon fuels. Ehsan and He (2015) reported synthesis of common cation heterostructure via modification of zinc oxide by zinc telluride (ZnTe) photocatalyst through a one-pot hydrothermal approach at a reaction temperature of 180°C. The fabricated heterostructure consisted of ZnO flower-like nanostructures (hundreds of nanometers for the rod length or sheet size of the petals and tens of nanometers for the corresponding diameter or thickness) exhibited photocatalytic capability for the reduction of carbon dioxide into methane under visible light irradiation ($\lambda \geq 420$ nm). The difference in the activity of materials with different morphology was explained by different exposed crystal planes of ZnO and different surface area (15.0 and 5.6 $m^2 g^{-1}$ for sheet-like petals and rod-like petals, respectively).

Various modified metal oxides were also prepared for efficient photocatalytic reduction of CO_2. These metal oxides were modified by different methods, and several new efficient heterogeneous catalysts (i.e., binary, ternary, and doped materials were used for reduction of CO_2). Song et al. (2015) synthesized $ZnFe_2O_4/TiO_2$ heterostructured photocatalysts with different mass percentages of $ZnFe_2O_4$ through the hydrothermal deposition method. The photocatalytic activities of the nanocomposites were tested by photocatalytic reduction of CO_2 in cyclohexanol under UV light (main wavelength at 360 nm) irradiation. It was shown that the main products were cyclohexanone and cyclohexyl formate. $ZnFe_2O_4/TiO_2$ nanocomposites showed much higher photocatalytic performance as compared with pure TiO_2 and $ZnFe_2O_4$ samples.

Shown et al. (2014) prepared graphene oxide (GO) decorated with copper nanoparticles (Cu/GO) and used these to enhance photocatalytic CO_2 reduction under visible light. A rapid one-pot microwave process was used to prepare the Cu/GO hybrids with various Cu contents. Metallic copper nanoparticles (4–5 nm in size) in the GO hybrid significantly enhanced the photocatalytic activity of GO, basically through the suppression of electron–hole pair recombination, reduction of bandgap of GO and modification of its work function. X-ray photoemission spectroscopy studies indicated a charge transfer from GO to Cu. Ehsan et al. (2014) demonstrated that ZnTe can be utilized as an efficient catalyst for the photoreduction of CO_2 into methane under visible light irradiation ($\geq$420 nm). The combination of ZnTe with $SrTiO_3$ increased the formation of CH_4 by efficiently promoting electron transfer from the conduction band of ZnTe to that of $SrTiO_3$ under visible light irradiation.

Yang and Jin (2014) synthesized Zn_2GeO_4 nanorods by a surfactant-assisted solution-phase route. It was thought that the cetyltetramethylammonium cations (CTA^+) preferentially adsorb on the planes of Zn_2GeO_4 nanorods, leading to preferential growth along the c-axis to form the Zn_2GeO_4 rods with larger aspect ratio and higher surface area. Hence, photocatalytic activity for photoreduction of CO_2 was improved. Jiang et al. (2014) prepared a series of novel microspheres of $CdIn_2S_4$ by hydrothermal process. $CdIn_2S_4$ was synthesized from l-cysteine, which exhibited higher photocatalytic activity for CO_2 reduction and has a potential application for catalysis under visible light. The mechanism of photocatalytic reduction of CO_2 in methanol over $CdIn_2S_4$ was also proposed. The narrow bandgap of the prepared catalyst promoted reduction of CO_2 to dimethoxymethane and methyl formate in methanol.

Li et al. (2014) synthesized a series of metal oxides (Ni/Zn/Cr layered double hydroxides [LDHs]) by coprecipitation and an annealing method at different temperatures. Their activities in photocatalytic reduction of CO_2 with H_2O vapor were tested at room temperature and atmospheric pressure. It was reported that the sample calcined at 500°C possessed the highest catalytic activity with respect to the formation of CH_4 and CO as the major products. It was attributed to the interaction of uniformly dispersed NiO, Cr_2O_3, $ZnCr_2O_4$, and $NiCr_2O_4$ with small grain size. Almeida et al. (2014) prepared pure and Cr(III) and Mo(V)-doped $BiNbO_4$ and $BiTaO_4$ by the citrate method. Pure $BiNbO_4$ and $BiTaO_4$ were obtained in triclinic phase at 600°C and 800°C, respectively. The metal doping was found to strongly influence the crystal structure as well as the photocatalytic activity of these oxides. They showed that Cr(III)-doped $BiTaO_4$ and $BiNbO_4$, in general, were more selective for hydrogen production, while Mo(V)-doped materials were more selective for CO_2 generation. $BiTaO_4$ showed higher photocatalytic activity than $BiNbO_4$ for hydrogen production as well as for CO_2 generation. A negligible change of conduction band minimum potential (CBM) was found for Mo(V)-doped materials, which indicates that there might be no improvement of the reduction power of the material following the substitutional doping. However there is a slight shift of the CBM potential, slightly increasing the reduction power in the case of Cr(III)-doped $BiNbO_4$. This effect was much stronger in the Cr(III)-doped $BiTaO_4$.

Wang et al. (2014) synthesized mesoporous Fe-doped CeO_2 catalysts with different Fe-doping concentrations through a nanocasting route using ordered mesoporous SBA-15 as the template. The samples were prepared by filling mesopores in silica template with a Fe-Ce complex precursor, followed by calcination and silica removal. Then their catalytic activity was tested for the photocatalytic reduction of CO_2 with H_2O under simulated solar irradiation. Fe species can effectively enhance photocatalytic performance of the reduction of CO_2 with H_2O, when compared with nondoped mesoporous CeO_2 catalyst.

In nature, plants use photosynthesis to convert sunlight and carbon dioxide into oxygen and carbohydrates. Plants use photosynthesis for producing food sources, and if one is able to mimic this process, this can create a clean and affordable supply of renewable energy. However, if CO_2 is to be used as the feedstock for the production of carbon-containing fuels, many scientific and engineering challenges have to be overcome first. These challenges include an efficient and inexpensive mechanism for capture of CO_2 at the energy-generating sites or even the atmosphere. The development of efficient, robust, and inexpensive catalysts for CO_2 reduction is one of the major scientific challenges. Finding ways to integrate CO_2 reduction catalysis with water oxidation half-reaction, all driven by the energy of sunlight presents another scientific and engineering challenge.

REFERENCES

Abe, R., K. Hara, K. Sayama, K. Domen, and H. Arakawa. 2000. Steady hydrogen evolution from water on eosin Y-fixed TiO_2 photocatalyst using a silane-coupling reagent under visible light irradiation. *J. Photochem. Photobiol. A: Chem.* 137: 63–69.

Abrahamsson, M. L. A., H. B. Baudin, A. Tran, C. Philouze, K. E. Berg, M. K. Raymond-Johansson, et al. 2002. Ruthenium-manganese complexes for artificial photosynthesis: Factors controlling intramolecular electron transfer and excited-state quenching reactions. *Inorg. Chem.* 41: 1534–1544.

Ahrens, M. J., L. E. Sinks, B. Rybtchinski, W. Liu, B. A. Jones, J. M. Giaimo, et al. 2004. Self-assembly of supramolecular light-harvesting arrays from covalent multi-chromophore perylene-3,4:9,10-bis(dicarboximide) building blocks. *J. Am. Chem. Soc.* 126: 8284–8294.

Albinsson, B., and J. Martensson. 2008. Long-range electron and excitation energy transfer in donor-bridge-acceptor systems. *J. Photochem. Photobiol. C. Rev.* 9: 138–155.

Alibabaei, L., M. K. Brennaman, M. R. Norris, B. Kalanyan, W. Song, M. D. Losego, et al. 2013. Solar water splitting in a molecular photoelectrochemical cell. *Proc. Natl. Acad. Sci. USA.* 110: 20008–20013.

Allakhverdiev, S. I. 2011. Recent progress in the studies of structure and function of photosystem II. *J. Photochem. Photobiol. B: Biol.* 104: 1–8.

Allakhverdiev, S. I., V. D. Kreslavski, V. Thavasi, S. K. Zharmukhamedov, V. V. Klimov, T. Nagata, et al. 2009. Hydrogen photoproduction by use of photosynthetic organisms and biomimetic systems. *Photochem. Photobiol. Sci.* 8: 148–156.

Almeida, C. G., R. B. Araujo, R. G. Yoshimura, A. J. S. Mascarenhas, A. F. Da Silva, C. M. Araujo, et al. 2014. Photocatalytic hydrogen production with visible light over Mo and Cr-doped $BiNb(Ta)O_4$. *Int. J. Hydrogen Energy.* 39:1220–1227.

Alstrum-Acevedo, J. H., M. K. Brennaman, and T. J. Meyer. 2005. Chemical approaches to artificial photosynthesis. *Inorg. Chem.* 44: 6802–6827.

Anthony F. C., and C. Critchley (Eds.). 2005. *Artificial Photosynthesis: From Basic Biology to Industrial Applications*, Wiley-VCH, Weinheim.

Aratani, N., D. Kim, and A. Osuka. 2009. Discrete cyclic porphyrin arrays as artificial light-harvesting antenna. *Acc. Chem. Res.* 42: 1922–1934.

Bak, T., J. Novotny, M. Rekas, and C. C. Sorrell. 2002. Photo-electrochemical hydrogen generation from water using solar energy. Material-related aspects. *Int. J. Hydrogen Energy.* 27: 991–1022.

Balzani, V., A. Credi, and M. Venturi. 1997. Photoprocesses. *Curr. Opin. Chem. Biol.* 1: 506–513.

Balzani, V., A. Credi, and M. Venturi. 2008. Photochemical conversion of solar energy. *ChemSusChem.* 1: 26–58.

Barber, J. 2009. Photosynthetic energy conversion: Natural and artificial. *Chem. Soc. Rev.* 38: 185–196.

Benniston, A. C. and A. Harriman. 2008. Artificial photosynthesis. *Mater. Today.* 11: 26–34.

Blakemore, J. D., N. D. Schley, D. Balcells, J. F. Hull, G. W. Olack, C. D. Incarvito, et al. 2010. Half-sandwich iridium complexes for homogeneous water-oxidation catalysis. *J. Am. Chem. Soc.* 132: 16017–16029.

Borgarello, E., K. Kalyanasundaram, D. Duonghong, and M. Graetzel. 1981b. Cleavage of water by visible-light irradiation of colloidal CdS solutions: Inhibition of photocorrosion by RuO_2. *Angew. Chem. Int. Ed.* 20: 987–988.

Borgarello, E., K. Kalyanasundaram, Y. Okuno, and M. Grätzel. 1981a. Visible light-induced oxygen generation and cyclic water cleavage sensitized by porphyrins. *Helv Chim. Acta.* 64: 1937–1942.

Brown, J. R., and W. F. Doolittle. 1997. Archaea and the prokaryote-to-eukaryote transition. *Microbiol. Mol. Biol. Rev.* 61: 456–502.

Casalot, L., and M. Rousset. 2001. Maturation of the [NiFe] hydrogenases. *Trends. Microbiol.* 9: 228–237.

Chiericato, G. Jr., C. R. Arana, C. Casado, I. Cuadrado, and H. D. Abruña. 2000. Electrocatalytic reduction of carbon dioxide mediated by transition metal complexes with terdentate ligands derived from diacetylpyridine. *Inorg. Chim. Acta.* 300–302, 32–42.

Cogdell, R. J., T. H. P. Brotosudarmo, A. T Gardiner, P. M Sanchez, and L. Cronin. 2010. Artificial photosynthesis—Solar fuels: Current status and future prospects. *Biofuels.* 1: 861–876.

Coleman, A., C. Brennan, J. G. Vos, and M. T. Pryce. 2008. Photophysical properties and applications of Re(I) and Re(I)–Ru(II) carbonyl polypyridyl complexes. *Coord. Chem. Rev.* 252: 2585–2595.

Collings, A. F., and C. Critchley (Eds.). 2005. *Artificial Photosynthesis: From Basic biology to Industrial Applications*. Wiley-VCH, Weinheim.

Concepcion, J. J., J. W. Jurss, M. K. Brennaman, P. G. Hoertz, A. O. T. Patrocinio, N. Y. Murakami Iha, et al. 2009. Making oxygen with ruthenium complexes. *Acc. Chem. Res.* 42: 1954–1965.

Corr, M. J., and J. A. Murphy. 2011. Evolution in the understanding of [Fe]-hydrogenase. *Chem. Soc. Rev.* 40: 2279–2292.

Danielsson, R., M. Suorsa, V. Paakkarinen, P. A. Albertsson, S. Styring, E. M. Aro, et al. 2006. Dimeric and monomeric organization of photosystem II. Distribution of five distinct complexes in the different domains of the thylakoid membrane. *J. Biol. Chem.* 281: 14241–14249.

Dinner, B. A., and G. T. Babcock. 1996. Structure, dynamics and energy conversion efficiency in photosystem II. In: *Oxygenic Photosynthesis: The Light Reactions*. D. R. Ort, C. F. Yocum, and I. F. Heichel (Eds.): Kluwer: The Netherlands, pp. 213–247.

Dogutan, D. K., R. McGuire, and D. G. Nocera. 2011. Electrocatalytic water oxidation by cobalt(III) hangman β-octafluoro corroles. *J. Am. Chem. Soc.* 133: 9178–9180.

Du, P., K. Knowles, and R. Eisenberg. 2008. A homogeneous system for the photogeneration of hydrogen from water based on a platinum(II) terpyridyl acetylide chromophore and a molecular cobalt catalyst. *J. Am. Chem. Soc.* 130: 12576–12577.

DuBois, M. R., and D. L. DuBois. 2008. The role of pendant bases in molecular catalysts for H_2 oxidation and production. *C.R. Chim.* 11: 805–817.

Ehsan, M. F., and T. He. 2015. In situ synthesis of ZnO/ZnTe common cation heterostructure and its visible-light photocatalytic reduction of CO_2 into CH_4. *Appl Catal B: Enviorn.* 166–167: 345–352.

Ehsan, M. F., M. N. Ashiq, F. Bi, Y. Bi, S. Palanisamy, and T. He. 2014. Preparation and characterization of $SrTiO_3$-ZnTe nanocomposites for the visible-light photoconversion of carbon dioxide to methane. *RSC Adv.* 4: 48411–48418.

Ellis, W. C., N. D. McDaniel, S. Bernhard, and T. J. Collins. 2010. Fast water oxidation using iron. *J. Am. Chem. Soc. 132*: 10990–10991.

Fukuzumi, S. 2008. Development of bioinspired artificial photosynthetic systems. *Phys. Chem. Chem. Phys.* 10: 2283–2297.

Gersten, S. W., G. J. Samuels, and T. J. Meyer. 1982. Catalytic oxidation of water by an oxo-bridged ruthenium dimer. *J. Am. Chem. Soc.* 104: 4029–4030.

Gholamkhass, B., H. Mametsuka, K. Koike, T. Tanabe, M. Furue, and O. Ishitani. 2005. Architecture of supramolecular metal complexes for photocatalytic CO_2 reduction: Ruthenium-rhenium bi- and tetranuclear complexes. *Inorg. Chem.* 44: 2326–2336.

Gloaguen, F. D. R., J. D. Lawrence, and T. B. Rauchfuss. 2001. Biomimetic proton reduction catalyzed by an iron carbonyl thiolate. *J. Am. Chem. Soc.* 123: 9476–9477.

Graetzel, M. (Ed.). 1983. *Energy Resources through Photochemistry and Catalysis*. Academic Press, New York.

Gust, D., and T. A. Moore. 1989. Mimicking photosynthesis. *Science*. 244: 35–41.

Gust, D., D. Kramer, A. Moore, T. A. Moore, and W. Vermaas. 2008. Engineered and artificial photosynthesis: Human ingenuity enters the game. *MRS Bull.* 33: 383–387.

Gust, D., T. A. Moore, and A. L. Moore. 2001. Mimicking photosynthetic solar energy transduction. *Acc. Chem. Res.* 34: 40–48.

Hambourger, M., G. F. Moore, D. M. Kramer, D. Gust, A. L. Moore, and T. A. Moore. 2009. Biology and technology for photochemical fuel production. *Chem. Soc. Rev.* 38: 25–35.

Hammarström, L., and S. Styring. 2008. Coupled electron transfers in artificial photosynthesis. *Philos. Trans. R. Soc. Lond. B Biol. Sci.* 363: 1283–1291.

Harriman, A., and J. P. Sauvage. 1996. Strategy for constructing photosynthetic models: Porphyrin-containing modules assembled around transition metals. *Chem. Soc. Rev.* 25: 41–48.

Hawecker, J., J. M. Lehn, and R. Ziessel. 1983. Efficient homogeneous photochemical hydrogen generation and water reduction mediated by cobaloxime or macrocyclic cobalt complexes. *Nouv. J. Chem.* 7: 271–277.

Helm, M. L., M. P. Stewart, R. M. Bullock, M. R. DuBois, and D. L. DuBois. 2011. A synthetic nickel electrocatalyst with a turnover frequency above 100,000 s^{-1} for H_2 production. *Science*. 333: 863–866.

Herek, J. L., W. Wohlleben, R. J. Cogdell, D. Zeidler, and M. Motzkus. 2002. Quantum control of energy flow in light harvesting. *Nature*. 417: 533–535.

Herrero, C., B. Lassalle-Kaiser, W. Leibl, A. W. Rutherford, and A. Aukauloo. 2008. Artificial systems related to light driven electron transfer processes in PSII. *Coord. Chem. Rev.* 252: 456–468.

Hoertz, P. G., Y. Kim, W. J. Youngblood, and T. E. Mallouk. 2007. Bidentate dicarboxylate capping groups and photosensitizers control the size of IrO_2 nanoparticle catalysts for water oxidation. *J. Phys. Chem. B.* 111: 6845–6856.

Hohmann-Marriott, M. F., and R. E. Blankenship. 2011. Evolution of photosynthesis. *Ann. Rev. Plant Biol.* 62: 515–548.

Hu, X., B. S. Brunschwig, and J. C. Peters. 2007. Electrocatalytic hydrogen evolution at low overpotentials by cobalt macrocyclic glyoxime and tetraimine complexes. *J. Am. Chem. Soc.* 129: 8988–8998.

Hu, X., B. M. Cossairt, B. S. Brunschwig, N. S. Lewis, and J. C. Peters. 2005. Electrocatalytic hydrogen evolution by cobalt difluoroboryl-diglyoximate complexes. *Chem. Commun.* 37: 4723–4725.

Huang, P., A. Magnuson, R. Lomoth, M. Abrahamsson, M. Tamm, L. Sun, et al. 2002. Photo-induced oxidation of a dinuclear $Mn_2^{II,II}$ complex to the $Mn_2^{III,IV}$ state by inter- and intramolecular electron transfer to Ru^{III} tris-bipyridine. *J. Inorg. Biochem.* 91: 159–172.

Hull, J. F., D. Balcells, J. D. Blakemore, C. D. Incarvito, O. Eisenstein, G. W. Brudvig, et al. 2009. Highly active and robust Cp* iridium complexes for catalytic water oxidation. *J. Am. Chem. Soc.* 131: 8730–8731.

Huynh, M. H. V., D. M. Dattelbaum, and T. J. Meyer. 2005. Exited state electron and energy transfer in molecular assemblies. *Coord. Chem. Rev.* 249: 457–483.

IEA (International Energy Agency). 2014. Key world energy statistics, IEA, pp. 1–82.

Ishida, H., K. Fukui, T. Ohta, K. Ohkubo, T. Tanaka, T. Terada, and T. Tanaka, 1990. Ligand effects of ruthenium 2,2′-bipyridine and 1,10-phenanthrolinecomplexes on electrochemical reduction of CO_2. *J. Chem. Soc. Dalton Trans.* 2155–2160.

Jiang, W., X. Yin, F. Xin, Y. Bi, Y. Liu, and X. Li. 2014. Preparation of $CdIn_2S_4$ microspheres and application for photocatalytic reduction of carbon dioxide. *Appl Surf. Sci.* 288: 138–142.

Kalyanasundaram, K. 1986. Luminescence and redox reactions of the metal-to-ligand charge-transfer excited state of tricarbonylchloro-(polypyridyl)rhenium(I) complexes. *J. Chem. Soc. Faraday Trans. II.* 82: 2401–2415.

Kanan, M. W., and D. G. Nocera. 2008. In situ formation of an oxygen-evolving catalyst in neutral water containing phosphate and Co^{2+}. *Science.* 321: 1072–1075.

Kanan, M. W., Y. Surendranath, and D. G. Nocera. 2009. Cobalt–phosphate oxygen-evolving compound. *Chem. Soc. Rev.* 38: 109–114.

Karunadasa, H. I., C. J. Chang, and J. R. Long. 2010. A molecular molybdenum-oxo catalyst for generating hydrogen from water. *Nature.* 464: 1329–1333.

Kaur-Ghumaan, S., L. Schwartz, R. Lomoth, M. Stein, and S. Ott, 2010. Catalytic hydrogen evolution from mononuclear iron(II) carbonyl complexes as minimal functional models of the [FeFe] hydrogenase active site. *Angew. Chem. Int. Ed.* 49: 8033–8036.

Kelley, R. F., S. J. Lee, T. M. Wilson, Y. Nakamura, D. M. Tiede, A. Osuka, et al. 2008. Intramolecular energy transfer within butadiyne-linked chlorophyll and porphyrin dimer-faced, self-assembled prisms. *J. Am. Chem. Soc.* 130, 4277–4284.

Khnayzer, R. S., V. S. Thoi, M. Nippe, A. E. King, J. W. Jurss, K. A. El Roz, et al. 2014. Towards a comprehensive understanding of visible-light photogeneration of hydrogen from water using cobalt(II) polypyridyl catalysts. *Energy Environ. Sci.* 7: 1477–1488.

Kiwi, J., K. Kalyanasundaram, and M. Graetzel. 1982. Visible light induced cleavage of water into hydrogen and oxygen in colloidal and microheterogeneous systems. *Struct. Bond.* 49: 37–125.

Kobayashi, K., T. Kikuchi, S. Kitagawa, and K. Tanaka. 2014. Selective generation of formamides through photocatalytic CO_2 reduction catalyzed by ruthenium carbonyl compounds. *Angew. Chem. Int. Ed.* 53: 11813–11817.

Kohl, S. W., L. L. Weiner, L. Schwartsburd, L. Konstantinovski, L. J. W. Shimon, Y. Ben-David, et al. 2009. Consecutive thermal H_2 and light-induced O_2 evolution from water promoted by a metal complex. *Science.* 324: 74–77.

Krishnan, C. V., and N. Sutin. 1981. Homogeneous catalysis of the photoreduction of water by visible light. 2. Mediation by a tris(2,2′-bipyridine)ruthenium(II)-cobalt(II) bipyridine system. *J. Am. Chem. Soc.* 103: 2141–2142.

Kumar, A., S. S. Sun, and A. Lees. 2010. Photophysics and photochemistry of organometallic rhenium diimine complexes. *Top. Organomet. Chem.* 29: 1–35.

Kyle, A. G., and P. K. Clifford. 2014. Recent studies of rhenium and manganese bipyridine carbonyl catalysts for the electrochemical reduction of CO_2. Chp-5. *Adv. Inorg. Chem.* 66: 163–185.

La Porte, N. T., D. B. Moravec, and M. D. Hopkins. 2014. Electron-transfer sensitization of H_2 oxidation and CO_2 reduction catalysts using a single chromophore. *Proc. Natl. Acad. Sci.* 111: 9745–9750.

Lalrempuia, R., N. D. McDaniel, H. Müller-Bunz, S. Bernhard, and M. Albrecht. 2010. Water oxidation catalyzed by strong carbene-type donor-ligand complexes of Iridium. *Angew. Chem. Int. Ed.* 49: 9765–9768.

Lehn, J. M., and R. Ziessel. 1982. Photochemical generation of carbon monoxide and hydrogen by reduction of carbon dioxide and water under visible light irradiation. *Proc. Natl. Acad. Sci. U. S. A.* 79: 701–704.

Leonard, C. J., L. Aravind, and E. V. Koonin. 1998. Novel families of putative protein kinases in bacteria and archaea: Evolution of the "eukaryotic" protein kinase superfamily. *Genome Res.* 8: 1038–1047.

Li, B. J., Z. J. Wu, C. Chen, W. F. Shangguan, and J. Yuan. 2014. Preparation and photocatalytic CO_2 reduction activity of Ni/Zn/Cr composite metal oxides. *J. Mol. Catal.* 28: 268–274.

Lukas, A. S., Y. Zhao, S. E. Miller, and M. R. Wasielewski. 2002. Biomimetic electron transfer using low energy excited states: A green perylene-based analogue of chlorophyll a. *J. Phys. Chem. B.* 106: 1299–1306.

Maeda, K., M. Eguchi, S. H. A. Lee, W. J. Youmgblood, H. Hata, and T. E. Mallouk. 2009. Photocatlytic hydrogen evolutionfrom hexaniobate nanoscrollsand calcium niobate nanosheetssensitized by ruthenium(II) bipyridyl complexes. *J. Phys. Chem. C.* 113: 7962–7969.

McConnell, I., G. Li, and G. W. Brudvig. 2010. Energy conversion in natural and artificial photosynthesis. *Chem. Biol.* 17: 434–447.

McDaniel, N. D., F. J. Coughlin, L. L. Tinker, and S. Bernhard. 2008. Cyclometalated iridium(III) aquo complexes: Efficient and tunable catalysts for the homogeneous oxidation of water. *J. Am. Chem. Soc.* 130: 210–217.

McEvoy, J. P., J. A. Gascon, V. S. Batista, and G. W. Brudvig. 2005. The mechanism of photosynthetic water splitting. *Photochem. Photobiol. Sci.* 4: 940–949.

Melis, A. 2009. Solar energy conversion efficiencies in photosynthesis: Minimizing the chlorophyll antennae to maximize efficiency. *Plant Sci.* 177: 272–280.

Merki, D., and X. Hu. 2011. Recent developments of molybdenum and tungsten sulfides as hydrogen evolution catalysts. *Energy Environ. Sci.* 4: 3878–3888.

Meyer, T. J. 1990. Intramolecular control of excited state electron and energy electron transfer. *Pure Appl. Chem.* 62:1003–1009.

Moore, T. A., D. Gust, P. Mathis, J. C. Mialocq, C. Chachaty, R. V. Bensasson, et al. 1984. Photodriven charge separation in a carotenoporphyrinquinone triad. *Nature.* 307: 630–632.

Morris, A. J., G. J. Meyer, and E. Fujita. 2009. Molecular approaches to the photocatalytic reduction of carbon dioxide for solar fuels. *Acc. Chem. Res.* 42: 1983–1994.

Mullins, C. S., and V. L. Pecoraro. 2008. Reflections on small molecule manganese models that seek to mimic photosynthetic water oxidation chemistry. *Coord. Chem. Rev.* 252: 416–443.

Nakamura, R., and H. Frei. 2006. Visible light-driven water oxidation by Ir oxide clusters coupled to single Cr centers in mesoporous silica. *J. Am. Chem. Soc.* 128: 10668–10669.

Naoki, A., D. Kim, and A. Osuka. 2009. Discrete cyclic porphyrin arrays as artificial light-harvesting antenna. *Acc. Chem. Res.* 42: 1922–1934.

Neri, G., J. J. Walsh, C. Wilson, A. Reynal, J. Y. C. Lim, X. Li, et al. 2015. A functionalised nickel cyclam catalyst for CO_2 reduction: Electrocatalysis, semiconductor surface immobilisation and light-driven electron transfer. *Phys. Chem. Chem. Phys.* 17: 1562–1566.

Nicolet, Y., C. Cavazza, and J. C. Fontecilla-Camps. 2002. Fe-only hydrogenases: Structure, function and evolution. *J. Inorg. Biochem.* 91: 1–8.

Nocera, D. G. 2012. The artificial leaf. *Acc. Chem. Res.* 45: 767–776.

Ogata, H., Y. Mizoguchi, N. Mizuno, K. Miki, S. Adachi, N. Yasuoka, et al. 2002. Structural studies of the carbon monoxide complex of [NiFe] hydrogenase from *Desulfovibrio vulgaris* Miyazaki F: Suggestion for the initial activation site for dihydrogen. *Am. Chem. Soc.* 124: 11628–11635.

Ort, D. R., X. Zhu, and A. Melis. 2011. Optimizing antenna size to maximize photosynthetic efficiency. *Plant Physiol.* 155: 79–85.

Peters, J. W. 1999. Structure and mechanism of iron-only hydrogenases. *Curr. Opin. Struct. Biol.* 9: 670–676.

Prashant, V. K., and D. Meisel. 2003. Nanoscience opportunities in environmental remediation. *Compt. Rend. Chimie.* 6: 999–1007.

Raebiger, J. W., J. W. Turner, B. C. Noll, C. J. Curtis, A. Miedaner, B. Cox, and D. L. DuBois. 2006. Electrochemical reduction of CO_2 to CO catalyzed by a bimetallic palladium complex. *Organometallics* 25: 3345–3351.

Rani, S., N. Bao, and S. C. Roy. 2014. Solar spectrum photocatalytic conversion of CO_2 and water vapour into hydrocarbons using TiO_2 nanoparticle membranes. *Appl. Surf. Sci.* 289: 203–208.

Razavet, M., V. Artero, and M. Fontecave. 2005. Proton electroreduction catalyzed by cobaloximes: Functional models of hydrogenases. *Inorg. Chem.* 44: 4786–4795.

Reithmeier, R. O., S. Meister, B. Rieger, A. Siebel, M. Tschurl, U. Heiz, et al. 2014. Mono- and bimetallic Ir(III) based catalysts for the homogeneous photocatalytic reduction of CO_2 under visible light irradiation. New insights into catalyst deactivation. *Dalton Trans.* 43: 13259–13269.

Röger, C., M. G. Müller, M. Lysetska, Y. Miloslavina, A. R. Holzwarth, and F. Würthner. 2006. Efficient energy transfer from peripheral chromophores to the self-assembled zinc chlorin rod antenna: A bioinspired light-harvesting system to bridge the "green gap." *J. Am. Chem. Soc.* 128: 6542–6543.

Romain, S., L. Vigara, and A. Llobet. 2009. Oxygen–oxygen bond formation pathways promoted by ruthenium complexes. *Acc. Chem. Res.* 42: 1944–1953.

Ryan, D. M., M. K. Coggins, J. J. Concepcion, D. L. Ashford, Z. Fang, L. Alibabaei, et al. 2014. Synthesis and electrocatalytic water oxidation by electrode-bound helical peptide chromophore-catalyst assemblies. *Inorg. Chem.* 53: 8120–8128.

Sayama, K., K. Mukasa, R. Abe, Y. Abe, and H. Arakawa. 2002. A new photocatalytic water splitting system under visible light irradiation mimicking a Z-scheme mechanism in photosynthesis. *J. Photochem. Photobiol. A: Chem.* 148: 71–77.

Sens, C., I. Romero, M. Rodriguez, A. Llobet, T. Parella, and J. Benet-Buchholz. 2004. A new Ru complex capable of catalytically oxidizing water to molecular dioxygen. *J. Am. Chem. Soc.* 126: 7798–7799.

Shown, I., H. C. Hsu, Y. C. Chang, C. H. Lin, P. K. Roy, A. Ganguly, et al. 2014. Highly efficient visible light photocatalytic reduction of CO_2 to hydrocarbon fuels by Cu-nanoparticle decorated graphene oxide. *Nano Lett.* 14: 6097–6103.

Small, Y. A., D. L. DuBois, E. Fujita, and J. T. Muckerman. 2011. Proton management as a design principle for hydrogenase-inspired catalysts. *Energy Environ. Sci.* 4: 3008–3020.

Song, G., F. Xin, and X. Yin. 2015. Photocatalytic reduction of carbon dioxide over $ZnFe_2O_4/TiO_2$ nanobelts heterostructure in cyclohexanol. *J. Colloid Interfacial Sci.* 442: 60–66.

Straight, S. D., G. Kodis, Y. Terazono, M. Hambourger, T. A. Moore, A. L. Moore, et al. 2008. Self-regulation of photoinduced electron transfer by a molecular nonlinear transducer. *Nat. Nanotechnol.* 3: 280–283.

Sun, L., H. Berglund, R. Davydov, T. Norrby, L. Hammarstro, P. Korall, et al. 1997. Binuclear ruthenium–manganese complexes as simple artificial models for photosystem II in green plants. *J. Am. Chem. Soc.* 119: 6996–7004.

Sun, L., L. Hammarström, B. Akermark, and S. Styring. 2001. Towards artificial photosynthesis: Ruthenium-manganese chemistry for energy production. *Chem. Soc. Rev.* 30: 36–49.

Sun, Y., J. P. Bigi, N. A. Piro, M. L. Tang, J. R. Long, and C. J. Chang. 2011. Molecular cobalt pentapyridine catalysts for generating hydrogen from water. *J. Am. Chem. Soc.* 133: 9212–9215.

Sutin, N., C. Creutz, and E. Fujita. 1997. Photo-induced generation of dihydrogen and reduction of carbon dioxide using transition metal complexes. *Comm. Inorg. Chem.* 19: 67–92.

Tagore, R., R. H. Crabtree, and G. W. Brudvig. 2008. Oxygen evolution catalysis by a dimanganese complex and its relation to photosynthetic water oxidation. *Inorg. Chem.* 47: 1815–1823.

Therrien, J. A., M. O. Wolf, and B. O. Patrick. 2014. Electrocatalytic reduction of CO_2 with palladium bis-N-hetercycliccarbene pincer complexes. *Inorg. Chem.* 53: 12962–12972.

Tseng, H. W., R. Zong, J. T. Muckerman, and R. Thummel. 2008. Mononuclear ruthenium(II) complexes that catalyze water oxidation. *Inorg. Chem.* 47: 11763–11773.

Umena, Y., K. Kawakami, J. R. Shen, and N. Kamiya. 2011. Crystal structure of oxygen-evolving photosystem II at a resolution of 1.9 Å. *Nature.* 473: 55–60.

Vlcek, A., and M. Busby. 2006. Ultrafast ligand-to-ligand electron and energy transfer in the complexes *fac*-$[Re^I(L)(CO)_3(bpy)]n^+$. *Coord. Chem. Rev.* 250: 1755–1762.

Wada, T., K. Tsuge, and K. Tanaka. 2000. Electrochemical oxidation of water to dioxygen catalyzed by the oxidized form of the bis(ruthenium-hydroxo) complex in H_2O. *Angew. Chem. Int. Ed.* 39: 1479–1482.

Wada, T., K. Tsuge, and K. Tanaka. 2001. Syntheses and redox properties of bis(hydroxoruthenium) complexes with quinone and bipyridine ligands. Water-oxidation catalysis. *Inorg. Chem.* 40: 329–337.

Wang, Y., F. Wang, Y. Chen, D. Zhang, B. Li, S. Kang, et al. 2014. Enhanced photocatalytic performance of ordered mesoporous Fe-doped CeO_2 catalysts for the reduction of CO_2 with H_2O under simulated solar irradiation. *Appl. Catal. B: Environ.* 147: 602–609.

Wasylenko, D. J., C. Ganesamoorthy, J. Borau-Garcia, and C. P. Berlinguette. 2011. Electrochemical evidence for catalytic water oxidation mediated by a high-valent cobalt complex. *Chem. Commun.* 47: 4249–4251.

Wenger, O. S. 2009. Long-range electron transfer in artificial systems with d(6) and d(8) metal photosensitizers. *Coord. Chem. Rev.* 253: 1439–1457.

Wiberg, J., L. Guo, K. Pettersson, D. Nilsson, T. Ljungdahl, J. Mårtensson, et al. 2007. Charge recombination versus charge separation in donor-bridge-acceptor systems. *J. Am. Chem. Soc.* 129: 155–163.

Wu, J. C. S., H. M. Lin, and C. L. Lai. 2005. Photo-reduction of CO_2 to methanol using optical-fiber photoreactor. *Appl. Catal. A: Gen.* 296: 194–200.

Xu, Y., T. Åkermark, V. Gyollai, D. Zou, L. Eriksson, L. Duan, et al. 2009. A new dinuclear ruthenium complex as an efficient water oxidation catalyst. *Inorg. Chem.* 48: 2717–2719.

Xu, Y., A. Fischer, L. Duan, L. Tong, E. Gabrielsson, B. Åkermark, et al. 2010. Chemical and light-driven oxidation of water catalyzed by an efficient dinuclear ruthenium complex. *Angew. Chem. Int. Ed.* 49: 8934–8937.

Yamashita, H., Y. Fujii, Y. Ichihashi, S. G. Zhang, K. Ikeue, D. R. Park, et al. 1998. Selective formation of CH_3OH in the photocatalytic reduction of CO_2 with H_2O on titanium oxides highly dispersed within zeolites and mesoporous molecular sieves. *Catal. Today.* 45: 221–227.

Yang, M., and X. Q. Jin. 2014. Facile synthesis of Zn_2GeO_4 nanorods toward improved photocatalytic reduction of CO_2 into renewable hydrocarbon fuel. *J. Cent. South Univ.* 21: 2837–2842.

Yin, Q., J. M. Tan, C. Besson, Y. V. Geletii, D. G. Musaev, A. E. Kuznetsov, et al. 2010. A fast soluble carbon-free molecular water oxidation catalyst based on abundant metals. *Science.* 328: 342–345.

Youngblood, W. J., S. H. A. Lee, K. Maeda, and T. E. Mallouk. 2009. Visible light water splitting using dye sensitized oxide semiconductors. *Acc. Chem. Res.* 42: 1966–1973.

Yuhas, B. D., C. Prasittichai, J. T. Hupp, and M. G. Kanatzidis. 2011. Enhanced electrocatalytic reduction of CO_2 with ternary $Ni-Fe_4S_4$ and $Co-Fe_4S_4$-based biomimetic chalcogels. *J. Am. Chem. Soc.* 133: 15854–15857.

Yui, T., Y. Tamaki, K. Sekizawa, and O. Ishitani. 2011. Photocatalytic reduction of CO_2: From molecules to semiconductors. *Topics Curr. Chem.* 303: 151–184.

Zhang, G., R. Zong, H. W. Tseng, and R. P. Thummel. 2008. Ru(II) Complexes of tetradentate ligands related to 2,9-di(pyrid-2′-yl)-1,10-phenanthroline. *Inorg. Chem.* 47: 990–998.

Zong, R., and R. P. Thummel. 2005. A new family of Ru complexes for water oxidation. *J. Am. Chem. Soc.* 127: 12802–12803.

11 Nanomaterials for Solar Energy

Mohammad Azad Malik, Sajid Nawaz Malik, and Asma Alenad

CONTENTS

11.1 Introduction 219
11.2 Binary Materials 220
    11.2.1 Copper Sulfide 220
    11.2.2 Iron Sulfide 222
    11.2.3 Lead Sulfide 223
    11.2.4 Tin Sulfide 225
11.3 I–III–VI Materials 226
    11.3.1 Copper Indium Disulfide ($CuInS_2$) 227
    11.3.2 Copper Indium Diselenide ($CuInSe_2$) 229
    11.3.3 Copper Indium Gallium Disulfide (CIGS) 231
    11.3.4 Copper Indium Gallium Diselenide (CIGSe) 232
11.4 Copper-Zinc-Tin Chalcogenides (CZTSSe) 235
11.5 Copper Iron Tin Sulfide and Related Materials 242
References 245

11.1 INTRODUCTION

Global energy consumption during 2010 was recorded as 524 quadrillion British thermal units (Btu). Various energy planning agencies have projected that this consumption will increase to 630 quadrillion Btu in 2020 and might rise to 820 quadrillion Btu in 2040 (U.S. Energy Information Administration [USEIA] 2013). This growth in energy consumption is driven by increasing growth in industrial activity in developing countries and changing lifestyles. Currently, most energy is obtained by the burning of fossil fuels, mainly coal, oil, and natural gas. Other major sources of energy are the nuclear reactors and hydroelectric power projects. The burning of fossil fuels results in the emission of greenhouse gases, especially CO_2 in the atmosphere, which has been identified as a major cause of global warming. Many scientific studies indicate that a temperature increase of 1.8°C–4.0°C may occur in the global temperature during the twenty-first century, which may cause serious and irreversible effects, such as the melting of polar ice, gigantic floods, and a rise in sea level (Intergovernmental Panel on Climate Change [IPCC] 2007). The disposal of nuclear fuels and nuclear security issues hinder the widespread use of nuclear energy. Large hydroelectric dams suffer from the issues of seasonal flow fluctuations and periods of draught. Furthermore, obtaining consent for more large hydroelectric sites is becoming increasingly difficult as such projects are opposed by local communities and environmentalists due to associated ecological disruptions (Sims 2008). Therefore, present energy supply trends are considered as being unsustainable for the long term.

Now the focus is shifting toward alternative energy technologies that are more viable, clean, efficient, and environmentally benign. The Sun is the most abundant source of energy. It produces enormous amounts of energy due to thermonuclear fusion reactions. According to one estimate, solar energy striking Earth in 1 h is more than the total energy consumed on the planet in a year

(Dhere 2007). Another calculation suggests that the solar energy falling on Earth in 30 sunny days is equivalent to the energy produced from all fossil fuels on Earth, either consumed or unused. If efficient and economical means of harvesting this energy could be developed, all energy needs could be satisfied by solar energy alone. Solar energy may, therefore, potentially ensure the transition toward a sustainable energy supply system for the twenty-first century.

Sunlight can be converted into useful energy by two routes: one is the solar thermal approach, whereby solar energy is converted to heat; the second is the solar photovoltaic approach, which uses semiconductor materials to generate electricity from solar radiation through the photoelectric effect. The first viable solar electricity-producing cell was demonstrated by Chapin, Pearson, and Fuller of Bell Telephone Laboratories in 1954 (Chapin et al. 1954). Since then this technology has shown a steady and substantial growth predominantly driven by government subsidies and incentives. Many policy papers on energy suggest that in 2030, solar photovoltaics may become a terawatt industry satisfying a considerable fraction of total global energy consumption (Bellemare 2006). Despite all this progress and inherent potentials of photovoltaics, the current share of photovoltaics in global energy production is very low; the cost per watt–peak of electricity produced is still expensive. This cost can be significantly reduced by market expansion, technological developments, improvement in solar module efficiencies, and utilization of superior production strategies (Fthenakis 2009).

At the heart of the solar cell lies light-absorbing semiconductor material that absorbs photons of light to produce electron and hole carriers via photovoltaic (PV) effect. The PV market is currently dominated by first-generation single-junction PV devices based on single or multicrystalline silicon wafers. However, cost per peak watt of electricity generated from a single-crystalline silicon PV technology is not low enough to make it economically viable, and about half of the cost involved is the material cost for silicon wafers. Second-generation PV technologies are thin-film single-junction devices using semiconductors that offer the key advantage of reduced material use and thus reduced associated costs, while maintaining the efficiencies comparable to or better than first-generation PV technologies. This approach is based on thin layers of semiconductors like cadmium telluride (CdTe), copper indium disulfide (CIS), copper indium gallium diselenide (CIGSe), and copper-zinc-tin chalcogenides (CZTSSe) deposited onto low-cost substrates like glass. These materials have high solar optical absorption coefficients (greater than $10^5\,cm^{-1}$), so film thickness is typically less than a micron and 100–1000 times lesser materials are required than silicon wafers–based PV. Furthermore, these technologies have the potential for upscaling by roll-to-roll manufacturing, thus offering even better economy of scales. Similarly, the third-generation photovoltaics encompass a range of evolving devices, upstarts, and wild ideas that have joined the race to achieve the performance and cost goals desired for solar PV (Kazmerski 2006). These technologies include polymer cells, biomimetics, quantum dot technologies, tandem/multijunction solar cells, hot carrier cells, upconversion and downconversion technologies, and solar thermal technologies. They may also include silicon nanostructures, modifying incident spectrum (concentration), use of excess thermal generation (caused by UV light) to enhance voltages or carrier collection, and use of infrared spectrum to produce electricity at night (Brown and Wu 2009).

The synthesis of nanocrystals, their properties, and their potential applications for solar energy have been reviewed (Ramasamy et al. 2012, 2013a, 2013b; Akhavan et al. 2012; Abermann 2013; Zhang et al. 2013a; Zhou et al. 2013; Aldakov et al. 2013; Fan et al. 2014; Azimi et al. 2014). This chapter will cover various technological developments, especially the use of semiconductor nanomaterials in solar photovoltaics and their overall impact on the growth of this technology.

11.2 BINARY MATERIALS

11.2.1 COPPER SULFIDE

Copper sulfide exists in a number of stoichiometric and nonstoichiometric forms; therefore, it has been an interesting material for fundamental studies. Its composition is based on nontoxic and

Earth-abundant elements, and it finds important applications in solar photovoltaics, bioimaging, and photocatalysis. Xiong and Zeng (2012) have reported the synthesis of multishelled copper sulfide hollow spheres. Their approach involved the synthesis of polyvinylpyrrolidone (PVP)-coated Cu_2O spheres by polyol method and the carrying out of ion exchange of these Cu_2O nanospheres using thiourea and sodium sulfide, thus producing multishelled Cu_2S hollow spheres. It was found that the optical bandgap of the Cu_2S hollow spheres varied upon increasing diameter of the spheres and was found to be up to 2.10, 1.49, and to 1.42 eV for single, double, and triple-shelled Cu_2S hollow spheres, respectively.

Synthesis of hierarchical hollow spheres of CuS at the interface of water and oil has been reported (Jiang et al. 2012a, 2012b). In a typical reaction, thioacetamide dissolved in water and copper naphthenate dissolved in dimethylbenzene were allowed to react at the interface for 24 h at room temperature. Amorphous hollow spheres were formed by the interfacial reaction at the interface of water and dimethylbenzene. Hierarchical CuS hollow spheres were subsequently obtained by autoclaving these amorphous spheres in ethanol at 60°C for 96 h.

Sun et al. (2012) have carried out the synthesis of polyhedral 26-facet Cu_7S_4 hollow cages by using the sacrificial template method. Cu_2O particles having 26 facets were synthesized as templates by reducing $Cu(CH_3COO)_2$ with glucose. The reaction of templates with Na_2S at room temperature resulted in the formation of Cu_2O/Cu_7S_4 core–shell particles. Selective removal of Cu_2O core using ammonia yielded Cu_7S_4 hollow cages.

Synthesis of hybrid $Ru-Cu_2S$ nanostructures having cage and nanonet-like morphologies has been reported by Vinokurov et al. (2012). Hybrid nanostructures are especially beneficial for solar cell applications as they offer better electron–hole separation. Han et al. (2012) have developed a one-pot colloidal method for the synthesis of $Cu_{1.94}S$-ZnS, $Cu_{1.94}S$-ZnS-$Cu_{1.94}S$, and $Cu_{1.94}S$-ZnS-$Cu_{1.94}S$-ZnS-$Cu_{1.94}S$ heteronanostructures. Copper iodide and zinc diethyldithiocarbamate complex in oleylamine were reacted to form heterostructures with screw-, dumbbell-, and sandwich-like morphologies.

A wide variety of single-source precursors have been synthesized and utilized for preparation of copper sulfide nanocrystals. Copper complexes of alkylxanthates, mercaptobenzothiazoles, thiobenzoates, dithiocarbamates, and dithiolates have been synthesized and used for the synthesis of copper sulfide nanocrystals with various sizes and morphologies. Abdelhady et al. (2012) have reported the use of 1,1,5,5-tetra-iso-propyl-2-thiobiuret complex of copper as a single-source precursor for the synthesis of copper sulfide nanocrystals in a continuous-flow microfluidic reactor. Spherical Cu_7S_4 nanocrystals with an average diameter of 6.7 ±1.6 nm, 10.8 ± 1.9 nm and 11.4 ± 2.4 nm were obtained at 170°C, 200°C, and 230°C, respectively. Thermolysis of a copper complex of S-methyl dithiocarbamate in different high boiling solvents at different temperatures has been used for the synthesis of copper sulfide nanocrystals (Bera and Seok 2012). High boiling solvents used in thermolysis include ethylene glycol, ethylenediamine, hydrazine hydrate, and hexamethylenediamine. Sobhani et al. (2012) have reported the use of [*bis*(thiosemicarbazide)copper(II)] chloride as a single-source precursor for the preparation of copper sulfide nanoparticles through hydrothermal approach. CuS nanoparticles synthesized by this method are 20–50 nm in size with an irregular morphology.

A low-temperature colloidal method has been used for the synthesis of ultrathin hexagonal copper sulfides nanosheets (Du et al. 2012a). Crystallographic phase of the nanoparticles was determined by x-ray diffraction to be the covellite CuS phase. It was found that the nanoparticles have a nanosheet-like morphology with a thickness of about 3.5 nm. CuS nanosheets obtained by this method showed an absorption peak at 465 nm and emission peaks at 418 and 445 nm.

A controllable solvothermal synthesis of CuS with hierarchical structures has been reported (Peng et al. 2014). Different morphologies like plates, nanoparticles, spheres, and nanoflowers have been obtained by optimally varying the reaction parameters. Electrochemical characterization results indicate that CuS structures exhibit remarkable capacitive performance. Flower-like CuS demonstrated a high specific capacitance (597 Fg^{-1}), excellent discharge rate, and a good stability, which reflect significant potential of this electrode material for use in super capacitor devices.

A simple wet chemical route for preparation of Cu_2S nanoneedles by a room temperature reaction of CuCl and Na_2S has been developed by Kumarakuru et al. (2014). Formation of Cu_2S nanoneedles takes place by the self-assembly of Cu_2S nanoparticles, whereas thioglycerol (TG) has been utilized as the capping agent. Formation of a polyphasic mixture containing different phases of copper sulfide as well as copper oxide and copper chloride was indicated in powder x-ray diffraction (p-XRD) studies. Deposition of copper sulfide thin films onto different substrates has been carried out by employing the thermal evaporation technique (Saadeldin et al. 2014). X-ray diffraction studies demonstrate that an orthorhombic chalcocite (γ-Cu_2S) phase has been deposited onto substrates. Atomic force microscopy revealed that thin films composed of nanoparticles have an average size of about 44 nm. Nanostructured assemblies of copper sulfide (CuS) have been synthesized by a facile, template-free route (Kundu and Pradhan 2014). It was found that the CuS nanoplates or nanoparticles underwent self-assembly to form either spheres or nanotubes. Detailed studies were carried out to elaborate the mechanism for the formation of nanotubes.

11.2.2 Iron Sulfide

Iron sulfides have recently attracted significant interest as solar absorber materials, mainly due to their suitable bandgap, high absorption coefficient, nontoxicity, and abundance in nature. Pyrite nanocrystals have emerged as a promising absorber material for large-scale production of solar photovoltaic devices, chiefly because of their Earth-abundant and nontoxic nature. Significant research interest has been devoted toward development of scalable routes for synthesis of phase-pure and shape-controlled colloidal pyrite nanocrystals. However, surface defects in pyrite crystals are a major detrimental factor for their use in solar cells. Iron sulfide nanocrystals having a variety of sizes and morphologies have been synthesized by employing various synthetic methods. Synthesis of FeS_2 nanoplates by injecting an organometallic precursor [$Fe(CO)_5$] into a solution containing oleylamine and sulfur has been reported (Kirkeminde et al. 2012). Nanoplates thus prepared had a lateral size of 150 nm and thickness around 30 nm. These irregular-shaped plates were composed mainly of hexagonal crystallites. The absorbance spectra of FeS_2 nanoplates grown for 180 min exhibited an excitonic peak at 895 nm (1.38 eV), which corresponded to direct bandgap of FeS_2. A hybrid solar cell constructed by blending a 1:1 ratio of FeS_2 nanoplates with P_3HT demonstrated an open-circuit voltage of 780 mV with a power conversion efficiency of 0.03%.

Steinhagen et al. (2012) have carried out the synthesis of iron sulfide (FeS_2) nanocrystals having a pyrite phase from solvent-based dispersions, or "solar paint," for fabrication of photovoltaic devices. Phase purity of the nanocrystals was demonstrated by the p-XRD and Raman spectroscopy. These nanocrystals were spray-deposited onto substrates to form absorber layers in devices with different architectures, including Schottky barrier, heterojunction, and organic/inorganic hybrid solar cells, to evaluate their suitability as a photovoltaic material. None of the devices exhibited a PV response, whereas the electrical conductivity of the nanocrystal films was about 4–5 S cm^{-1}. This lack of PV response may be attributed to the highly conductive surface-related defects in pyrite nanocrystals.

The synthesis of single-crystalline cubic iron pyrite (FeS_2) nanowires has been carried out by thermal sulfidation of steel foil (Cabán-Acevedo et al. 2012). Isolated nanocubes had a size of ~150 nm, whereas size of the dendrites was about 40 nm, and they were composed of smaller particles of ~10 nm. It was observed that the sizes of both the nanocubes and nanodendrites increased by increasing the reaction time. The pyrite nanowires have length greater than 2 µm and a 4–10 nm diameter. Electrical transport measurements showed the pyrite nanowires to be highly p-doped, with an average resistivity of 0.18 ± 0.09 Ω cm and carrier concentrations of the order of 10^{21} cm^{-3}. Bandgap calculated by optical measurement was found to be around 0.9 eV.

Macpherson and Stoldt (2012) have reported the synthesis of pyrite nanocubes by reaction of $FeCl_2$ and elemental sulfur in different alkylamines. The nanocrystals obtained after initial heating at 250°C had a random oblate shape, whereas the first stage of growth at 200°C resulted in the formation of nanocubes with large size distribution. If the reaction was further carried out to allow a second growth stage, the size distribution of the nanocubes was improved. The FeS$_2$ nanocubes prepared by this method had a lateral size of 37 ± 11 nm. Absorption measurements of the nanocubes showed an indirect bandgap around 1.1 eV along with two excitonic transitions at 1.9 and 3.0 eV.

Morrish et al. (2012) have reported the preparation of FeS$_2$ nanoparticles through plasma-assisted sulfurization of Fe$_2$O$_3$ nanorods. Nanorods of Fe$_2$O$_3$ having an approximate size of 150 nm were deposited by chemical bath deposition using FeCl$_3$ and NaNO$_3$ on fluorine-doped tin oxide (FTO) glass plates. Formation of the marcasite phase was somewhat suppressed by prolonged sulfurization of the F$_2$O$_3$ nanorods; however, complete eradication could not be achieved. The bandgap (direct) of obtained FeS$_2$ was found to be 1.2 eV.

Beal et al. (2011) have reported the synthesis of greigite (Fe$_3$S$_4$) nanoparticles with spherical morphology and a size of 6.5 ± 0.5 nm by hot-injection approach. Magnetic properties of these Fe$_3$S$_4$ nanoparticles were compared with magnetic properties of similar size Fe$_3$O$_4$ nanocrystals. Greigite (Fe$_3$S$_4$) nanoparticles showed saturation magnetization of 12 emu g^{-1} at 10 K and blocking temperature around ~50 K.

Greigite (Fe$_3$S$_4$) nanoparticles have been prepared by vapor–solid interaction using a laterally resolving ultrahigh vacuum multimethod instrument (Bauer et al. 2014). Bandgaps of the FeS nanoparticles synthesized using ethylene glycol, ethylenediamine, and ammonia were found to be 3.13 eV, 3.02 eV, and 2.75 eV, respectively (Maji et al. 2012). Photocatalytic activity of FeS nanoparticles was determined by carrying out methylene blue degradation experiments; it was found to have better activity than commercial TiO$_2$.

Cummins et al. (2013) have reported the synthesis of phase-pure iron sulfide nanowires through sulfurization of hematite nanowire arrays. Transmission electron microscope (TEM) images showed hollow iron sulfide nanotubes with diameters in the range of 100–300 nm and wall thicknesses around 60 nm. The average length of the nanotubes was around 3 μm. Beal et al. (2012) have recently reported the synthesis of Fe$_{1-x}$S and Fe$_3$S$_4$ nanocrystals by reaction of elemental sulfur with Fe(acac)$_2$ in oleylamine at 200°C for 4 h. Thin sheets of Fe$_3$S$_4$ were formed by following these conditions, whereas the same reaction carried out at above 300°C for 30 min resulted in formation of Fe$_{1-x}$S nanocrystals. The nanocrystals possessed a hexagonal plate and prism-like morphologies with an average size of 70 nm. Hexagonal iron sulfide (Fe$_7$S$_8$) nanoflowers have been developed by Wang et al. (2013) from thermal decomposition of ferric hexadecylxanthate at 260°C without any solvent or inert gas protection, forming Fe$_7$S$_8$ nanoflowers.

Li et al. (2014) have carried out a systematic study of optical and electronic properties of pyrite nanocrystal thin films. They had used a variety of ligands having different anchor and bridging groups. It was found that the anchor group of ligand mainly controls the optical absorption. The conductivity and photoconductivity of the nanocrystals are however, controlled by combined effects of anchor and bridging groups. Zhu et al. (2014) have investigated the effect of reaction conditions and the local chemical environment on the shape, composition, and crystallographic phase of iron pyrite NCs synthesized by using the hot-injection approach. Different morphologies were obtained by varying the solvents, concentration of reactants, and capping ligand (either trioctylphosphine oxide or 1,2-hexanediol). These morphologies include short, branched, and chromosome-like rods having ~10 nm diameter and 20–30 nm length as well as quasi-cubic NC agglomerates of 200 nm. The as-synthesized iron pyrite NCs can be dispersed well in chloroform, chlorobenzene, toluene, and hexane, and thus are promising in solution-processable photovoltaic applications.

11.2.3 LEAD SULFIDE

Lead sulfide is a narrow-bandgap semiconductor that has been extensively studied. Its direct bandgap and large exciton Bohr radius make it a suitable material for various solar photovoltaic devices. A variety of techniques have been employed for the synthesis of PbS nanocrystals. Khan et al. (2013a) have used a hot-injection approach for the synthesis of nearly monodispersed PbS nanoparticles with tunable bandgap. It was found that the size of the nanocrystals could be controlled by suitably varying the injection temperature and the reaction time. For example, nanoparticles grown at 110, 120, 130, 150, and 160°C had a mean size of 2.3 ± 0.18, 2.7 ± 0.19, 3.5 ± 0.23, 5.5 ± 0.25, and 10 ± 0.3 nm, respectively. Bandgap values calculated from UV-Vis absorption spectra were found to vary from ~1.7 eV for size of 2.3 ± 0.18 nm and ~0.6 eV for 10 ± 0.3 nm. Heterojunction solar cells fabricated using these nanocrystals have demonstrated an open-circuit voltage of 235 mV for 10 ± 0.3 nm and 386 mV for 2.3 ± 0.18 size nanoparticles. The device constructed using nanoparticles having bandgap of 1.2 eV exhibited the highest short-circuit current of 1.67 mA cm^{-2}.

Synthesis of spherical PbS nanoparticles from $PbCl_2$ and elemental sulfur using the hot-injection approach has been reported (Nakashima et al. 2013). The experimental procedure involved the injection of an oleylamine solution of sulfur into the reaction vessel containing $PbCl_2$ and oleylamine preheated at 70°C and 80°C, and the reaction was continued for different time periods ranging from 1 to 25 min. The mean diameter of the particles grown at 70°C and 80°C was estimated to be 4.47 ± 0.92 nm and 4.53 ± 0.57 nm for 1 min reaction, and 5.30 ± 0.52 nm and 6.27 ± 0.46 nm for 15 min of reaction time. A red shift in the photoluminescence wavelength of the nanoparticles from 1221 to 1288 nm was observed for the nanoparticles grown at 70°C from 1 to 15 min, whereas the nanoparticles grown at 80°C demonstrated a red shift from 1282 to 1370 nm.

A wet chemical reaction using lead acetate and thiourea was used for the synthesis of self-supporting arrays of quadrangular PbS nanopyramids (Hu et al. 2013a). The Scherrer equation was used to calculate the crystallite size which was estimated to be about 150 nm. Scanning electron microscope (SEM) images showed large sheets of 3–15 μm. These sheets were formed as a result of self-assembly of 100–200 nm quadrangular nanopyramids.

Lead sulfide nanocrystals with a starfish like morphology were obtained by the reaction of lead acetate and thioacetamide in the presence of cetyltrimethylammonium bromide (CTAB) and sodium dodecyl sulfate (SDS) at 80°C (Li et al. 2013d). The selected area electron diffraction (SAED) pattern from the star-shaped PbS crystals imaged along the [111] zone axis indicated that the arms of the star-shaped PbS have grown along the [100] direction. The influence of various reaction parameters like reaction temperature and lead source on the morphology of PbS nanocrystals has been investigated.

Pan et al. (2013) have reported the synthesis of PbS quantum dots (QDs) in a flow reactor for applications in solar cells. It was shown that the PbS quantum dots produced in the flow reactor had a comparable performance to those synthesized in a batch reaction. A dual-temperature-stage flow reactor was used to carry out the synthesis of PbS nanoparticles with optimal results. Precursor A used in this method was composed of lead oxide, oleic acid (OA), and octadecene (ODE), whereas precursor B contained *bis*(trimethylsilyl) sulfide (TMS) and ODE. The two precursors were mixed together under nitrogen. The temperature of the reaction was controlled by thermocouples. The reaction between the precursors at elevated temperature leads to the formation of PbS nuclei which act as seed and subsequently grow to form PbS quantum dots. These quantum dots are finally isolated by precipitation using acetone and are redispersed in toluene.

Multiple exciton generation (MEG) in PbS, and a PbS_xSe_{1-x} alloy have been studied (Midgett et al. 2013). It was observed that a linear decrease in MEG efficiency occurs for both PbS and $PbSxSe_{1-x}$ alloyed dots with an increase in diameter within the strong confinement regime. Synthesis of PbS nanoparticles has also been carried out as part of a composite material for use in solar cells and solar water splitting (Kawawaki and Tatsuma 2013). For solar cell application, the PbS nanoparticles are used in conjunction with plasmonic gold nanoparticles to serve as light-harvesting antennae, whereas PbS nanoparticles have been decorated with Al-doped ZnO (AZO) nanorod arrays for solar water-splitting applications (Hsu et al. 2013a).

A controlled synthesis of the high-quality flower-shaped PbS nanostructures by A solvothermal approach using propylene glycol as solvent and a thio Schiff-base (2-(benzylideneamino)-benzenethiol) as a new sulfur source has been reported (Arani and Niasari 2014). It was demonstrated that the good-quality, flower-shaped PbS nanoparticles having nearly uniform morphology and size could be synthesized. Furthermore, the influence of various reaction parameters like solvents, reaction time, temperature, and concentration of sulfurizing agent on morphology and size of the nanocrystals has been investigated.

Colloidal PbS quantum dots having very small size possess significant potential for increasing the open-circuit voltages of quantum dot–based solar cells due to their large energy gap. A low-temperature synthesis of ultra-small red light–emitting PbS QDs from organometallic precursors in the presence of 1,2-dichloroethane has been developed by Reilly et al. (2014). Average size and optical properties of the PbS quantum dots have been determined using HRTEM and optical spectroscopy. HRTEM images revealed that the PbS quantum dots have an average size of 1.6 nm with a standard deviation of 0.2 nm.

PbS nanoparticle films have been deposited electrochemically onto ITO glass substrates (Mocanu et al. 2014). Deposition experiments resulted in the formation of different morphologies of PbS nanostructures in the presence of different water-soluble polymers like polyacrylic acid (PAA), PVP, polyvinyl alcohol (PVA), and poly(2-acrylamido-2-methylpropane sulfonic acid) (PAMPSA). SEM images revealed that at least two different morphologies of PbS nanostructures are deposited for all samples. The role of polymers in determining the shape of PbS nanostructures has been explained.

McPhail and Weiss (2014) have investigated the reaction mechanism for the chemical reaction of elemental sulfur with 1-octadecene (ODE). They have described the factors that induce a change in morphology of the PbS quantum dots from cubic to hexapodal during synthetic reaction between the S/ODE precursor and lead(II) oleate. Hexapodal geometry is obtained when the organosulfur ligands bind to the growing QDs causing a preferential growth at the ⟨100⟩ faces rather than at ⟨111⟩ faces. It was also suggested that S/ODE can be used more reliably by decreasing the reaction temperature and time for dissolution of the sulfur.

11.2.4 Tin Sulfide

Tin sulfide belongs to the family of IV–VI semiconductors. Three forms of tin sulfide exist, which include SnS, SnS_2, and Sn_2S_3. The bandgaps of SnS, SnS_2, and Sn_2S_3 are 1.3, 2.18, and 0.95 eV, respectively. Significant research attention has been devoted to explore the use of tin sulfide in solar photovoltaic devices as an absorber based on Earth-abundant elements and having a low toxicity. SnS has been reported to be either a p-type or n-type conductor depending on the tin content. Furthermore, its conductivity might also change upon heat treatment. SnS_2 is an n-type semiconductor, whereas Sn_2S_3 possesses highly anisotropic conduction behavior.

The synthesis of nanocrystalline tin sulfide is mostly carried out with a view to synthesize the SnS phase, having a quantum confinement effect. The synthesis of SnS nanocrystals having three different morphologies, such as cubes, spherical polyhedral, and nanosheets, has been reported by Biacchi et al. (2013). Spherical polyhedral nanoparticles synthesized by this method had a mean size of 9.7 ± 1.5 nm. SnS nanoparticles with cubic morphology were obtained by following the same procedure but carrying out injection of SnS at 170°C and maintaining the reaction temperature between 165°C and 170°C for 1 h. Resulting nanocubes had an average edge length of 11.5 ± 1.9 nm. A slightly modified procedure for the preparation of SnS nanosheets has also been reported by the same group. They have used a thermolytic reaction of tin acetate and elemental sulfur dissolved in oleylamine. Thus-prepared nanosheets had a lateral dimension of about 270 ± 50 nm. Investigations into the photocatalytic activity of these SnS nanocrystals against methylene blue dye showed that the SnS with nanocube morphology had the highest photocatalytic activity.

Diethylthiocarbamate complex of tin has been used as a single-source precursor for the phase-controlled synthesis of nanosized SnS, SnS_2, and SnS/SnS_2 heterostructures (Hu et al. 2013b). Thermolysis of the complex in oleylamine at 320°C for 30 min yielded SnS nanocrystals, whereas nanocrystals with SnS_2 phase were obtained by using 300 μL CS_2 in addition to the precursor during thermolysis. Liang et al. (2013) have reported the synthesis of SnS nanocrystals by thermal decomposition of SnO and elemental sulfur in oleic acid and oleylamine at different reaction temperatures.

The size of the nanocrystals increased from 50 to 200 nm when the decomposition temperature was increased from 150°C to 210°C. Similarly, morphology of the nanocrystals was changed from spherical to sheet-like morphology by increasing the volume of oleic acid used in the reaction. A hydrothermal reaction of $SnCl_2.2H_2O$, hydrazine, and thiourea at 180°C for 23 h has been used for the preparation of SnS nanorods (Iqbal et al. 2013). The lengths of thus-prepared nanorods varied from 1 to 2 μm, whereas the width was estimated to be about 80 nm. It was demonstrated that as-prepared SnS nanorods had a maximum hydrogen absorption value of 0.73 wt%, demonstrating their potential as a hydrogen storage material.

Yan et al. (2013a) have carried out microwave-assisted hydrothermal synthesis of CuS, ZnS, and SnS nanocrystals and compared it with conventional hydrothermal synthesis. SnS nanoparticles prepared by both processes, conventional hydrothermal heating and microwave hydrothermal heating, had nearly the same morphology; however, microwave-assisted heating proved to be much quicker (15 min heating period) than conventional heating (4 h).

Sonochemical preparation of SnS nanoparticles exhibiting quantum confinement has been reported (Azizian-Kalandaragh et al. 2013). In a typical reaction, an aqueous solution of tin chloride, sodium sulfide, and triethanolamine was sonicated for 2 h at room temperature by a high-intensity ultrasonic transducer. Polydispersed nanoclusters having particle sizes smaller than 100 nm were thus obtained. The bandgap value of the SnS nanoparticles was determined to be 1.74 eV, indicating the presence of the quantum confinement effect.

Rath et al. (2014) have proposed a fabrication scheme for improving the device performance of low-cost chalcogenide-based solar cells. They have developed an In_2S_3 layer having embedded SnS quantum dots. Their fabrication scheme involves incorporation of this intermediate bandgap layer in copper indium sulfide (CIS) cells for improving the current generation and efficiency. They have also proposed measures to optimally cap the surface of quantum dots for effective passivation of defects and to protect the quantum dots from doping from the environment.

11.3 I–III–VI MATERIALS

Polycrystalline copper indium/gallium sulfur/selenium ($CuIn_{1-x}Ga_xS_{1-y}Se_y$) thin films have attracted considerable research and industrial interest among chalcogenide semiconductor photoabsorber materials because of their high optical absorption coefficient (10^5 cm^{-1}), tunable bandgap energy, long-term photoirradiation stability, and higher efficiency as compared to other absorber materials. CIGS thin-film solar cells have already exceeded 20% power conversion efficiency and are approaching the performance of silicon solar cells. A power conversion efficiency of 20.4% has been reported for CIGS-based thin-film solar cells, whereas solar devices based on potassium-doped CIGS have demonstrated an efficiency of 20.8%. Typical configuration of a CIGS-based solar device is glass substrate/Mo back contact/CIGS-based absorber layer/CdS or ZnS buffer layer/TCO window layer/antireflection coating layer (MgF_2).

The most critical process that governs the cost and performance of CIGS solar cells is the deposition of the photoabsorber layer. So far, the CIGS thin films with a suitable composition profile, phase purity, and grain structure have generally been deposited by the use of vacuum-based techniques like co-evaporation or sputtering followed by post-treatment operations like high-temperature annealing. However, these methods cannot be scaled up for mass production and require huge capital investment on vacuum and deposition equipment. Nonvacuum methods, in contrast, offer the possibility of faster, scaled-up, continuous, roll-to-roll manufacturing and a low initial capital cost. The scarcity of materials, such as In and Ga, has resulted in increased costs and has raised concerns about the availability of these metals for large-scale deployment of CIGS-based solar cells. Therefore, efficient processes making more judicious use of such materials are required to reduce the material costs and to overcome the barriers to commercialization. These methods rely on solution-based deposition of CIGS nanoparticles or molecular precursors at low temperature followed by post-treatment at higher temperature.

Nanomaterials for Solar Energy

$CuInS_2$, $CuInSe_2$, $CuInGaS_2$ (CIGS), and $CuInGaSe_2$ (CIGSe) are the leading I–III–VI$_2$ materials finding application in solar photovoltaics. During the past two decades, extensive research studies have been undertaken to develop controllable synthesis of I–III–VI$_2$ materials as nanocrystals. These nanocrystals are then used for low-cost, solution-based nonvacuum deposition of the photoabsorber layer for thin-film solar cells. Subsequent sections will deal with the synthesis of nanoparticles of these materials, which can be used as inks for fabrication of a photoabsorber layer in solar cells.

11.3.1 Copper Indium Disulfide ($CuInS_2$)

Copper indium disulfide ($CuInS_2$) is an important ternary chalcogenide material of the I–III–VI class of compound semiconductors, owing to its direct bandgap value of 1.5 eV closely corresponding with the solar spectrum, excellent photoirradiation stability, and high absorption coefficient ($>10^5$). $CuInS_2$ was regarded as the material of choice for thin-film solar photovoltaics (Yan et al. 2013b). Solar cells based on its quaternary selenium analogue $CuIn_xGa_{1-x}Se_2$ have already demonstrated submodule power conversion efficiency of 20.4% (Jackson et al. 2011). Theoretical calculations suggest power conversion efficiencies of 27%–32% for $CuInS_2$-based solar cells (Connor et al. 2009). However, recombination losses in the space charge region have limited the practical efficiencies to only 13% (Zhong et al. 2008). Other important applications of $CuInS_2$ nanostructures include in bioimaging (Liu et al. 2013c), for H_2 evolution from water (Tsuji et al. 2006), in light-emitting diodes (Chen et al. 2013b), and as counterelectrode material for dye-sensitized solar cells (Yao et al. 2013).

Nanoparticles of $AgInS_2$ and $CuInS_2$ have been synthesized by hot injection at 270°C using copper acetate, indium acetate, and silver acetate as the metal sources (Nadar et al. 2013). Dodecanethiol and elemental sulfur were used as sulfide precursors, whereas oleylamine was used as surfactant. It was demonstrated that the formation of the desired material is slow in dodecanethiol, probably due to its dual role as surfactant and sulfur source. Binary phases could be traced in the reaction products even after 4 h of reaction. In contrast, the samples synthesized using elemental sulfur had no/little traces of intermediate binary phases such as β-In_2S_3. Generalized phase-controlled synthesis of zinc blende and wurtzite $CuInS_2$, Cu_2SnS_3, and Cu_2ZnSnS_4 nanoparticles has been carried out using a wet chemical reaction (Chang and Waclawik 2013). Synthesis of $CuInS_2$ involved the use of copper iodide, indium acetate, and dodecanethiol in either octadecene or oleylamine. It was observed that the nature of the coordinating solvent and temperature control the evolution of the crystallographic phase of the nanoparticles. Triangular pyramidal nanocrystals of zinc blende phase with an average size of 9.3 ± 0.5 nm were formed when noncoordinating octadecene was used as solvent, whereas wurtzite $CuInS_2$ nanoparticles were obtained when octadecene was replaced by oleylamine. The optical bandgap of zinc blende and wurtzite $CuInS_2$ was found to be 1.39 and 1.50 eV, respectively. A similar effect of the ligand on the phase of nanoparticles was also observed in preparation of Cu_2SnS_3 and Cu_2ZnSnS_4 nanoparticles.

Chen et al. (2013a) have demonstrated the *in situ* growth of $CuInS_2$ nanocrystals onto nanoporous TiO_2 film by solvothermal treatment. They have thoroughly investigated the effect of the precursor's concentration on the morphology of as-grown nanostructures. The $CuInS_2$ film formed on the nonconductive glass side was removed by scraping, whereas the film on the nanoporous TiO_2 side was repeatedly washed with deionized water and absolute ethanol. Highly ordered potato chip–like arrays were formed with the 0.01 M and 0.03 M concentration of $InCl_3$ used in the reaction, whereas the formation of flower-shaped structures with an average diameter 3 μm was found to cover the entire FTO/compact-TiO_2/nanoporous-TiO_2 substrate when the concentration of $InCl_3$ was increased to 0.1 M. The pores of nanoporous-TiO_2 film were also filled by $CuInS_2$ nanoparticles. This later film was used to fabricate a heterojunction solar cell of configuration FTO/TiO_2/CIS/P_3HT/PEDOT:PSS/Au which yielded a power conversion efficiency of 1.4%. Further optimization of the $CuInS_2$ layer as well as cell structure might improve the efficiency of the cell.

Guo et al. (2013) have reported the preparation of inorganic ligand–capped $CuInS_2$ nanocrystals by a ligand exchange reaction using organic ligand (dodecanethiol and oleylamine)–capped $CuInS_2$ nanoparticles and $(NH_4)_2S$. These sulfide (S^{2-})-capped nanoparticles were subsequently used as aqueous ink to fabricate counterelectrodes by drop casting the ink onto cleaned FTO glass substrates. Thin films were obtained by drying at room temperature and sintering at 500°C for 30 min in an argon environment. These films were used as counterelectrode in dye-sensitized solar cells (DSSCs). The power conversion efficiency of the DSSCs significantly improved from 0.35% to 6.32% with the use of inorganic ligand–capped $CuInS_2$ nanocrystals as counterelectrode in the DSSC. A further improvement to 6.49% was observed when the nanocrystals were sintered at 500°C. This efficiency value is comparable to that of platinum (Pt) counterelectrodes and demonstrates a possibility of substituting costly Pt electrodes with low-cost $CuInS_2$ nanocrystals–based counterelectrodes.

Metastable wurtzite and zinc blende forms of $CuInS_2$ nanoparticles were synthesized by the one-pot reaction of copper-thiourea precursors having chloride, sulfate, and nitrate counterions and indium sulfate/indium acetate in ethylene glycol (Gusain et al. 2013). Optical bandgap of the wurtzite $CuInS_2$ nanoparticles was found to be 1.4 eV. Studies were carried out to determine the effect of incorporation of Ga^{3+} and Fe^{3+} ions into a $CuInS_2$ lattice. It was observed that the wurtzite structure was retained by replacing small amounts of In^{3+} by Ga^{3+} atoms. Higher loading of Ga^{3+} atoms into $CuInS_2$ results in the formation of mixed chalcopyrite and wurtzite phases. The bandgap value calculated from the UV-visible diffuse reflectance spectrum for the iron-doped samples exhibited a decrease from 1.40 to 1.05 eV.

The spray pyrolysis technique has been used to deposit polycrystalline thin films composed of $CuInS_2$ nanocrystals with the size range of 40–60 nm onto glass substrates with a bandgap of 1.55 eV. A solar device fabricated by using these films demonstrated a power conversion efficiency of 7.60% (Khan et al. 2013b). Wurtzite $CuInS_2$ nanowires have been synthesized by means of Ag_2S nanocrystals catalyzed growth in a solution-phase reaction (Li et al. 2013a). Typical synthesis was carried out by using diethyldithiocarbamate complexes of silver, copper, and indium. Optical and photoelectrical measurements of thin films prepared by drop casting as-synthesized nanowires showed promising photoresponse characteristics. Bandgap of the nanowires was calculated to be 1.5 eV. A similar solution-based approach has also been used for the synthesis of single-crystalline wurtzite ternary $CuInS_2$ and quaternary semiconductor $CuIn_xGa_{1-x}S_2$ nanoribbons (Li et al. 2013b). It was observed that $Cu_{1.75}S$ nanocrystals formed in the initial reaction stage serve as a catalyst for anisotropic growth of the nanoribbons. The optical bandgaps of the as-synthesized $CuIn_xGa_{1-x}S_2$ nanoribbons could be varied from 1.44 to 1.91 eV by varying the Ga concentration.

Monodispersed $CuInS_2$ nanopompons and hierarchical nanostructure have been synthesized by a solvothermal route using Cu_2O and $In(OH)_3$ as metal precursors, thioacetic acid as sulfur source, and ammonia (Liu et al. 2013d). These films were subsequently used as counterelectrodes of dye-sensitized solar cells. A power conversion efficiency of up to 4.8% demonstrated good catalytic activity of as-synthesized nanoflake films as counterelectrodes in DSSCs. Yang et al. (2013) have also used the same method using different concentrations of $CuSO_4 \cdot 5H_2O$ and $InCl_3 \cdot H_2O$ metal precursors and CH_3CSNH_2 as the sulfur source in ethanol as the solvent. SEM imaging revealed that the as-deposited films are composed of vertically aligned nanosheets. TEM images show that the nanosheets consist of a large number of crystal grains. Yao et al. (2013) have synthesized $CuInS_2$ nanocrystals capped by organic ligand dodecanethiol to prevent aggregations and to control the size and morphology. A complete phase transfer was observed by mixing and stirring the two solutions, thus yielding S^{2-}-capped $CuInS_2$ nanoparticles in the polar formamide phase. Thin films of both the organic-capped as well inorganic S^{2-}-capped $CuInS_2$ nanocrystals were prepared. These films were then used as counterelectrodes in DSSCs. It was observed that significant improvement occurs in the optoelectronic properties of $CuInS_2$ nanoparticles by replacement of organic surfactant with all inorganic S^{2-} ligand.

Highly luminescent $CuInS_2$ nanoparticles were synthesized by a noninjection method (Yu et al. 2013). Reaction conditions were optimized by using different copper and indium compounds as metal sources, varying molar ratios of Cu, In, and SDPP, and employing different solvents/capping ligands like octadecene, oleic acid, and oleylamine. The $CuInS_2$ nanocrystals showed high crystallinity with a mean diameter of 3.4 ± 0.4 nm.

The $CuInS_2$ particles were synthesized by using $Cu(NO_3)_2$ and $In(NO_3)_3$ as the metal precursors and $CS(NH_2)_2$ as the sulfur source in an aqueous solution at 160°C for 2 h (Yidong et al. 2013). The particles thus prepared were deposited as thin films onto the quartz substrate by a spin-coating process. Optical bandgap energy of the films was determined to be 1.40 eV. A heating-up synthesis approach using $Cu(acac)_2$, $In(acac)_3$, elemental sulfur, and octadecene was used (Dierick et al. 2014). TEM images showed that the as-synthesized chalcopyrite nanocrystals had a quasi-spherical morphology with a mean diameter of ~8 nm. No peaks from the binary phases were detected in the p-XRD pattern, and the bandgap of the nanocrystals was found to be 1.54 eV.

Deng et al. (2014) have recently reported a generalized strategy for the synthesis of a variety of ternary metal sulfides with controlled size, morphology, crystallographic phase, and stoichiometric composition. Various materials synthesized by this approach include orthorhombic Cu_3BiS_3 nanosheets and nanoparticles, orthorhombic $Cu_4Bi_4S_9$ nanowires and nanoribbons, wurtzite $CuInS_2$ nanopencils, cubic $AgBiS_2$ nanocubes, orthorhombic Ag_8SnS_6 nanoparticles, and orthorhombic Cu_3SnS_4 nanorods.

11.3.2 Copper Indium Diselenide ($CuInSe_2$)

$CuInSe_2$ nanowires were synthesized by a solution–liquid–solid (SLS) growth process catalyzed by gold–bismuth core–shell nanoparticles (Wooten et al. 2009). $CuInSe_2$ nanowires have also been grown by thermal decomposition of single-source precursor $[(PPh_3)_2Cu(\mu-SePh)_2In(SePh)_2]$ and by using multiple source precursors. It was observed that the morphology and stoichiometric composition of the nanowires obtained were strongly influenced by the chemical nature of the precursors Use of single-source precursor $[(PPh_3)_2Cu(\mu-SePh)_2In(SePh)_2]$ leads to the formation of single-crystalline, straight, and stoichiometric $CuInSe_2$ nanowires, whereas optimization of the reaction parameters was required to grow good-quality $CuInSe_2$ nanowires by using multiple source precursor systems.

Thin films of $CuIn(Se,S)_2$ have been fabricated by using a solution-processed, hydrazine-based approach (Liu et al. 2009). $CuInSe_2$ films were deposited onto various substrates including Mo-coated glass and thermally oxidized silicon wafers by spin coating without carrying out annealing in a toxic Se- or S-containing environment. $CuInSe_2$ films with good crystallinity could be obtained by simple heat treatment in an inert environment. It was possible to tailor bandgap of the absorber layer by varying the amount of sulfur in the film. Solar cells having configuration glass/Mo/CIS/CdS/i-ZnO/ITO were constructed using thus-prepared hydrazine-processed $CuInSe_2$ absorber layer, which demonstrated a power conversion efficiency of 12.2% under AM 1.5 illumination. Xu et al. (2010) have developed a facile chemical approach for preparation of morphology-controlled CuSe, $CuInSe_2$ nanowire, and $CuInSe_2/CuInS_2$ core–shell nanocable bundles. Cubic $Cu_{2-x}Se$ nanowire bundles were first synthesized and used as a self-sacrificial template for preparation of hexagonal CuSe nanowire with a diameter of about 10–15 nm and lengths up to hundreds of micrometers. It was observed that the smaller copper ions diffuse outward from the interior to the surface of nanowires and subsequently react with sulfur and indium ions to form a $CuInS_2$ shell over $CuInSe_2$ core. These nanocable bundles exhibited strong optical absorption in UV–vis region.

Ahn et al. (2012) have fabricated a $CuInSe_2$ particles–based absorber layer for solar cells using binary nanoparticles as precursors through a nonvacuum route. $CuInSe_2$ absorber thin films were obtained by selenization of the nanoparticle-coated film for 30 min. It was observed that nonuniform growth of the particles occurred, leaving large voids in the final films when these films are subjected to selenization. These voids are detrimental to the performance of the solar cell as they acted as short-circuiting paths in the solar cells. An additional solution-filling treatment was applied to the binary precursor film to mitigate this issue, and a conversion efficiency of up to 1.98% from a $CuInSe_2$ film selenized at 430°C was obtained.

Malik et al. (2011) have synthesized $CuInSe_2$, $CuGaSe_2$, and $CuIn_{1-x}Ga_xSe_2$ (CIGS) nanoparticles by thermolysis of the diisopropyldiselenophosphinatometal complexes $M_x[^iPr_2PSe_2]_n$ (M = Cu(I), In(III), Ga(III); n = 1, 3) in HDA/TOP at 120°C –210°C or 250°C. The diameters of the nanoparticles for $CuInSe_2$, $CuGaSe_2$, and $CuIn_{0.7}Ga_{0.3}Se_2$ were found to be 4.9 ± 0.6 nm (at 180°C), 13.5 ± 2.9 nm (at 250°C), and 14 ± 2.22 nm (at 250°C), respectively. They had previously reported the preparation of $CuInSe_2$ nanoparticles by a two-step reaction using CuCl, $InCl_3$, and TOPSe in TOPO as the coordinating solvent and ligand (Malik et al. 1999).

A simple heating-up of reactants approach has been described for high-yield synthesis of the chalcopyrite phase of quaternary $CuIn(S_{1-x}Se_x)_2$ nanoparticles (Chiang et al. 2011). Gradual heating of the reactants to a temperature of 265°C and maintaining this temperature led to the formation of $CuIn(S_{1-x}Se_x)_2$ nanoparticles with controllable S/Se ratios. Tuning of the bandgap energies of the nanocrystals from 0.98 to 1.46 eV was also feasible.

Stolle et al. (2012) have compared the photovoltaic response of the absorber layer based on $CuInSe_2$ nanocrystals capped with organic ligand (oleylamine) and inorganic ligands like metal chalcogenide–hydrazinium complexes (MCCs), S^{2-}, HS^-, and OH^-. Thin-film solar cells were fabricated with $CuInSe_2$ nanocrystals capped with different capping ligands used as absorber layer. It was observed that the PV device based on MCC ligand-capped $CuInSe_2$ nanocrystal demonstrated a power conversion efficiency of 1.7% under AM 1.5 illumination, while the PV device based on the oleylamine-capped $CuInSe_2$ nanocrystals exhibited 1.6% under similar conditions. Hsin et al. (2011) have demonstrated the feasibility for transformation of phase change material In_2Se_3 to $CuInSe_2$ through a solid-state reaction. This study therefore represented a unique practical process for transformation of the phase change material In_2Se_3 to the solar energy material $CuInSe_2$ and the preparation of the nanoheterostructures composed of $In_2Se_3/CuInSe_2$ for use in future nanodevices and solar cells.

A method based on microwave irradiation assisted chemical reaction for preparation of multiphase $CuInSe_2$ nanoparticles from copper acetate, indium acetate, and elemental selenium has been reported (Ahn et al. 2010). This method also featured formation of a metastable CuSe phase that enabled fabrication of a solution-processed, crack-free, crystalline, and high-performance $CuInSe_2$ absorber layer for a solar cell. Solar cells with configuration glass/Mo/CISe/CdS/i-ZnO/n-ZnO/Al were then fabricated by using either CuSe phase–free or CuSe phase–containing $CuInSe_2$ nanoparticles to assess the role of the CuSe phase on the preparation of a device-quality absorber layer. It was demonstrated that a $CuInSe_2$ absorber layer fabricated from CuSe containing $CuInSe_2$ multiphase nanoparticles showed a power conversion efficiency of 8.2%, while open-circuit voltage, short-circuit current density, and fill factor were measured as 0.44 V, 33.7 mA cm^{-2}, and 55%, respectively.

A simple method for colloidal preparation of highly luminescent $CuInSe_2$ nanocrystals through a silylamide-promoted reaction has been reported (Yarema et al. 2013). $CuInSe_2$ nanocrystals of average sizes ranging from 2.7 and 7.9 nm and small size distribution were prepared by suitably varying the reaction temperature and time.

A unique approach for activating the elemental selenium for use in the synthesis of metal selenide nanocrystals via the solvothermal method has been described (Zhang et al. 2014). Higher reactivity of Se^{2-} ions than zero-valent Se atoms also contributes toward improvement of reaction kinetics. Oleksak et al. (2014) have used a one-pot microwave-assisted solvothermal reaction for the synthesis of $CuInSe_2$ nanoparticles. The reaction was typically carried out by using a combination of precursors, which strongly absorb microwave energy, and low microwave–absorbing solvents like tri-n-octylphosphine (TOP) and oleic acid.

Lim et al. (2013) have deposited $CuInSe_2$ thin films from nanoparticle precursors synthesized by a solution-based colloidal approach. Absorber films with ca. 20% more density were obtained by cold-isostatic pressing (CIP). It was demonstrated that thus formed $CuInSe_2$ thin films had improved microstructure, lower porosity, a more uniform surface morphology, and a relatively thinner $MoSe_2$ layer. A significant increase in photovoltaic performance of the solar cells fabricated by using these films was observed. A threefold increase in the average efficiency, from 3.0% to 8.2%, was also demonstrated.

$CuInSe_2$ nanoparticles have been synthesized by using a two-step, continuous-flow, solar microreactor (Kreider et al. 2014). Radiative heat from simulated, concentrated solar radiation was used as a faster heating source, which served to reduce the reaction duration. Synthesis of both chalcopyrite and sphalerite phases of $CuInSe_2$ nanoparticles was carried out by suitably altering the nucleation temperature and residence time in the solar microreactor. It was observed that the formation of the chalcopyrite phase of $CuInSe_2$ nanoparticles is favored by the higher nucleation temperatures and longer residence times. A continuous method has also been reported for synthesis of $CuInSe_2$ nanoparticles in a microtubular reactor (Kim et al. 2014). Monodispersed colloidal $CuInSe_2$ nanoparticles thus formed were used as nanocrystal ink for deposition of $CuInSe_2$ thin films. The shape of the nanocrystals gradually transformed from spherical to hexagonal to trigonal with increasing In or Se content. $CuInSe_2$ nanoparticles synthesized at a high temperature possessed trigonal morphologies with chalcopyrite crystallographic structures. Utilization of these inks for solar photovoltaics was verified by fabricating a lab-scale device with 1.9% efficiency under AM 1.5 G illumination.

A hybrid ink has been prepared from copper-rich CuSe nanoparticles and an indium precursor solution to form $CuInSe_2$ thin films for solar cell applications (Cho et al. 2014). PV devices fabricated by using these films demonstrated a power conversion efficiency of 5.04% as compared to an efficiency of 1.04% for normally synthesized copper-rich CuSe nanoparticles. This observation confirmed that the Cu-MEA complex had a strong influence on the performance of $CuInSe_2$-based solar cells produced with the hybrid ink process.

11.3.3 Copper Indium Gallium Disulfide (CIGS)

Pan et al. (2009) have reported the synthesis of quaternary $Cu_{1.0}Ga_xIn_{2-x}S_{3.5}$ and $Cu_{1.0}In_xTl_{2-x}S_{3.5}$ nanocrystals by hot-injection approach. TEM images showed nearly monodispersed $Cu_{1.0}Ga_xIn_{2-x}S_{3.5}$ nanocrystals with an average diameter of 6.2 nm. Bandgap of alloyed $Cu_{1.0}Ga_xIn_{2-x}S_{3.5}$ nanocrystals, as determined by UV-vis absorption spectroscopy, could be tailored in the range of 1.43 to 2.42 eV depending upon the composition.

$CuInGaS_2$ (CIGS) nano-ink has been prepared using the hot-injection approach (Singh et al. 2012). CIGS nanocrystals were obtained by precipitation with ethanol and centrifugation at 10,000 rpm for 20 min. Nano-ink prepared by dispersing CIGS nanocrystals in toluene is used to fabricate thin films by drop casting onto glass substrates. Optical bandgaps of $CuInS_2$ thin films were found to be 1.57 eV, which was slightly higher than bulk $CuInS_2$ bandgap because of the small size of nanocrystals (~5 nm). CIGS thin films exhibited a bandgap of 1.65 eV as a result of Ga substitution.

An et al. (2014) have reported a simple strategy based on paste coating and sulfurization for fabrication of CIGS thin films for photovoltaic applications. The films prepared using this paste demonstrated good photoelectrical characteristics; however, multiple coating and drying cycles were required to deposit thin films of the desired thickness. Another paste was prepared by using ethanol as solvent and ethyl cellulose as organic binder. Solar devices with configuration Al/Ni/ZnO:Al/i-ZnO/CdS/CIGS/Mo-coated glass were fabricated to compare the performance of both pastes. The solar cell fabricated by using the CIGS thin films deposited from these pastes showed an improved efficiency (4.66%) over those fabricated using CIGS thin-film deposited from thicker paste only (2.90%).

Deposition of CIGSSe absorber layers by electrospraying a propylene glycol solution of copper(II), In(III), and Ga(III) nitrates as metal precursors has been reported (Yoon et al. 2014). The bandgap energy of the absorber layer was found to be 1.15 eV. Solar cells with Mo/CIGSSe/CdS/i-ZnO/n-ZnO/Ni/Al structure have been fabricated. The solar cell demonstrated a power conversion efficiency of 4.63% with an open-circuit voltage of 410 mV, a short-circuit current of 21 mA cm^{-2}, and a fill factor of 0.5337 for an active area of 0.46 cm^2.

Harvey et al. (2013) have reported the colloidal preparation of CIGS nanoparticles with a stoichiometric composition of $Cu_{0.8}In_{0.7}Ga_{0.3}Se_2$. The nanocrystals were used as ink for the deposition of thin films onto Mo-coated glass using the spray deposition process. More uniform and thicker CIGSe films were obtained by multiple cycles of ink deposition and sintering. A power conversion efficiency, as high as 7%, has been achieved by optimal use of this multiple deposition and sintering approach.

Coughlan et al. (2013) have carried out an extensive study on the synthesis and shape evolution of $CuIn_xGa_{1-x}S_2$ nanocrystals and nanorods. These nanocrystals form ternary chalcopyrite $CuInS_2$ nanoparticles, whereas incorporation of Ga into the lattice formed quaternary CIGS nanoparticles. Nanorods with a controllable aspect ratio from 1.8 to 3.3 have been synthesized using these aliphatic amines.

By varying the initial amount of metal precursors, a full range of stoichiometric compositions from $CuGaS_2$ to $CuInS_2$, through intermediate $CuIn_xGa_{1-x}S_2$ with tunable x were synthesized. The bandgap of the as-synthesized nanocrystals significantly varied with the composition, and an increase in bandgap energy from 1.48 to 2.2 eV was observed with increasing gallium composition. Furthermore, the average size of the nanoparticles could be varied from 13 to 19 nm by employing different reaction temperatures ranging from 230°C to 290°C. A ligand exchange reaction carried out to replace organic capping ligand with the inorganic S^{2-} capping ligand resulted in only 50% exchange. Thin films prepared by using these partially ligand exchanged nanocrystals exhibited high conductivity.

A solution-based approach has been used for the synthesis of wurtzite phase $CuIn_xGa_{1-x}S_2$ nanocrystals with the value of x ranging from 0 to 1 (Wang et al. 2011). As-synthesized $CuInS_2$ nanocrystals had a black color, whereas $CuIn_{0.5}Ga_{0.5}S_2$ and $CuGaS_2$ nanocrystals had dark red and dark yellow colors, respectively. The TEM images revealed the formation of bullet-like $CuInS_2$ nanocrystals having a uniform size with 16 nm width and about 35 nm length. The morphology changed to rod-like for the CIGS composition $CuIn_{0.75}Ga_{0.25}S_2$ with a shorter length (Figure 11.1). In general, the length of nanorods decreased with an increase in Ga concentration. When octadecene was used as solvent, the morphology changed from nanobullets to nanorods, nanospheres, and nanotadpoles. X-ray diffraction studies showed that the nanocrystals had a wurtzite crystallographic structure. The bandgap of as-synthesized $CuIn_xGa_{1-x}S_2$ nanocrystals increased with increasing Ga composition from 1.53 eV for $CuInS_2$ to 2.48 eV for $CuGaS_2$. Due to the relatively large size of nanocrystals, no broadening of bandgap due to quantum confinement was observed.

11.3.4 Copper Indium Gallium Diselenide (CIGSe)

Tang et al. (2008) have reported the synthesis of monodispersed, high-quality $CuGaSe_2$, $CuInSe_2$, and $Cu(InGa)Se_2$ nanoparticles. It was demonstrated that the nanoparticle size can be tuned by carefully choosing the reaction temperature, whereas the desired stoichiometric composition can be attained by suitably adjusting the precursor concentrations. As-grown $CuGaSe_2$ and $CuInSe_2$ nanocrystals had bandgap energy of 1.68 and 1.01 eV, respectively.

Panthani et al. (2008) introduced the concept of using nanoparticles-based ink for fabrication of an absorber layer in solar cells. They have reported the synthesis of chalcopyrite $CuInS_2$ and $CuIn_xGa_{1-x}Se_2$ (CIGSe) by arrested precipitation in solution. As-synthesized CIGSe nanocrystals had a mean diameter ranging from 5 to 25 nm as revealed by TEM analyses. It was demonstrated that the relative atomic ratio (In/Ga) in the CIGS nanocrystals could be controlled by suitably varying the ratio of In/Ga precursors used in the reaction. The bandgap of CIGSe nanocrystals showed significant variation with change in stoichiometric composition of the nanocrystals. Uniform, crack-free films of micrometer thickness were fabricated by using as-prepared nanocrystal ink. Prototype photovoltaic devices were constructed to demonstrate a proof of concept for a reproducible photovoltaic response.

Nanomaterials for Solar Energy

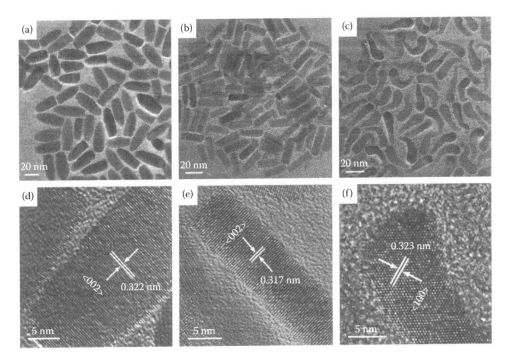

FIGURE 11.1 TEM images of (a) bullet-like $CuInS_2$ nanocrystals synthesized in OLA, (b) rod-like $CuIn_{0.75}Ga_{0.25}S_2$ nanocrystals synthesized in ODE, and (c) tadpole-like $CuGaS_2$ nanocrystals synthesized in ODE; HRTEM images of individual (d) bullet-like, (e) rod-like, and (f) tadpole-like nanocrystals. (Adapted from Wang, Y.-H. A., X. Zhang, N. Bao, B. Lin, and A. Gupta, *J. Am. Chem. Soc.*, 133, 11072–11075, 2011. With permission.)

Preparation of $CuIn_{1-x}Ga_x(S_{1-y}Se_y)_2$ nanocrystals with controllable stoichiometric composition (x, y ranging from 0 to 1) and graded bandgaps has been reported (Chang et al. 2011). Stoichiometric composition of the $CuIn_{1-x}Ga_x(S_{1-y}Se_y)_2$ nanocrystals was successfully controlled by using variable ratios of In and Ga as well as S and Se reactants in the reaction. It was observed that tuning of optical bandgap from 0.98 to 2.40 eV was possible by suitably varying the composition of CIGSSe nanocrystals.

Jiang et al. (2012c) have used colloidal nanocrystals capped with metal chalcogenide complexes as soluble precursors for $CuInSe_2$, CIGSe, and CZTS materials for solar cell applications. TEM and p-XRD studies demonstrated that the ligand exchange reaction proceeds without altering the phase and morphology of $Cu_{2-x}Se$ nanocrystals. Thin films deposited from metal chalcogenide complex of $Cu_{2-x}Se$ by spin coating or spray coating were subsequently annealed at 500°C, and solid-state reactions resulted in complete transformation of $In_2Se_4^{2-}$-capped $Cu_{2-x}Se$ nanocrystals into $CuInSe_2$ thin films without any cracks. The bandgap of the as-deposited $CuInSe_2$ thin films was found to be 1.01 eV. This approach has also been used to deposit high-quality semiconductor thin films of CIGSe and CZTS.

A new versatile solution-based strategy for *in situ* preparation of metal–organic molecular precursors based on butyldithiocarbamic acid has been reported (Wang et al. 2012). Carbon disulfide and 1-butylamine undergo a reaction producing butyldithiocarbamic acid, which reacts with a variety of metal oxides and hydroxides to form thermally degradable metal–organic precursors. This approach has been used for the fabrication of a CIGSSe-based absorber layer in thin-film solar cells. SEM images showed compact and dense film without any noticeable cracks. Solar cells fabricated with the configuration glass/Mo/CIGSSe/CdS/i-ZnO/ITO/Al demonstrated a power conversion efficiency of 8.8%. It was suggested that optimization of film thickness and bandgap can further improve performance of the solar cells.

Ahmadi et al. (2012) have carried out a systematic study to investigate the growth mechanism and evolution pathways of quaternary $CuIn_{0.5}Ga_{0.5}Se_2$ nanocrystals in a hot coordinating solvent. A mixture of binary $Cu_{2-x}Se$ and ternary $CuGaSe_2$ nanocrystals is formed. After 40 min of reaction, a small amount of CIGSe was observed to form by reaction of $Cu_{2-x}Se$, $CuGaSe_2$ nanocrystals and amorphous In_2Se_3 nuclei. At 60 min, a biphasic $CuIn_xGa_{1-x}Se_2$ mixture with variable x values was formed which underwent subsequent transformation to monophasic $CuIn_{0.5}Ga_{0.5}Se_2$ after another 60 min.

Synthesis of $CuIn_{0.7}Ga_{0.3}Se_2$ (CIGSe) nanopowders has been carried out by microwave irradiation (Seelaboyina et al. 2013). CIGSe nanopowders thus synthesized had particle size ranging from 20 to 80 nm, whereas the bandgap was found to be 1.1 eV. CIGSe nanopowders were obtained by centrifugation at 5000–10,000 rpm, which was dried at 100°C–200°C for 4–8 h. Finally, CIGSe ink was prepared by ball milling the powder in isopropanol and 1,2-propanediol for 2–10 h. X-ray diffraction studies showed that the CIGSe nanopowders had a chalcopyrite crystallographic structure corresponding to standard ICDD pattern 035-1102 for $CuIn_{0.7}Ga_{0.3}Se_2$. SEM, TEM, and dynamic light scattering (DLS) results showed particle size to be less than 100 nm, whereas elemental mapping confirmed uniform distribution of constituent elements throughout the powder particles.

A process for nonvacuum deposition of CIGSe absorber layers from hydroxide-containing particles and subsequent selenization by Se vapors has been reported (Uhl et al. 2013). Another type of CIGSe thin film was prepared by sequential deposition of multiple layers by varying precursor inks (In + Ga/Cu/In + Ga) and repeating the drying and deposition steps. These films were also selenized at 550°C. Any residual Cu–Se phases were removed by etching of the selenized films with KCN. CIGSe absorber formed from deposition of the inks mixture was mainly porous; however, sequential deposition of inks resulted in improved sintering properties. *In situ* XRD measurements showed that $CuSe_2$, $CuSe$, and $Cu_{2-x}Se$ are formed as binary phases at 220°C to 350°C in the stacked film, while only $Cu_{2-x}Se$ was detected from 270°C to 440°C in the case of mixed precursor ink deposition. Power conversion efficiencies up to 4.8% and 5.8% were achieved with the absorber layer deposited from mixed inks and stacked deposition, respectively, despite the porosity and large compositional gradients in thin films.

Synthesis of $CuIn_{0.7}Ga_{0.3}Se_2$ nanocrystals by a facile heating-up process using metal chloride salts, selenium, octadecene, and oleylamine has been reported (Hsu et al. 2013b). It was observed that with the supersaturation, formation of CuSe nanocrystals occurs first, whereas amorphous indium selenide and gallium selenide nuclei are later formed on the surfaces of CuSe nanocrystallites at the lower temperature and shorter reaction time. Finally, the amorphous indium selenide and gallium selenide nanocrystals react with CuSe crystallites at high temperature (~250°C), yielding thermodynamically stable chalcopyrite CIGSe nanocrystallites.

Li et al. (2013c) have reported the synthesis of ternary and quaternary $CuIn_{1-x}Ga_xSe_2$ nanocrystals by an organo-alkali–assisted diethylene–glycol solution-based approach. Monophasic, well-dispersed, and granular chalcopyrite nanocrystals with controlled stoichiometry ($0 \leq x \leq 1$) and a size ranging from 10 to 20 nm were synthesized using the hot-injection approach. The best results were obtained by using triethylenetetramine with the additional range of 2–4 vol% into a diethylene glycol–based solution. The In/Ga atomic ratios of the as-synthesized $CuIn_{1-x}Ga_xSe_2$ nanocrystals were very similar to those in the feeding precursor solutions in the case of TETA-assisted diethylene glycol synthesis. Tuning of the optical bandgap from 1.05 to 1.7 eV was feasible by controlled variation of the In/Ga ratio in $CuIn_{1-x}Ga_xSe_2$ nanocrystals.

A microwave-assisted solvothermal synthetic method has been used for the preparation of multiphase CIGSe nanoparticles (Seo et al. 2014). By using this approach, composition-controlled synthesis of CIGSe materials with Cu to (In + Ga) atomic ratio varying from 0.6 to 1.0 has been carried out, and dense absorber layers have been fabricated. Finally, a photovoltaic device based on an as-deposited CIGSe absorber layer was fabricated, and its performance parameters were investigated.

A ligand exchange method has been used where readily synthesized $CuInSe_2$ and $CuIn_{1-x}Ga_xSe_2$ (CIGSe) nanocrystals capped with organic oleylamine are converted to 1-ethyl-5-thiotetrazole–capped nanocrystals (Lauth et al. 2014). The ligand exchange reaction proceeds with no deleterious effect on the properties of nanoparticles, and colloidal stability of the chalcopyrite materials is preserved in the process. Measurement of the current–voltage characteristics of thus-obtained nanocrystal films before and after thermolysis of ligand were carried out in the dark and under illumination. It was observed that the conductivity of trigonal pyramidal $CuInSe_2$ nanocrystals demonstrated an increase by four orders of magnitude for ligand-free nanocrystal films. Similarly, a two orders of magnitude improvement in the photoconductivity of the CIGS nanocrystal film was observed.

A one-pot solution-based approach has been used for the synthesis of monophasic, chalcopyrite $CuIn_{1-x}Ga_xSe_2$ nanoparticles with controlled stoichiometric composition ($0 \leq x \leq 1$) (Han et al. 2014). The effects of the Ga/In + Ga atomic ratio were determined on crystallographic structure, shape, and optical properties of the as-prepared $CuIn_{1-x}Ga_xSe_2$ nanoparticles. The SEM images revealed polydispersed size and shape of the nanoparticles, and nanoparticles with higher Ga content showed a pronounced tendency toward agglomeration. Bandgap energy of the as-synthesized nanocrystals exhibited an increase from 1.00 to 1.68 eV with increasing Ga/In + Ga atomic ratio.

Roux et al. (2013) have developed a new ink-based process for fabrication of an efficient chalcopyrite $CuIn_{1-x}Ga_xSe_2$-based absorber for use in thin-film solar cells. Inks prepared by using as-synthesized nanoparticles could be safely coated under ambient conditions to form thin films by commercially available techniques like doctor blading. Annealing of the precursor thin films under Se vapors in a primary vacuum formed a functional absorber layer for thin-film solar cells. $CuIn_{1-x}Ga_xSe_2$ thin films formed by employing this simple, two-step process exhibited strong mechanical adhesion. Solar cells fabricated using these absorbers demonstrated a power conversion efficiency higher than 7%.

11.4 COPPER-ZINC-TIN CHALCOGENIDES (CZTSSe)

Considerable research attention has been given toward $Cu_2ZnSn(S,Se)_4$ (CZTSSe) system due to the potential of these materials for the production of low-cost solar cells with good power conversion efficiency (PCE). This material has a bandgap ranging from 1.0 to 1.5 eV, with high optical absorption ($>10^4$ cm) and is based on low-cost, nontoxic, and Earth-abundant elements. Therefore, it has emerged as a promising material that can lead to both economically and ecologically sustainable production of solar cells without any issue of scarcity of constituent elements. PCE of CZTSSe-based thin films has already reached 12% during the last few years. Metal chalcogenide–based materials can tolerate a higher degree of either structural or electronic defects and, hence, can be deposited in a cost-effective manner by employing non-vacuum-based deposition approaches. In the case of CZTSSe, non-vacuum-based solution processing has provided device performances that are even better than those achieved through vacuum-based approaches. Processing of the CZTSSe must be controlled to form a pure kesterite phase only, and this can be achieved only within a limited stoichiometric range with [Cu/(Zn + Sn)] = 0.8–0.9 and Zn/Sn = 1.1–1.4. Deviation from this stoichiometry leads to the formation of binary phases (Cu, Sn, and Zn chalcogenides) and/or ternary phases (copper tin chalcogenides). Due to the volatile nature of constituent elements and binary phases, phase control becomes somewhat difficult, especially in deposition techniques involving ultrahigh vacuum and high temperatures. Furthermore, in solution-based approaches, phase formation occurs in the liquid phase, thus allowing better diffusivity for constituting elements in order to overcome the barriers of activation energy and form a thermodynamically stable phase. Therefore, solution-based approaches providing targeted stoichiometric composition at relatively lower temperatures have demonstrated better control over the formation of monophasic CZTSSe. Device performance is also dependent on defect properties of the CZTSSe. A dependable synthetic strategy is vital for controlling the defect profile of the final photoabsorber film, and nanoink-based approaches provide a facile way of incorporating extrinsic dopants for the optimization of defect structure and passivation of grain boundaries.

The first step in a CZTSSe nanocrystals–based approach is obviously the synthesis of high-quality kesteritic nanocrystals. These nanocrystals along with additives are then deposited as thin films by a variety of commercially available methods like spin coating, doctor blading, inkjet printing, and so on. The final step involves post-treatment procedures like annealing in a sulfur/selenium environment. This section will mainly focus on various synthetic strategies for obtaining high-quality CZTSSe nanocrystals for use as inks in solar devices.

The first-ever synthesis of Cu_2ZnSnS_4 nanocrystals and their use in the fabrication of solar cells was reported by Guo et al. (2009). The resulting CZTS nanocrystals had sizes ranging from 15 to 25 nm and a bandgap of 1.5 eV. These nanocrystals were drop coated onto Mo-coated soda lime glass substrates and then selenized at 500°C to form $Cu_2ZnSnS_ySe_{1-y}$ (CZTSSe). Photovoltaic devices fabricated using these CZTSSe films demonstrated a power conversion efficiency of 0.74% under AM 1.5 illumination. Riha et al. (2009) carried out synthesis of homogeneous and nearly monodispersed CZTS nanocrystals by hot-injection approach. The mean diameter of the thus-prepared nanocrystals was found to be 12.8 ± 1.2 nm, and they had a bandgap of 1.5 eV. Using the same method, Guo et al. (2010) synthesized CZTS nanocrystals with copper-poor and zinc-rich stoichiometry. Hexanthiol-dispersed nanocrystals were deposited as 1 μm–thick film by knife coating onto Mo-coated soda lime glass. The films after post-treatment processes had a Cu/(Zn + Sn) = 0.79 and Zn/Sn = 1.11 composition. Solar cells fabricated by using these films showed 7.2% power conversion efficiency under one sun illumination.

Riha et al. (2011) extended their work to the compositionally controlled synthesis of phase-pure $Cu_2ZnSn(S_{1-x}Se_x)_4$ nanocrystals (with $0 \leq x \leq 1$). Compositional control of the chalcogenide ratios was obtained by sonicating the targeted amounts of S and Se along with sodium borohydride and oleylamine to balance S and Se reactivities. A probe into lattice parameters and bandgap energies of the composition tunable CZTSSe was carried out to investigate the effect of Se on material properties of finally deposited film. With an increase in the amount of Se, lattice parameters showed an increase, bandgap was slightly decreased, and the electrical conductivity of the nanocrystals increased due to more pronounced grain growth and passivation of grain boundaries.

A facile colloidal synthesis of CZTS nanocrystals was carried out using metal dithiocarbamic acid salt $[Cu_2ZnSn(S_2CNEt_2)_{10}]$ as the precursor, oleylamine as the activation agent, and oleic acid as the capping agent, respectively (Zou et al. 2011). EDX analyses showed that the CZTS nanocrystals had a sulfur-poor stoichiometric composition; ratio of Cu:Zn:Sn:S being 26:14:18:42 precisely. Oleylamine balanced the reactivity of cationic constituents, while oleic acid bound to the surface of growing CZTS nanocrystals.

Guo et al. (2012) also prepared CZTGeSSe nanocrystals by partial substitution of tin precursor with germanium precursor. The bandgap energy of the CZTGeSSe nanocrystals was found to be 1.09 eV. The device fabricated using these nanocrystals as the absorber layer demonstrated 8.4% PCE without the use of any antireflection coating.

Synthesis of CZTS nanocrystals with the average nanocrystal diameter ranging from 2 to 7 nm was carried out by heating a stoichiometric mixture of copper, zinc, and tin diethyl dithiocarbamate complexes (Khare et al. 2011). It was observed that the size of the nanocrystals could be tuned by varying the growth temperature and the amount of oleylamine used, whereas the reaction duration had no effect on the size of the nanocrystals. Nanocrystals with average diameters of 2.0 and 2.5 nm showed a significant shift in their optical absorption spectra. Similarly, colloidal preparation of CZTS nanocrystals using copper ethylxanthate, zinc ethylxanthate, and tin(IV) chloride has been reported by Liu et al. (2011). CZTS nanocrystals had a mean diameter of 15.6 ± 2.0 nm and showed a tendency to grow larger with increased reaction durations. Stoichiometric composition of the nanocrystals was determined by EDX analyses, which showed Cu:Zn:Sn:S ratios to be 1.81:1.17:0.95:4.07. A bandgap of the nanocrystals estimated from their optical absorption spectra was found to be 1.5 eV.

Good-quality monodispersed CZTS nanoparticles having thermodynamically stable kesterite and wurtzite phases have been synthesized via a simple, one-pot, low-cost solution method (Cattley et al. 2013). The nanoparticles prepared at 140°C had the tetragonal CZTS crystallographic structure, whereas those grown at higher temperature had a wurtzite crystallographic phase as revealed by high-resolution transmission electron microscopy (HRTEM), selected area electron diffraction (SAED), and XRD analyses.

Spindle-shaped CZTS nanoparticles of the kesterite phase have been prepared by hot-injection approach using corresponding metal chlorides as metal precursors and thiourea as the sulfur source (Wei et al. 2012). The nanoparticles had an average length of 22.5 ± 2.0 nm and average width of 13.9 ± 1.5 nm. The bandgap of the nanocrystals was found to be 1.54 eV. Formation of spindle-like morphology and narrow size distribution was attributed to the gradual release of sulfur from thiourea into solution, resulting in homogeneous nucleation and growth.

A hot-injection approach involving the injection of concentrated metal precursors (copper(II) acetylacetonate, zinc acetate, and tin(IV) acetate) and elemental sulfur solution in oleylamine into preheated triphenylphosphate (TPP) as capping ligand has been reported (Kim et al. 2013). It was observed that the TPP ligand surrounding the CZTS nanocrystals decomposes more readily than conventionally employed capping ligands such as oleylamine, thus giving a better absorber layer for solar cell applications. As-synthesized CZTS nanocrystals had an average diameter of 14.5 ± 4.6 nm, whereas the bandgap was calculated as 1.5 eV. Photovoltaic performance of the CZTS absorber layer was determined by fabricating a solar device with a configuration Mo-coated SLG/CZTS/CdS/i-ZnO/indium tin oxide (ITO)/patterned Ni/Al grid. The device showed a power conversion efficiency of 3.6% without any Se treatment.

Cao et al. (2013) have reported the synthesis of CZTSSe nanocrystals through hot-injection approach using copper(II) acetate, zinc acetylacetonate, tin(IV) acetate, elemental sulfur, selenium, and oleylamine. $Cu_2ZnSn(S_{1-x}Se_x)_4$ nanocrystals ($x = 0$, 0.2, 0.5, 0.85, and 1.0) were prepared and used as counterelectrodes in quantum dot–sensitized solar cells (QDSSCs). It was observed that all of the QDSSCs using CZTSSe as counterelectrodes demonstrated superior current–voltage characteristics than the Pt-based QDSSCs, regardless of the selenium amount in CZTSSe. The QDSSC with CZTSSe ($x = 0.5$) counterelectrode showed the highest energy conversion efficiency of 3.01%, which was higher than that (1.24%) obtained using platinum counterelectrode.

The CZTS nanocrystals have been prepared by the hot-injection method using corresponding diethyldithiocarbamate complexes of copper, zinc, and tin in oleylamine at 280°C for 10 min (Chernomordik et al. 2014a). Thin films based on nanocrystals were prepared by drop casting the colloidal CZTS suspension onto soda lime glass and quartz substrates. The effect of various parameters like annealing temperature, annealing time, and sulfur vapor pressure on the evolution of microstructure of CZTS nanocrystal films was thoroughly investigated.

Guan et al. (2014a) have reported the synthesis of CZTS nanocrystals by a hot-injection method. It was observed that crystalline CZTS and Cu_2SnS_3 and amorphous ZnS were formed after injection of sulfur–oleylamine into the mixture solution. The reaction of ZnS with Cu_2SnS_3 forms CZTS during sulfur annealing at a high temperature. Therefore, in case of a much higher Zn/Sn ratio, the excess ZnS cannot be eliminated and appears as an impurity phase. When the Zn/Sn precursor ratio was reduced to 0.6, a relatively pure CZTS phase was obtained.

Synthesis of CZTSe nanoparticles by the hot-injection approach using bis-(triethylsilyl)selenide $[(Et_3Si)_2Se]$ as the selenium source has been reported (Jin et al. 2014). Near stoichiometric and monophasic nanocrystals with average size ranging from 25 to 30 nm were prepared. The presence of secondary and ternary phases in the material was ruled out by XPS and Raman spectroscopy measurements. The prepared CZTSe nanocrystals had an optical bandgap of 1.59 eV.

Steinhagen et al. (2009) have synthesized the CZTS nanocrystals through a high-temperature arrested precipitation approach using oleylamine (OLA) as the coordinating solvent. EDX showed average composition of the as-prepared nanocrystals to be $Cu_{2.08}Zn_{1.01}Sn_{1.20}S_{3.70}$, and average diameter of the crystals was found to be 10.6 ± 2.9 nm. The bandgap of the nanocrystals was calculated to be 1.3 eV. Solar PV devices were fabricated by deposition of CZTS thin films through spray coating the toluene dispersion of nanocrystals. No annealing or other post-treatment steps were performed. Thus, fabricated devices with Au/CZTS/CdS/ZnO/indium tin oxide (ITO) configuration showed 0.23% PCE under AM 1.5 illumination.

Thermal reactions of metal acetate and elemental sulfur in hot oleylamine were used to prepare CZTS nanoparticles (Kameyama et al. 2010). It was observed that the nanocrystals deposited at 240°C or higher temperature have pure kesterite phase, whereas those deposited at 180°C or below showed broad peaks originating from the CuS phase. CZTS nanoparticles had an average size of 5.6 nm, and bandgap of the nanocrystals was determined to be 1.5 eV.

Due to improved device quality by partial replacement of indium in $CuInSe_2$ with a lower atomic number gallium to form CIGSe, tin atoms in CZTS were partially replaced by smaller group IV element germanium to give $Cu_2Zn(Sn_{1-x}Ge_x)S_4$ (Ford et al. 2011). Alloying germanium in CZTS nanocrystals opened up a new window for optimizing the bandgap by suitably adjusting the Ge/(Sn + Ge) ratio. CZTGSSe nanocrystals thus synthesized were polydispersed ranging in size from 5 to 30 nm (Figure 11.2). The bandgap of the as-synthesized nanocrystals was found to increase with increasing Ge content in the CZTGSSe lattice. The solar cells with SLG/Mo/CZGSSe/CdS/i-ZnO/ITO/Ni-Al architecture showed a power conversion efficiency 6.8% for CZTGS nanocrystals synthesized with Ge/(Ge + Sn), Cu/(Zn + Sn + Ge), and Zn/(Sn + Ge) ratios of 0.7, 0.8, and 1.2, respectively.

Phase-pure wurtzite CZTS nanocrystals were synthesized by Li et al. (2012a). It was found that the use of dodecanethiol alone in the reaction leads to the formation of coexisting wurtzite and kesterite phases. However, when some amount of oleylamine was added, the reaction environment was changed, and only the wurtzite phase was obtained. Rice-like CZTS nanocrystals had a mean diameter ranging from 10 to 40 nm depending on the reaction duration. The preparation of wurtzite CZTS nanocrystals based on arrested precipitation has been reported, which utilizes acetates of copper, zinc, and tin as a metal source—diethanolamine as solvent and thiourea as sulfur source (Li et al. 2012b). It was reported that the initial reaction at 160°C and the amount of thiourea are key factors controlling the selective growth of CZTS nanocrystals as the wurtzite phase. Nanocrystals thus formed had a mean diameter of 10 ± 1.1 nm with a bandgap of 1.56 eV.

The synthesis of CZTS and CZTSe nanoparticles has also been carried out by using metal halide salts CuI, $ZnCl_2$, and SnI_4 and sulfur/selenium (Rath et al. 2012). CZTS nanocrystals grown at

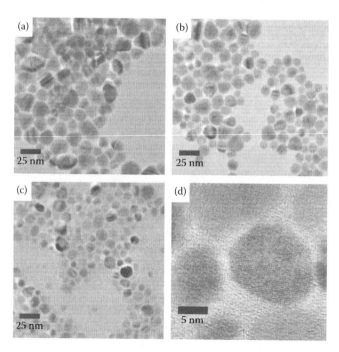

FIGURE 11.2 TEM images of (a) CZTS, (b) CZTGS, and (c) CZGS. (d) HR-TEM of CZGS nanocrystals. (Adapted from Ford, G. M., Q. Guo, R. Agrawal, and H. W. Hillhouse, *Chem. Mater.*, 23, 2626–2629, 2011. With permission.)

210°C had an average size ranging from 7 to 10 nm. The EDX analysis showed near stoichiometric composition for CZTS nanoparticles, whereas the Se analogue prepared using the same methodology had no zinc and was found to be Cu_2SnSe_3. Yang et al. (2012) have reported the synthesis of CZTS nanocrystals and their utilization as a potential thermoelectric material. The as-prepared CZTS nanocrystals had a mean diameter of 10.6 ± 1.9 nm, and bandgap was found to be 1.51 eV. The nanocrystals were then compressed into compact pellets by spark plasma sintering (SPS) and hot press. A significant enhancement in Seebeck coefficient and decrease in thermal conductivity in comparison with bulk crystals was observed in electrical and thermal measurements between 300 and 700 K. It was also observed that doping of CZTS nanocrystals with extra copper significantly increased the electrical conductivity and decreased the thermal conductivity.

Chesman et al. (2013a) have reported a gram-scale noninjection synthesis of CZTS nanocrystals. Nanocrystals were grown using CuI, $SnCl_4 \cdot 5H_2O$, zinc ethyl xanthate, and dodecanethiol in addition to oleylamine at 250°C for 30 min. It was observed that due to different decomposition temperatures of sulfur precursors. Morphology of the nanocrystals was predominantly triangular, and the mean size was in the range of 5.5–7.8 nm. Optical bandgap of CZTS nanocrystals ranged from 1.45 to 1.55 eV. This one-pot, noninjection synthetic strategy was extended by using CS_2 to form oleyldithiocarbamate and dodecyltrithiocarbonate *in situ* as the sulfur source (Chesman et al. 2013b). It was observed that the size of the nanocrystals was 1.6 ± 0.4 nm at 138°C and it grew to 5.0 ± 0.1 in 10 min period after decomposition of dodecanethiol at 250°C. After 30 min reaction at 250°C, the average size of the crystallites was recorded as 7.4 ± 0.1 nm.

Liao et al. (2013) have reported a simple, one-pot, noninjection strategy for large-scale synthesis of wurtzite CZTS nanocrystals. The bandgap of the CZTS nanocrystals was 1.5 eV, while the average size was 13 nm. The nanocrystals had somewhat bullet- or leaf-like morphology. High-quality w-CZTS nanocrystals demonstrated high crystallinity, a monophasic nature, a uniform composition, and narrow size distribution.

A phosphine-free approach for the synthesis of high-quality CZTSe nanocrystals in organic solvents has been reported (Liu et al. 2012).

Wurtzite CZTS nanorods have been synthesized using commercially available cationic precursors (Thompson et al. 2013). Clear variation in the composition of rods along their axes was observed, and a Cu-rich end and a Zn-rich end could be identified. The copper precursor had the most reactivity and the highest rate of nucleation followed by the zinc precursor, while tin had the least reactivity. The aspect ratio of the CZTS was increased by decreasing the initial loading of the most reactive Cu precursor. In this case, three metal precursors nucleate at more comparable rates; thus, longer nanorods with more homogeneous elemental composition along their axes were formed. Increasing the initial loading of Cu precursor results in a low aspect ratio (formation of dots) and/or phase segregation (yielding the binary phases).

Synthesis of CZTS nanocrystals by thermolysis of single-source precursors of copper, zinc, and tin diethyldithiocarbamates has been reported (Chernomordik et al. 2014b). CZTS nanocrystals having mean size between 2 and 40 nm were synthesized by varying the growth temperature between 150°C and 340°C. The kesterite phase of the nanocrystals was confirmed by x-ray diffraction, Raman spectroscopy, and transmission electron microscopy, whereas energy dispersive x-ray spectroscopy confirmed stoichiometric composition of the CZTS nanocrystals. Nanocrystals having 2 nm size had a bandgap of 1.67 eV exhibiting strong quantum confinement, whereas larger nanocrystals had a bandgap of 1.5 eV.

Todorov et al. (2010) used a hydrazine-based nonvacuum, slurry-coating method. The slurry (or ink) used for deposition of CZTSSe was composed of a Cu–Sn chalcogenide (S or S–Se) solution in hydrazine and particle-based Zn-chalcogenide precursors, $ZnSe(N_2H_4)$ or $ZnS(N_2H_4)$ formed *in situ* upon the addition of Zn powder. The CZTSSe obtained had a zinc-rich and copper-poor stoichiometric composition with ratios Cu/(Zn/Sn) = 0.8 and Zn/Sn = 1.22. This material was deposited onto Mo-coated glass for fabrication of a solar device, which demonstrated a PCE of 9.6%. A similar technique was used to prepare 2.5 μm thick CZTSSe thin films with S/(S + Se) ratio of 0.4 ± 0.1

(Barkhouse et al. 2012). Deposition of CdS buffer, a ZnO window layer, and indium tin oxide (ITO) was carried out by using CBD and RF magnetron sputtering. An antireflection MgF_2 coating was deposited using electron beam evaporation. The finished device demonstrated a power conversion efficiency of 10.1% under AM 1.5 illumination conditions. CZTSe-based thin-film solar absorbers with some percentage of Sn atoms replaced by smaller isoelectronic Ge atoms were also prepared. These films were prepared using the hydrazine-based mixed particle-solution approach pioneered by this group. They observed an increase in bandgap of the absorber layer from 1.08 to 1.15 eV with 40% substitution of Sn atoms with Ge in CZTSe thin film. Devices fabricated by using this absorber layer showed a power conversion efficiency of 9.1% and a high open-circuit voltage.

An optical-design approach has been reported, which resulted in improvement of the short-circuit current and power conversion efficiency of CZTSSe solar cells (Winkler et al. 2014). They have optimized the thickness of the upper-device layers in a way that allows maximum transmission into the CZTSSe absorber. This design approach was based on optical modeling of idealized planar devices with a semi-empirical approach for treating the impact of surface roughness. Effectiveness of the new device architecture based on thinner CdS and TCO layers was experimentally demonstrated by fabricating a solar device with an overall power conversion efficiency of 12.0%—a new record in CZTSSe-based devices.

Solvothermal synthesis of spherical CZTS nanoparticles using ethylene glycol, copper (II) chloride dehydrate, zinc (II) chloride and tin (IV) chloride tetrahydrate, and thiourea and PVP was reported by Zhou et al. (2011a). Nanocrystals of about 100–150 nm size agglomerated to form spheres of kesteritic CZTS having bandgap of 1.48 eV. The same group has also reported the preparation of hierarchal flower-like CZTS nanostructures through a similar solvothermal approach (Zhou et al. 2011b). The effect of various synthesis parameters such as reaction temperature, reaction duration, precursor concentration, and amount of PVP on size and morphology of the final CZTS nanostructures was investigated. Formation of flower-like morphology was attributed to a nucleation-dissolution–recrystallization mechanism during crystal growth.

The synthesis of surfactant-free, solvent-redispersible CZTS nanoparticles based on high-temperature polycondensation reactions has been reported (Zaberca et al. 2012). Thiourea acted both as complexing agent to inhibit growth of the nanoparticles and also as surfactant to aid redispersion of nanocrystals in polar solvents. Films of the CZTS absorber were formed on Mo-coated glass by dip coating from a concentrated dispersion of nanocrystals with intermediate heat treatment at 400°C. Crack- and defect-free sintered films formed from surfactant-free CZTS can be a promising option for low-cost solar cells.

Arul et al. (2013) have reported the solvothermal synthesis of CZTS nanospheres with an average diameter of 3.26 nm. Small CZTS nanospheres exhibited a bandgap of 1.84 eV, significantly larger than the bandgap of bulk CZTS due to strong quantum confinement effects. An organic photovoltaic cell fabricated by using these CZTS nanospheres showed a PCE of 0.952%. Mali et al. (2013) have reported a simple, single-step synthesis of CZTS nanocrystals from corresponding metal halides, thiourea, and PVP. Nanoparticles thus obtained had a tetragonal crystallographic structure with size ranging from 4 to 7 nm. Synthesis of less than 10 nm CZTS nanocrystals by solvothermal reaction at 180°C for 15 h has also been reported by Pal et al. (2013).

The CZTSe nanoparticles with a mean diameter of 200–300 nm have been prepared using a one-step reaction without using any surfactant or template (Du et al. 2012b). The CZTSe nanoparticles were drop-casted onto FTO substrate and used as counterelectrode (CE) in dye-sensitized solar cells (DSSCs), which demonstrated an efficiency of 3.85%. A one-step, solvothermal treatment–based approach has been used for *in situ* deposition of CZTS thin films onto stainless steel and FTO glass substrates (Zhai et al. 2014). Characterization of the resulting material has been done by p-XRD, SEM, TEM, UV-vis spectroscopy, and Raman spectroscopy. The effects of temperature, reaction duration, and ratio of Cu/Zn/Sn precursors on the formation of CZTS nanocrystal films have been investigated. Phase-pure CZTS thin films were obtained at a temperature of 250°C or higher. The use of flexible substrates for deposition of CZTS thin films is attractive for industrial applications.

Jiang et al. (2012d) have reported a metastable orthorhombic phase of CZTS prepared through a hydrothermal approach using $SnCl_2$, H_2O, $ZnCl_2$, $CuCl_2.2H_2O$, and thiocarbamide in a water-ethylenediamine mixture at 200°C. Nanocrystals thus obtained had a crystal structure based on a double-wurtzite cell having space group $Pmn2_1$. The bandgap of the orthorhombic CZTS nanocrystal was 1.45 eV. Annealing of the nanocrystals at 500°C resulted in phase transformation, yielding the tetragonal kesterite phase of CZTS.

CZTS nanoparticles powder has been prepared by using corresponding metal halides and $Na_2S.9H_2O$ (Mariama et al. 2013). An ethylene glycol suspension of milled CZTS nanoparticles was deposited onto 1 mm FTO coated glass by spin coating. The bandgaps of thin films deposited after 18, 12, and 6 cycles of spin coating were found to be 1.42, 1.57, and 1.67 eV, respectively. Photoelectrical characterization showed a short-circuit current of 23.8 mA cm^{-2}, open-circuit voltage as 0.394 V, fill factor as 28.78%, and efficiency as ~2.25%. Synthesis of phase-pure kesterite CZTS nanocrystals with controllable size between 3 and 10.5 nm has been carried out using inorganic metal salts copper (II) acetate monohydrate, zinc (II) nitrate, and tin (II) chloride and thiourea by using the hydrothermal method (Liu et al. 2013a). The average size of the nanocrystals with the reaction duration of 6, 12, 24, and 48 h at 180°C were 3.0, 4.8, 6.5, and 10.5 nm, respectively. Bandgaps for CZTS nanocrystals with an average size of 10.5, 6.5, 4.8, and 3 nm were found to be 1.48, 1.52, 1.61, and 1.89 eV, respectively. A slight blue shift in bandgap of the 4.8 nm CZTS nanocrystals was observed, while a significantly large blue shift was observed for the 3 nm CZTS nanocrystals due to the quantum confinement effect. The same group has also reported the hydrothermal synthesis of ~5 nm CZTS nanocrystals with an optical bandgap of 1.47 eV (Liu et al. 2013b).

Synthesis of good-quality CZTS nanocrystals by using a water-based one-step hydrothermal method has been reported, and the influence of the sulfur precursor used in a hydrothermal reaction has been investigated (Tiong et al. 2014). It was observed that the sulfur precursor significantly affects the phase purity and crystal structure of the CZTS material. The use of organic sulfur precursors like thioacetamide and thiourea gave a mixture of kesterite phase and wurtzite crystal phase. However, pure kesterite CZTS was obtained by using Na_2S as the sulfur source.

The CZTS nanocrystals–based ink has been prepared using a microwave-assisted synthesis (Flynn et al. 2012). This method is capable of producing CZTS nanoparticles at significantly lower temperatures and in lesser reaction duration. The CZTS nanocrystals were purified by removing the unreacted species through centrifugation from de-ionized water and ethanol alternatively. The TEM images showed polydispersed spherical particles having an average diameter of 7.2 ± 1.2 nm; SEM images showed spherical clusters formed by the agglomeration of smaller nanoparticles. The bandgap of the as-prepared nanoparticles was found to be 1.5 eV. Analyses using p-XRD and Raman spectrometry confirmed that the CZTS nanocrystals correspond with the kesterite phase. Solar cells fabricated by using these crystals showed a maximum power conversion efficiency of 0.25%. Shin et al. (2012) have also reported the synthesis of CZTS nanocrystals through irradiation of aqueous solutions of copper acetate, zinc acetate, tin chloride, and thioacetamide by microwave energy of 700 W for 10 min.

A rapid and facile deposition of CZTS thin films from a homogeneous precursor solution directly onto conductive films has been carried out via thermolysis of precursors by microwave heating (Knutson et al. 2014). The conductive films strongly absorb microwave energy, which causes rapid heating to a sufficiently high temperature for the decomposition of precursors and deposition of CZTS on the conductive layer. By using this approach, thin films having thicknesses of 1–3 mm can be deposited in a very short time. Microwave annealing of the as-deposited CZTS films in solvent resulted in an increase of the crystallite size, while no formation of impurity phases was observed.

The synthesis of kesteritic CZTS nanocrystals using a continuous-flow mesofluidic reactor has been reported (Flynn et al. 2013). Ethylene glycol was used as the solvent, and the composition of the CZTS nanocrystals was controlled by judiciously varying the parameters, such as precursors concentrations, reaction temperature, and residence time. The growth of CZTS nanocrystals from binary metal sulfide precursors was studied in detail. The average size of the nanocrystals obtained was 5.4 ± 2.0 nm.

11.5 COPPER IRON TIN SULFIDE AND RELATED MATERIALS

The crystal structure of copper iron tin sulfide, Cu_2FeSnS_4 (CFTS), was first reported by Brockway (1934). Bernardini et al. (1990, 2000) have investigated the crystallographic structure and magnetic properties of bulk single-crystalline CFTS. Synthesis of CFTS nanoparticles was reported by hydrothermal reaction (Gui et al. 2004). The CFTS nanoparticles were formed as black precipitate. Studies using p-XRD showed the monophasic nature of the stannite material, and TEM images revealed that the size of crystallites ranged from 10–20 nm. It was observed that formation of a white colloidal precursor is important for subsequent conversion to the final multinary sulfide (CFTS).

Zhang et al. (2012) have reported the synthesis of wurtzite CFTS nanocrystals by hot-injection approach. The CFTS nanocrystals thus obtained had a mean size of approximately 20 nm. The x-ray diffraction pattern of the nanocrystals synthesized at 210°C was similar to that of the simulated pattern for the wurtzite phase of CFTS, whereas those grown at higher temperatures exhibited zinc blende structure.

The stoichiometric composition of the CFTS nanocrystals grown at 210°C was slightly iron deficient; however, it was improved by increasing the reaction temperature. The bandgap of the nanocrystals grown at 210°C, 240°C, 270°C, and 310°C was found to be 1.54, 1.53, 1.51, and 1.49 eV, respectively.

Zhang et al. (2013b) have reported the preparation of wurtzite Cu_2CoSnS_4 via a simple thermal decomposition approach. Nanocrystals thus formed had a nanorod-like morphology with average length of 32 ± 2.0 nm and average width of 16 ± 1.5 nm. It was demonstrated that the stable stannite crystallographic phase of Cu_2CoSnS_4 nanocrystals can be obtained by using higher reaction temperatures or by postsynthesis annealing at above 400°C. Wurtzite nanocrystals exhibited a bandgap of 1.58 eV.

Photoelectrical characterization of thin films based on Cu_2CoSnS_4 nanocrystals showed a photoresponse indicating their potential for use in thin-film solar cells. A flexible chemical approach for preparation of colloidal CFTS nanocrystals by thermal reaction of copper (II) acetate, iron (II) chloride, tin (II) chloride, and sulfur in oleylamine has been reported (Li et al. 2012c). The p-XRD pattern of the as-synthesized CFTS nanocrystals corresponded with the stannite CFTS. The average size of the as-synthesized nanocrystals, as revealed by TEM images, was 15–25 nm, which increased to 20–40 nm after annealing. The bandgap was found to be 1.33 eV.

The synthesis of zinc blende CFTS nanocrystals by hot injection has also been reported by Liang et al. (2012a). The p-XRD pattern of the as-prepared nanocrystals corresponded with the simulated pattern for zinc blende CFTS with space group *F-43 m* and unit cell parameter a = 5.429 Å. Nanocrystals had a spherical morphology with an average diameter of 7.5 nm. Magnetic studies by superconducting quantum interference device (SQUID) revealed that the CFTS nanocrystals thus obtained had ferromagnetic behavior, and bandgap was measured to be 1.1 eV. Liang and Wei (2012) have also reported the synthesis of zinc blende CFTS nanocrystals using the same reaction strategy; however, the nanocrystals had a lower bandgap of 0.92 eV.

A microwave irradiation–based method for the preparation of quaternary CFTS nanotubes with an outer diameter of 400–800 nm and a thickness of 100–200 nm from corresponding metal halide precursors has been reported (Lunhong and Jing 2012). Benzyl alcohol was used as the microwave-absorbing solvent in this nonaqueous reaction. As-prepared CFTS nanotubes exhibited strong absorption in the visible region and had a bandgap of 1.71 eV. Ai and Jiang (2012) have also reported the synthesis of hierarchal porous quaternary CFTS hollow microspheres by a nonaqueous microwave reaction. These microstructures showed excellent adsorption capacity. Furthermore, these CFTS hollow microstructures demonstrated superb ability to remove organic pollutant from water.

The synthesis of CFTS nanocrystals by hot-injection approach has been reported by Yan et al. (2012). CFTS nanocrystals grown for 1 h had a triangular and spherical morphology with a mean diameter of 13.2 ± 1.1 nm. Raman spectroscopy and p-XRD confirmed phase-pure deposition of CFTS material, as no peaks assignable to Cu_2SnS_3 and FeS were observed. The EDX measurements revealed that the CFTS nanocrystals had a sulfur-rich stoichiometry $Cu_2Fe_{1.05}Sn_{1.06}S_{4.69}$. The band gap was found to be 1.28 ± 0.02 eV. The CFTS nanocrystals thus obtained were coated directly onto ITO substrates and annealed at 350°C to form thin-film electrodes. Photoelectrochemical characterization revealed p-type conductivity of the CFTS films with clear photoelectrochemical response and good photostability. Values of the photocurrent density obtained for this system ranged from 11 to 13 mA cm^{-2} without any optimization.

The formation of highly crystalline and phase-pure quaternary chalcogenide (Cu_2CoZnS_4, Cu_2ZnSnS_4) particles by a high-temperature route in molten KSCN at 400°C has been reported (Benchikri et al. 2012). Metal oxides like Cu_2O, SnO_2, CoO, and ZnO were used as metal precursors. It was demonstrated that the chemical homogeneity of various oxide precursors Cu_2O–MO–(M = Zn or Co)–SnO_2_KSCN played a crucial role in the formation of a pure quaternary chalcogenide structure. The nucleation rate of the quaternary chalcogenide particles was also affected remarkably by the chemical homogeneity. These results thus highlight the ability to synthesize highly pure and highly crystallized CZTS particles with controlled primary crystallite sizes.

A general strategy for the synthesis of quaternary semiconductor nanocrystals of Cu_2CoSnS_4, Cu_2FeSnS_4, Cu_2NiSnS_4, and Cu_2MnSnS_4 based on solvothermal reaction has been reported by Cui et al. (2012a). Cu_2CoSnS_4 and Cu_2FeSnS_4 nanocrystals possessed a zinc blende crystallographic structure, whereas Cu_2NiSnS_4, and Cu_2MnSnS_4 had a wurtzite structure as revealed by p-XRD measurements. The TEM images demonstrate that the Cu_2CoSnS_4 and Cu_2FeSnS_4 nanocrystals had somewhat spherical morphology with a mean diameter of 5.6 ± 1.4 nm and 7.5 ± 1.5 nm, respectively. Cu_2NiSnS_4, and Cu_2MnSnS_4 nanocrystals exhibited a nail-like and rod-like morphology. Cu_2FeSnS_4, Cu_2CoSnS_4, Cu_2MnSnS_4, and Cu_2NiSnS_4 nanocrystals possessed optical bandgaps of 1.19, 1.25, 1.28, and 1.49 eV, respectively. Photoelectrical measurements of the Cu_2MnSnS_4 thin films prepared by spin coating a concentrated solution of Cu_2MnSnS_4 nanorod exhibited a photoresponse behavior, thus showing that these Cu_2MnSnS_4 nanorods can potentially be used for fabrication of low-cost PV solar cells.

Synthesis of Cu_2CoSnS_4 nanoparticles through low temperature, simple, and one-pot hydrothermal process has been reported (Murali and Krupanidhi 2013). The p-XRD studies showed that the nanoparticles had a wurtzite crystallographic phase. A size ranging from 20 to 60 nm was observed for Cu_2CoSnS_4 nanoparticles prepared under different conditions of reaction temperature, reaction time, and surfactant. An unusual red-edge effect was exhibited by CCTS nanoparticles in all the samples. Hybrid devices fabricated using these nanocrystals showed their potential use in solar cells, photodetectors, and light-emitting diodes.

Park et al. (2013) have synthesized Cu_2FeSnS_4 nanoparticles with zinc blende crystal structure. The TEM images revealed that the CFTS nanocrystals are well dispersed, having spherical morphology. Average size of the nanocrystals varied from 11 to 15 nm, whereas the nanocrystals grew to a size of 30–40 nm after annealing at 400°C for 15 min. Counterelectrodes for DSSC were fabricated by spin coating the CFTS nanocrystals on FTO glass at 1000 rpm for 20 s, and the desired thickness of the film was obtained after seven cycles of spin coating. It was demonstrated that the DSSC based on CFTS as counterelectrode had efficiency comparable to that of DSSCs based on Pt-based counterelectrode.

The $Cu_2(Fe_xZn_{1-x})SnS_4$ nanocrystals with x ranging from 0 to 1 have been synthesized by using the hot-injection approach in order to tune the bandgap of Cu_2ZnSnS_4 nanocrystals (Huang et al. 2013). Nearly monodispersed $Cu_2(Fe_xZn_{1-x})SnS_4$ nanocrystals with sizes ranging between 10 and 20 nm were shown by TEM images. Raman spectroscopy and p-XRD studies demonstrated that the Cu_2ZnSnS_4 and Cu_2FeSnS_4 crystallize as kesterite and stannite phases, respectively, and no binary or ternary phases were formed. Shifting of the Raman peaks toward higher frequencies with increasing Zn content was also reported. It was observed that the bandgap of the nanocrystals can be tailored from 1.25 eV for $x = 0$ to 1.5 eV for $x = 1$ by adjusting the composition variable x. Improvement in the photoelectrical conversion performance was also observed when $x = 0.4$.

Synthesis of CFTS nanoparticles by the solvothermal route using N,N-dimethylformamide (DMF) as solvent has been reported (Jiang et al. 2013). Peaks in the x-ray diffraction pattern of as-synthesized nanocrystals could be indexed to the stannite phase. The TEM images revealed the formation of larger (100–200 nm) particles with good crystallinity. Bandgap was determined to be 1.28 eV.

Most recently, the synthesis of Cu_2FeSnS_4 particles having a flower-like morphology has been carried out by using a microwave irradiation method (Guan et al. 2014b). Characterization of as-prepared CFTS nanoparticles was carried out by p-XRD, SEM, TEM, and UV-vis-NIR spectroscopy. The CFTS nanoparticles possessed a stannite structure. The SEM images show flower-like morphology, whereas TEM images revealed that the individual particles had a spherical structure. The bandgap of the CFTS particles was found to be 1.52 eV, indicating its potential for solar cell applications.

Shibuya et al. (2014) have carried out first principles calculations to investigate the substitution of Zn atoms by Fe atoms into the CZTS lattice to form the $Cu_2(Zn,Fe)SnS_4$ solid solution in order to tune the lattice parameters and bandgap energy. It was demonstrated that by the incorporation of Fe atoms, a phase transition from kesterite (Zn-rich) to stannite (Fe-rich) occurred at the Fe/Zn ratio of 0:4. An increase in bandgap and slight decrease in lattice volume were observed with increasing Fe concentrations in the solid solution. A linear bandgap bowing was observed for each phase, which resulted in blue-shifted photoabsorption for alloys having Fe-rich composition due to the confinement of the conduction states. Similarly, Khadka and Kim (2014) have reported the fabrication of polycrystalline $Cu_2Zn_zFe_{1-z}SnS_4$ (CZFTS) thin films by the chemical spray pyrolysis process followed by sulfurization. It was found that the postsulfurized CZFTS films possess morphology, structure, and optoelectronic features that make them potential photoabsorbers in thin-film photovoltaics. Raman spectroscopy and p-XRD studies clearly showed transition from the stannite to the kesterite phase with the increase of zinc content in the CZFTS alloy. A parabolic increasing trend was observed in the bandgap of postsulfurized CZFTS films, and it was possible to tune the bandgap from 1.36 ± 0.02 to 1.51 ± 0.02 eV with increasing Zn content ($0 \leq z \leq 1$).

The synthesis of Cu_2MnSnS_4 nanocrystals having zinc blende and wurtzite structure has been carried out using a hot-injection approach (Liang et al. 2012b). It was found that the reactivity of the sulfur precursor used in the reaction controls the crystal structure of Cu_2MnSnS_4 nanocrystals. A more stable wurtzite nanostructure was obtained by using less reactive thiourea as the sulfur source. The TEM images revealed that the average size of triangular zinc blende Cu_2MnSnS_4 nanocrystals was 9.3 nm, whereas wurtzite Cu_2MnSnS_4 was composed of hexagonal nanoplates having diameter of about 23.1 nm. The EDX measurements indicated that the nanocrystals had a stoichiometric composition closely matching the theoretical composition of Cu_2MnSnS_4. The bandgap of the nanocrystals was found to be 1.1 eV, indicating their potential for use in the manufacture of low-cost thin-film solar cells.

The Cu_2CdSnS_4 nanorods were synthesized by the solvothermal method (Cui et al. 2012b). Then, p-XRD studies indicated a wurtzite crystallographic structure of the nanorods. The TEM images showed that the Cu_2CdSnS_4 nanorods have an average diameter and length of about 5.7 and 26 nm, respectively. The EDX measurements confirmed a relative atomic ratio of 2:1:1:4 for Cu, Cd, Sn, and S atoms. The bandgap of the as-synthesized nanorods was determined to be 1.4 eV. Photoelectrical measurements of the spin-casted films demonstrated that the Earth-abundant Cu_2CdSnS_4 nanorods have remarkable potential as a low-cost material for thin-film solar cells. Solvothermal synthesis of Cu_2CdSnS_4 nanocrystals has also been reported by other research groups (Cao et al. 2012).

This chapter highlights some of the representative semiconductor nanomaterials for solar photovoltaics. Nanomaterials find applications in solar cells owing to their unique optoelectronic properties, large surface area, and high optical absorption coefficient and exceptional mechanisms like quantum confinement effects. Ternary and quaternary metal chalcogenide nanocrystals of I–III–VI$_2$

and I_2–II–IV–VI$_4$ family of semiconductors have demonstrated good power conversion efficiencies (PCEs), thus proving their potential for commercialization. Current research efforts are focused on semiconductors based on Earth-abundant and nontoxic elements. However, this field still faces challenges due to contamination from binary materials and the large variety of compositional phases. It has been established that defect states of the nanocrystals strongly influence the performance of final devices; therefore, theoretical modeling of the defects in semiconductor systems should be considered. Recent studies involving doping of the semiconductors have shown significant improvement in their absorption/emission and electronic properties. Similarly, the use of small molecular ligands or simple ions for surface passivation can have an insulating effect. Future development efforts will emphasize greener and lower temperature-processing strategies to obtain higher yield nanomaterials offering good PCE. The challenges associated with developing size, morphology, phase, and stoichiometry control during synthesis, and understanding the fundamental characteristics as well as the industrial-level scaled-up manufacturing issues will be optimally addressed during the next decade.

REFERENCES

Abdelhady, A. L., M. A. Malik, and P. O'Brien. 2012. High-throughput route to $Cu_{2-x}S$ nanoparticles from single molecular precursor. *Mater. Sci. Semicond. Proc.*, 15: 218–221.

Abermann, S. 2013. Non-vacuum processed next generation thin film photovoltaics: Towards marketable efficiency and production of CZTS based solar cells. *Solar Energy*, 94: 37–70.

Ahmadi, M., S. S. Pramana, L. Xi, C. Boothroyd, Y. M. Lam, and S. Mhaisalkar. 2012. Evolution pathway of CIGSe nanocrystals for solar cell applications. *J. Phys. Chem. C.*, 116: 8202–8209.

Ahn, S., C. Kim, J. H. Yun, J. Gwak, S. Jeong, B.-H. Ryu, et al. 2010. $CuInSe_2$ (CIS) thin film solar cells by direct coating and selenization of solution precursors. *J. Phys. Chem. C.*, 114: 8108–8113.

Ahn, S., K. Kim, A. Cho, J. Gwak, J. H. Yun, K. Shin et al. 2012. $CuInSe_2$ (CIS) thin films prepared from amorphous Cu–In–Se nanoparticle precursors for solar cell application. *ACS Appl. Mater. Interfaces*, 4: 1530–1536.

Ai, L., and J. Jiang. 2012. Hierarchical porous quaternary Cu-Fe-Sn-S hollow chain microspheres: Rapid microwave nonaqueous synthesis, growth mechanism, and their efficient removal of organic dye pollutant in water. *J. Mater Chem.*, 22: 20586–20592.

Akhavan, V. A., B. W. Goodfellow, M. G. Panthani, C. Steinhagen, T. B. Harvey, C. J. Stolle, et al. 2012. Colloidal CIGS and CZTS nanocrystals: A precursor route to printed photovoltaics. *J. Solid State Chem.*, 189: 2–12.

Aldakov, D., A. Lefrancois, and P. Reiss. 2013. Ternary and quaternary metal chalcogenide nanocrystals: Synthesis, properties and applications. *J. Mater. Chem. C.*, 1: 3756–3776.

An, H. S., Y. Cho, S. J. Park, H. S. Jeon, Y. J. Hwang, D.-W. Kim, et al. 2014. Cocktails of paste coatings for performance enhancement of $CuInGaS_2$ thin-film solar cells. *ACS Appl. Mater. Interfaces*, 6: 888–893.

Arani, M. S., and M. S. Niasari. 2014. A facile and reliable route to prepare flower shaped lead sulfide nanostructures from a new sulfur source. *J. Indust. Eng. Chem.*, 20: 3141–3149.

Arul, N. S., D. Y. Yun, D. U. Lee, and T. W. Kim. 2013. Strong quantum confinement effects in kesterite Cu_2ZnSnS_4 nanospheres for organic optoelectronic cells. *Nanoscale*, 5: 11940–11943.

Azimi, H., Y. Hou, and C. J. Brabec, 2014. Towards low-cost, environmentally friendly printed chalcopyrite and kesterite solar cells. *Energy Environ. Sci.*, 7: 1829–1849.

Azizian-Kalandaragh, Y., A. Khodayari, Z. Zeng, C. Garoufalis, S. Baskoutas, and L. Gontard. 2013. Strong quantum confinement effects in SnS nanocrystals produced by ultrasound-assisted method. *J. Nanopart. Res.*, 15: 1388.

Barkhouse, D. A. R., O. Gunawan, T. Gokmen, T. K. Todorov, and D. B. Mitzi. 2012. Device characteristics of a 10.1% hydrazine-processed $Cu_2ZnSn(Se,S)_4$ solar cell. *Prog. Photovolt.: Res. Appl.*, 20: 6–11.

Bauer, E., K. L. Man, A. Pavlovska, A. Locatelli, T. O. Mentes, M. A. Nino et al. 2014. Fe_3S_4 (Greigite) formation by vapor-solid reaction. *J. Mater. Chem. A.*, 2: 1903–1913.

Beal, J. H. L., P. G. Etchegoin, and R. D. Tilley. 2012. Synthesis and characterisation of magnetic iron sulfide nanocrystals. *J. Solid State Chem.*, 189: 57–62.

Beal, J. H. L., S. Prabakar, N. Gaston, G. B. Teh, P. G. Etchegoin, G. Williams, et al. 2011. Synthesis and comparison of the magnetic properties of iron sulfide spinel and iron oxide spinel nanocrystals. *Chem. Mater.*, 23: 2514–2517.

Bellemare, B. 2006. Solar boiling over. *UtiliPoint International Article*, August 9.

Benchikri, M., O. Zaberca, R. El Ouatib, B. Durand, F. Oftinger, A. Balocchi, et al. 2012. A high temperature route to the formation of highly pure quaternary chalcogenide particles. *Mater. Lett.*, 68: 340–343.

Bera, P., and S. I. Seok. 2012. Nanocrystalline copper sulfide of varying morphologies and stoichiometries in a low temperature solvothermal process using a new single-source molecular precursor. *Solid State Sci.*, 14: 1126–1132.

Bernardini, G. P., P. Bonazzi, M. Corazza, F. Corsini, G. Mazzetti, L. Poggi, et al. 1990. New data on the Cu_2FeSnS_4-Cu_2ZnSnS_4 pseudobinary system at 750 degree and 550 degree C. *Eur. J. Mineral*, 2: 219–225.

Bernardini, G. P., D. Borrini, A. Caneschi, F. Di Benedetto, D. Gatteschi, S. Ristori, et al. 2000. EPR and SQUID magnetometry study of Cu_2FeSnS_4 (stannite) and Cu_2ZnSnS_4 (kesterite). *Phys. Chem. Miner.*, 27: 453–461.

Biacchi, A. J., D. D. Vaughn, and R. E. Schaak. 2013. Synthesis and crystallographic analysis of shape-controlled SnS nanocrystal photocatalysts: Evidence for a pseudotetragonal structural modification. *J. Am. Chem. Soc.*, 135: 11634–11644.

Brockway, L. O. 1934. The crystal structure of stannite, Cu_2FeSnS_4. *Z. Kristallogr.*, 89: 434–441.

Brown, G. F., and J. Wu. 2009. Third generation photovoltaics. *Laser Photonics Rev.*, 3: 394–405.

Cabán-Acevedo, M., M. S. Faber, Y. Tan, R. J. Hamers, and S. Jin. 2012. Synthesis and properties of semiconducting iron pyrite (FeS_2) nanowires. *Nano Lett.*, 12: 1977–1982.

Cao, M., L. Li, W. Z. Fan, X. Y. Liu, Y. Sun, and Y. Shen. 2012. Quaternary Cu_2CdSnS_4 nanoparticles synthesized by a simple solvothermal method. *Chem. Phys. Lett.*, 534: 34–37.

Cao, Y., Y. Xiao, J.-Y. Jung, H.-D. Um, S.-W. Jee, H. M. Choi, et al. 2013. Highly electrocatalytic $Cu_2ZnSn(S_{1-x}Se_x)_4$ counter electrodes for quantum-dot-sensitized solar cells. *ACS Appl. Mater. Interfaces*, 5: 479–484.

Cattley, C. A., C. Cheng, F. S. M. Airclough, L. M. Droessler, N. P. Young, J. H. Warner, et al. 2013. Low temperature phase selective synthesis of Cu_2ZnSnS_4 quantum dots. *Chem. Commun.*, 49: 3745–3747.

Chang, J., and E. R. Waclawik. 2013. Controlled synthesis of $CuInS_2$, Cu_2SnS_3 and Cu_2ZnSnS_4 nano-structures: Insight into the universal phase-selectivity mechanism. *Cryst. Eng. Comm.*, 15: 5612–5619.

Chang, S.-H., M.-Y. Chiang, C.-C. Chiang, F.-W. Yuan, C.-Y. Chen, B.-C. Chiu, et al. 2011. Facile colloidal synthesis of quaternary $CuIn_{1-x}Ga_x(S_ySe_{1-y})_2$ (CIGSSe) nanocrystal inks with tunable band gaps for use in low-cost photovoltaics. *Energy Environ. Sci.*, 4: 4929–4932.

Chapin, D. M., C. S. Fuller, and G. L. Pearson. 1954. A new silicon p-n junction photocell for converting solar radiation into electrical power. *J. Appl. Phys.*, 25: 676–677.

Chen, B., H. Zhong, M. Wang, R. Liu, and B. Zou. 2013b. Integration of $CuInS_2$-based nanocrystals for high efficiency and high colour rendering white light-emitting diodes. *Nanoscale*, 5: 3514–3519.

Chen, Z., M. Tang, L. Song, G. Tang, B. Zhang, L. Zhang, et al. 2013a. In situ growth of $CuInS_2$ nanocrystals on nanoporous TiO_2 film for constructing inorganic/organic heterojunction solar cells. *Nanoscale Res. Lett.*, 8: 354.

Chernomordik, B. D., A. E. Béland, D. D. Deng, F. L. F. Rancis, and E. S. Aydil. 2014a. Microstructure evolution and crystal growth in Cu_2ZnSnS_4 thin films formed by annealing colloidal nanocrystal coatings. *Chem. Mater.*, 26: 3191–3201.

Chernomordik, B. D., A. E. Béland, N. D. Trejo, A. A. Gunawan, D. D. Deng, K. A. Mkhoyan, et al. 2014b. Rapid facile synthesis of Cu_2ZnSnS_4 nanocrystals. *J. Mater. Chem. A*, 2: 10389–10395.

Chesman, A. S. R., N. W. Duffy, S. Peacock, L. Waddington, N. A. S. Webster, and J. J. Jasieniak. 2013a. Non-injection synthesis of Cu_2ZnSnS_4 nanocrystals using a binary precursor and ligand approach. *RSC Adv.*, 3: 1017–1020.

Chesman, A. S. R., J. Van Embden, N. W. Duffy, N. A. S. Webster, and J. J. Jasieniak. 2013b. In situ formation of reactive sulfide precursors in the one-pot, multigram synthesis of Cu_2ZnSnS_4 nanocrystals. *Cryst. Growth Design*, 13: 1712–1720.

Chiang, M.-Y., S.-H. Chang, C.-Y. Chen, F.-W. Yuan, and H.-Y. Tuan. 2011. Quaternary $CuIn(S_{1-x}Se_x)_2$ nanocrystals: Facile heating-up synthesis, band gap tuning, and gram-scale production. *J. Phys. Chem. C*, 115: 1592–1599.

Cho, A., H. Song, J. Gwak, Y.-J. Eo, J. H. Yun, K. Yoon, et al. 2014. A chelating effect in hybrid inks for non-vacuum-processed $CuInSe_2$ thin films. *J. Mater. Chem. A*, 2: 5087–5094.

Connor, S. T., C.-M. Hsu, B. D. Weil, S. Aloni, and Y. Cui. 2009. Phase transformation of biphasic Cu_2S–$CuInS_2$ to monophasic $CuInS_2$ nanorods. *J. Am. Chem. Soc.*, 131: 4962–4966.

Coughlan, C., A. Singh, and K. M. Ryan. 2013. Systematic study into the synthesis and shape development in colloidal $CuIn_xGa_{1-x}S_2$ nanocrystals. *Chem. Mater.*, 25: 653–661.

Cui, Y., G. Wang, and D. Pan. 2012b. Synthesis and photoresponse of novel Cu_2CdSnS_4 semiconductor nanorods. *J. Mater. Chem.*, 22: 12471–12473.

Cui, Y., R. Deng, G. Wang, and D. Pan. 2012a. A general strategy for synthesis of quaternary semiconductor Cu_2MSnS_4 (M = Co^{2+}, Fe^{2+}, Ni^{2+}, Mn^{2+}) nanocrystals. *J. Mater. Chem.*, 22: 23136–23140.

Cummins, D. R., H. B. Russell, J. B. Jasinski, M. Menon, and M. K. Sunkara. 2013. Iron sulfide (FeS) nanotubes using sulfurization of hematite nanowires. *Nano Lett.*, 13: 2423–2430.

Deng, M., S. Shen, Y. Zhang, H. Xu, and Q. Wang. 2014. A generalized strategy for controlled synthesis of ternary metal sulfide nanocrystals. *New J. Chem.*, 38: 77–83.

Dhere, N. G. 2007. Toward GW/year of CIGS production within the next decade. *Sol. Energy Mater. Solar Cells*, 91: 1376–1382.

Dierick, R., B. Capon, H. Damm, S. Flamee, P. Arickx, E. Bruneel, et al. 2014. Annealing of sulfide stabilized colloidal semiconductor nanocrystals. *J. Mater. Chem. C*, 2: 178–183.

Du, Y.-F., J.-Q. Fan, W.-H. Zhou, Z.-J. Zhou, J. Jiao, and S.-X. Wu. 2012b. One-step synthesis of stoichiometric $Cu_2ZnSnSe_4$ as counter electrode for dye-sensitized solar cells. *ACS Appl. Mater. Interfaces*, 4: 1796–1802.

Du, Y., Z. Yin, J. Zhu, X. Huang, X.-J. Wu, Z. Zeng, et al. 2012a. A general method for the large-scale synthesis of uniform ultrathin metal sulphide nanocrystals. *Nat. Commun.*, 3: 1177.

Fan, F.-J., L. Wu, and S.-H. Yu, 2014. Energetic I-III-VI_2 and I_2-II-IV-VI_4 nanocrystals: Synthesis, photovoltaic and thermoelectric applications. *Energy Environ. Sci.*, 7: 190–208.

Flynn, B., I. Braly, P. A. Glover, R. P. Oleksak, C. Durgan, and G. S. Herman, 2013. Continuous flow mesofluidic synthesis of Cu_2ZnSnS_4 nanoparticle inks. *Mater. Lett.*, 107: 214–217.

Flynn, B., W. Wang, C.-H. Chang, and G. S. Herman. 2012. Microwave assisted synthesis of Cu_2ZnSnS_4 colloidal nanoparticle inks. *Physica Status Solidi (A)*, 209: 2186–2194.

Ford, G. M., Q. Guo, R. Agrawal, and H. W. Hillhouse. 2011. Earth abundant element $Cu_2Zn(Sn_{1-x}Ge_x)S_4$ nanocrystals for tunable band gap solar cells: 6.8% efficient device fabrication. *Chem. Mater.*, 23: 2626–2629.

Fthenakis, V. 2009. Sustainability of photovoltaics: The case for thin-film solar cells. *Renew. Sust. Energy Rev.*, 13: 2746–2750.

Guan, H., Y. Shi, B. Jiao, X. Wang, and F. Yu. 2014b. Flower-like Cu_2FeSnS_4 particles synthesized by microwave irradiation method. *Chalcogenide Lett.*, 11: 9–12.

Guan, Z., W. Luo, and Z. Zou. 2014a. Formation mechanism of ZnS impurities and their effect on photoelectrochemical properties on a Cu_2ZnSnS_4 photocathode. *Cryst. Eng. Comm.*, 16: 2929–2936.

Gui, Z., R. Fan, X. Chen, Y. Hu, and Z. Wang. 2004. A new colloidal precursor cooperative conversion route to nanocrystalline quaternary copper sulfide. *Mater. Res. Bull.*, 39: 237–241.

Guo, J., X. Wang, W.-H. Zhou, Z.-X. Chang, X. Wang, Z.-J. Zhou, et al. 2013. Efficiency enhancement of dye-sensitized solar cells (DSSCs) using ligand exchanged $CuInS_2$ NCs as counter electrode materials. *RSC Adv.*, 3: 14731–14736.

Guo, Q., G. M. Ford, W.-C. Yang, C. J. Hages, H. W. Hillhouse, and R. Agrawal. 2012. Enhancing the performance of CZTSSe solar cells with Ge alloying. *Sol. Energy Mater. Solar Cells*, 105: 132–136.

Guo, Q., G. M. Ford, W.-C. Yang, B. C. Walker, S. E. A. Tach, H. W. Hillhouse, et al. 2010. Fabrication of 7.2% efficient CZTSSe solar cells using CZTS nanocrystals. *J. Am. Chem. Soc.*, 132: 17384–17386.

Guo, Q., H. W. Hillhouse, and R. Agrawal. 2009. Synthesis of Cu_2ZnSnS_4 nanocrystal ink and its use for solar cells. *J. Am. Chem. Soc.* 131: 11672–11673.

Gusain, M., P. Kumar, and R. Nagarajan. 2013. Wurtzite $CuInS_2$: Solution based one pot direct synthesis and its doping studies with non-magnetic Ga^{3+} and magnetic Fe^{3+} ions. *RSC Adv.*, 3: 18863–18871.

Han, S.-K., M. Gong, H.-B. Yao, Z.-M. Wang, and S.-H. Yu. 2012. One-pot controlled synthesis of hexagonal-prismatic $Cu_{1.94}S$-ZnS, $Cu_{1.94}S$-ZnS-$Cu_{1.94}S$, and $Cu_{1.94}S$-ZnS-$Cu_{1.94}S$-ZnS-$Cu_{1.94}S$ heteronanostructures. *Angew. Chem. Int. Ed.*, 51: 6365–6368.

Han, Z., D. Zhang, Q. Chen, R. Hong, C. Tao, Y. Huang, et al. 2014. Synthesis of single phase chalcopyrite $CuIn_{1-x}Ga_xSe_2$ ($0 \leq x \leq 1$) nanoparticles by one-pot method. *Mater. Res. Bull.*, 51: 302–308.

Harvey, T. B., I. Mori, C. J. Stolle, T. D. Bogart, D. P. Ostrowski, M. S. Glaz, et al. 2013. Copper indium gallium selenide (CIGS) photovoltaic devices made using multistep selenization of nanocrystal films. *ACS Appl. Mater. Interfaces*, 5: 9134–9140.

Hsin, C.-L., W.-F. Lee, C.-T. Huang, C.-W. Huang, W.-W. Wu, and L.-J. Chen. 2011. Growth of $CuInSe_2$ and $In_2Se_3/CuInSe_2$ nano-heterostructures through solid state reactions. *Nano Lett.*, 11: 4348–4351.

Hsu, C.-H., C.-H. Chen, and D.-H. Chen. 2013a. Decoration of PbS nanoparticles on Al-doped ZnO nanorod array thin film with hydrogen treatment as a photoelectrode for solar water splitting. *J. Alloys Compds.*, 554: 45–50.

Hsu, W.-H., H.-I. Hsiang, C.-T. Chia, and F.-S. Yen. 2013b. Controlling morphology and crystallite size of $Cu(In_{0.7}Ga_{0.3})Se_2$ nano-crystals synthesized using a heating-up method. *J. Solid State Chem.*, 208: 1–8.

Hu, H., C. Deng, K. Zhang, M. Sun, H. Xuan, and S. Kong. 2013a. Novel two-dimensional lead sulfide quadrangular pyramid-aggregated arrays with self-supporting structure prepared at room temperature. *Mater. Sci. Semicond. Proc.*, 16: 1566–1572.

Hu, X., G. Song, W. Li, Y. Peng, L. Jiang, Y. Xue, et al. 2013b. Phase-controlled synthesis and photocatalytic properties of SnS, SnS_2 and SnS/SnS_2 heterostructure nanocrystals. *Mater. Res. Bull.*, 48: 2325–2332.

Huang, C., Y. Chan, F. Liu, D. Tang, J. Yang, Y. Lai, et al. 2013. Synthesis and characterization of multicomponent $Cu_2(Fe_xZn_{1-x})SnS_4$ nanocrystals with tunable band gap and structure. *J. Mater. Chem. A*, 1: 5402–5407.

Intergovernmental Panel on Climate Change. 2007. *IPCC Fourth Assessment Report (AR4)*. IPCC.

Iqbal, M. Z., F. Wang, M. Y. Rafique, S. Ali, M. H. Farooq, and M. Ellahi. 2013. Hydrothermal synthesis, characterization and hydrogen storage of SnS nanorods. *Mater. Lett.*, 106: 33–36.

Jackson, P., D. Hariskos, E. Lotter, S. Paetel, R. Wuerz, R. Menner, et al. 2011. New world record efficiency for $Cu(In,Ga)Se_2$ thin-film solar cells beyond 20%. *Prog. Photovolt.: Res. Appl.*, 19: 894–897.

Jiang, C., J.-S. Lee, and D. V. Talapin. 2012c. Soluble precursors for $CuInSe_2$, $CuIn_{1-x}Ga_xSe_2$, and $Cu_2ZnSn(S,Se)_4$ based on colloidal nanocrystals and molecular metal chalcogenide surface ligands. *J. Am. Chem. Soc.*, 134: 5010–5013.

Jiang, D., W. Hu, H. Wang, B. Shen, and Y. Deng. 2012a. Controlled synthesis of hierarchical CuS architectures by a recrystallization growth process in a microemulsion system. *J. Mater. Sci.*, 47: 4972–4980.

Jiang, D., W. Hu, H. Wang, B. Shen, and Y. Deng. 2012b. Synthesis, formation mechanism and photocatalytic property of nanoplate-based copper sulfide hierarchical hollow spheres. *Chem. Eng. J.*, 189–190: 443–450.

Jiang, H., P. Dai, Z. Feng, W. Fan, and J. Zhan. 2012d. Phase selective synthesis of metastable orthorhombic Cu_2ZnSnS_4. *J. Mater. Chem.*, 22: 7502–7506.

Jiang, X., W. Xu, R. Tan, W. Song, and J. Chen. 2013. Solvothermal synthesis of highly crystallized quaternary chalcogenide Cu_2FeSnS_4 particles. *Mater. Lett.*, 102–103: 39–42.

Jin, C., P. Ramasamy, and J. Kim. 2014. Facile hot-injection synthesis of stoichiometric $Cu_2ZnSnSe_4$ nanocrystals using bis(triethylsilyl) selenide. *Dalton Trans.*, 43: 9481–9485.

Kameyama, T., T. Osaki, K.-I. Okazaki, T. Shibayama, A. Kudo, S. Kuwabata, et al. 2010. Preparation and photoelectrochemical properties of densely immobilized Cu_2ZnSnS_4 nanoparticle films. *J. Mater. Chem.*, 20: 5319–5324.

Kawawaki, T., and T. Tatsuma. 2013. Enhancement of PbS quantum dot-sensitized photocurrents using plasmonic gold nanoparticles. *Phys. Chem. Chem. Phys.*, 15: 20247–20251.

Kazmerski, L. L. 2006. Solar photovoltaics R & D at the tipping point: A 2005 technology overview. *J. Electron Spectros. Relat. Phenomena* 150(2–3): 105–135.

Khadka, D. B., and J. Kim. 2014. Structural transition and band gap tuning of $Cu_2(Zn,Fe)SnS_4$ chalcogenide for photovoltaic application. *J. Phys. Chem. C*, 118: 14227–14237.

Khan, A. H., U. Thupakula, A. Dalui, S. Maji, A. Debangshi, and S. Acharya. 2013a. Evolution of long range bandgap tunable lead sulfide nanocrystals with photovoltaic properties. *J. Phys. Chem. C*, 117: 7934–7939.

Khan, M. A. M., S. Kumar, and M. S. Al-Salhi. 2013b. Synthesis and characteristics of spray deposited $CuInS_2$ nanocrystals thin films for photovoltaic applications. *Mater. Res. Bull.*, 48: 4277–4282.

Khare, A., A. W. Wills, L. M. Ammerman, D. J. Norris, and E. S. Aydil. 2011. Size control and quantum confinement in Cu_2ZnSnS_4 nanocrystals. *Chem. Commun.*, 47: 11721–11723.

Kim, K.-J., R. P. Oleksak, C. Pan, M. W. Knapp, P. B. Kreider, G. S. Herman, et al. 2014. Continuous synthesis of colloidal chalcopyrite copper indium diselenide nanocrystal inks. *RSC Adv.*, 4: 16418–16424.

Kim, Y., K. Woo, I. Kim, Y. S. Cho, S. Jeong, and J. Moon. 2013. Highly concentrated synthesis of copper-zinc-tin-sulfide nanocrystals with easily decomposable capping molecules for printed photovoltaic applications. *Nanoscale*, 5: 10183–10188.

Kirkeminde, A., B. A. Ruzicka, R. Wang, S. Puna, H. Zhao, and S. Ren. 2012. Synthesis and optoelectronic properties of two-dimensional FeS_2 nanoplates. *ACS Appl. Mater. Interfaces*, 4: 1174–1177.

Knutson, T. R., P. J. Hanson, E. S. Aydil, and R. L. Penn. 2014. Synthesis of Cu_2ZnSnS_4 thin films directly onto conductive substrates via selective thermolysis using microwave energy. *Chem. Commun.*, 50: 5902–5904.

Kreider, P. B., K.-J. Kim, and C.-H. Chang. 2014. Two-step continuous-flow synthesis of $CuInSe_2$ nanoparticles in a solar microreactor. *RSC Adv.*, 4: 13827–13830.

Kumarakuru, H., M. J. Coombes, J. H. Neethling, and J. E. Westraadt. 2014. Fabrication of Cu_2S nanoneedles by self-assembly of nanoparticles via simple wet chemical route. *J. Alloys Compd*, 589: 67–75.

Kundu, J., and D. Pradhan. 2014. Controlled synthesis and catalytic activity of copper sulfide nanostructured assemblies with different morphologies. *ACS Appl. Mater. Interfaces*, 6: 1823–1834.

Lauth, J., J. Marbach, A. Meyer, S. Dogan, C. Klinke, A. Kornowski, et al. 2014. Virtually bare nanocrystal surfaces: Significantly enhanced electrical transport in $CuInSe_2$ and $CuIn_{1-x}Ga_xSe_2$ thin films upon ligand exchange with thermally degradable 1-ethyl-5-thiotetrazole. *Adv. Funct. Mater.*, 24: 1081–1088.

Li, C., E. Ha, W.-L. Wong, C. Li, K.-P. Ho, and K.-Y. Wong. 2012b. A facile arrested precipitation method for synthesis of pure wurtzite Cu_2ZnSnS_4 nanocrystals using thiourea as a sulfur source. *Mater. Res. Bull.*, 47: 3201–3205.

Li, J., Z. Jin, T. Liu, J. Wang, D. Wang, J. Lai, et al. 2013c. Ternary and quaternary chalcopyrite $Cu(In_{1-x}Ga_x)Se_2$ nanocrystals: Organoalkali-assisted diethylene glycol solution synthesis and band-gap tuning. *Cryst. Eng. Comm.*, 15: 7327–7338.

Li, L., X. Liu, J. Huang, M. Cao, S. Chen, Y. Shen, et al. 2012c. Solution-based synthesis and characterization of Cu_2FeSnS_4 nanocrystals. *Mater. Chem. Phys.*, 133: 688–691.

Li, M., W.-H. Zhou, J. Guo, Y.-L. Zhou, Z.-L. Hou, J. Jiao, et al. 2012a. Synthesis of pure metastable wurtzite CZTS nanocrystals by facile one-pot method. *J. Phys. Chem. C*, 116: 26507–26516.

Li, Q., C. Zou, L. Zhai, L. Zhang, Y. Yang, X. A. Chen, et al. 2013a. Synthesis of wurtzite $CuInS_2$ nanowires by Ag_2S-catalyzed growth. *Cryst. Eng. Comm.*, 15: 1806–1813.

Li, Q., L. Zhai, C. Zou, X. Huang, L. Zhang, Y. Yang, et al. 2013b. Wurtzite $CuInS_2$ and $CuIn_xGa_{1-x}S_2$ nanoribbons: Synthesis, optical and photoelectrical properties. *Nanoscale*, 5: 1638–1648.

Li, W., T. Dittrich, F. Jäckel, and J. Feldmann. 2014. Optical and electronic properties of pyrite nanocrystal thin films: The role of ligands. *Small*, 10: 1194–1201.

Li, Y., Y. Hu, H. Zhang, Y. Shen, and A. Xie. 2013d. Controlled synthesis of bionic microstructures PbS crystals by mixed cationic/anionic surfactants. *Russ. J. Phys. Chem.* 87A: 1239–1245.

Liang, B. Y. J., Y.-M. Shen, S.-C. Wang, and J.-L. Huang. 2013. The influence of reaction temperatures and volume of oleic acid to synthesis SnS nanocrystals by using thermal decomposition method. *Thin Solid Films*, 549: 159–164.

Liang, X. L., and X. H. Wei. 2012. Synthesis and characterization of Cu_2FeSnS_4 semiconductor nanocrystals with zincblende structure. *Adv. Mater. Res.*, 512–515: 2019–2022.

Liang, X., P. Guo, G. Wang, R. Deng, D. Pan, and X. Wei. 2012b. Dilute magnetic semiconductor Cu_2MnSnS_4 nanocrystals with a novel zincblende and wurtzite structure. *RSC Adv.*, 2: 5044–5046.

Liang, X., X. Wei, and D. Pan. 2012a. Dilute magnetic semiconductor Cu_2FeSnS_4 nanocrystals with a novel zincblende structure. *J. Nanomater.*, 708648.

Liao, H.-C., M.-H. Jao, J.-J. Shyue, C. Y.-F. Hen, and W.-F. Su. 2013. Facile synthesis of wurtzite copper-zinc-tin sulfide nanocrystals from plasmonic djurleite nuclei. *J. Mater. Chem. A*, 1: 337–341.

Lim, Y. S., H.-S. Kwon, J. Jeong, J. Y. Kim, H. Kim, M. J. Ko, et al. 2013. Colloidal solution-processed $CuInSe_2$ solar cells with significantly improved efficiency up to 9% by morphological improvement. *ACS Appl. Mater. Interfaces*, 6: 259–267.

Liu, L., R. Hu, W.-C. Law, I. Roy, J. Zhu, L. Ye, et al. 2013c. Optimizing the synthesis of red- and near-infrared $CuInS_2$ and $AgInS_2$ semiconductor nanocrystals for bioimaging. *Analyst*, 138: 6144–6153.

Liu, W., B. Guo, C. Mak, A. Li, X. Wu, and F. Zhang. 2013b. Facile synthesis of ultrafine Cu_2ZnSnS_4 nanocrystals by hydrothermal method for use in solar cells. *Thin Solid Films*, 535: 39–43.

Liu, W. C., B. L. Guo, X. S. Wu, F. M. Zhang, C. L. Mak, and K. H. Wong. 2013a. Facile hydrothermal synthesis of hydrotropic Cu_2ZnSnS_4 nanocrystal quantum dots: Band-gap engineering and phonon confinement effect. *J. Mater. Chem. A*, 1: 3182–3186.

Liu, W., D. B. Mitzi, M. Yuan, A. J. Kellock, S. J. Chey, and O. Gunawan. 2009. 12% efficiency $CuIn(Se,S)_2$ photovoltaic device prepared using a hydrazine solution process. *Chem. Mater.* 22: 1010–1014.

Liu, Y., D. Yao, L. Shen, H. Zhang, Z. X. Hang, and B. Yang. 2012. Alkylthiol-enabled Se powder dissolution in oleylamine at room temperature for the phosphine-free synthesis of copper-based quaternary selenide nanocrystals. *J. Am. Chem. Soc.*, 134: 7207–7210.

Liu, Y., G. M.e, Y. Yue, Y. Sun, Y. Wu, X. Chen, et al. 2011. Colloidal Cu_2ZnSnS_4 nanocrystals generated by a facile route using ethylxanthate molecular precursors. *Phys. Status Solidi (RRL)*, 5: 113–115.

Liu, Y., Y. Xie, H. Cui, W. Zhao, C. Yang, Y. Wang, et al. 2013d. Preparation of monodispersed $CuInS_2$ nano-pompons and nanoflake films and application in dye-sensitized solar cells. *Phys. Chem. Chem. Phys.*, 15: 4496–4499.

Lunhong, A., and J. Jing. 2012. Self-sacrificial templating synthesis of porous quaternary Cu–Fe–Sn–S semiconductor nanotubes via microwave irradiation. *Nanotechnology*, 23: 495601.

Macpherson, H. A., and C. R. Stoldt. 2012. Iron pyrite nanocubes: Size and shape considerations for photovoltaic application. *ACS Nano*, 6: 8940–8949.

Maji, S. K., A. K. Dutta, P. Biswas, D. N. Srivastava, P. Paul, A. Mondal, et al. 2012. Synthesis and characterization of FeS nanoparticles obtained from a dithiocarboxylate precursor complex and their photocatalytic, electrocatalytic and biomimic peroxidase behavior. *Appl. Catal. A: Gen.*, 419–420: 170–177.

Mali, S. S., H. Kim, C. S. Shim, P. S. Patil, and C. K. Hong. 2013. Polyvinylpyrrolidone (PVP) assisted single-step synthesis of kesterite Cu_2ZnSnS_4 nanoparticles by solvothermal process. *Phys. Status Solidi (RRL)*, 7: 1050–1054.

Malik, M. A., P. O'Brien, and N. Revaprasadu. 1999. A novel route for the preparation of CuSe and $CuInSe_2$ nanoparticles. *Adv. Mater.*, 11: 1441–1444.

Malik, S. N., S. Mahboob, N. Haider, M. A. Malik, and P. O'Brien. 2011. A colloidal synthesis of $CuInSe_2$, $CuGaSe_2$ and $CuIn_{1-x}Ga_xSe_2$ nanoparticles from diisopropyldiselenophosphinatometal precursors. *Nanoscale*, 3: 5132–5139.

Mariama, C. S., W. Lingling, and Z. Xintong. 2013. Easy hydrothermal preparation of Cu_2ZnSnS_4 (CZTS) nanoparticles for solar cell application. *Nanotechnology*, 24: 495401.

McPhail, M. R., and E. A. Weiss. 2014. Role of organosulfur compounds in the growth and final surface chemistry of PbS quantum dots. *Chem. Mater.*, 26: 3377–3384.

Midgett, A. G., J. M. Luther, J. T. Stewart, D. K. Smith, L. A. Padilha, V. I. Klimov, et al. 2013. Size and composition dependent multiple exciton generation efficiency in PbS, PbSe, and PbS_xSe_{1-x} alloyed quantum dots. *Nano Lett.*, 13: 3078–3085.

Mocanu, A., E. Rusen, A. Diacon, and A. Dinescu. 2014. Hierarchical nanostructures of PbS obtained in the presence of water soluble polymers. *Powder Technol.*, 253: 237–241.

Morrish, R., R. Silverstein, and C. A. Wolden. 2012. Synthesis of stoichiometric FeS_2 through plasma-assisted sulfurization of Fe_2O_3 nanorods. *J. Am. Chem. Soc.*, 134: 17854–17857.

Murali, B., and S. B. Krupanidhi. 2013. Facile synthesis of Cu_2CoSnS_4 nanoparticles exhibiting red-edge-effect: Application in hybrid photonic devices. *J. Appl. Phys.*, 114: 144312.

Nadar, L., N. Destouches, N. Crespo-Monteiro, R. Sayah, F. Vocanson, S. Reynaud, et al. 2013. Multicolor photochromism of silver-containing mesoporous films of amorphous or anatase TiO_2. *J. Nanopart. Res.*, 15: 1–10.

Nakashima, S., K. Kikushima, and K. Mukai. 2013. Infrared emitting property and spherical symmetry of colloidal PbS quantum dots. *J. Crystal Growth*, 378: 537–541.

Oleksak, R. P., B. T. Flynn, D. M. Schut, and G. S. Herman. 2014. Microwave-assisted synthesis of $CuInSe_2$ nanoparticles in low-absorbing solvents. *Phys. Status Solidi. (A)*, 211: 219–225.

Pal, M., N. R. Mathews, R. S. Gonzalez, and X. Mathew. 2013. Synthesis of Cu_2ZnSnS_4 nanocrystals by solvothermal method. *Thin Solid Films*, 535: 78–82.

Pan, D., X. Wang, Z. H. Zhou, W. Chen, C. Xu, and Y. Lu, 2009. Synthesis of quaternary semiconductor nanocrystals with tunable band gaps. *Chem. Mater.*, 21: 2489–2493.

Pan, J., A. A. O. El-Ballouli, L. Rollny, O. Voznyy, V. M. Burlakov, A. Goriely, et al. 2013. Automated synthesis of photovoltaic-quality colloidal quantum dots using separate nucleation and growth stages. *ACS Nano*, 7: 10158–10166.

Panthani, M. G., V. Akhavan, B. Goodfellow, J. P. Schmidtke, L. Dunn, A. Dodabalapur, et al. 2008. Synthesis of $CuInS_2$, $CuInSe_2$, and $Cu(In_xGa_{1-x})Se_2$ (CIGS) nanocrystal "inks" for printable photovoltaics. *J. Am. Chem. Soc.*, 130: 16770–16777.

Park, J.-Y., J. H. Noh, T. N. Mandal, S. H. Im, Y. Jun, and S. I. Seok. 2013. Quaternary semiconductor Cu_2FeSnS_4 nanoparticles as an alternative to Pt catalysts. *RSC Adv.*, 3: 24918–24921.

Peng, H., G. Ma, J. Mu, K. Sun, and Z. Lei. 2014. Controllable synthesis of CuS with hierarchical structures via a surfactant-free method for high-performance supercapacitors. *Mater. Lett.*, 122: 25–28.

Ramasamy, K., V. L. Kuznetsov, K. Gopal, M. M. A. Alik, J. Raftery, P. P. Edwards, et al. 2013b. Organotin dithiocarbamates: Single-source precursors for tin sulfide thin films by aerosol-assisted chemical vapor deposition (AACVD). *Chem. Mater.*, 25: 266–276.

Ramasamy, K., M. A. Malik, and P. O'Brien. 2012. Routes to copper zinc tin sulfide Cu_2ZnSnS_4 a potential material for solar cells. *Chem. Commun.*, 48: 5703–5714.

Ramasamy, K., M. A. Malik, N. Revaprasadu, and P. O'Brien, 2013a. Routes to nanostructured inorganic materials with potential for solar energy applications. *Chem. Mater.*, 25: 3551–3569.

Rath, J. K., C. Prastani, D. E. Nanu, M. Nanu, R. E. I. Schropp, A. Vetushka, et al. 2014. Fabrication of SnS quantum dots for solar-cell applications: Issues of capping and doping. *Physica Status Solidi (B)*, 251: 1309–1321.

Rath, T., W. Haas, A. Pein, R. Saf, E. Maier, B. Kunert, et al. 2012. Synthesis and characterization of copper zinc tin chalcogenide nanoparticles: Influence of reactants on the chemical composition. *Sol. Energy Mater. Solar Cells*, 101: 87–94.

Reilly, N., M. Wehrung, R. A. O'Dell, and L. Sun. 2014. Ultrasmall colloidal PbS quantum dots. *Mater. Chem. Phys.*, 147: 1–4.

Riha, S. C., B. A. Parkinson, and A. L. Prieto. 2009. Solution-based synthesis and characterization of Cu_2ZnSnS_4 nanocrystals. *J. Am. Chem. Soc.*, 131: 12054–12055.

Riha, S. C., B. A. Parkinson, and A. L. Prieto. 2011. Compositionally tunable $Cu_2ZnSn(S_{1-x}Se_x)_4$ nanocrystals: Probing the effect of Se-inclusion in mixed chalcogenide thin films. *J. Am. Chem. Soc.*, 133: 15272–15275.

Roux, F., S. Amtablian, M. Anton, G. Besnard, L. Bilhaut, P. Bommersbach, et al. 2013. Chalcopyrite thin-film solar cells by industry-compatible ink-based process. *Sol. Energy Mater. Solar Cells*, 115: 86–92.

Saadeldin, M., H. S. Soliman, H. A. M. Ali, and K. Sawaby. 2014. Optical and electrical characterizations of nanoparticle Cu_2S thin films. *Chin. Phys. B*, 23: 046803.

Seelaboyina, R., M. Kumar, A. Venkata Madiraju, K. Taneja, A. K. Keshri, S. Mahajan, et al. 2013. Microwave synthesis of copper indium gallium (di)selenide nanopowders for thin film solar applications. *J. Renew. Sust. Energy*, 5: 031608.

Seo, Y.-H., Y. Jo, Y. Choi, K. Yoon, B.-H. Ryu, S. Ahn, et al. 2014. Thermally-derived liquid phase involving multiphase $Cu(In,Ga)Se_2$ nanoparticles for solution-processed inorganic photovoltaic devices. *RSC Adv.*, 4: 18453–18459.

Shibuya, T., Y. Goto, Y. Kamihara, M. Matoba, K. Yasuoka, L. A. Burton, et al. 2014. From kesterite to stannite photovoltaics: Stability and band gaps of the $Cu_2(Zn,Fe)SnS_4$ alloy. *Appl. Phys. Lett.*, 104: 021912.

Shin, S. W., J. H. Han, C. Y. Park, A. V. Moholkar, J. Y. Lee, and J. H. Kim. 2012. Quaternary Cu_2ZnSnS_4 nanocrystals: Facile and low cost synthesis by microwave-assisted solution method. *J. Alloys Compds.*, 516: 96–101.

Sims, R. E. H. 2008. Hydropower, geothermal, and ocean energy. *MRS Bull.*, 33: 389–395.

Singh, S., S. K. Samji, and M. S. R. Rao. 2012. Synthesis and characterisation of $CuInGaS_2$ nano-ink for photovoltaic applications. *J. Exp. Nanosci.*, 8: 320–325.

Sobhani, A., M. Salavati-Niasari, and S. M. Hosseinpour-Mashkani. 2012. Single-source molecular precursor for synthesis of copper sulfide nanostructures. *J. Cluster Sci.*, 23: 1143–1151.

Steinhagen, C., T. B. Harvey, C. J. Stolle, J. Harris, and B. A. Korgel. 2012. Pyrite nanocrystal solar cells: Promising, or fool's gold? *J. Phys. Chem. Lett.*, 3: 2352–2356.

Steinhagen, C., M. G. Panthani, V. Akhavan, B. Goodfellow, B. Koo, and B. A. Korgel. 2009. Synthesis of Cu_2ZnSnS_4 nanocrystals for use in low-cost photovoltaics. *J. Am. Chem. Soc.*, 131: 12554–12555.

Stolle, C. J., M. G. Panthani, T. B. Harvey, V. A. Akhavan, and B. A. Korgel, 2012. Comparison of the photovoltaic response of oleylamine and inorganic ligand-capped $CuInSe_2$ nanocrystals. *ACS Appl. Mater. Interfaces*, 4: 2757–2761.

Sun, S., D. Deng, C. Kong, X. Song, and Z. Yang. 2012. Twins in polyhedral 26-facet Cu_7S_4 cages: Synthesis, characterization and their enhancing photochemical activities. *Dalton Trans.*, 41: 3214–3222.

Tang, J., S. Hinds, S. O. Kelley, and E. H. Sargent. 2008. Synthesis of colloidal $CuGaSe_2$, $CuInSe_2$, and $Cu(InGa)Se_2$ nanoparticles. *Chem. Mater.*, 20: 6906–6910.

Thompson, M. J., T. P. A. Ruberu, K. J. Blakeney, K. V. Torres, P. S. Dilsaver, and J. Vela. 2013. Axial composition gradients and phase segregation regulate the aspect ratio of Cu_2ZnSnS_4 nanorods. *J. Phys. Chem. Lett.*, 4: 3918–3923.

Tiong, V. T., Y. Zhang, J. Bell, and H. Wang. 2014. Phase-selective hydrothermal synthesis of Cu_2ZnSnS_4 nanocrystals: The effect of the sulphur precursor. *Cryst. Eng. Comm.*, 16: 4306–4313.

Todorov, T. K., K. B. Reuter, and D. B. Mitzi. 2010. High-efficiency solar cell with earth-abundant liquid-processed absorber. *Adv. Mater.*, 22: E156–E159.

Tsuji, I., H. Kato, and A. Kudo. 2006. Photocatalytic hydrogen evolution on ZnS–$CuInS_2$–$AgInS_2$ solid solution photocatalysts with wide visible light absorption bands. *Chem. Mater.*, 18: 1969–1975.

Uhl, A. R., M. Koller, A. S. Wallerand, C. M. Fella, L. Kranz, H. Hagendorfer, et al. 2013. $Cu(In,Ga)Se_2$ absorbers from stacked nanoparticle precursor layers. *Thin Solid Films*, 535: 138–142.

U.S. Energy Information Administration, U.S. Department of Energy. 2013. *Int. Energy Outlook*. Washington, DC.

Vinokurov, K., J. E. Macdonald, and U. Banin. 2012. Structures and mechanisms in the growth of hybrid Ru–Cu_2S nanoparticles: From cages to nanonets. *Chem. Mater.*, 24: 1822–1827.

Wang, G., S. Wang, Y. Cui, and D. Pan. 2012. A novel and versatile strategy to prepare metal–organic molecular precursor solutions and its application in $Cu(In,Ga)(S,Se)_2$ solar cells. *Chem. Mater.*, 24: 3993–3997.

Wang, X., W. Zhou, Z. Chang, Z. Zhou, and S. Wu. 2013. Solvent-free synthesis of hexagonal iron sulfide nanoflowers. *Chin. J. Chem.*, 31: 983–986.

Wang, Y.-H. A., X. Zhang, N. Bao, B. Lin, and A. Gupta. 2011. Synthesis of shape-controlled monodisperse wurtzite $CuIn_xGa_{1-x}S_2$ semiconductor nanocrystals with tunable band gap. *J. Am. Chem. Soc.*, 133: 11072–11075.

Wei, M., Q. Du, D. Wang, L. W. Iu, G. Jiang, and C. Zhu. 2012. Synthesis of spindle-like kesterite Cu_2ZnSnS_4 nanoparticles using thiorea as sulfur source. *Mater. Lett.*, 79: 177–179.

Winkler, M. T., W. Wang, O. Gunawan, H. J. Hovel, T. K. Todorov, and D. B. Mitzi. 2014. Optical designs that improve the efficiency of $Cu_2ZnSn(S,Se)_4$ solar cells. *Energy Environ. Sci.*, 7: 1029–1036.

Wooten, A. J., D. J. Werder, D. J. Williams, J. L. Casson, and J. A. Hollingsworth. 2009. Solution–liquid–solid growth of ternary Cu–in–Se semiconductor nanowires from multiple- and single-source precursors. *J. Am. Chem. Soc.*, 131: 16177–16188.

Xiong, S., and H. C. Zeng. 2012. Serial ionic exchange for the synthesis of multishelled copper sulfide hollow spheres. *Angew. Chem. Int. Ed.*, 51: 949–952.

Xu, J., C.-S. Lee, Y.-B. Tang, X. Chen, Z.-H. Chen, W.-J. Zhang, et al. 2010. Large-scale synthesis and phase transformation of CuSe, $CuInSe_2$, and $CuInSe_2/CuInS_2$ core/shell nanowire bundles. *ACS Nano.*, 4: 1845–1850.

Yan, C., C. Huang, J. Yang, F. Liu, J. Liu, Y. Lai, et al. 2012. Synthesis and characterizations of quaternary Cu_2FeSnS_4 nanocrystals. *Chem. Commun.*, 48: 2603–2605.

Yan, X., E. Michael, S. Komarneni, J. R. Brownson, and Z.-F. Yan. 2013a. Microwave- and conventional-hydrothermal synthesis of CuS, SnS and ZnS: Optical properties. *Ceram. Int.*, 39: 4757–4763.

Yan, Z., Y. Zhao, M. Zhuang, J. Liu, and A. Wei. 2013b. Solvothermal synthesis of $CuInS_2$ powders and $CuInS_2$ thin films for solar cell application. *J. Mater. Sci.: Mater. Electronics*, 24: 5055–5060.

Yang, H., L. A. Jauregui, G. Zhang, Y. P. Chen, and Y. Wu. 2012. Nontoxic and abundant copper zinc tin sulfide nanocrystals for potential high-temperature thermoelectric energy harvesting. *Nano Lett.*, 12: 540–545.

Yang, J., C. Bao, J. Zhang, T. Yu, H. Huang, Y. Wei, et al. 2013. In situ grown vertically oriented $CuInS_2$ nanosheets and their high catalytic activity as counter electrodes in dye-sensitized solar cells. *Chem. Commun.*, 49: 2028–2030.

Yao, R. Y., Z. J. Zhou, Z. L. Hou, X. Wang, W. H. Zhou, and S. X. Wu. 2013. Surfactant-free $CuInS_2$ nanocrystals: An alternative counter-electrode material for dye-sensitized solar cells. *ACS Appl. Mater. Interfaces* 5: 3143–3148.

Yarema, O., D. Bozyigit, I. Rousseau, L. Nowack, M. Yarema, W. Heiss, et al. 2013. Highly luminescent, size- and shape-tunable copper indium selenide based colloidal nanocrystals. *Chem. Mater.*, 25: 3753–3757.

Yidong, Z., H. Weiwei, and J. Huimin. 2013. Hydrothermal fabrication of chalcopyrite-type $CuInS_2$ film and their optical properties. *Physica Scripta*, 88: 015705.

Yoon, H., S. H. Na, J. Y. Choi, M. W. Kim, H. Kim, H. S. An, et al. 2014. Carbon- and oxygen-free Cu(InGa)(SSe)$_2$ solar cell with a 4.63% conversion efficiency by electrostatic spray deposition. *ACS Appl. Mater. Interfaces*, 6: 8369–8377.

Yu, K., P. Ng, J. Ouyang, M. B. Zaman, A. Abulrob, T. N. Baral, et al. 2013. Low-temperature approach to highly emissive copper indium sulfide colloidal nanocrystals and their bioimaging applications. *ACS Appl. Mater. Interfaces*, 5: 2870–2880.

Zaberca, O., F. Oftinger, J. Y. Chane-Ching, L. Datas, A. Lafond, P. Puech, et al. 2012. Surfactant-free CZTS nanoparticles as building blocks for low-cost solar cell absorbers. *Nanotechnology*, 23: 185402.

Zhai, X., H. Jia, Y. Zhang, Y. Lei, J. Wei, Y. Gao, et al. 2014. In situ fabrication of Cu_2ZnSnS_4 nanoflake thin films on both rigid and flexible substrates. *Cryst. Eng. Comm.*, 16: 6244–6249.

Zhang, J., S. Zhang, H. Zhang, Y. Zhang, Z. Zheng, and Y. Xiang. 2014. Activated selenium for promoted formation of metal selenide nanocrystals in solvothermal synthesis. *Mater. Lett.*, 122: 306–308.

Zhang, Q., E. Uchaker, S. L. Candelaria, and G. Cao. 2013a. Nanomaterials for energy conversion and storage. *Chem. Soc. Rev.*, 42: 3127–3171.

Zhang, X., N. Bao, B. Lin, and A. Gupta. 2013b. Colloidal synthesis of wurtzite Cu_2CoSnS_4 nanocrystals and the photoresponse of spray-deposited thin films. *Nanotechnology*, 24: 105706.

Zhang, X., N. Bao, K. Ramasamy, Y.-H. A. Wang, Y. Wang, B. Lin, et al. 2012. Crystal phase-controlled synthesis of Cu_2FeSnS_4 nanocrystals with a band gap of around 1.5 eV. *Chem. Commun.*, 48: 4956–4958.

Zhong, H., Y. Zhou, M. Ye, Y. He, J. Ye, C. He, et al. 2008. Controlled synthesis and optical properties of colloidal ternary chalcogenide $CuInS_2$ nanocrystals. *Chem. Mater.*, 20: 6434–6443.

Zhou, H., W.-C. Hsu, H.-S. Duan, B. Bob, W. Yang, T.-B. Song, et al. 2013. CZTS nanocrystals: A promising approach for next generation thin film photovoltaics. *Energy Environ. Sci.*, 6: 2822–2838.

Zhou, Y.-L., W.-H. Zhou, Y.-F. Du, M. Li, and S.-X. Wu. 2011a. Sphere-like kesterite Cu_2ZnSnS_4 nanoparticles synthesized by a facile solvothermal method. *Mater. Lett.*, 65: 1535–1537.

Zhou, Y.-L., W.-H. Zhou, M. Li, Y.-F. Du, and S.-X. Wu. 2011b. Hierarchical Cu_2ZnSnS_4 particles for a low-cost solar cell: Morphology control and growth mechanism. *J. Phys. Chem. C*, 115: 19632–19639.

Zhu, L., B. J. Richardson, and Q. Yu. 2014. Controlled colloidal synthesis of iron pyrite FeS_2 nanorods and quasi-cubic nanocrystal agglomerates. *Nanoscale*, 6: 1029–1037.

Zou, C., L. Zhang, D. Lin, Y. Yang, Q. Li, X. Xu, et al. 2011. Facile synthesis of Cu_2ZnSnS_4 nanocrystals. *Cryst. Eng. Comm.*, 13: 3310–3313.

12 Other Solar Cells

Rakshit Ameta

CONTENTS

12.1 Introduction 253
12.2 Plasmonic Solar Cells 253
12.3 Hybrid Solar Cells 255
12.4 Biohybrid Solar Cell 258
12.5 Perovskite Solar Cells 259
12.6 Miscellaneous Solar Cells 260
References 262

12.1 INTRODUCTION

Solar cells can be divided into three different generations. The first-generation solar cells are made from crystalline semiconductor wafers (200–300 µm), and 90% of the solar cell market is based on these first-generation solar cells. Approximately 40% of the cost of a solar module is due to thick silicon wafers. Second-generation solar cells are based on thin-film (1–2 µm) technology. This film is deposited on low-cost substrates such as glass, plastic, or stainless steel. The main focus of these solar cells is lowering the amount of material used. They contain a variety of semiconductors like cadmium telluride, copper indium diselenide, and so on, as well as amorphous and polycrystalline silicon. A major limitation of thin-film solar cells is their ineffective absorbance near bandgap, in particular, for the indirect bandgap semiconductor silicon. Therefore, it is important to trap light inside the solar cell in order to increase the absorbance. Third-generation solar cells are presently investigated with the goal to increase the efficiency using second-generation SCs.

Most of the commercial solar cells are made from a refined, highly purified silicon crystal (silicon wafer). These silicon solar cells and complex production processes involve a high cost, and therefore, have generated interest in alternative technologies.

12.2 PLASMONIC SOLAR CELLS

Plasmonic solar cells are a class of photovoltaic devices that converts light into electricity by using plasmons. This is a type of thin-film solar cell normally 1–2 µm thick. Low-cost substrates such as glass, plastic, or steel can be used in place of costly silicon to fabricate these cells. The major problem for thin-film solar cells is limited absorption as compared to thicker solar cells. Plasmonic cells can have more absorption by scattering light using metal nanoparticles excited at their surface plasmon resonance (Catchpole and Polman 2008). This permits light to be absorbed more directly without a relatively thick additional layer. It was found that Raman scattering can be increased by using metal nanoparticles, which makes more photons available to excite surface plasmons and, in turn, electrons, which travel through the thin-film solar cells, thus generating a current.

Photocurrent enhancement was observed by Stuart and Hall (1998) using silver nanoparticles, while Westphalen et al. (2000) reported an enhancement for silver clusters incorporated into indium tin oxide and zinc-phthalocyanine solar cells. Enhanced efficiencies for ultrathin-film organic solar cells were also observed using 5 nm–diameter silver nanoparticles (Rand et al. 2004). Gold nanoparticles were also used for scattering and absorption of light on doped silicon resulting in 80% enhancements (Schaadt et al. 2005). Derkacs et al. (2006) also used gold nanoparticles on thin-film silicon, giving 8% on conversion efficiency.

Another method of utilizing surface plasmons for harvesting solar energy is to have a thin film of silicon and a thin layer of metal deposited on the lower surface. The light will travel through the silicon and generate surface plasmons on the interface of the silicon and metal. This generates electric fields inside of the silicon, because electric fields do not travel very far into metals. If the electric field is strong enough, electrons produced a photocurrent. The thin film of metal must be of nanometric-sized grooves, which act as waveguides for the incident light to excite more photons in the silicon thin film (Ferry et al. 2008).

When a photon strikes the substrate of a solar cell, an electron and hole pair is generated. Once the electrons and holes are separated, they will try to recombine. If the electrons can be collected prior to this recombination, then they can be utilized as a current in an external circuit. In a solar cell, it is important to keep a balance by minimizing this recombination, which required thin layers and the absorption of more photon, which needs thicker-layer Tanabe (2009).

The basic functioning of plasmonic solar cells depends on scattering and absorption of light due to the deposition of metal nanoparticles. Silicon does not absorb light appreciably. Therefore, more light is to be scattered across the surface in order to increase the absorption. It has been found that metal nanoparticles help in scattering the incident light across the surface of the silicon. Surface plasmon resonance depends mainly on the density of free electrons in the particle. The order of densities of electrons for different metals (aluminum and silver in ultraviolet, and gold and copper in visible light) corresponds to the resonance. This resonant frequency can be shifted by variation of the dielectric constant for the embedding medium.

Presently, first-generation solar cells do not exceed efficiencies of about 30%, while third-generation cells can be expected to achieve efficiencies up to 40%–60%. However, in second-generation cells, required material needs have been reduced through the use of thin-film technology. The goal of third-generation solar cells is to increase the efficiency of second-generation solar cells (thin film) using materials found abundantly on Earth.

Standridge et al. (2009) fabricated titanium dioxide–based dye-sensitized solar cells by incorporating corrosion-protected silver nanoparticles as plasmonic optical elements of the photoelectrode. There was an enhancement of the dye extinction when plasmonic particles are present, and they increase on decreasing TiO_2 thickness.

Au NPs were produced by ultrafast laser ablation in liquids and were incorporated on the devices in a single-walled nanotube (SWNT)/poly(3,4-ethylene-dioxythiophene):poly(4-styrenesulfonate)/NP/poly(3-hexylthiophene):[6,6]-phenyl-C61-butyric acid methyl ester/Al configuration (Kymakis et al. 2011). The incorporation of NPs leads to a power conversion efficiency improvement of 70%. This increase has been attributed to the improved photocurrent and fill factor due to an enhanced exciton generation rate of the photoactive layer caused by localized surface plasmon resonances of the conduction electrons within the NPs.

Ding et al. (2011) reported that on addition of 22 vol% of core–shell particles to a 5 μm thick TiO_2 film, the energy conversion efficiency of DSSCs increases from 2.7% to 4.0%, in spite of a more than 20% decrease in the amount of dyes adsorbed on the composite films. Plasmonic structures of fluorine-doped tin oxide (FTO)/TiO_2/NPs-Ag and FTO/NPs-Ag/TiO_2 electrodes were fabricated and used in DSSC (Lin et al. 2012). An enhancement of 60% in photocurrent as well as improvement in photovoltage was observed. Xu et al. (2012) observed that Au@PVP not only adds to chemical stability to iodide/triiodide electrolyte, but also to the adhesiveness to dye molecules. They obtained a power conversion efficiency (PCE) enhancement of 30% from 3.3% to 4.3% with incorporation of Au@PVP NPs.

Plasmonic core–shell nanoparticles (PCSNPs) can function as nanoantenna and will improve the efficiency of dye-sensitized solar cells. It was observed that PCSNPs with a thinner shell mainly enhance the current, whereas particles with a thicker shell improve the voltage (Liu et al. 2013). Synthesis of TiO_2 branched nanorod arrays (TiO_2 BNRs) with plasmonic Au nanoparticles attached on the surface was reported by Su et al. (2013). These Au/TiO_2 BNR composites exhibit high photocatalytic activity in photoelectrochemical water splitting. The Au/TiO_2 BNRs achieved the highest efficiency of ~1.27%, which indicated elevated charge separation and transportation efficiencies. The high PEC performance is mainly due to the plasmonic effect of Au nanoparticles, which enhances the visible light absorption, together with the large surface area, efficient charge separation, and high carrier mobility of the TiO_2 BNRs.

Zhang et al. (2014) fabricated silver nanoprisms (AgNPs) on indium tin oxide (ITO) by immersing it in AgNPs solution for a series of immersion times. Performance of the device was the best when the immersion time was 30 min, corresponding to AgNPs coverage of 68%. Under this condition, the device showed short-circuit current density of 10.10 mA cm^{-2} (18% improved), and power conversion efficiency of 3.88% (23% improved). Poly(3,4-ethylenedioxythiophene) (PEDOT) films incorporating gold nanoparticles have been used for counterelectrodes in dye-sensitized solar cells by Koussi-Daoud et al. (2014). An increase in efficiency of about 130% has been reported as compared to the use of PEDOT alone.

A sandwich-structured CdS-Au-TiO_2 nanorod array was used as the photoanode in a photoelectrochemical cell for hydrogen generation via the splitting of water (Li et al. 2014a). The gold nanoparticles sandwiched between the TiO_2 nanorod and the CdS quantum dot layer play a dual role in enhancing the solar-to-chemical energy conversion efficiency. Jiao et al. (2015) used plasmonic Ag nanoparticles with vertically aligned N-doped TiO_2 nanotube arrays showing improved photoelectrochemical performance. The N-doped anatase nanotube arrays were fabricated, and cubic silver nanoparticles (diameter 5 nm) were deposited at room temperature on TiO_2 nanotubes without organic additives. The plasmonic Ag/N-doped TiO_2 composites are favorable for the separation for photoelectron–hole pairs and increasing electron transfer resulting in enhanced photoelectrochemical performance.

Recently, Wang et al. (2015) reported the feasibility of simultaneous production of hydrogen and electricity as well as contaminant removal from actual urban wastewater within a dye-sensitized photoelectrochemical cell (DSPC). The photoanode in the DSPC was a novel nanostructured plasmonic Ag/AgCl@chiral TiO_2 nanofiber (Ag and AgCl nanoparticles supported on chiral TiO_2 nanofibers). The electrolyte was actual wastewater, with added 17-β-ethinylestradiol and copper.

12.3 HYBRID SOLAR CELLS

Hybrid solar cells combine advantages of both organic and inorganic semiconductors. Hybrid photovoltaics have organic materials that consist of conjugated polymers that absorb light as the donor and transport holes (Milliron et al. 2005). Inorganic materials in hybrid cells are used as the acceptor and electron transporters in the structure. The hybrid photovoltaic devices have a potential not only for low cost but also for scalable solar power conversion.

An organic material is mixed with a high electron transport material to form the photoactive layer in hybrid solar cells (Shaheen et al. 2005). Assemblage of two materials together in a heterojunction-type photoactive layer will have a greater power conversion efficiency than a single material (Saunders and Turner 2008). One of these materials acts as the photon absorber as well as exciton donor, while the other material facilitates dissociation of exciton at the junction. Charge is transferred and then separated after an exciton created in the donor is delocalized on a donor–acceptor complex (Sariciftci et al. 1993).

The average distance an exciton can diffuse through a material before annihilation by recombination is called *exciton diffusion length*. This is relatively short in polymers—that is, 5–10 nm (Ginger and Greenham 1999). Excitons generated within this length close to an acceptor would contribute to the photocurrent.

Increased efficiency can be achieved by increasing the interfacial surface area between the organic and inorganic material to facilitate charge separation and by controlling the nanoscale lengths. The charges then remain separate and move toward the electrode before these recombine. The three main nanoscale structures used are

- Mesoporous inorganic films infused with electron-donating organics
- Alternating inorganic–organic lamellar structures
- Nanowire structures

Mesoporous films have been used for a high-efficiency hybrid solar cell. The structure of mesoporous thin-film solar cells usually includes a porous inorganic material that is saturated with an organic surfactant, which absorbs light and transfers electrons to the inorganic semiconductor. The electrons are then transferred to the electrode. Herman et al. (2011) used alternating layers of organic and inorganic compounds, which are controlled through electrodeposition-based self-assembly. The lamellar structure and periodicity of the alternating organic-inorganic layers can be controlled through solution chemistry. Larger organic surfactants that absorb more visible radiation must be deposited between the layers of electron-accepting inorganics to obtain cells with practical efficiencies.

Nanostructure-based solar cells using nanowires or nanotubes of inorganics surrounded by electron-donating organics utilizing self-organization processes have been developed. They offer the advantages of directed charge transport and controlled phase separation between donor and acceptor materials (Weickert et al. 2011). The nanowire-based morphology offers reduced internal reflection, facile strain relaxation, and increased defect tolerance. The efficiencies of nanowire-based solar cells have been increased with time. Now, it seems that they are one of the most promising nanoscale solar hybrid technologies (Garnett et al. 2011).

Hybrid cell efficiency must be increased to start large-scale manufacturing. Three factors are likely to affect efficiency:

- Bandgap should be reduced to absorb longer-wavelength (red) photons, which contain a significant fraction of the energy in solar insolation. Current organic photovoltaics have shown 70% quantum efficiency for shorter wavelength (blue) photons.
- Contact resistance between each layer in the device should be minimized to offer higher fill factor and power conversion efficiency.
- Charge carrier mobility should be increased to allow the photovoltaics to have thicker active layers while minimizing the recombination of the carrier and keeping the resistance of the device low.

The size of nanoparticle control creates quantum confinement and allows for the tuning of optoelectronic properties, such as bandgap and electron affinity. The cells also have a large surface area to volume ratio, which presents more area for charge transfer to occur (Wu et al. 2005).

The photoactive layer can be created by mixing nanoparticles into a polymer matrix. Solar devices based on polymer–nanoparticle composites mostly resemble polymer solar cells. Hole mobilities are greater than electron mobilities, so the polymer phase is used to transport holes, while nanoparticles transport electrons to the electrode. A large interfacial area between the polymer phase and the nanoparticles is needed, which is achieved by dispersing the particles throughout the polymer matrix. However, the nanoparticles need to be interconnected to form percolation networks for electron transport. Efficiency is also affected by aspect ratio, geometry, and volume fraction of the nanoparticles.

The nanoparticle bandgap should be tuned so that it matches the corresponding polymer. The nanoparticles involved are typically colloids, and these are stabilized in solution by ligands, which decrease efficiency because they impede interaction between the donor and nanoparticle acceptor and decrease electron mobility. Some success has been achieved by Saunders (2012) by exchanging the initial ligands for pyridine or another short-chain ligand.

Hybrid solar cells based on dye-sensitized solar cells are fabricated by dye-absorbed inorganic materials and organic materials. The most preferred inorganic material is TiO_2 because it is a low-cost material, is easily synthesized, and acts as an n-type semiconductor due to the donor-like oxygen vacancies. Therefore, molecular sensitizers (dyes) are attached to the titania surface to collect a greater portion of the solar spectrum. A photon is absorbed by a sensitizer molecule layer, which induces electron injection into the conduction band of titania, thus resulting in flow of current. A shorter diffusion length in such solar cells decreases the solar-to-energy conversion efficiency. A variety of organic materials are attached to the titania to increase this diffusion length (or carrier lifetime).

Some supramolecular or multifunctional sensitizers have been used to enhance carrier diffusion length (Moser 2005). A dye chromophore has been modified by the addition of secondary electron donors. Minority carriers (holes) diffuse to the attached electron donors to recombine. Therefore, electron–hole recombination is retarded by the physical separation between the dye–cation moiety and surface of TiO_2. This results in an enhanced carrier diffusion length, and as a consequence, the carrier lifetime is also increased.

A dye-sensitized mesoporous film of TiO_2 can be used for making photovoltaic cells. Such a solar cell is called a *solid-state dye-sensitized solar cell*. The pores in mesoporous TiO_2 thin film (2–50 μm) are filled with any solid hole-conducting material, such as p-type semiconductors or organic hole-conducting material. Liquid electrolyte is changed with a solid charge transport material in these cells; however, the process of electron–hole generation and recombination remains the same. Here, electrons are injected from photoexcited dye into the conduction band of titania, and holes are transported by a solid charge transport electrolyte to an electrode. Attempts have been made to obtain organic materials that have a high solar-to-energy conversion efficiency in dye-synthesized solar cells using mesoporous titania thin film (Lancelle-Beltron et al. 2006).

Sofos et al. (2009) reported a nanostructured lamellar structure that provides an ideal design for bulk heterojunction solar cells. This cell has ZnO and some small conducting organic molecules, which co-assemble into alternating layers of organic and inorganic components. This highly organized structure is stabilized by π–π stacking between the organic molecules, which allows for conducting pathways in both organic and inorganic layers. The thicknesses of the layers are within the exciton diffusion length, thus minimizing recombination among charge carriers. Initial photoconductivity measurements have exhibited higher values as compared with other organic, hybrid, and amorphous silicon photoconductors. Such systems may prove to be promising and efficient hybrid photovoltaic devices.

Bai et al. (2012) reported a method to improve the efficiency of the promising CNT-Si solar cells to about 10% by hydrogen peroxide doping. It forms hybrid solar cells consisting of numerous CNT-Si heterojunction cells and CNT-Si-H_2O_2 photoelectrochemical cells in parallel by hydrogen peroxide doping. The quantum efficiency of the H_2O_2-doped solar cell was higher than that of the original cell.

Wang and Tang (2012) reported a photoelectrochemical biofuel cell (PEBFC) generating electrical energy directly from sunlight and biomass. This cell had a natural chlorophyll-sensitized titanium dioxide film photoanode and Pt black cathode. The incident photon-to-current efficiency (IPCE) was found to be 8.4%. Xia et al. (2011) synthesized mesoporous CdS spheres with large surface areas and ordered pore-size distribution. This electrode has a conversion efficiency of 2.39%. The good performance, low cost, and straightforward fabrication method made this mesoporous CdS material promising for the development of effective photoelectrochemical cells. A hybrid heterojunction and solid-state photoelectrochemical solar cell based on graphene woven fabrics (GWFs) and silicon was fabricated by Li et al. (2014b), and significant improvement in power conversion efficiency was achieved (11%).

Kong et al. (2014) reported a hybrid photoelectrode by using single-crystalline rutile TiO_2 nanowires (NWs) inlaid with anatase TiO_2 nanoparticles (NPs). This NW-NP electrode exhibited

6.2% conversion efficiency, which corresponds to an almost 48% improvement over the efficiency of the nanoparticle-dye sensitized solar cell (NP-DSC). This may be attributed to the synergistic effects of the enhanced light confinement, charge collection, and dye loading. A novel photoanode consisting of CdS/CdSe cosensitized sea urchin–shaped ZnO/TiO$_2$-based nano/micro hybrid heterostructures offers better light scattering over the entire visible frequency domain as well as better separation of photogenerated charge carriers. A remarkable photoconversion efficiency of 6.4% was observed (Ali et al. 2014).

A hybrid photovoltaic/photoelectrochemical (PV/PEC) water-splitting device with a benchmark solar-to-hydrogen conversion efficiency of 5.2% has been reported (Han et al. 2014). This cell consists of a gradient-doped tungsten-bismuth vanadate (W:BiVO$_4$) photoanode and a thin-film silicon solar cell. Hybrid TiO$_2$-nanoparticle (NP)/TiO$_2$-SiO$_2$ (TS) composites were prepared using rice straw biotemplates and then were used as photoelectrodes in dye-sensitized solar cells (Jin et al. 2014). The hybrid TiO$_2$-NP/TS composite showed a maximum power conversion efficiency of 5.81%.

12.4 BIOHYBRID SOLAR CELL

A biohybrid solar cell is made up of a combination of organic matter (photosystem I) and inorganic matter. These solar cells have been made by a team of researchers at Vanderbilt University. This team used the photosystem I (a photoactive protein complex located in the thylakoid membrane) to recreate the natural process of photosynthesis to obtain a greater efficiency in solar energy conversion. These biohybrid solar cells present a new type of renewable energy (Ciesielskia et al. 2010; Yehezkeli et al. 2012).

Multiple layers of this photosystem I gather photonic energy and convert it into chemical energy, generating a current. This cell consists of similar nonorganic materials found in other solar cells, with the only difference that the injected photosystem I complexes are introduced and gathered for some days in the gold layer. The layers of photosystem I are made visible and appear as a thin green film. This thin green film helps in improving the energy conversion, prompting attempts to incorporate and improve different technologies. Spinach was used as the source for photosystem I. Thylakoid membranes were isolated and purified to separate photosystem I from this membrane. This resulted in improved electrical current compared to that of other solar cells (almost a thousand times greater).

Biohybrid solar cells are advantageous because they convert solar energy to electricity with almost 100% efficiency. This conversion is quite efficient as compared to 40% efficiency in traditional solar cells. Producing a biohybrid solar cell is relatively low cost because extracting the protein from spinach and other plants is much cheaper as compared to the cost of metals required to produce other solar cells. Although the efficiency of the biohybrid cells is much higher, there are many disadvantages. Traditional solar cells produce more power than that currently being achieved by biohybrid cells. Also, the life span of a biohybrid solar cell is short—a few weeks to about 9 months.

Qian et al. (2010) reported a solar-driven microbial photoelectrochemical cell (MPC) that can produce sustainable energy through coupling the microbial catalysis of biodegradable organic matter with solar energy conversion. This MPC consists of a p-type cuprous oxide nanowire-arrayed photocathode and an electricigen (*Shewanella oneidensis* MR-1)-colonizing anode. A substantial current generation of 200 µA was achieved from the MPC device based on the synergistic effect of the bioanode and photocathode at zero bias under illumination of 20 mW cm^{-2}.

Current photovoltaic and photocatalytic systems are almost entirely based on semiconductor materials. Efforts are being made to develop semiconductor/biomolecular composites for converting sunlight to electricity and fuels (Li et al. 2011). Light-induced cis/trans isomerization in the family of merocyanine (MC) dyes offers a recyclable proton pumping ability, which can be potentially used in hybrid bioelectronic devices (Tayebi et al. 2014). Lipid molecules play a critical role in stabilizing the dye in a membrane structure for practical use in energy devices. A major modification in the cell was introduced by eliminating the I_2/I^- electrolyte, resulting in a twofold increase in the open-circuit voltage as compared with that of the conventional cell. The charging time was also reduced by approximately four orders of magnitude.

12.5 PEROVSKITE SOLAR CELLS

A perovskite solar cell is a type of solar cell that includes a perovskite absorber, most commonly a hybrid organic–inorganic lead or tin halide–based material, as the light-harvesting active layer. Perovskite absorber materials are extremely cheap to produce and simple to manufacture (e.g., methylammonium or formamidinium lead halide). Solar cell efficiencies of such devices have increased from 3.8% (Kojima et al. 2009) to 20.1%. It is an advancing solar technology. Their high efficiencies and production costs make perovskite solar cells an commercially attractive option.

The name *perovskite solar cell* was derived from perovskite structure. The unit cell of the most commonly employed perovskite absorber in solar cells is methylammonium lead trihalide ($CH_3NH_3PbX_3$, where X I$^-$, Br$^-$, Cl$^-$). It has a bandgap between 2.3 and 1.57 eV depending on the content of halide. Formamidinium lead trihalide ($H_2NCH_3NH_3PbX_3$) is another newer material, which shows promise, with a bandgap between 2.23 and 1.48 eV. This minimum bandgap is closer to the optimal for a single-junction cell than methylammonium lead trihalide; therefore, it should be capable of higher efficiencies (Eperon et al. 2014). Lead in perovskite materials may be replaced by tin in perovskite absorber, $CH_3NH_3SnI_3$, and this cell shows a power-conversion efficiency of more than 6% (Noel et al. 2014).

Perovskite solar cells hold an advantage over traditional silicon solar cells in the simplicity of their processing. Traditional silicon cells require expensive, multistep processes with high temperatures (>1000°C) and vacuums in special clean-room facilities to produce high-purity silicon wafers. Organic–inorganic perovskite material can be manufactured with simpler wet chemistry and processing techniques. Most notably, methylammonium and formamidinium lead trihalides have been prepared using a variety of solvent techniques and vapor deposition techniques. Lead halide and methylammonium iodide can be dissolved in solvent and spin coated onto a substrate in solution processing. Simple solution processing results in the development of voids, platelets, and other defects in the layer, which may decrease the efficiency of a solar cell. Lee et al. (2012) mentioned that rapid improvement of perovskite solar cells has made them the new generation of the photovoltaics world with high promise. There is a huge potential for more efficient perovskite solar cells to reach in excess of 20% power conversion efficiency.

Zhang et al. (2013) reported enhanced photocurrent in perovskite solar cells utilizing plasmonic core–shell (gold–silica) nanoparticles. Although electrical power conversion efficiencies around 17% have been obtained in organic–inorganic lead halide perovskite solar cells, the potential toxicology of lead is a problem. Tin may be a suitable substitute. While organic–inorganic tin halide perovskites have shown good semiconducting behavior, the instability of tin in its 2+ oxidation state is an existing challenge. Completely lead free, a $CH_3NH_3SnI_3$ perovskite solar cell possessing a mesoporous TiO_2 has shown efficiency over 6%.

In the last 2–3 years, a new class of solar cell, which is based on mixed organic–inorganic halide perovskites, has emerged. The first efficient solid-state perovskite cells were reported in 2012, they attained 16.2% energy appreciable conversion efficiencies by 2013, and they were claimed to achieve between 17.9% and 19.3% efficiencies last year. Green et al. (2014) has reviewed this field.

Organic–inorganic perovskites, such as $CH_3NH_3PbX_3$ (X = I, Br, Cl), have emerged as attractive absorber materials for the fabrication of low-cost and high-efficiency solar cells. The n-type contact was modified with a self-assembled fullerene monolayer, and electron transfer is switched on. Both n-type and p-type heterojunctions with the perovskite are active in driving the photovoltaic operation. The fullerene-modified devices achieve up to 17.3% power conversion efficiency with significantly reduced hysteresis.

Perovskite-based solar cells using $CH_3NH_3PbI_{3-x}Cl_x$ with different hole-transporting materials were fabricated (Giacomo et al. 2014). 2,2′,7,7′-Tetrakis-(N,N-di-p-methoxyphenylamine)-9,9′-spirobifluorene (Spiro-OMeTAD) has been compared with poly(3-hexylthiophene-2,5-diyl) (P3HT). A power conversion efficiency of 9.3% was obtained by tuning the energy level of P3HT and optimizing the fabrication of the device. This is the highest reported efficiency for a solar cell using P3HT.

Bai et al. (2014) reported high-performance planar heterojunction perovskite solar cells. PCE up to 15.9% has been achieved in this cell. Successful fabrication of highly efficient, stable, and reproducible planar heterojunction $CH_3NH_3PbI_{3-x}Cl_x$ solar cells was made using improved cathode interface bilayer-structured electron-transporting interlayers of [6,6]-phenyl-C61-butyric acid-methyl ester (PCBM)/ZnO. They also demonstrated that this simple planar structure is promising for large-scale devices.

Moisture is assumed to be detrimental to organometal trihalide perovskite, as excess water can damage the crystallinity of the perovskite structure. A growth mode for thermal annealing of the perovskite precursor film in a humid environment (e.g., ambient air) has been reported (You et al. 2014) to greatly improve the film quality, grain size, carrier mobility, and lifetime. This method produces devices with maximum power conversion efficiency of 17.1% and fill factor of 80%, revealing a promising route to achieve high-quality perovskite polycrystalline films with superior optoelectronic properties that can pave the way toward efficient photovoltaic conversion.

Different lengths of rutile TiO_2 nanowires with wide-open space for effective material filling were used as photoanodes by Jiang et al. (2014) for perovskite solar cells with an efficiency of 11.7%. A uniform and pinhole free hole-blocking layer is necessary for high-performance perovskite-based thin-film solar cells. Wu et al. (2014) investigated the effect of nanoscale pinholes in compact TiO_2 layers on the performance of solar cells. It was shown that TiO_2 compact layers fabricated using atomic layer deposition (ALD) contain a much lower density of nanoscale pinholes than layers obtained by spin coating and spray pyrolysis methods. This TiO_2 layer acts as an efficient hole-blocking layer in perovskite solar cells and gives a power conversion efficiency of 12.56%.

Inorganic–organic lead halide perovskite materials appear quite promising for the next generation of solar devices due to their high power conversion efficiency. The highest efficiencies reported for these cells so far have been obtained mainly with methylammonium lead halide materials. A narrow bandgap, but relatively unstable formamidinium lead iodide ($FAPbI_3$) with methylammonium lead bromide ($MAPbBr_3$) has been used as the light-harvesting unit in a bilayer solar cell architecture. Incorporation of $MAPbBr_3$ into $FAPbI_3$ has stabilized the perovskite phase of $FAPbI_3$ and improved the power conversion efficiency of the solar cell to more than 18% (Jeon et al. 2015).

12.6 MISCELLANEOUS SOLAR CELLS

A polymer solar cell is a type of flexible solar cell made with polymers. Polymer solar cells include organic solar cells and are also called *plastic solar cells*. These polymer solar cells are lightweight as compared to silicon-based devices, disposable, inexpensive to fabricate, flexible, and have relatively less adverse environmental impact. These cells have the potential to exhibit transparency, suggesting applications in windows, walls, flexible electronics, and so on, but there are some disadvantages of polymer solar cells. They offer about one-third of the efficiency of hard materials and may undergo substantial photochemical degradation. Although polymer solar cells have some stability problems, these cells are low cost and more efficient. Polymer solar cells were able to achieve over 10% efficiency via a tandem structure.

Carbon nanotubes (CNTs) have high electron and thermal conductivity, robustness, and flexibility. The CNTs have been used as either the photoinduced exciton carrier transport medium impurity within a polymer-based photovoltaic layer or as the photoactive (photon–electron conversion) layer. Electron-accepting impurities must be added to the photoactive region to increase the photovoltaic efficiency. Dissociation of the exciton pair can be accomplished by the CNT matrix by incorporating those into the polymer. The high surface area (~1600 m^2/g) of CNTs offers a good opportunity for exciton dissociation (Cinke et al. 2002). Separated carriers within the polymer–CNT matrix are transported by the percolation pathways of adjacent CNTs, providing the means for high carrier mobility and efficient charge transfer. The factors of performance of CNT–polymer hybrid photovoltaics are low as compared to those of inorganic photovoltaics.

Metal nanoparticles may be applied to the exterior of CNTs to increase the exciton separation efficiency. The metal provides a higher electric field at the CNT–polymer interface, thus accelerating the exciton carriers to transfer them more effectively to the CNT matrix. The CNT may be used as an add-in material to increase carrier transport and also as the photoactive layer (Chen et al. 2008). The single-walled CNT (SWCNT) is a potentially attractive material for photovoltaic applications because of its unique structural and electrical properties. This has high electric conductivity and shows ballistic carrier transport, thus greatly decreasing carrier recombination. The bandgap of the SWCNT is inversely proportional to the diameter of the tube. Therefore, SWCNT may show multiple, direct bandgaps.

Several challenges must be addressed for CNT to be used in photovoltaic applications. CNT degrades in an oxygen-rich environment with time. The passivation layer required to prevent CNT oxidation may reduce the optical transparency of the electrode region; hence, the photovoltaic efficiency is lower. The dispersion of CNT within the polymer photoactive layer is another problem. It should be well dispersed within the polymer matrix to form charge-transfer-efficient pathways between the excitons and the electrode (Somani et al. 2008).

The CNT is lacking to form a p-n junction, due to the difficulty of doping certain segments. A p-n junction creates an internal built-in potential, which provides a pathway for efficient carrier separation. Energy band bending using two electrodes of different work functions may solve this problem. The oxidation of CNT will create another difficulty. Oxidized CNTs have a tendency to become more metallic, therefore limiting their use as a photovoltaic material (Collins et al. 2000).

A quantum dot solar cell utilizes quantum dots (QDs) as the absorbing photovoltaic material. Bulk materials such as silicon, CuInGaSe$_2$, or CdTe are replaced by quantum dots, because their bandgaps are tunable across a wide range of energy levels by changing the size of QDs. The bandgap is fixed by the choice of material. This property makes quantum dots attractive for multijunction solar cells, where a variety of materials may be used to improve the efficiency by harvesting multiple portions of the solar spectrum. In such solar cells, 8.7% efficiency has been reported by Chuang et al. (2014).

The idea of using quantum dots for achieving high efficiency was first noted by Barnham and Duggan (1990). Quantum dots are semiconducting particles. Their sizes have been reduced to less than exciton Bohr radius, and as such, electron energies that can exist within them become finite, much like energies in an atom due to quantum mechanics considerations. Quantum dots have also been referred to as *artificial atoms*. They can be grown over a range of sizes, allowing them to have a variety of bandgaps without changing the underlying material or construction techniques (Baskoutas and Terzis 2006). This tuning of bandgap is achieved by varying the duration of synthesis or temperature, and this ability makes quantum dots desirable for solar cells. Single-junction solar cells lead sulfide (PbS) CQDs have bandgaps that can be tuned into the far-infrared, which is very difficult to achieve with traditional materials modification. Half of the solar energy reaching the Earth is in the infrared region, and mostly in the near-infrared region. A quantum dot solar cell makes this infrared energy as accessible as any other (Sargent 2005).

In the future, many new types of solar cells will be developed, or modifications of existing generations of solar cells will be made, leading to easy fabrication, higher efficiency, lower cost, increased durability, and lighter weight.

REFERENCES

Ali, Z., I. Shakir, and D. J. Kang. 2014. Highly efficient photoelectrochemical response by sea-urchin shaped ZnO/TiO$_2$ nano/micro hybrid heterostructures co-sensitized with CdS/CdSe. *J. Mater. Chem. A*, 2: 6474–6479.

Bai, S., Z. Wu, X. Wu, Y. Jin, N. Zhao, Z. Chen, et al. 2014. High-performance planar heterojunction perovskite solar cells: Preserving long charge carrier diffusion lengths and interfacial engineering. *Nano Res.*, 7: 1749–1758.

Bai, X., H. Wang, J. Wei, Y. Jia, H. Zhu, K. Wang, et al. 2012. Carbon nanotube-silicon hybrid solar cells with hydrogen peroxide doping. *Chem. Phys. Lett.*, 533: 70–73.

Barnham, K. W. J., and G. Duggan. 1990. A new approach to high-efficiency multi-band-gap solar cells. *J. Appl. Phys.*, 67: 3490.

Baskoutas, S., and A. F. Terzis. 2006. Size-dependent band gap of colloidal quantum dots. *J. Appl. Phys.*, 99: 013708.

Catchpole, K. R., and A. Polman. 2008. Plasmonic solar cells. *Opt. Express*, 16: 21793–21800.

Chen, C., Y. Lu, E. S. Kong, Y. Zhang, and S.-T. Lee. 2008. Nanowelded carbon-nanotube-based solar microcells. *Small*, 4: 1313–1318.

Chuang, C.-H. M., P. R. Brown, V. Bulović, and M. G. Bawendi. 2014. Improved performance and stability in quantum dot solar cells through band alignment engineering. *Nat. Mater.*, 13: 796–801.

Ciesielskia, P. N., F. M. Hijazib, A. M. Scott, C. J. Faulkner, L. Beard, K. Emmett, et al. 2010. Photosystem I-based biohybrid photoelectrochemical cells. *Biosource Technol.*, 101: 3047–3053.

Cinke, M., J. Li, B. Chen, A. Cassell, L. Delzeit, J. Han, et al. 2002. Pore structure of raw and purified HiPco single-walled carbon nanotubes. *Chem. Phys. Lett.*, 365: 69–74.

Collins, P. G., K. Bradley, M. Ishigami, and A. Zettl. 2000. Extreme oxygen sensitivity of electronic properties of carbon nanotubes. *Science*, 287: 1801–1804.

Derkacs, D., S. H. Lim, P. Matheu, W. Mar, and E. T. Yu. 2006. Improved performance of amorphous silicon solar cells via scattering from surface plasmon polaritons in nearby metallic nanoparticles. *Appl. Phys. Lett.*, 89: 093103.

Ding, B., B. J. Lee, M. Yang, H. S. Jung, and J.-K. Lee. 2011. Surface-plasmon assisted energy conversion in dye-sensitized solar cells. *Adv. Energy Mater.*, 1: 415–421.

Eperon, G. E., S. D. Stranks, C. Menelaou, M. B. Johnston, L. M. Herz, and H. J. Snaith. 2014. Formamidinium lead trihalide: A broadly tunable perovskite for efficient planar heterojunction solar cells. *Energy Environ. Sci.*, 7: 982–988.

Ferry, V. E., L. A. Sweatlock, D. Pacifici, and H. A. Atwater. 2008. Plasmonic nanostructure design for efficient light coupling into solar cells. *Nano Lett.*, 8: 4391–4397.

Garnett, E. C., M. L. Brongersma, Y. Cui, and M. D. McGehee. 2011. Nanowire solar cells. *Ann. Rev. Mater. Res.*, 41: 269–295.

Giacomo, F. D., S. Razza, F. Matteocci, A. D'Epifanio, S. Licoccia, T. M. Brown, et al. 2014. High efficiency CH$_3$NH$_3$PbI$_{(3-x)}$Cl$_x$ perovskite solar cells with poly(3-hexylthiophene) hole transport layer. *J. Power Sources*, 251: 152–156.

Ginger, D. S., and N. C. Greenham. 1999. Photoinduced electron transfer from conjugated polymers to CdSe nanocrystals. *Phys. Rev. B*, 59: 10622–10629.

Green, M. A., A. Ho-Baillie, and H. J. Snaith. 2014. The emergence of perovskite solar cells. *Nat. Photonics*, 8: 506–514.

Han, L., F. F. Abdi, R. Vande Krol, R. Liu, Z. Huang, H.-J. Lewerenz, et al. 2014. Efficient water-splitting device based on a bismuth vanadate photoanode and thin-film silicon solar cells. *ChemSusChem.*, 7: 2832–2838.

Herman, D. J., J. E. Goldberger, S. Chao, D. T. Martin, and S. I. Stupp 2011. Orienting periodic organic–inorganic nanoscale domains through one-step electrodeposition. *ACS Nano.*, 5: 565–573.

Jeon, N. J., J. H. Noh, W. S. Yang, Y. C. Kim, S. Ryu, J. Seo, et al. 2015. Compositional engineering of perovskite materials for high-performance solar cells. *Nature*, 517: 476–480.

Jiang, Q., X. Sheng, Y. Li, X. Feng, and T. Xu. 2014. Rutile TiO$_2$ nanowire-based perovskite solar cells. *Chem. Commun.*, 50: 14720–14723.

Jiao, J., J. Tang, W. Gao, D. Kuang, Y. Tong, and L. Chen. 2015. Plasmonic silver nanoparticles matched with vertically aligned nitrogen-doped titanium dioxide nanotube arrays for enhanced photoelectrochemical activity. *J. Power Sources*, 274: 454–470.

Jin, E. M., J.-Y. Park, K.-J. Hwang, H.-B. Gu, and S. M. Jeong. 2014. Biotemplated hybrid TiO$_2$ nanoparticle and TiO$_2$-SiO$_2$ composites for dye-sensitized solar cells. *Mater. Lett.*, 131: 190–193.

Kojima, A., K. Teshima, Y. Shirai, and T. Miyasaka. 2009. Organometal halide perovskites as visible-light sensitizers for photovoltaic cells. *J. Am. Chem. Soc.*, 131: 6050–6051.

Kong, E.-H., Y.-H. Yoon, Y.-J. Chang, and H. M. Jang. 2014. Hybrid photoelectrode by using vertically aligned rutile TiO_2 nanowires inlaid with anatase TiO_2 nanoparticles for dye-sensitized solar cells. *Mater. Chem. Phys.*, 143: 1440–1445.

Koussi-Daoud, S., D. Schaming, P. Martin, and J.-C. Lacroix. 2014. Gold nanoparticles and poly(3,4-ethylenedioxythiophene) (PEDOT) hybrid films as counter-electrodes for enhanced efficiency in dye-sensitized solar cells. *Electrochim. Acta*, 125: 601–660.

Kymakis, E., E. Stratakis, E. Koudoumas, and C. Fotakis. 2011. Plasmonic organic photovoltaic devices on transparent carbon nanotube films. *IEEE Trans. Electron. Devices*, 58: 860–864.

Lancelle-Beltran, E., P. Prené, C. Boscher, P. Belleville, P. Buvat, and C. Sanchez. 2006. All-solid-state dye-sensitized nanoporous TiO_2 hybrid solar cells with high energy-conversion efficiency. *Adv. Mater.*, 18: 2579–2582.

Lee, M. M., J. Teuscher, T. Miyasaka, and H. J. Snaith. 2012. Efficient hybrid solar cells based on meso-superstructured organometal halide perovskites. *Science*, 338: 643–647.

Li, C., F. Wang, and J. C. Yu. 2011. Semiconductor/biomolecular composites for solar energy applications. *Energy Environ. Sci.*, 4: 100–113.

Li, J., S. K. Cushing, P. Zheng, T. Senty, F. Meng, A. D. Bristow, et al. 2014a. Solar hydrogen generation by a CdS-Au-TiO_2 sandwich nanorod array enhanced with au nanoparticle as electron relay and plasmonic photosensitizer. *J. Am. Chem. Soc.*, 136: 8438–8449.

Li, X., X. Zang, X. Li, M. Zhu, Q. Chen, K. Wang, et al. 2014b. Hybrid heterojunction and solid-state photoelectrochemical solar cells. *Adv. Energy Mater.*, 4: 1400224.

Lin, S.-J., K.-C. Lee, J.-L. Wu, and J.-Y. Wu. 2012. Plasmon-enhanced photocurrent in dye-sensitized solar cells. *Solar Energy*, 86: 2600–2605.

Liu, W.-L., F.-C. Lin, Y.-C. Yang, C.-H. Huang, S. Gwo, M. H. Huang, et al. 2013. The influence of shell thickness of Au@TiO_2 core-shell nanoparticles on the plasmonic enhancement effect in dye-sensitized solar cells. *Nanoscale*, 5: 7953–7962.

Milliron, D. J., I. Gur, and A. P. Alivisatos. 2005. Hybrid organic–nanocrystal solar cells. *MRS Bull.*, 30: 41–44.

Moser, J. 2005. Solar cells: Later rather than sooner. *Nature Mater.*, 4: 723–724.

Noel, N. K., S. D. Stranks, A. Abate, C. Wehrenfennig, S. Guarnera, A.-A. Haghighirad, et al. 2014. Lead-free organic–inorganic tin halide perovskites for photovoltaic applications. *Energy Environ. Sci.*, 7: 3061–3068.

Qian, F., G. Wang, and Y. Li. 2010. Solar-driven microbial photoelectrochemical cells with a nanowire photocathode. *Nano Lett.*, 10: 4686–4691.

Rand, B. P., P. Peumans, and S. R. Forrest. 2004. Long-range absorption enhancement in organic tandem thin-film solar cells containing silver nanoclusters. *J. Appl. Phys.*, 96: 7519.

Sargent, H. E. 2005. Infrared quantum dots. *Adv. Mater.*, 17: 515–522.

Sariciftci, N. S., L. Smilowitz, A. J. Heeger, and F. Wudl. 1993. Semiconducting polymers (as donors) and buckminsterfullerene (as acceptor): Photoinduced electron transfer and heterojunction devices. *Synth. Met.*, 59: 333–352.

Saunders, B. R. 2012. Hybrid polymer/nanoparticle solar cells: Preparation, principles and challenges. *J. Colloid Interface Sci.*, 369: 1–15.

Saunders, B. R., and M. L. Turner. 2008. Nanoparticle-polymer photovoltaic cells. *Adv. Colloid Interface Sci.*, 138: 1–23.

Schaadt, D. M., B. Feng, and E. T. Yu. 2005. Enhanced semiconductor optical absorption via surface plasmon excitation in metal nanoparticles. *Appl. Phys. Lett.*, 86: 063106.

Shaheen, S. E., D. S. Ginley, and G. E. Jabbour. 2005. Organic-based photovoltaics. *MRS Bull.*, 30: 10–19.

Sofos, M., J. Goldberger, D. A. Stone, J. E. Allen, Q. Ma, D. J. Herman, et al. 2009. A synergistic assembly of nanoscale lamellar photoconductor hybrids. *Nature Mater.*, 8: 68–75.

Somani, P. R., S. P. Somani, and M. Umeno. 2008. Application of metal nanoparticles decorated carbon nanotubes in photovoltaics. *Appl. Phys. Lett.*, 93: 033315.

Standridge, S. D., G. C. Schatz, and J. T. Hupp. 2009. Distance dependence of plasmon-enhanced photocurrent in dye-sensitized solar cells. *J. Am. Chem. Soc.*, 131: 8407–8409.

Stuart, H. R., and D. G. Hall. 1998. Island size effects in nanoparticle-enhanced photodetectors. *Appl. Phys. Lett.*, 73: 3815.

Su, F., T. Wang, R. Lv, J. Zhang, P. Zhang, J. Lu, et al. 2013. Dendritic Au/TiO_2 nanorod arrays for visible-light driven photoelectrochemical water splitting. *Nanoscale*, 5: 9001–9009.

Tanabe, K. 2009. A review of ultrahigh efficiency III-V semiconductor compound solar cells: Multijunction tandem, lower dimensional, photonic up/down conversion and plasmonic nanometallic structures. *Energies*, 2: 504–530.

Tayebi, L., M. Mozafari, R. El-Khouri, P. Rouhani, and D. Vashaee. 2014. Energy harvesting capability of lipid-merocyanine macromolecules: A new design and performance model development. *Photochem. Photobiol.*, 90: 517–521.

Wang, D., Y. Li, G. L. Puma, C. Wang, P. Wang, W. Zhang, et al. 2015. Dye-sensitized photoelectrochemical cell on plasmonic Ag/AgCl @ chiral TiO_2 nanofibers for treatment of urban wastewater effluents, with simultaneous production of hydrogen and electricity. *Appl. Catal. B: Environ.*, 168–169: 25–32.

Wang, K., and J. Tang. 2012. Natural chlorophyll-sensitized nanocrystalline TiO_2 as photoanode of a hybrid photoelectrochemical biofuel cell. *Adv. Mater. Res.*, 496: 399–402.

Weickert, J., R. B. Dunbar, W. Wiedemann, H. C. Hesse, and L. Schmidt-Mende. 2011. Nanostructured organic and hybrid solar cells. *Adv. Mater.*, 23: 1810–1828.

Westphalen, M., U. Kreibig, J. Rostalski, H. Lüth, and D. Meissner. 2000. Metal cluster enhanced organic solar cells. *Sol. Energy Mater. Solar Cells*, 61: 97–105.

Wu, X. L., J. Y. Fan, T. Qiu, X. Yang, G. G. Siu, and P. K. Chu. 2005. Experimental evidence for the quantum confinement effect in 3C-SiC nanocrystallites. *Phys. Rev. Lett.*, 94: 026102.

Wu, Y., X. Yang, H. Chen, K. Zhang, C. Qin, J. Liu, et al. 2014. Highly compact TiO_2 layer for efficient hole-blocking in perovskite solar cells. *Appl. Phys. Express*, 7: 052301.

Xia, C., N. Wang, and X. Kim. 2011. Mesoporous CdS spheres for high-performance hybrid solar cells. *Electrochim. Acta*, 56: 9504–9507.

Xu, Q., F. Liu, W. Meng, and Y. Huang. 2012. Plasmonic core-shell metal-organic nanoparticles enhanced dye-sensitized solar cells. *Opt Express*, 20: A898–A907.

Yehezkeli, O., R. Tel-Vered, J. Wasserman, A. Trifonov, D. Michaeli, R. Nechushtai, et al. 2012. Integrated photosystem II-based photoelectrochemical cells. *Nature Commun.*, 3: 742.

You, J., Y.-(M.) Yang, Z. Hong, T.-B. Song, L. Meng, Y. Liu, et al. 2014. Moisture assisted perovskite film growth for high performance solar cells. *Appl. Phys. Lett.*, 105: 183902.

Zhang, Q., W. J. Qin, H. Q. Cao, L. Y. Yang, and S.-G. Yin. 2014. Efficiency enhancement of organic solar cells with process-optimized silver nanoprisms. *Appl. Mech. Mater.*, 598: 327–330.

Zhang, W., M. Saliba, S. D. Stranks, Y. Sun, X. Shi, U. Wiesner, et al. 2013. Enhancement of perovskite-based solar cells employing core-shell metal nanoparticles. *Nano Lett.*, 13: 4505–4510.

Index

A

Absorption photon-to-current efficiency (APCE), 145, 146
Air mass ratio (m), 8
3-Aminopropyltriethoxysilane (APTES), 78
APCE, *see* Absorption photon-to-current efficiency (APCE)
Artificial atoms, 261
Artificial photosynthesis, 10, 41, 155, 163, 187, 195
    approaches in, 190–196
    basic components of, 190–191
    catalysis for CO_2 reduction and fuel production, 208–213
    catalytic proton reduction and hydrogen evolution
        light-driven catalysis for hydrogen production, 206–207
        proton reduction catalyst by hydrogenases, 204–206
    catalytic water oxidation, 199–200
        Iridium- based catalyst, 202–203
        Mn-, Co-, and Fe-based catalysts, 203–204
        Ruthenium-based catalysts, 200–202
    charge separation by molecular donor–acceptor systems, 197–198
    heterogeneous artificial photosynthesis systems, 195–196
    homogeneous artificial photosynthesis systems, 191–195
    light absorption, 196–197
    light-driven catalytic water splitting, 199
    natural photosynthesis to, 188–190
    for production of oxygen and hydrogen, 191
Au NPs, 254, 255

B

Band bending, 21, 166
    in n-doped semiconductor, 19
    in p-doped semiconductor, 19, 20
Band engineering, 182
Bandgaps, reducing through doping, 159
Becquerel, Alexandre-Edmond, 86
Becquerel effect, 115
Bell Laboratories, 1, 57, 86, 220
Bilayer cells, 14, 69–71
Bilayer organic photovoltaic cells, 69–71, 72
Biohybrid solar cells, 14, 258–259
Biohydrogen, 148
Biological hydrogen production, 148
Biomass waste, of wine industries, 154
Bromophenol red-EDTA system, 124
Bulk heterojunction (BHJ), 71–75
    active layer system for OPV, 72
    hybrid photovoltaic cells, 72–73
    morphology of, 72, 75
    nonfullerene-based, 73
    p-type perylene diimide in, 73
    solution-processed, 73
    solvent-processes, 73–74

C

Carbon dioxide (CO_2), 12–13
    in atmosphere, 173
    conversion of, to fuels, 173–174
    emissions, 2
    mechanism of photocatalytic reduction of, 174–176
    molecule, characteristics of, 174
    operation conditions, 181–182
        pressure effect, 181
        reactor-type effect, 182
        reductant effect, 181
        temperature effect, 181
    outlook, 182–183
    photocatalysts for reduction of, 177–181
        metal oxide photocatalysts, 179
        nonoxide semiconductor photocatalysts, 179–181
        TiO_2 and related titanium-containing solids, 177–179
    reduction and fuel production, 208–213
Carbon nanotubes (CNTs), 74–75, 100, 101, 260–261
Cascade heterojunction organic solar cells, 75
Catalytic proton reduction and hydrogen evolution
    light-driven catalysis for hydrogen production, 206–207
    proton reduction catalyst by hydrogenases, 204–206
Catalytic water oxidation, 199–204
    Iridium-based catalyst, 202–203
    Mn-, Co-, and Fe-based catalysts, 203–204
    Ruthenium-based catalysts, 200–202
Cathode interfacial material (CIM), 67
Cell efficiency, 36, 99
    hybrid, 256
    of photogalvanic cells, 117
Charge
    carrier mobilities, 64, 67
    collection, 60
    separation, 59, 197–198
    transfer, 59–60
Charge-transfer complex, 59
Chlor-alkali plants, 153
Chlorophyll, 9, 10, 145, 189, 192, 196
Cis-genes, 205
CNT-Si solar cells, 257
Composite photocatalyst, 163
Concentrated solar power (CSP), 143, 149–150
Concentration cell, 117, 118
Conduction band (CB), 10, 11, 13, 17, 26, 41
Conjugative polymers, 63
Copolymers, 73, 74
Copper indium diselenide ($CuInSe_2$), 86, 229–231
Copper indium disulfide ($CuInS_2$), 227–229
Copper indium gallium diselenide (CIGSe), 232–235

Copper indium gallium disulfide (CIGS), 231–232
Copper iron tin sulfide and related materials, 242–245
Copper sulfide, 220–222
Copper-zinc-tin chalcogenides (CZTSSe), 235–242
    nanocrystals–based approach, 235–236
    nanoparticles, 240–241
    power conversion efficiency of, 240
    processing of, 235
    synthesis of, 237–239
Counterelectrodes, 12, 100–101, 243
CSP, see Concentrated solar power (CSP)
Cu_2CoSnS_4, 243, 244
Cu_2FeSnS_4, 243
$Cu_2(Fe_xZn_{1-x})SnS_4$ nanocrystals, 243–244
Cu_2MnSnS_4, synthesis of, 244
Cyanine dyes, 62

D

Dendrimers, 63, 65
Depletion layer, 21, 25
Differential cell, 117, 118
Donor-bridge-acceptor (D-B-A) system, 193, 194
Doped semiconductors, 17, 98–99; see also Semiconductors
DSPC, see Dye-sensitized photoelectrochemical cell (DSPC)
DSPECs, see Dye-sensitized photoelectrosynthesis cells (DSPECs)
DSSC, see Dye-sensitized solar cell (DSSC)
Dyes, 62–63, 91–92
    photogalvanic cell and, 119–124
    quantum dots (QDs) and, 157–159
Dye-sensitized photoelectrochemical cell (DSPC), 255
Dye-sensitized photoelectrosynthesis cells (DSPECs), 39, 41
Dye-sensitized solar cells (DSSCs), 10, 85, 195
    applications of, 105
    basic parts of, 10–11
    characterization, 88
    components of, 89
    counterelectrode, 100–101
    electrodes
        doped semiconductors, 98–99
        modified semiconductors, 97
        naïve semiconductors, 95–97
        other nanomaterials, 99–100
    electrolytes, 11, 93–95
        liquid, 93–94
        other, 94–95
        polymer gel, 94
    fabrication of, 101
    four energy levels, 11
    heterojunction three-dimensional fabrication of, 86
    incident photon-to-current conversion efficiency, 89
    nanocrystalline semiconductor working electrode, 95–100
    photo-to-electric conversion efficiency of, 99–100
    potential of, 100
    principle, 87
    sensitizer, 89
        dyes, 91–92
        metal complexes, 89–91
        natural pigments, 93
    types of, 101
        quantum dot DSSC, 104–105
        quasi-solid state DSSC, 103–104
        solid-state DSSC, 101–103

E

Electrochemical aspect, of hydrogen, 164–167
Electrochemical cell, 56, 132; see also Photoelectrochemical (PEC) cells
Electrochemical photovoltaic cells, 31–32
Electrodes
    doped semiconductors, 98–99
    modified semiconductors, 97
    naïve semiconductors, 95–97
    other nanomaterials, 99–100
Electrolysis, of water, 143, 148–149
    solarothermal, 150
    thermal, 150–151
Electrolytes, component in DSSCs, 11, 93
    liquid, 11, 93–94
    other, 94–95
    polymer gel, 94
    quasi-solid-state, 11
    solid, 11
Energy, 1; see also Solar energy
    crisis, 2
    global, use and production, 187–188
    renewable, 2–3, 7, 86, 140–143, 187, 208
Energy Conversion Devices, 166
Engineered solid solutions, 159
Excitons, 13, 14
    diffusion, 58–59
    exciton diffusion length, 255

F

Fe(II)-β-diketonate/thionine, 123
Fermi level (E_f), 18, 19, 25, 164
    definition, 17
    in intrinsic semiconductor, 17, 22
Fe-thionine photogalvanic cell, 122
Fill factor (FF), 31, 61, 66, 75, 88
First-generation solar cells, 86, 253, 254
Formaldehyde, 13
Formamidinium lead iodide ($FAPbI_3$), 260
Formic acid, 208
Fossil fuels, 2, 4, 5, 143, 147–148, 155, 187, 219
Fuel cells, 14, 168, 208
    hydrocarbon waste of high-temperature, 153–154
    hydrogen and, 141, 142, 143
Fuel production, catalysis for CO_2 reduction and, 208–213
Fullerene, 15, 35, 63, 64, 70, 73, 74, 259
β-Functionalized porphyrin sensitizers, 32

G

Graphene, 67–68, 166–167
Graphene oxide (GO), 44, 47, 76, 212
Graphene quantum dots (GQDs), 104–105
Gratzel, Michal, 10
Gratzel cell, see Dye-sensitized solar cell (DSSC)

Index

H

Heterocyclic dyes, 124
Heterogeneous artificial photosynthesis systems, 195–196
Heterogeneous photocatalysis, 26
High-resolution transmission electron microscopy (HRTEM), 66
High-temperature fuel cells, hydrocarbon waste of, 153–154
Homogeneous artificial photosynthesis systems, 191–195
Homogeneous photocatalysis, 26
Homogenous water oxidation, 201
Honda FCX concept car, 140–141
HRTEM, *see* High-resolution transmission electron microscopy (HRTEM)
Hybrid solar cells, 14–15, 77, 255–258
    dye-sensitized solar cells, 257
    nanoscale structures, 255–256
    organic material, 255
    photoactive layer, 256
Hybrid tandem photovoltaic cell, 76–77
Hydrocarbon fuels, 43–44, 147, 211
Hydrogen, 12, 13, 139–141
    biological production, 148
    as by-product of other chemical processes, 152–154
        chlor-alkali plants, 153
        high-temperature fuel cells, hydrocarbon waste of, 153–154
        oil refinery, catalytic reforming of, 153
        waste biomass of wine industries, 154
    chemical fuel, 141
    concentrating solar thermal power, 149–150
    eco-friendly production of, 140
    economy, 168
    electrochemical aspect, 164–167
    electrolysis of water, 148–149
    as energy carrier, 144
    evolution and catalytic proton reduction and, 204–207
    fuel cells, 140, 141, 143
    generation, technology for, 147–154
    as key solution, 167–168
    Kvaerner carbon black and hydrogen process, 148
    photocatalysts, modification of, 157
        composite photocatalyst, 163
        engineered solid solutions, 159
        photocatalyst, nanostructuring of, 162–163
        photoelectrochemical water splitting, 163–164
        photosensitization, 157–159
        reducing bandgaps through doping, 159
        surface plasmon resonance, plasmonic nanostructures with, 161–162
        water splitting, macromolecular systems for, 159–161
    photocatalytic and photoelectrocatalytic hydrogen production, 151–152
    photocatalytic hydrogen generation, 155–156
    photocatalytic water splitting, mechanisms of, 156–157
    photochemical hydrogen generation, 144–147
    photogeneration of, 13
    as renewable energy, 140
    as safe energy source, 142
    as sustainable and clean energy source, 142
    as synthetic fuel, 13
    thermal electrolysis of water, 150–151

Hydrogenases
    catalytic action of, 206
    definition, 204
    proton reduction catalyst by, 204–206
Hydrogen generation, 140, 142–144
    light-driven catalysis for, 206–207
    photochemical, 144–147
    resources for, 142–143
    technology for, 147–152
        biological production, 148
        concentrating solar thermal power, 149–150
        electrolysis of water, 148–149
        Kvaerner carbon black and hydrogen process, 148
        photocatalytic and photoelectrocatalytic, 151–152
        thermal electrolysis of water, 150–151

I

IEA, *see* International Energy Agency (IEA)
Incident photon-to-current conversion efficiency (IPCE), 61, 89, 145–146, 257
Infrared (IR) radiation, 7, 59
Inorganic semiconductors, 14, 56, 115, 255, 256
Insulator, electrical conductivity of, 17
Internal combustion engines (ICEs), 168
Internal reforming, 153–154
International Energy Agency (IEA), 2, 143
Intrinsic semiconductor, 17–18
Inverted organic photovoltaic cells, 77–78
Inverted structure, 77
Inverted tandem solar cell, 77
IPCE, *see* Incident photon-to-current efficiency (IPCE)
Iridium-based catalyst, 202–203
Iron-iron hydrogenases, 204
Iron sulfide, 222–223
Iron–thionine photogalvanic cell, 115, 120–121

K

Kvaerner carbon black and hydrogen process, 148

L

Langley (L), 8
Lanthanides, 66
Lead sulfide, 223–225
LHE, *see* Light-harvesting efficiency (LHE)
Light absorption (LA), 46, 160, 161, 188, 192, 196–197, 209
Light-driven catalytic water splitting, 199
Light-harvesting efficiency (LHE), 146
Liquid crystals, 65–66
Liquid electrolytes, 11, 93–94
Liquid-junction PEC cell, 31
Liquid-junction solar cells, *see* Photogalvanic cells

M

Macromolecular systems for, water splitting, 159–161
Manganese-molybdenum-diethyldithiocarbamate complex, 128
MEG, *see* Multiple exciton generation (MEG)
Metal-based electrolytes, 95
Metal complexes, 89–91

free dyes, 91–92
other metal complexes, 90–91
Ru metal complex dyes, 90
Metal-free organic dyes, 92
Metal oxide photocatalysts, 179
Metal-to-ligand charge-transfer (MLCT) transitions, 35, 196, 208
Methylammonium lead bromide (MAPbBr$_3$), 260
Microbial photoelectrochemical cell (MPC), 258
Mn-, Co-, and Fe-based catalysts, 203–204
Modified semiconductors, 97
Molecular donor–acceptor systems, charge separation by, 197–198
Molecular light-harvesting systems, 193
Mott–Schottky equation, 22
MPC, see Microbial photoelectrochemical cell (MPC)
Multiple exciton generation (MEG), 224

N

Naïve semiconductors, 95–97
Nanocrystalline semiconductor working electrode, 95–100
doped semiconductors, 98–99
modified semiconductors, 97
naïve semiconductors, 95–97
other nanomaterials, 99–100
Nanomaterials for solar energy, 219–220
binary materials, 220
copper sulfide, 220–222
iron sulfide, 222–223
lead sulfide, 223–225
tin sulfide, 225–226
copper-zinc-tin chalcogenides, 235–242
copper iron tin sulfide and related materials, 242–245
I–III–VI materials, 226–227
copper indium diselenide, 229–231
copper indium disulfide, 227–229
copper indium gallium diselenide, 232–235
copper indium gallium disulfide, 231–232
Natural dye pigments, 92
Natural photosynthesis, 9
to artificial photosynthesis, 188–190
at molecular level, 193
n-Doped semiconductor, 17, 19, 165; see also p-doped semiconductor
effect of pH on potential drop in, 24
effect on band bending and Fermi level on, 25
photoexcitation of, 19
redox couple in vacuum, 20–21
reverse band-bending, 19
space charge layer in, 21
in vaccum, 23
n-DSSCs, 10
Nevada Policy Research Institute (NPRI), 3
n-GaAs electrodes, 32, 34
NiO photocathode–based dye-sensitized solar cell, 99
Nonoxide semiconductor photocatalysts, 179–181
n-p-Type cells, 20

O

OECD, see Organisation for Economic Co-operation and Development (OECD)
Oil refinery, catalytic reforming of, 153
Oil shocks, 2
Oligomers, 63
Open-circuit voltage (V_{oc}), 38, 60, 64–65, 68, 73, 88, 122
Operation conditions, for reduction of CO_2, 181–182
pressure effect, 181
reactor-type effect, 182
reductant effect, 181
temperature effect, 181
Organic cell, electrochemical cell vs., 56
Organic chromophore dyes, 91–92
Organic dyes, 62–63
Organic light-emitting devices (OLEDs), 58
Organic materials, for photovoltaic applications, 61–62
Organic photovoltaic cells, 13–14, 55–78
advantages of, 56
basic processes, 58–60
charge collection, 60
charge separation, 59
charge transfer, 59–60
exciton diffusion, 58–59
photon absorption, 58
characteristics, 60–61
history, 57–58
hybrid tandem photovoltaic cell, 76–77
introduction, 55–57
inverted, 77–78
inverted tandem solar cell, 77
materials, 61–65
dyes, 62–63
liquid crystals, 65–66
other materials, 66–68
pigments, 62
small molecules, oligomers, polymers, and dendrimers, 63–65
MoS_2 with, 67
tandem solar cells, 75–76
types of, 68–75
bilayer, 69–71
heterojunctions, 71–75
single layer, 68–69
Organic semiconductors, 56, 57, 70
Organisation for Economic Co-operation and Development (OECD), 142
Oxide photocatalysts, 163

P

PbS quantum dots, 224–225
PCSNPs, see Plasmonic core–shell nanoparticles (PCSNPs)
p-Doped semiconductor, 18, 19; see also n-doped semiconductor
band bending in, 19, 20
electrolyte with a redox couple, 23
redox couple in vacuum, 22
reverse band-bending, 19, 20
space charge layer in, 22
p-DSSC, 10
PEBFC, see Photoelectrochemical biofuel cell (PEBFC)
Perovskite solar cells, 14, 15, 259–260
Phenozine dyes, 123
Photocatalysis, 26–28, 175
definition, 26
example, 145

Index

heterogeneous, 26
homogeneous, 26
various probabilities of reactions, 26–27
Photocatalysts, 145
　composite, 163
　modification of, 157
　　composite photocatalyst, 163
　　engineered solid solutions, 159
　　photocatalyst, nanostructuring of, 162–163
　　photoelectrochemical water splitting, 163–164
　　photosensitization, 157–159
　　reducing bandgaps through doping, 159
　　surface plasmon resonance, plasmonic nanostructures with, 161–162
　　water splitting, macromolecular systems for, 159–161
　nanostructuring of, 162–163
　for reduction of CO_2, 177–181
　　mechanism, 174–176
　　metal oxide photocatalysts, 179
　　nonoxide semiconductor photocatalysts, 179–181
　　TiO_2 and related titanium-containing solids, 177–179
Photocatalytic cells, 30, 39, 42–47
Photocatalytic hydrogen generation, 47, 155–156
Photocatalytic water splitting
　analogy of photosynthesis with, 145
　for hydrogen production, 42, 47, 156–157
　mechanisms of, 156–157
Photochemical conversion modes, 9–15
Photochemical hydrogen generation, 144–147
Photochemical reaction, 117–119
Photoconductivity, 18, 57, 62, 63, 257
Photoconversion efficiency (η), 42, 47, 146–147
Photocurrent enhancement, 75, 253–254
Photoelectrochemical biofuel cell (PEBFC), 257
Photoelectrochemical (PEC) cells, 12, 196
　applications of, 12
　classification, 30–31
　　photocatalytic cells, 42–47
　　photoelectrosynthetic cells, 30–31, 39–42
　　regenerative PEC solar cells, 30, 31–39
　commercial use of, 29
　definition, 29
　energy production by, 29
　semiconductor–liquid-junction-based, 30
　as storage cells, 47–48
　titanium dioxide in, 12
Photoelectrochemical (PEC) reactions, 163
Photoelectrochemical water splitting, 163–164
Photoelectrochemistry, 12, 24–26, 122, 131
Photoelectrolysis, of water, 129
Photoelectrolytic solar cells, 39–42
Photoelectrosynthetic cells, 30–31, 39–42
Photogalvanic cells, 12, 115, 116, 124, 128
　cell efficiency, 117
　complexes, 124–129
　definition, 115
　dyes, 119–124
　at electrode, 11
　examples, 125–126
　functioning of, 120
　introduction, 115–117
　miscellaneous, 129–132
　operation
　　construction, 128–129

　　photochemical determinants of, 124–125
　　of rubidium anion, 126–127
　　requirements for efficient, 117–119
　　theory of the operation, 120
Photogalvanic device functions, 12, 115, 132
Photogalvanovoltaic (PGV) cell, 130
Photoinitiated electron collection (PEC) process, 160
Photon absorption, 58
Photonic flux density, 9
Photoredox process, 116
Photosensitization, 34, 124, 157–159
Photosynthesis, 9–10, 119, 145,155, 163, 258
　artificial, see Artificial photosynthesis
　natural, see Natural photosynthesis
　reactions, 188–189
Photosystem I (PS I), 14, 188–189, 190, 258
Photosystem II (PS II), 14, 188–189, 191, 198
Photovoltaic cells, 3, 13–14, 55–56, 57, 130–131
　bilayer organic, 69–71, 72
　electrochemical, 31–32
　hybrid tandem, 76–77
　inverted organic, 77–78
　organic, see Organic photovoltaic cells
　silicon-based, 56
　single-layer organic, 68–69
Photovoltaic (PV) effect, 220
Photovoltaic/photoelectrochemical (PV/PEC), 258
Pigments, 62
　natural, 92
Planar donor–acceptor heterojunction, 69
Planar donor–acceptor hydrogen, 14
Plasmonic core–shell nanoparticles (PCSNPs), 254–255
Plasmonic nanostructures, 161–162
Plasmonic solar cells, 15, 253–255
Plasmon resonance energy transfer (PRET), 161
Plastic solar cell (PSC), 15, 63, 260
Poly(adipic acid pentaerythritol ester) (PAAPE), 104
Poly(3-hexylthiophene) (P3HT), 71, 78
Poly(vinyl carbazole) (PVK), photoconductivity of, 63
Polymer gel electrolytes, 94, 103
Polymers, 14, 15, 35, 57, 62, 63–65, 71–72, 73
Polymer solar cell, 260
Polythiophene, 63, 68
Porphyrin quinine (P–Q) system, 197–198
Power conversion efficiency (PCE, η), 14, 32, 38, 57, 61, 70, 73, 105, 260
Power point, 88
PRINT (pattern replication in nonwetting templates), 58
Proton reduction catalyst, by hydrogenases, 204–206
PSt-bpy-Ru complex, 32–33
Pt electrode, 100, 101
Push–pull conjugated copolymers, synthesis of, 73

Q

Quantum dot dye-sensitized solar cells (QD-DSSC), 104–105
Quantum dots, 3, 11, 261
　CdS, 104
　colloidal, 76, 77
　dyes and, 157–159
　graphene, 104–105
　PbS, 224–225
　solar cell, 261

Quantum efficiency (QE), 61, 145, 147
Quasi-solid-state DSSC (QSS-DSSC), 11, 103–104
Quasi-solid-state electrolytes, 11

R

Regenerative PEC solar cells, 30, 31–39
Renewable energy resources, 2–3, 7, 86, 140–143, 187, 208
Reverse water gas shift (RWGS) reaction, 208
Ru(II) complex dyes, 90, 91, 92, 102, 202
Ruthenium-based catalysts, 200–202

S

Schottky-type cells, 20
Schottky-type junction, 21
Second-generation solar cells, 86, 253, 254
Selenium, photoconductive effect in, 57
Semiconductors
    electrochemistry of, 17–24
    electron distribution, 17, 18
    Fermi level, 17, 18
    material classification, 17, 18
    n-doped semiconductors, *see* n-doped semiconductors
    p-doped semiconductor, *see* p-doped semiconductors
Sensitization, 119, 157
Sensitizer, 89
    dyes, 91–92
    metal complexes, 89–91
        other metal complexes, 90–91
        Ru metal complex dyes, 90
    metal complex free dyes, 91–92
    natural pigments, 93
Short-circuit current (*isc*), 60–61
Silicon-based photovoltaic cells, 56
Silicon solar cells, 3, 226
Silver nanoprisms (AgNPs), 255
Single-layer organic photovoltaic cells, 68–69
Single-site photocatalysts, 177–178
Single-walled carbon nanotube (SWNT), in bulk heterojunction, 75
Solar cells, 1, 3, 86
    biohybrid, 14, 258–259
    cascade heterojunction organic, 75
    characteristics, 60–61
    dye-sensitized, *see* Dye-sensitized solar cells (DSSCs)
    first-generation, 86, 253, 254
    hybrid, *see* Hybrid solar cells
    inverted tandem, 77
    perovskite, 14, 15, 259–260
    photoelectrolytic, 39–42
    plasmonic, 15, 253–255
    plastic, 15, 63, 260
    polymer, 260
    quantum dot dye-sensitized solar cells, 104–105
    regenerative PEC, 30, 31–39
    second-generation, 86, 253, 254
    silicon, 3, 226
    solid-state dye-sensitized, 257
    tandem, *see* Tandem solar cells
    third-generation, 86, 253, 254
Solar constant, 7
Solar electricity, 55–56, 220
Solar energy, 1, 2, 85–86, 219–220
    advantages, 3–4
    conversion efficiency, 39–40
    disadvantages, 4
    future, 4–5
    in generating electricity, 2–3, 4
    nanomaterials for, 219
        copper indium diselenide, 229–231
        copper indium disulfide, 227–229
        copper indium gallium diselenide, 232–235
        copper indium gallium disulfide, 231–232
        copper iron tin sulfide and related materials, 242–245
        copper sulfide, 220–222
        copper-zinc-tin chalcogenides, 235–242
        iron sulfide, 222–223
        lead sulfide, 223–225
        tin sulfide, 225–226
    scenario, 7–9
    for water breaking, 145
Solar fuel, 9, 10, 195
Solar influx, 8
Solar radiation, 1, 55, 85
    annual mean of, 8
    conversion into electrical energy, 58
        charge collection, 60
        charge separation, 59
        charge transfer, 59–60
        exciton diffusion, 58–59
        photon absorption, 58
    unit of, 8
Solar spectrum, 7–8, 194, 196
Solar-to-electricity conversion efficiency, 33, 92
Solar-to-hydrogen (STH) conversion efficiency, 145, 146
Solid-state dye-sensitized solar cell, 101–103, 257
Solid-state electrolytes, 11, 91, 93, 103
Solid-state photogalvanic dye-sensitized solar cells, 124
Space charge layer, *see* Depletion layer
Spectral distribution, of sunlight, 8–9
Spray pyrolysis technique, 228
Steam reforming, 208
Stille polymerization reaction, 72
Successive ionic layer adsorption and a reaction (SILAR) process, 37
Sunlight, 3, 7
    diffusion of, 8
    spectral distribution of, 8–9
Surface plasmon resonance (SPR), 161–162

T

Tandem solar cells, 10, 75–76
    colloidal quantum dots and, 76
    interlayer, 76
    multilayer, 76
    single-layer, 75–76
Thermal electrolysis, of water, 150–151
    three-step cycle, 151
    two-step cycles, 150–151
Thiazine dyes, 120
Thin-film solar cells, 253
Thionine-EDTA system, for solar energy conversion, 122
III–V materials, 3
Third-generation solar cells, 86, 253, 254
Tin sulfide, 225–226

Index

Titanium dioxide (TiO_2), 10, 12, 96, 156
    coated SnO_2 hybrid nanorod, 99
    Fe-doped titania, 98
    MgO-coated, 97
    nanofibers, 100
    nitrogen-doped, 98
    photocatalytic activity of, 177
    photoreduction of CO_2 by using, 177
    titanium-containing solids and, 177–179
    for water splitting, 156
Trans-genes, 205
Transition-metal catalysts, 194, 210

U

Ultraviolet (UV) radiation, 7, 8, 67, 178

V

Valence band (VB), 11, 17, 22, 26, 157
Vanadium oxides (VO_x), 77
Visible (VIS) radiation, 7

W

Waste biomass, of wine industries, 154

Water
    electrolysis of, 148–149
    splitting
        by Ir complex, 202
        light-driven catalytic, 199
        macromolecular systems for, 159–161
        phenomena, 156–157
        photoelectrochemical, 163–164
    thermal electrolysis of, 150–151
Water electrolysis cell, 145, 149
Water oxidation, 44, 189, 199
    catalytic, 40, 41, 42, 199–204
        Ruthenium-based catalysts, 200–202
        Iridium-based catalyst, 202–203
        Mn-, Co-, and Fe-based catalysts, 203–204
    homogenous, 201
Wine industries, waste biomass of, 154

X

Xanthene dyes, 123

Z

ZnO nanorod, 99, 104